CONSERVATION BIOLOGY

CONSERVATION BIOLOGY

The Science of Scarcity and Diversity

Edited by Michael E. Soulé

SCHOOL OF NATURAL RESOURCES
UNIVERSITY OF MICHIGAN

SINAUER ASSOCIATES, INC. • PUBLISHERS
Sunderland, Massachusetts

THE COVER
The black rhinoceros (*Diceros bicornis*)—endangered by man—otherwise
symbolizes the best in interspecific relationships. Not beloved of hunters
because of their loud alarm calls, the red-billed ox pecker (*Buphagus
erythrorhynchus*, with a yellow eye-ring) and the yellow-billed ox pecker
(*Buphagus africanus*) not only warn their mammalian hosts of approaching danger, but also rid them of ticks and other parasites. Another tick
eater, the cattle egret (*Bulbulcus ibis*), feeds on insects disturbed by the
passage of large mammals. (Courtesy of Patricia D. Moehlman)

CONSERVATION BIOLOGY: THE SCIENCE OF SCARCITY AND DIVERSITY
Copyright © 1986 by The Regents of the University of Michigan
For information address Sinauer Associates Inc., Sunderland, Massachusetts
01375.

Library of Congress Cataloging in Publication Data

Conservation biology

  Bibliography: p.
  Includes index.
  1. Nature conservation.   2. Population biology.
  3. Ecology.   I. Soulé, Michael E..
  QH75.C664   1986   574.5   86-1902
  ISBN 0-87893-794-3
  ISBN 0-87893-795-1 (pbk.)

Printed in U.S.A.

5 4

Contents

23 Restoration, Reclamation, and Regeneration of Degraded or
    Destroyed Ecosystems   465
    JOHN CAIRNS, JR.

24 The Design of a Nature Reserve System for Indonesian
    New Guinea   485
    JARED DIAMOND

25 Intrinsic Value: Will the Defenders of Nature Please Rise?   504
    ARNE NAESS

    Acknowledgments   517

    Literature Cited   521

    Index   571

Preface

This is the second book with the title "Conservation Biology." It differs from the first one (*Conservation Biology: An Evolutionary-Ecological Perspective*, M.E. Soulé and B.A. Wilcox, eds., 1980, Sinauer Associates, Sunderland, MA) by having a different subtitle, and all 25 contributed chapters are new. It is about 25 percent longer, and about 90 percent of its authors are new.

Much has happened to conservation biology in the six-year interval between these two books. Many people now refer to themselves as conservation biologists; there is a new Society for Conservation Biology, and a new journal is being established. Partly a cause and partly as an effect of this movement, there is growing concern among biologists about biological diversity, genetic resources, and extinction.

Some things have improved and some have worsened during these six years. But in spite of rampant population growth, acid precipitation, changes in the distribution of atmospheric gases (in the wrong direction), massive deforestation, and all the other global ecological disasters, I believe there is a sense that something can be done, and that scientists can and must play an important role in preventing the erasure of the planet's biological print. One reason for wary optimism is that society has learned some lessons from the gross environmental mismanagement of the last four decades.

One objective of this book is to provide an up-to-date synthesis of conservation biology. If this is a nearly impossible task now, it will be even more so in the future, as information and ideas continue to multiply. Other objectives include the encouragement of communication among all sectors of the conservation community, and the engendering of a sense of purpose and excitement in students and professionals.

The book is divided into six sections, each preceded by a short "overview." Rather than syntheses, these short introductions touch on some of the commonalities and differences in the chapters. They may also serve to facilitate application of the principles discussed and to assist students in working through some of the more technical or controversial topics.

The advantages of multi-authored volumes include breadth of coverage, expertise, and perspective. The disadvantages include uneven style and content. I have done what I can to ameliorate the latter problem and to present the information in a logical sequence. I hope you enjoy it, but more importantly, I hope that it makes you a more informed and effective member of the conservation biology community.

This book, like most human endeavors, would have been impossible without the enthusiastic cooperation of many people and organizations. The rallying point for the project was the Second Conference on Conservation Biology, sponsored by the School of Natural Resources, University of Michigan, in May, 1985, and supported by generous contributions from the National Science Foundation, World Wildlife Fund—U.S., UNESCO, the International Union for the Conservation of Nature and Natural Resources (IUCN), World Wildlife Fund International, and the New York Zoological Society.

Some of the many people who have made this project possible are David Hales, Jim Crowfoot, April Oja, and Tanya Bernard of the University of Michigan; and William Conway, Paul Ehrlich, Lynn Greenwalt, David Hales, Thomas Lovejoy, Peter Raven, Daniel Simberloff, and John Terborgh of the Conference Organizing Committee. In addition, I wish to extend special thanks to some others who have made important contributions of time, energy, advice, and aid, including Buff Bohlen, Brand Brickman, Larry Gilbert, Michael Gilpin, Jeffrey McNeely, Kenton Miller, Anatola Ondricek, Jane Robertson, Hal Salwasser, Andy Sinauer, Carol Wigg, Bruce Wilcox, and especially the authors. Finally, I have to thank numerous creatures because:

> They gently came to this mud mind,
>   clear footprints left, synaptically enshrined.
> Each intimate mind fossil erases duality;
>   each mud-fossil embrace defies mortality.

MICHAEL E. SOULÉ
March 5, 1986

The Contributors

Fred W. Allendorf, Department of Zoology, University of Montana, Missoula, Montana

R. O. Bierregaard, Jr., World Wildlife Fund—U.S., 1255 23rd Street NW, Washington, D.C.

K. S. Brown, Jr. Department of Zoology, Universidade Estadual de Campinas, São Paulo, Brazil

John Cairns, Jr., University Center for Environmental Studies and Department of Biology, Virginia Polytechnic Institute and State University, Blacksburg, Virginia

Sara Cairns, Department of Neurobiology and Behavior, Cornell University, Ithaca, New York

Martin L. Cody, Department of Biology, University of California, Los Angeles, California

David C. Culver, Department of Ecology and Evolutionary Biology, Northwestern University, Evanston, Illinois

Jared Diamond, Department of Physiology, University of California Medical School, Los Angeles, California

Theresa Dillon, Section of Ecology and Systematics, Cornell University, Ithaca, New York

Andrew P. Dobson, Department of Biology, Princeton University, Princeton, New Jersey

Robin B. Foster, Department of Botany, Field Museum of Natural History, Chicago, Illinois

Alwyn H. Gentry, Missouri Botanical Gardens, St. Louis, Missouri

Hubert Gillet, Laboratoire d'Ethnobotanique, Museum National d'Histoire Naturelle, 57 rue Cuvier, Paris, France

Michael E. Gilpin, Department of Biology, University of California, San Diego, California

L. H. Harper, Department of Biological Sciences, State University of New York, Albany, New York

Paul H. Harvey, Department of Zoology, University of Oxford, Oxford, England

B. G. Hatcher, Marine Biological Laboratory, University of Western Australia, North Beach, W.A., Australia

M. B. Hays, World Wildlife Fund—U.S., 1255 23rd Street NW, Washington, D.C.

Stephen P. Hubbell, Department of Biology, University of Iowa, Iowa City, Iowa

Daniel H. Janzen, Department of Biology, University of Pennsylvania, Philadelphia, Pennsylvania

R. E. Johannes, CSIRO Marine Laboratories, Hobart, Tasmania, Australia

Carl F. Jordan, Institute of Ecology, University of Georgia, Athens, Georgia

Robb F. Leary, Department of Zoology, University of Montana, Missoula, Montana

F. Thomas Ledig, Institute of Forest Genetics, U.S.D.A. Forest Service, Pacific Southwest Forest and Range Experiment Station, 1960 Addison Street, Berkeley, California

Henri Nöel Le Houérou, CEPE—Louis Emberger, Centre National de la Recherché Scientifique, B.P. 5051, Montpellier, France

Thomas E. Lovejoy, World Wildlife Fund—U.S., 1255 23rd Street NW, Washington, D.C.

Anna Marie Lyles, Department of Biology, Princeton University, Princeton, New Jersey

J. R. Malcolm, Department of Zoology, University of Florida, Gainesville, Florida

Robert M. May, Department of Biology, Princeton University, Princeton, New Jersey

Charles H. McLellan, Centre for Environmental Technology, Imperial College, London, England

Norman Myers, Consultant in Environment and Development, Upper Meadow, Old Road, Headington, Oxford, England

Arne Naess, Institutt for Filosofi, University of Oslo, Oslo, Norway

Stuart L. Pimm, Department of Zoology and Graduate Program in Ecology, University of Tennessee, Knoxville, Tennessee

A. H. Powell, 228 Apache Street, Tavernier, Florida

G. V. N. Powell, National Audubon Society Research Department, 115 Indian Mound Trail, Tavernier, Florida

C. E. Quintela, Museum of Zoology, Louisiana State University, Baton Rouge, Louisiana

Deborah Rabinowitz, Section of Ecology and Systematics, Cornell University, Ithaca, New York

Katherine Ralls, Department of Zoological Research, National Zoological Park, Smithsonian Institution, Washington, D.C.

A. B. Rylands, National Institute for Amazon Research (INPA), C.P. 478, 69000 Manaus, Brazil

H. O. R. Schubart, National Institute for Amazon Research (INPA), C.P. 478, 69000 Manaus, Brazil

Harald Sioli, Max Planck Institut für Limnologie, Postfach 165, D-232 Plon, West Germany

Michael E. Soulé, School of Natural Resources, University of Michigan, Ann Arbor, Michigan

Alan R. Templeton, Department of Biology, Washington University, St. Louis, Missouri

John Terborgh, Department of Biology, Princeton University, Princeton, New Jersey

David S. Wilcove, The Nature Conservancy, 1800 North Kent Street, Arlington, Virginia

CHAPTER 1

CONSERVATION BIOLOGY AND THE "REAL WORLD"

Michael E. Soulé

A good field biologist is often more concerned with what is missing in a situation than with what is present. Similarly, the perspicacious reader will ask what was left out of this book. It is a healthy sign that the answer is "a lot." Entire disciplines and many topics could not be included, simply because the boundaries and activities of the science keep expanding. Ideally, we would also have preferred to include many more contributions written from the perspectives of people who work with nature reserves and zoological and botanic gardens. In addition, there could have been much more about captive breeding, about introductions, about genetic engineering, about restoration and rehabilitation of degraded habitats, and about management conundrums and insights. In fact, most of these areas and topics have had recent conferences of their own, and there are several excellent books and publications either recently published or in press (Mooney, 1985; Regal, 1985; Ralls and Ballou, in press; Jordan and Gilpin, in press). One glaring exception is the interface between theory and practice, especially from the viewpoint of the manager of parks and forest reserves. This deserves its own project, although many of the authors (especially in Sections V and VI) have been involved at various levels with management agencies and managers.

Another area that was regretfully omitted is integrated agro-ecosystems—the study of the traditional knowledge and methods in rural life that are gentler on the landscape than is high subsidy, highly mechanized agriculture. We could not have done justice to this field in the space available. It also deserves a book of its own.

The line had to be drawn somewhere, and with the exception of the last two sections, it was drawn around the "core" areas, including biogeography, ecology, systematics, genetics, and behavior (Sections I through IV).

Section V takes us from the realm of theory and experiment (in more or less natural communities) to the realm of human impact. This section is by no means an attempt to provide a comprehensive overview of the status of all of the planet's sensitive or threatened habitats; it is merely a sampling of some of the most obvious (tropical forests, the Sahel) and some of the most neglected (cave faunas, tropical marine and freshwater habitats). The status of some other habitats is referred to in Chapters 7 (Mediterranean climates), 8 (tropical montane regions), and 14 (islands).

Section VI examines several of the interfaces between conservation biology and society (Chapters 23 and 24), and between conservation biology and ethics (Chapter 25). Thus, this volume steps gingerly into the "real world." Many of the authors also appear to be suggesting that the perennial reluctance of scientists to discuss matters of ethics may imperil the very organisms and processes they hold most dear.

The omission of any subject, regardless of how arbitrary it may appear, is certainly not meant to imply any limits to the content of conservation biology. This discipline should attract and penetrate every field that could possibly benefit and protect the diversity of life.

WHAT IS THE "REAL WORLD"?

The term "real world" appears in the title of this chapter, and it is always part of the banter during conferences and workshops when the subject matter is conservation. This is because the ultimate test of conservation biology is the application of its theories in actual management situations. Like many colloquialisms, however, the term is somewhat problematic. To most of us, I think, the term "real world" means the part of our lives that involves face-to-face interactions with others and with their desires, priorities, and prejudices; it is the world "out there," the world of politics and economics, and all the vagaries of human nature that we associate with these areas.

But the term "real world" implies that other worlds are less than real. It begs the question, for example, of the "internal world" of subjective experience, and the "middle world" of cognition, experiment, and discourse. Such distinctions are oversimplified, but this one is so ingrained that attack is futile. However, it is important to bear in mind that there *is* traffic bridging these so-called worlds. Events and prevailing philosophies in the real world influence and determine our private intellectual lives. Therefore the activity of these private worlds, including models and theories, are in part the products of learning and conditioning in the social-cultural milieu. Just as the idea of a "star wars" defense shield had to await the development of nuclear missiles, so the idea of conservation biology had to await a serious threat to biological diversity.

Mission-oriented crisis disciplines such as conservation biology straddle the frontier between these worlds. Perhaps a better metaphor is that they are like a shuttle bus going back and forth, with a cargo of ideas, guidelines, and empirical results in one direction, and a cargo of issues, problems, criticism, constraints, and changed conditions in the other.

As I will suggest, however, none of these worlds is at bottom any more real or important in a social or physical sense than any other. Each is part of a larger whole. When this fact is ignored, we generate disciplinary hierarchies that inhibit progress in solving conservation problems. Conservation biology will succeed to the degree that its theoreticians, practitioners, and users acknowledge the larger context in which they exist, and to the degree that they respect one another's roles, contributions, and problems. Anyone wishing to cultivate such an attitude of respect, or wishing to counter a narrow, chauvinistic definition of the discipline need only consider the history and diverse intellectual tributaries of conservation biology. This is the subject of the following section.

THE ORIGINS OF CONSERVATION BIOLOGY

There used to be a popular saying in taxonomic circles that a species was "good" (real) when enough systematists agreed that it was good. Consensus is an important part of any human enterprise.

Consensus can also define a discipline. Disciplines are not logical constructs; they are social crystallizations which occur when a group of people agree that association and discourse serve their interests. Conservation biology began when a critical mass of people agreed that they were conservation biologists. There is something very social and very human about this realization.

When and why did conservation biology become a discipline? The preliminatry groundwork occurred in the early decades of this century with the emergence of such disciplines as forestry, fisheries, and wild-life management. The history of the current crystallization has been documented by Brussard (1985). It seems to have happened in the early years of this decade. "Why" is a little harder to explain, but I think it is possible to identify a few of the important tributaries that fed this social and scientific confluence.

One of these was intellectual. The 60s and 70s saw an explosion of theories and data in community ecology and island biogeography. By the end of this period, there were many people who were testing these ideas in conservation programs.

At the same time, there was a perception of a widening gulf separating modern population biology (including the above fields) from the more traditional disciplines in the natural resources. Biology departments were fragmenting into modern versus traditional, botany versus zoology. Departments and schools of conservation, forestry, fisheries, wildlife management, and similar "applied" fields had become institutionally isolated from each other and from the "mother" scientific disciplines. This inhibited the natural diffusion and application of new ideas and information from one discipline to another and into agency and management circles. Isolation leads to alienation, alienation to hostility. Maybe some reconciliation was overdue.

Social trends may also have contributed to the origins of conservation biology. In the late 70s, many academics in the "pure" sciences wanted very much to contribute to conservation—to be "relevant." Altruism was part of it. The environmental movement had begun. Care for the planet, appropriate and harmonious lifestyles, and a healthy environment were the lenses that focused the energies of a generation. Even today, many university students wish to find a career that promises idealism along with security. But another part of it was a defensive reaction; many of us were finding that development projects were obliterating our research sites, whether they were tropical forests in Costa Rica or vacant fields and canyons next to the campus.

Still another factor has been the extinction crisis. The probable disappearance of the majority of species during the next 50 to 100 years is—or certainly should be—a matter of great concern. This concern will grow as the fruits of poor planning and greed, particularly in the tropics, begin to foul the lives of more and more hungry and deprived people, and to destroy the environments in which they live. As first hundreds, then thousands, and finally millions of species are threatened with extinction over the next century, there will be a crescendo of protest from those to whom "right to life" is more than just a chauvinistic movement by and for *Homo sapiens*.

Finally, the current interest in conservation biology owes much to the intellectual leadership and vision of individuals in the zoo and herbarium world, in some government agencies, and in some private foundations. Some of these people have been posing challenging questions (such as "how big must a population be to minimize the risk of extinction?"). Others have generously supported research, conferences, and publications. From all of these sources, conservation biology was born. Its current polymorphism reflects this.

The idea of conservation biology seems to convey several things at once, including scholarship, a common purpose, and the potential for making a significant personal contribution to the world. For students and established scientists alike, conservation biology seems to represent a community of commitment, and something of value to identify with. The famous quotation of Victor Hugo is relevant: "There is one thing stronger than all the armies in the world: and that is an idea whose time has come."

I referred above to the tendency for suspicion and hostility to grow out of disciplinary fission and isolation. A related tendency is the inevitable tension generated by anything new. To the extent that conservation biology is perceived as a new field, and as enthusiasm for it grows, it will become threatening to some people. It need not, and it can minimize conflict if it can avoid repeating the errors of the past. These errors are elitism and isolation.

Elitism lurks whenever a field has a strong academic foothold. Whether the root of elitism is arrogance from within the ivory tower or fear from without, it is always a danger. There is no hiding the fact that much of the current interest in conservation biology is occurring within academic circles, because it is. But it would be a mistake to lose sight of the fact that the origins of the current movement can be traced back to Thoreau, Pinchot, and Leopold, to Darwin and Elton, to Marsh and Carson, to Teddy Roosevelt and Ashoka, and to Lao Tzu and Saint Francis. More recently its predecessors and contemporaries include many practitioners of ecology, biogeography, systematics, genetics, evolution, epidemiology, sociobiology, forestry, fisheries, wildlife biology, and of the auxiliary sciences of agronomy, veterinary science, resource economics and policy, ethnobiology, and environmental ethics.

Another, related, danger is isolation—elitism's child. If conservation biology becomes isolated in the mental world of academia, it will be of little use. Its prescriptions will not be informed by the real-world problems of the managers, by the actual circumstances of the people who are most involved and affected, and by the knowledge and experience of other consultants (agency employees, extension workers, social scientists, etc.). On the other hand, it is just as important that managers make an effort to be informed about the real world of science as it is that the scientist make an effort to be informed about the real world of management. The only thing that is really real is the whole thing.

THE DILEMMAS OF GOING PUBLIC

Avoiding the social and psychological pitfalls just described will be the greatest social challenge faced by conservation biologists. Simply, most of us received little or no training in the skills needed for functioning in the real worlds of institutions, government, and management. Many academic scientists, in particular, have had no training or preparation for "going public." Speaking out takes courage. Here I will explore two dilemmas associated with speaking out. In different ways, both are concerned with the problem of insufficient data, or, "when is it time to go public?"

The Nero dilemma

In many situations conservation biology is a crisis discipline (Soulé, 1985). In crisis disciplines, in contrast to "normal" science, it is sometimes imperative to make an important tactical decision before one is confident in the sufficiency of the data. In other words, the risks of non-action may be greater than the risks of inappropriate action. Warfare is the epitome of a crisis discipline. On a battlefield, if you observe a group of armed men stealthily approaching your lines, you are justified in taking precautions, which may include firing on the men. To a remote and objective observer, however, the behavior of the men does not prove that they are enemy troops. Alternative possibilities include (1) they are "your" troops returning from a mission; (2) they are forces belonging to a third, neutral, or friendly, force; (3) they are illusory manifestations of hysteria. Of course, this objective observer, like most pure scientists, is free to entertain such alternative hypotheses, because being wrong is not a matter of life and death for him or her. Except in a few highly competitive fields (which, indeed, are war-like), there are no severe penalties for "fiddling" with such ideas for as long as one likes—or until Rome is in ashes.

Here we are dealing with the problem of provisional validity. In pure science, the idea of provisional validity is often enveloped by the term "working hypothesis." In the world of management, however, the situation rarely permits the time-consuming testing of all the relevant working hypotheses. Rather, the "best" hypothesis must be chosen and implemented. As an example, the evidence presented in Chapters 4 and 5 supports, in my opinion, the inference that any significant loss of genetic variation is likely to reduce fitness significantly. Therefore, such genetic erosion should be avoided in managed species or populations, especially in small ones where the loss of genetic variation is chronic and irreversible. Strictly speaking, such a statement is a working hypothesis because it probably lacks sufficient empirical backing to satisfy many experts, and because there are several categorical exceptions (as there are with all biological "laws"). The possible risks of choosing a bad hypothesis, however, must be weighed against doing nothing.

In general, the provisional nature of a hypothesis is no impediment to its operational validity, as long as there are no contradictory, equally supported, guidelines. To ignore such a hypothesis may endanger a species that is being managed, and it exposes the manager to the criticism of reviewers and peers. This is because the manager exists in the real-time world of risks and benefits, not in the null-time world of pure science. In conservation, dithering and endangering are often linked.

The "bottom line" dilemma

Another dimension to the problem of "speaking out" has to do with the issue of complexity. Facing this problem also requires courage. Biologists, including managers of wildlands, are painfully aware of this complexity in organisms and ecosystems, but many people are not, and nonscientists are often unprepared to deal with the conditional, qualified, diffuse, and nonlinear nature of Nature. Stuart Pimm (Chapter 14) points out, for instance, that engineers and politicians, who are always looking for "the bottom line," will not be pleased with community ecologists, in part because the latter are unable to provide a straightforward, one-line statement about community structure and stability.

One solution is to simplify to the point of meaninglessness. A better one is to educate nonbiologists at every opportunity about the nature of complex systems. Simply put, the "bottom line" for complex, nonlinear systems is usually "it depends."

One reason why "it depends" is diversity. Communities and ecosystems are like individuals—no two are alike. Therefore, the quest for a simple bottom line on such issues as stability, extinction thresholds, effects of atmospheric pollution, effects of overharvesting or species introductions, effects of eutrophication, etcetera, is a quest for a phantom by an untrained mind. Qualification of generalizations is inherent in the subject matter of biology, just as it is in the social sciences. It benefits no one to be defensive about the "bottom line" issue. Just as in social issues, there is rarely an answer in conservation biology that is both simple and politically appealing.

One perennial solution to this dilemma is the call for more research. This slogan is often abused when it is used as an excuse to delay action. But it can usually be justified; we can always use more research. But this brings us back to the Nero dilemma—"more research" is sometimes a dangerous luxury when the world is coming down around our heads.

Courage in education and communication

Dealing with emotionally and ethically complex material often requires courage. It is natural to back away from "nonscientific" or "messy" matters, leaving such issues to journalists, ethics committees, popularists, and "nature lovers." For example, there are probably conservation biologists who would be reluctant to lecture their students on how to love nature. In their defense, one could argue that loving nature is not a scientific subject, that ardor reveals subjectivity, and, above all, that a biologist is not an expert on loving nature.

This is rubbish. Everyone knows that ardor and enthusiasm are inspiring, and that love of subject matter is infectious. Regarding the issue of expertise, no one has more expertise on loving nature than the professional naturalist or manager who has spent his or her career (or lifetime) studying and admiring plants and animals.

Arne Naess (Chapter 25) points out that it is very important that experts be heard to espouse the intrinsic value of species and nature, and that they try to communicate their spontaneous, inner experience of their unique insights. Admittedly, most of us find it hard, at first, to overcome our lifelong conditioning and inhibitions. But we don't have to go around kissing flowers and beetles. We should, however, let our students and audiences know that nature is more than just a collection of useful reseach tools or genetic resources for industry and agriculture.

Not that there is anything bad about utilitarian or instrumentalist arguments for nature protection, especially when talking to economists and some politicians. But these arguments are dry, and besides, they have no shortage of effective spokespersons. More to the point, I agree with Naess that most people respond better to appeals that come directly and spontaneously from one's private experience of nature, including our anguish as we witness its slow death. Who is more capable than biologists to spread the word that it is wrong to terminate evolutionary lines, and that it is wrong to wipe out entire communities?

It is often tempting to leave speeches to the politicians, to leave morality to the priests, and to leave ethics to philosophers; however, few of them can speak with authority and familiarity about the exquisite detail and amazing diversity of life.

Students and other people respond well to honesty and emotion. Emotions are as real as cognition, and emotional appeals are often effective in the real world. (Just look at television commercials.) I don't mean that appeals to emotions and intuition should always replace rational and materialistic arguments, or that one true ethical structure exists that can provide answers to all complex problems and conflicts. It is simply that it is approaching "high noon," and we should use every (ethical) tool at our disposal to minimize the damage to this planet.

In this context, it is important to note that many of the authors in this volume touch on ethical issues. Several of them imply or suggest that the "right" of humans to produce as many babies as they like, or as God wills, is arrogant, callous, and selfish. Other conflicts about values are pointed out. These include the rights of wolves versus those of sheep in Scandinavia, the benefits for wildlife (and ultimately for humans) of the tsetse fly versus the rights of cattle and contemporary African pastoralists, and the rights of rivers and traditional human cultures in Amazonia versus those of dams and developers.

Here again, biologists should ask for the help of professionals in other disciplines, including philosophers, anthropologists, sociologists, and community development workers. Otherwise, we risk espousing solutions that are theoretically robust, but socially and politically naive.

PROGRESS AND THE FUTURE

Conservation biology has made significant strides recently. Much of value has been learned about the utility and limitations of island biogeography, about area effects, edge effects, genetic management, disease, the results of removals and introductions, endemicity, rarity, etcetera. Much of this information is already being applied in the design and management of natural areas, in captive breeding programs, and in other conservation projects. These are encouraging tidings. But they are not enough.

One major impediment is the virtual absence of any formal training in this field. There are no degree programs, and resources for training and research, in relative terms, are negligible. The rate of technology transfer is too slow, and the communications between our colleagues "on the ground" and colleagues in the academy is often poor at best. And last, but certainly not least, there is pitifully little funding for research. Our principal social challenges in the years to come will be:

1. To strengthen the financial and institutional supports for conservation biology, in both its research and educational roles.
2. To strengthen the contacts with related, natural resource disciplines.
3. To strengthen contacts with the management community itself.

The necessary research

What research problems need addressing next? I suppose that the number is infinite, but certain ones stand out to me as especially important and challenging. Before mentioning these, however, it might be helpful to note that just as the genetics of nature conservation is the genetics of scarcity (Frankel and Soulé, 1981), so *conservation biology is the biology of scarcity*. The conservation biologist is called in when an ecosystem, habitat, species, or population is subject to some kind of artificial limitation—usually a reduction of space and numbers.

Scarcity is an obvious issue when there is gross habitat destruction or when a species suffers a great reduction in population size from poaching or overexploitation. These are examples of quantitative scarcity. But there is also qualitative scarcity—as when air pollutants or hazardous chemicals degrade a habitat without necessarily reducing its area.

Problems of scarcity can be examined at various levels, from that of the local population, through the metapopulation, the entire species, to the community and the ecosystem. But three levels of complexity and problems are clearly identifiable. The first level is the *population*; the principle issue is population viability (or extinction), often referred to as the minimum viable population (MVP) problem (Chapter 2).

I propose that the MVP problem is one of the most seminal issues in population biology, let alone in conservation biology. Not only does the problem of viability require the simultaneous consideration of demography, genetics, behavior, and ecology, but it forces us, as biologists, to come face to face with the ultimate inquisitor. Shaking his finger in our faces, he demands, "If you say that you are a population biologist, and if you say that population biology is a genuine science, then tell me how I can predict the length of time that a population will persist?" Until we can answer such questions (at least within an order of magnitude), the maturity of conservation biology (and population biology) and its status as a predictive science will be in serious doubt.

Challenging problems also exist at higher levels of biological organization. One of the obvious problems at the *community* level is the question of community structure, and the relative weights of stochastic versus deterministic forces affecting such structure. The specific questions include:

1. What is community equilibrium?
2. Is it ever achieved?
3. Is it more or less likely to exist in the humid tropics than elsewhere?
4. Are there multiple stable states or equilibria?

The skeptic might ask how such arcane questions could ever be germane to the day-to-day management of wildlands and preserves. In answer we might appeal to such nebulous rationalizations as "we cannot manage what we don't understand." A better answer would be to point to real issues where a more mature theory would help us prescribe better guidelines. For example, if communities are often saturated or supersaturated with species, it is harder to justify trying to prevent the extinction of rare species; there could be ecological and economic arguments for allowing the community to "settle" into a lower, but more stable, level of diversity. On the other hand, if it is common for there to be a relatively large random element in community composition, and if species are not packed very tightly, then there might be less concern about the impacts of reintroductions and the introductions of "analog species" to replace lost elements; furthermore there would be less justification for permitting any extinctions.

Pressing multidisciplinary research problems also abound at the level of the *ecosystem* and biosphere. An incomplete list of these problems includes the greenhouse effect, acid rain and symptomatically similar forms of atmospheric pollution, the biological and climatic consequences of nuclear war, and the climatic, economic, and ecological effects of deforestation and desertization. In addition, we would like to predict the effects that damming most of the world's large rivers will have on climate, commerce, the oceans, and biological communities.

Finally, we cannot ignore people. For example, the implementation of "biosphere reserves" as sites for the harmonious coexistence for humans and nature (UNESCO-UNEP, 1984) depends on both a good grasp of the local biology and on the enthusiastic support of the indigenous peoples. In fact, the survival of many natural biological communities is going to require the creative cooperation of biologists, social scientists, and politicians, especially in the tropics. It won't be long before many conservation biologists are spending more time at community meetings than in the field or laboratory.

Clearly, there is no scarcity of issues—only a scarcity of time and the resources to address them.

CODA

The planetary tragedy is also a personal tragedy to those scientists who feel compelled to devote themselves to the rescue effort. It is painful to witness so much termination.

The other side of the coin is that most humans want to contribute something of social value, and want to be recognized for that contribution. It is the fortunate scientist who can do this without compromising standards and without taking time off from his or her career.

Conservation biologists are such people. Not many scientists, during the few decades of their careers, are able to commit their minds and labors to an epochal task like saving the planet.

It will be a long struggle; many generations of conservation biologists have been and will be conscripted. In the process there will be as much pettiness and jealousy as always, but this will not be a serious impediment if we keep in mind that no one individual or subgroup can ultimately accomplish much on its own. Complementarity is inherent in conservation biology, not only because it is multidisciplinary, but also because it is just one of several fields that are essential for the protection of natural systems in places where humans are part of the landscape.

This book is dedicated to the students who will come after, who will witness the worst and accomplish the most.

THE FITNESS AND VIABILITY

OF POPULATIONS

BIOLOGICAL PRINCIPLES, RULES OF APPLICATION

What are the rules for maintaining the fitness of individuals and populations, and what are the biological principles upon which these rules are based? These are fundamental questions in conservation biology. The chapters in this section offer answers to these questions. Equally important, they suggest avenues for further research.

One of the propositions of conservation biology is that populations have demographic and genetic thresholds below which nonadaptive, random forces prevail over adaptive, deterministic forces (Soulé, 1985). For example, genetic drift can overwhelm natural selection in small populations. Similarly, small populations can suddenly succumb to the effects of demographic stochasticity, even when all the individuals are quite robust. Gilpin and Soulé (Chapter 2) examine these and other consequences of smallness; they describe how random, deleterious processes lead to vicious cycles, and how these "vortices" can interact and exacerbate one another. Their chapter asks more questions than it answers, but it provides a map to some of the most difficult and challenging terrain in the conceptual territory of population vulnerability.

A perennial issue in conservation biology is the optimum level of genetic relatedness of potential mates. Imagine a continuum of relationships, beginning at one end with an individual and extending to a family of related individuals, to a patch of related families, to a local population, and continuing to a metapopulation, the species, a cluster of sibling species, and so on. Now we ask, "Where along this continuum is the best place from which to draw potential breeding partners?" In other words, what is the optimum degree of relatedness for the maximization of fitness? In one way or another, all of the chapters in this section deal with this problem. As a first approximation, based on a huge body of data and practical experience, we hardly need to qualify the statement that the breeding of close relatives produces unhappy results (inbreeding depression) in outbreeding species.

At the other extreme, hybridization between individuals from genetically distinct gene pools, especially if there are obvious differences in chromosomal structure and number, causes outbreeding depression, which may be manifest in lowered viability or fertility. These black and white extremes leave a lot of gray in the middle.

INBREEDING DEPRESSION

If close inbreeding is really as bad as the biological literature suggests, then we might expect the evolution of physiological and behavioral mechanisms that allow organisms to avoid breeding with close relatives. Based on an extensive survey, Ralls, Harvey, and Lyles (Chapter 3) demonstrate the ubiquitousness of such mechanisms in animals. Many mammals apparently eschew breeding altogether if the only available partners are siblings, parents, or offspring. Why? Such behavior makes evolutionary sense only if close inbreeding is tantamount to the waste of reproductive effort—if, in other words, the prognosis for inbred offspring is very poor, and the effort to produce and rear them is simply not worth it.

It is one thing to have a rule that says "inbreeding is bad." It is another to have a rule that says "the gradual loss of genetic variation is bad." The latter rule implies that *homozygosity* is bad, even if it occurs without breeding between close relatives. Evidence for such a rule has been accumulating for decades. In this book the zoological literature is reviewed by Allendorf and Leary (Chapter 4), and the tree literature by Ledig (Chapter 5). The rule is sustained; relatively heterozygous individuals are, in general, better off than relatively homozygous individuals. Of course there are exceptions.

Does this mean that populations lacking detectable heterozygosity are doomed? Not necessarily in the short run, but in the long run such populations cannot evolve, so their future is bleak unless they are large enough to regenerate a modicum of variation. Three such cases are discussed in this section: the cheetah (Chapter 4), the Torrey pine (Chapter 5), and the collared lizard (Chapter 6). The former two species show serious disabilities that are probably attributable to homozygosity for deleterious alleles; the data are not in on the lizard. All three cases lack sympatric congeners; hence it is possible that their continued existence depends on the absence of competition and other environmental challenges.

There is a serious bias when comparing the average fitness of homozygous populations with that of genetically more variable ones. The reason is that some fraction, probably a large one, of such homozygous populations is invisible—the fraction that has already gone extinct. The surviving populations are probably not a random sample; it is likely that they constitute the fraction with the least genetic load to begin with.

Even so, the cheetah and Torrey pine examples suggest that the shedding of genetic load (of deleterious recessive alleles) does not automatically accompany the loss of most genetic variation. Rather, theory and experience both lead us to the rule that genetic drift is often accompanied by the fixation of deleterious alleles. Theory also shows (Ledig, Chapter 5) the astronomical odds against the accidental fixation of the best alleles at each locus. In addition, genetically depauperate populations are robbed of whatever contribution to fitness is made by overdominance.

OUTBREEDING DEPRESSION

If inbreeding is bad and heterozygosity is good, should we conclude that the optimum mate is the most distantly related one? Templeton (Chapter 6) explains the genetic and ecological reasons why the answer to this is "no"—the various manifestations of outbreeding depression. Outbreeding depression is one of the areas of population biology where semantic and definitional problems obscure the small body of real information. In addition, this subject area is burdened by mythic, subconscious currents that occasionally rise to the surface in formulations such as "racial purity," "pure types," and similar projections of troop and tribal atavism.

Some of the less anthropocentric terms used to "explain" the phenomena underlying outbreeding depression are "coadaptation," "epistasis," "genetic integration," "harmonious interaction," and "internal balance." The breakdown of coadaptation is sometimes referred to as "negative heterosis," and is manifested by "developmental instability" and by a decline in "fitness." Some of these terms (such as epistasis) have clear genetic meanings, but even these are context-dependent. A few remarks might be helpful, especially in light of the differing uses of the term "coadaptation" by Ledig (Chapter 5) and Templeton (Chapter 6).

Ledig uses the term in its traditional sense (see Futuyma, 1979 for a discussion). In this sense, however, the term is a conceptual receptacle for all developmental–organismic, integrative phenomena. An organism *must* be "harmoniously integrated" or "coadapted" merely to undergo mitosis, ontogeny, and reproduction. At the very least, the parts of an organism must be compatible enough so they don't work at cross purposes. Mere existence in nature is a sufficient, if trivial, demonstration of this kind of "coadaptation." Outbreeding depression is an equally trivial demonstration of "coadaptation," because if two biological entities are different enough to be recognized as different, it should be anticipated that their parts will not be fully compatible.

If the traditional use of coadaptation is to be retained, we should probably refer to it as "general" or "universal" coadaptation, because it is a catch-all category that includes nonepistatic and epistatic coadaptation, as well as many kinds of regulatory and integrative processes at every level from the gene to social behavior.

Returning to the practical problem, does the genetic basis of coadaptation differ sufficiently between populations of a species so that outbreeding depression should be a concern of managers? Templeton (Chapter 6) sheds some strong light on this question. First he points out that a major component of outbreeding depression is the specific, local adaptations of populations, and that the mixture of these adaptations that occurs in hybrids is known to be nonadaptive in some cases.

Next Templeton narrows the field of discourse. He provides an operational definition for one class of coadaptation, "intrinsic coadaptation." It exists when the descendants of a particular cross have reduced fitness even though the parental lineages have identical chromosomal architecture (linkage relationships). The apparent mechanism for such outbreeding depression is the breakup by recombination of adaptive linkage relations. Internal coadaptation can be produced by either (a) alleles that improve the efficiency of interaction between loci or their products (positive epistasis), (b) alleles that improve efficiency in different noninteracting, independent processes, or (c) a combination of these two phenomena. The former kind of internal coadaptation is a functional example of a "coadapted gene complex" as the term is used in the older literature in evolutionary biology.

Templeton believes that such intrinsic coadptation is rare in outcrossing species, at least those with even modest levels of gene flow. Of course, intrinsic coadaptation is only a subset of relevant integrative phenomena, and there often exist other kinds of coadaptation (some involving chromosomal rearrangements) that might reduce fitness in the descendants of crosses between distantly related individuals.

It has long been held that natural selection should increase the functional integration of linked alleles in populations. Templeton shows that this is not always possible. Nor does a gradual improvement in fitness, whether produced by internal coadaption or by something else, imply anything about epistasis. For example, a population will be more fit if both disease resistance and vision are improved. But these systems may be completely independent genetically and biochemically; the improvements may be in nonoverlapping sets of genes.

In other words, fitness does not necessarily depend on epistasis or on linkage. Conversely, outbreeding depression is not proof of epistasis or of internal coadaptation. Ledig (Chapter 5) points out that the biochemical and morphological apparatus of photosynthesis may differ among species in the same genus, and that hybrids may be physiologically inferior to the parents. His point is that the "components" of different species may manifest degrees of incompatibility. But the degree of incompatibility is unrelated to the "internal harmony" of the parents. Analogously, the windshield wiper blade of a Ford is more likely to fit on the wiper of another Ford than on the wiper of a Toyota, but this observation cannot be used to predict anything about the engineering standards of Fords or Toyotas.

PATTERNS OF HETEROZYGOSITY

This is not the place to review the decades-long debates that have raged around the topic of heterozygosity. Such questions as, "Is heterozygosity deleterious, neutral, or selected?"; "Is its level in a species related to population size, to ecological amplitude, or to niche width?"; and, "Is it maintained by overdominance, marginal overdominance, or dominance?" are fascinating intellectually, but are not particularly fruitful in terms of applications. Nevertheless, I must refer to Ledig's hypothesis (Chapter 5) that the high levels of heterozygosity occurring in many trees is the end product of a positive feedback cycle driven by genetic load. If Ledig is correct, then scale effects—that is, longevity and size (cell number)—force most trees to adopt an outcrossing breeding system, this being the "cheapest" way for them to minimize the costs of the large load of deleterious mutations that their size and life histories compel them to bear. This is a pregnant hypothesis.

MANAGEMENT IMPLICATIONS

Most wildlands managers can't choose the species in their sanctuary, reserve, or park. Their limited resources must be used to defend what is present. This usually means protection and restoration of habitat on the one hand, and population management of certain indicator or critical species on the other. Often, as with large herbivores, the two are interdependent.

The five chapters in this Section speak mostly to the latter category of problems: maintaining the vigor and evolutionary potential of specific target species. These will be the species that are most vulnerable to environmental perturbation, or are so small that random birth and death events and genetic drift threaten them. The number of species in this set of vulnerable populations will depend on many factors—especially the size of the reserve, because the larger the reserve, the smaller the proportion of species that will have marginal population sizes. If choices must be made among candidate species, then the ecological role of the species is perhaps the most important criterion. Large predators and other keystone or mutualistic species such as some trees (Chapter 15) and pollinators (Chapters 19 and 21) are automatic targets.

If "Nature knows best" and Nature abhors close inbreeding (Chapter 3), then managers should too. Every rule has its exceptions, however, including this one—especially when a past history of inbreeding requires extreme remedies, even the purging of certain recessive deleterious genes from a group (Chapter 6).

Chapters 4 and 5 suggest that any measurable increase in homozygosity is likely to extract a cost in the currencies of immediate fitness and long-term adaptability. But even more dangerous than inbreeding depression is the kind of "depression" that afflicts some agencies when it appears that a population is too small to save. The point is that even when numbers are very low recovery is possible, given the will and the resources. Theory indicates that most genetic variation (though not the rare alleles) can be saved from even a handful of individuals. In such cases, it is usually financial resources, not genetic resources, that is the limiting factor.

Finally, genetic systems are too heterogeneous to allow generalizations about the dangers of outbreeding depression in specific cases. Managers of threatened or endangered species, whether the species are captive or managed in the wild, must be aware of the hazards discussed above and in the following chapters. The problems range from meiotic disturbances and sterility in the F_1 generation to quantitative reduction in developmental homeostasis, viability, local adaptation, and host recognition, to inappropriate timing of reproduction (Templeton, Chapter 6). Anticipating all the potential problems may be impossible, but chromosomal surveys and autecological studies could eliminate most sources of failure.

M.S.

CHAPTER 2

MINIMUM VIABLE POPULATIONS: PROCESSES OF SPECIES EXTINCTION

Michael E. Gilpin and Michael E. Soulé

The term "minimum viable population" has come into vogue, possibly because of an injunction from the Congress of the United States (National Forest Management Act of 1976) to the U.S. Forest Service to maintain "viable populations" of all native vertebrate species in each National Forest. The term implies that there is some threshold for the number of individuals, or some multivariate set of thresholds and limits, that will insure (at some acceptable level of risk) that a population will persist in a viable state for a given interval of time. This chapter introduces the term "population vulnerability analysis" (PVA) for analyses that estimate minimum viable populations (MVPs). That is, MVP is the product, and PVA the process.

Earlier investigations of the MVP problem, including MacArthur and Wilson (1967), Richter-Dyn and Goel (1972), and Leigh (1975), emphasized a demographic approach, in which the expected lifetime of a population was the objective. Their work was based on birth and death branching processes, and they found that there were critical "floors" for size, below which the population would quickly go extinct. A second, more recent body of work (for reviews see Frankel and Soulé, 1981, and Schoenwald-Cox et al., 1983) has focused on genetic aspects of the population extinction question; again, the findings support the existence of critical factors of population size and population structure, below which inbreeding and loss of selectable variation become a problem for the continued survival of the population.

19

A paper by Mark Shaffer (1981) appears to have been the first to take an overall systems perspective. He began by distinguishing deterministic extinction from chance or stochastic extinction. Then he distinguished four separate forces, or kinds of variation, that independently contribute to population extinction. The first two, demographic stochasticity and genetic stochasticity, were just mentioned. The second two, identified by Shaffer, were environmental stochasticity (environmental shocks received by all members of a population), and catastrophes. Despite this comprehensive view, however, Shaffer used only a combination of demographic and environmental stochasticities in his study of the grizzly bear (Shaffer, 1983).

Estimating MVPs is complex. Efforts to consider all four effects jointly and, beyond these, to consider the consequences of habitat fragmentation, have been slow to develop. As elaborated in this chapter, the probability of extinction cannot be pegged to population size alone. In addition, each situation will have a set of "minima" (Salwasser et al., 1984; Soulé, in press), depending on the life history of the species (population), the temporal and spatial distribution of its resources, and its level of genetic variation. In other words, there will be no "magic number," no single MVP that is universally applicable to all species.

These considerations might appear to undermine the idea of MVPs, but we would not agree that the term should be abandoned. First, it is already widely used, and it focuses attention on a critical problem. Second, its disappearance or repudiation would have deleterious effects on conservation. Third, it conveys three important concepts:

1. It defines the single species population as the unit of study.[1]
2. The term "viability" stresses that we are concerned with the persistence of the population over some relatively long temporal interval.
3. The idea of "minimum" suggests that there are critical aspects of the "population-in-its-environment," whether involving its total size, its distribution, or some feature of its genetics, that govern its probabilistic decay from existence to extinction.

In this chapter we take a conceptual, pluralistic overview of the process of species extinction. We do this to provide a framework for the integration of ecological, population dynamic, and population genetic models that must be brought to bear on this vital issue. We base our

---

[1] But, as will be discussed later in the chapter, this unit may not apply to systems of patch populations (i.e., "metapopulations").

efforts on the seminal analysis of Shaffer (1981), but we go beyond Shaffer in two important ways. We not only identify the components of population vulnerability analysis (PVA), but we examine the feedback loops by which the decay in one factor (such as population size) can exacerbate not only itself but also the behavior of other factors (such as inbreeding and fragmentation).[2]

A general conceptual model of the viable population problem is important to a number of concerns in conservation biology. Captive populations must be kept above the MVP that assures retention of genetic variation and fitness. Ecological preserves must be large enough to provide for the minimum viability of their important species. MVP also plays a role in the emerging discipline of ecosystem restoration, because it may govern the composition and timing of the reintroduction of extirpated species. The following general model of viability and its decay is intended to be a guide for further work, not a solution, *per se,* of the MVP problem.

THE DYNAMICS OF SPECIES VIABILITY

Species extinction is a systems phenomenon, involving the interaction of processes and states. In the remainder of this chapter, we will attempt to disentangle some of the major processes that lead to species extinction. At the very least, our exercise provides a checklist for assessing the completeness of a PVA for a particular species. Our approach conceptually integrates the more exact and quantitative models for the demographic, environmental, and genetic mechanisms that contribute to species extinction (Soulé, in press).

For reasons that will become apparent, we view PVA to be based on three interacting fields, the states of which are constantly changing and interacting. "Population phenotype" (PP) includes all of the physical, chemical and biological manifestations of the population. A second field, the environment (E), is the context. It includes all aspects of the abiotic and biotic factors that influence the population. Together, these two fields determine the third field: the "population structure and fitness" (PSF). This is the field in which the dynamic consequences of the interactions of population phenotype and the environment are manifested. Table 1 lists the components of each of these fields, and Figure 1 represents their overlaps and interactions schematically.

[2] We see population vulnerability analysis as a touchstone for the validity of much of the extant theory in population biology. PVA implicitly asks whether current theory is sufficiently accurate and comprehensive to make long-term predictions as to the probable persistence of populations.

TABLE 1. Components of the three fields of population vulnerability analysis (PVA).

Field	Components of the field[a]

Population phenotype (PP)	Morphology:
	  Variation of sizes, shapes, and patterns
	  Geographic and temporal variation
	Physiology:
	  Metabolism
	  Metabolic efficiency
	  Reproduction
	  Disease resistance
	Behavior (intra- and interspecific):
	  Courtship and breeding
	  Social behavior
	  Interspecific interactions
	Behavior (distribution):
	  Dispersal
	  Migration
	  Habitat selection

Environment (E)	Habitat quantity
	Habitat quality:
	  Abundance (density) of resources
	  Abundance of interacting species
	  Patterns of disturbance (duration, frequency,
	   severity, and spatial scale of disturbances)

Population structure and fitness (PSF)	Dynamics of spatial distribution:
	  Patch distribution
	  Metapopulation structure and fragmentation
	Age structure
	Size structure
	Sex ratio
	Saturation density
	Growth rate (r)
	Variance of r:
	  Individual
	  Within patches
	  Between patches

hard one

[a]The categories listed for each field are meant to be suggestive, not comprehensive.

Population phenotype

The population phenotype field can be divided into four sections. Although these sections are somewhat artificial, they have heuristic value. The first section of this field is the most concrete; it is the *morphology* section. Here we are dealing with the tangible, physical,

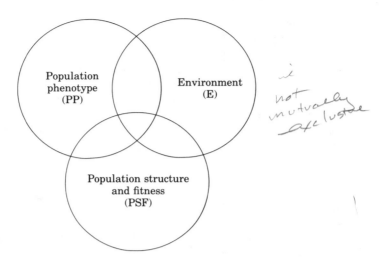

FIGURE 1. The three interacting fields of analysis in population biology: population phenotype (PP), environment (E), and population structure and fitness (PSF). See text for complete discussion.

static representations of size, volume, color, texture, and shape of morphological structures. These are things that can be stored in museum cabinets.

The other three sections or components of the PP field are dynamic; they are the physiological and behavioral processes of individuals. The physiology section is *metabolism* in the broadest sense, including nutrition, coordination, reproductive processes such as gametogenesis, and many aspects of disease resistance. A third section includes sexual, courtship, breeding, and social *behaviors*. It also includes the behavioral aspects of interactions with other species (prey, predators and competitors). The fourth section includes those behaviors (migration, taxes, dispersal) that govern the choice of and *movement* between different habitats.

Environment

It is useful to examine the environment field from two perspectives—*quality* and *quantity*, because a deterioration of either can extinguish a population. While both quantity and quality contribute to the number of individuals that will be found in the associated region, each does it in a different way (though this distinction may collapse under certain circumstances, as discussed below). For terrestrial species, "quantity" refers to the amount of habitat available to the population, everything else being equal. Environment or habitat quantity scales the total population size and may influence aspects of its distribution, including how it responds to a perturbation of environment quality.

Environment quality comprises everything extrinsic that determines the adaptedness of the species including the relative fitness of the individuals. It interacts with the population phenotype to establish the population density or carrying capacity. Quality includes the states of the physical environment, the abundance of resources (food, nutrients, shelter, mutualists, breeding sites), and the kinds and numbers of interacting species (competitors, predators, herbivores, and disease organisms). Environmental quality also has dynamic components; it includes the patterns of change in all of these factors. In other words, quality depends on disturbance dynamics as much as it does on average conditions.

From the viewpoint of population vulnerability analysis, the disturbance regime where a population lives is often the most important aspect of the environment. This is because many of the populations that are subject to PVA will be restricted to island-like habitats such as parks and refuges, and they will be unable to escape to other refugia when the environment deteriorates (Peters and Darling, 1985). Therefore, a PVA must anticipate the "worst-case" eventualities for each targeted population in each place.

The complex nature of environmental quality notwithstanding, a good naturalist is often able to summarize his or her habitat knowledge of a species, using such categories as "marginal" to "central," or "good, medium, and bad." Such subjective judgments about environment quality will often be our best and only guides to the relative merits of particular sites.

Population structure and fitness

Population phenotype and the environment interact to produce what we call PSF. These are the measurable aspects of population structure and fitness: the population's age structure, sex ratio, size structure, and the distribution of individuals over time and space. In this field we also include the dynamic changes in these variables, which give the population growth rate (r) and the saturation density of the population.

DETERMINISTIC AND STOCHASTIC EXTINCTIONS

The goal of a population vulnerability analysis is to establish a minimum viable population that reduces the risk of an extinction to an acceptable level. In this quest, it helps to classify extinctions into the two kinds mentioned above, deterministic and stochastic. Deterministic extinctions are those that result from some inexorable change or force from which there is no hope of escape. Deforestation and glaciation are two such forces for many species of trees. For monkeys, the disappearance of all the trees in a region is such a force. For the species-specific parasites of monkeys, the death of all the monkeys is such a force. *Are oil spills deterministic or stochastic* [handwritten annotation]

A deterministic extinction occurs when something essential is removed (such as space, shelter, or food), or when something lethal is introduced (such as too many cats or hydrogen ions). Deterministic extinctions can be linked. For example, the extermination of the dodo bird by humans doomed a Mauritian plant species to extinction (Temple, 1977), because only by passage through the gut of the dodo could the seeds of this plant be primed for germination. Loss of habitat has already been a major factor in the extinction of a number of tropical plant species (Gentry, Chapter 8). The total extinction of a species due to habitat loss, however, can only occur (1) when the species is a local endemic, (2) when it is already on the verge of extinction, or (3) when humans or global events cause wholesale habitat destruction.

Stochastic extinctions are those that result from normal, random changes or environmental perturbations. Usually such perturbations thin a population but do not destroy it; once thinned, however, the population is at an increased risk from the same or from a different kind of random event. The smaller a population, the greater its vulnerability to such perturbations. Also, the shorter the interval between such events, the more likely the population will be pushed over the brink before it can recover to a safe size.

The following section will emphasize stochastic extinctions, because the processes involved are more subtle and more difficult to observe and defend against. As with most dichotomies, however, this one can be misleading. Many extinctions are the result of a deterministic event that brings the population into a size range where rather frequent or probable stochastic events can easily terminate it. For example, habitat destruction or overharvesting will reduce a population to the point where a stochastic extinction is inevitable (see Soulé and Simberloff, 1986, for examples).

THE FOUR EXTINCTION VORTICES

We are proposing that any environmental change can set up positive feedback loops of biological and environmental interactions that have further negative impacts on the population, possibly leading to its extinction. We refer to these event trains as "extinction vortices." The first question is, "What are the changes that can bring a population under the influence of one of these vortices?"

Figure 2 attempts to capture such events—where a major loss of habitat causes reductions in population size, *N,* and in population distribution, *D*.[3] The consequences of reductions in *N* and *D* are shown at the bottom of Figure 2. One possible outcome is the immediate (deterministic) extinction of the population. If the population survives, it will probably be more fragmented and will suffer an increase in demographic stochasticity, and an increased chance of stochastic extinction.

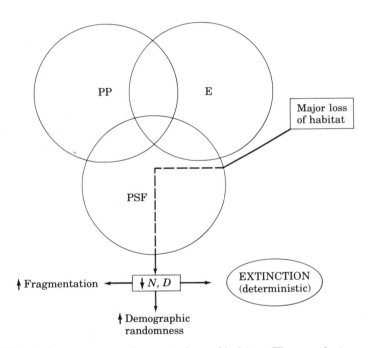

FIGURE 2. Consequences of a major loss of habitat. The population suffers reductions in population size *(N)* and in distribution *(D)*. One possible outcome is the immediate deterministic extinction of the population. Even if the population survives, it will be increasingly vulnerable to stochastic extinction. (↑ = increase; ↓ = decrease.)

[3] For the purposes of this chapter we characterize the population distribution with a single variable, *D*, where this is a summary statistic that describes the intensity of the fragmentation—i.e., the number of patches and the degree of their isolation. We recognize that fragmentation has both genetic and population dynamic consequences (Gilpin, in press a) and that its description may require more than a scalar variable.

Demographic stochasticity

"Demographic stochasticity" is often the immediate precursor of extinction. It is defined as the chance variation in individual birth and death. It is assumed to be independent for each individual. In a small population, extinction can occur accidentally (randomly) because of high death rates or low birth rates. A population is not safe from such a chance failure in recruitment until it has a large number of members. For an organism that reproduces by fission, such as a microorganism, MacArthur and Wilson (1967) found a critical limit of 10 individuals. For a dioecious organism with a more complex life history, including a prereproductive period, Shaffer (1981) obtained numbers in the range of 50. Whenever a population is drawn down to this size, it is in danger of falling prey to demographic stochasticity.

Very small populations are therefore in extreme jeopardy. In Chapter 12, Lovejoy et al. show that 10-ha patches of Amazonian forest are large enough to contain only one or two groups of red howler monkeys *(Alouatta seniculus)*, with up to about 10 individuals in each group. Such groups are not likely to persist for more than a few generations, according to population biology theory.

Environmental stochasticity

There are two general routes to the domain of strong demographic stochasticity. The first is a decrease in habitat quantity. The second is disturbance, or a deterioration in environmental quality. Most changes in quality fall under the heading of environmental stochasticity.

"Environmental stochasticity" is the random series of environmental changes. In models, a frequent assumption about environmental variation is that all individuals in a patch feel these perturbations in an equivalent way, and that regardless of the initial population size, a series of negative blows can reduce the population by orders of magnitude and bring it to a state where demographic stochasticity can take hold. The distribution of the species in space, however, may tend to ameliorate this, because a large range may introduce ecological heterogeneity that buffers the action of the perturbations.

Figure 3 adds two additional state variables. It suggests that a perturbation can reduce r (for example, by reducing per capita resources, and via mortality by affecting age structure). It also shows that the genetic effective population size, N_e, is likely to be reduced.

Both effects—a lower N_e and increased demographic stochasticity—have consequences or "outputs," to use systems theory terminology. The output of a lower N_e is an increase in genetic drift and inbreeding,

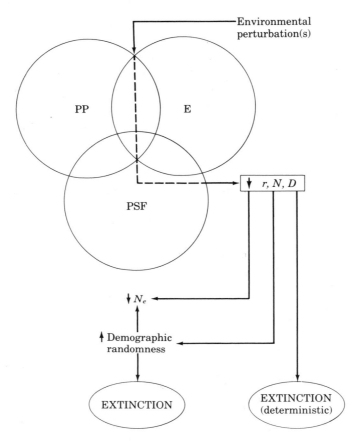

FIGURE 3. The effects of one or more environmental perturbations. These can affect the population's growth rate *(r)* and also reduce the genetic effective population size *(N_e)*.

while the output of increased demographic stochasticity is manifested as an increase in the variance of the population growth rate, denoted *Var (r)*.

Var (r) has been employed to calculate extinction probabilities. This study was pioneered by Feller (1957) and was developed further by MacArthur and Wilson (1967), Richter-Dyn and Goel (1972), Leigh (1981), and Goodman (in press). *Var (r)* should not be viewed as a simple manifestation of any one ecological factor. Rather, it is the distillation of all of the forces acting on and through a population. We place *Var (r)* under PSF, because it is neither part of the population phenotype, *per se,* nor is it part of the environment. Rather, it is a statistic that depends on the interaction of these two fields over time.

These outputs lead to further changes of state in the field of PP—where they can modify the population's genetics and phenetics—and then to the field of PSF, where population dynamics are affected, with the possibility of deleterious effects on N, N_e, r, and $Var\ (r)$. We distinguish four such loops: the R, D, F, and A vortices. It should also be emphasized that the pathways of responses we are sketching can be traveled multiple times and can become complex and interconnected.

The R vortex

The R vortex is triggered when chance lowerings of N and increases of $Var\ (r)$ make the population vulnerable to further disturbances, in turn reducing N further and increasing $Var\ (r)$. That is, the severity of the impact of a disturbance may be exacerbated by the current states of N and r, with a series of otherwise similar disturbances having progressively more serious consequences on the population (Figure 4). An example of this would be a case where the early disturbances alter the age structure of a population in such a way as to make it more vulnerable to subsequent disturbances. Or, an alteration

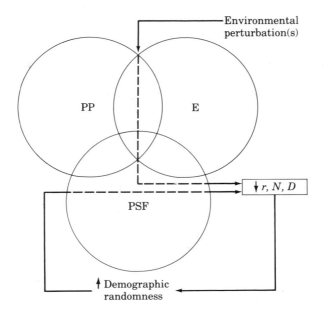

FIGURE 4. The demographic (R) vortex is triggered when chance lowerings of N and increases of $Var\ (r)$ make the population vulnerable to further disturbances, which in turn reduce N further and increase $Var\ (r)$ still more.

of the sex ratio away from 50:50 could lead to even greater variation in birth and death rates (due, for example, to increased difficulty in finding mates). The effects of low N over a number of generations can, in a sense, be cumulative. That is, the probability that a small population will be extinguished by demographic stochasticity alone increases at a rate that is greater than the linear sum of the per generation probabilities of extinction.

The D Vortex

A lowering of N and an increase in $Var\ (r)$ can alter the spatial distribution of a population and can increase the patchiness of its distribution (Figure 5). Fragmentation has a number of detrimental implications. First, because the probability of extinction of a local patch varies inversely with the population size on the patch (Gilpin and Diamond, 1976), more fragmented distributions are likely to in-

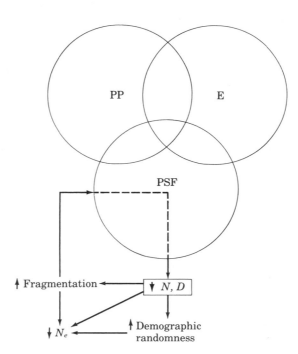

FIGURE 5. The fragmentation (D) vortex. The effects of the R vortex can alter the spatial distribution of a population, introducing or increasing fragmentation. More fragmented distributions increase the likelihood of local extinctions, as well as having detrimental effects on N_e.

crease further the rate of local extinction, further exacerbating the problem of isolation.

Second, and less obviously, fragmentation with turnover of patch populations has profound negative consequences for N_e, potentially producing effective population sizes that are orders of magnitude lower than the actual total census count (Maruyama and Kimura, 1980; Gilpin, in press b).

The F vortex

A decrease in N_e can have far-reaching consequences for a population, especially if the low N_e persists for many generations. If severe enough, such reduced sizes can lead to the initiation of two genetically-based vortices. Both vortices are the consequences of increased genetic drift and the loss of heterozygosity and genetic variance.

The F vortex is predicated on the well-established connection between inbreeding and inbreeding depression (Frankel and Soulé, 1981; Ralls and Ballou, 1983) and also on the frequent, direct relationship between individual heterozygosity and individual fitness (Allendorf and Leary, Chapter 4; Ledig, Chapter 5). Inbreeding depression and the loss of heterozygosity probably undermine most components of the population phenotype, including metabolic efficiency, growth rate, reproductive physiology, and disease resistance. In turn, these affect the schedule of births and deaths that determine r in the PSF field. These effects are relatively "hard," or absolute, because their impact is more or less independent of the environment. Lower r and N further reduce N_e, completing this feedback loop (Figure 6), and increasing the probability of extinction via this vortex and the others.

The A vortex

The last vortex is also the consequence of genetic drift and loss of genetic variance, but it manifests itself differently. Genetic drift can affect the precision with which selection can "tune" a population to its environment. Any process or event that decreases N_e will also reduce the efficacy of stabilizing and directional selection, in turn causing an increasing and accelerating lack of fit between the population phenotype (PP) and the environment it faces (E). This reduces r and N still further, draws the population still deeper into a vicious cycle, and exacerbates all of the other vortices at the same time.

For example, a reduction in fitness (from either the F or A vortices, or both) may increase fragmentation, with all of the derivative effects mentioned above. In part, this may occur because marginal habitats or patches become submarginal as fitness deteriorates. Second, and

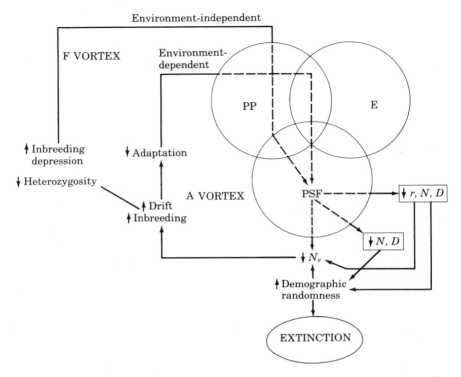

FIGURE 6. The inbreeding (F) and adaptation (A) vortices. Both are the consequences of increased genetic drift and the loss of heterozygosity and genetic variance. The F vortex reflects inbreeding depression and the effects of loss of heterozygosity on phenotype viability (metabolic efficiency, growth rate, reproductive efficiency, disease resistance, etc.) and eventually on r. The A vortex reflects that a decrease in N_e brought about by genetic drift will affect the ability of a population to adapt to its environment, causing an increasing lack of "fit" between the population phenotype (PP) and the environment (E).

probably more important, loss of genetic variance may make it increasingly unlikely for a population to track environmental changes genetically (Frankel and Soulé, 1981; Futuyma, 1983). The A vortex has the longest time scale, meaning that the other vortices will often lead to extinction before the population suffers serious genetic erosion and loss of adaptation.

To briefly illustrate the interaction of these positive feedback vortices, we use the case of the howler monkey mentioned earlier in this chapter and described in Chapter 12. Assuming that the monkeys are all confined to those 10-ha patches of forest, the consequences include an overall decline in population size. Each unit of the fragmented population is now isolated. The small sizes of these units mean that they are prone to severe demographic stochasticity and a very high probability of extinction. Small population size and isolation also produce a low effective size and a high level of inbreeding and loss of heterozygosity. In addition, edge effects and other characteristics of the fragmented habitat change the selective pressures on the monkeys. Assuming that one generation is equivalent to one cycle through a vortex, the longer that a population of monkeys survives, the more it cycles through the F and A vortices. Each cycle further reduces its fitness, increasing mortality and decreasing natality. This exacerbates demographic stochasticity (the R vortex), lowering N_e still further. As populations go extinct, the probability of gene flow and recolonization of empty patches is further reduced, exacerbating the D vortex as well as the F and A vortices.

SUMMARY

The foregoing treatment of population vulnerability analysis is general, but we hope it has been exhaustive at the level of systems identification. But even if this systems model is comprehensive, it is far from being a universal protocol for the estimation of minimum viable populations in specific cases.

One reason for this lack of precision is that species will differ in their vulnerability to the four vortices. To some extent, it is possible to predict the vulnerability to these vortices using taxonomic, ecological, and body size categories. For example, small-bodied, high-r insects, fish and rodents—especially those with relatively long-range juvenile dispersal—are unlikely to suffer from the F and A vortices. Usually their total population sizes and their rates of gene flow are great enough to avoid the loss of genetic variation. Besides, their local populations are relatively susceptible to capture by the R vortex because r and $Var (r)$ are positively correlated. In other words, extinctions of patch populations of small annual species are likely to occur before the F and A vortices are fully engaged (for example see Ehrlich, 1983).

At the other extreme, large, iteroparous organisms can persist at low population levels for generations, because of longevity and because they are physiologically buffered from short-term environmental changes. Therefore, trees and large vertebrates have relatively high probabilities of being sucked into the F and A vortices.

Lineages have been episodically exposed to one or more of these vortices during their evolution, and therefore may have evolved some kind of "resistance" to them. For example, the high dispersal rates and high r characteristic of many small organisms may be evolutionary adaptations that minimize the threats of frequent fragmentation (D vortex) and local extinctions (R vortex). At the other extreme, it is conceivable that large-bodied, high trophic level species with chronically low N_e might evolve some resistance to inbreeding depression, such as allelic and physiological redundancy. To our knowledge, however, little evidence exists for such adaptations.

We hope that further elaboration and improvement of this systems approach will lend itself to the management of rare or endangered populations. At this point, our approach can be used as a conceptual checklist for those responsible for the viability of particular species, and as a guide to gathering the sorts of data that may be most useful in assessing the connectedness of the scheme we have sketched.

SUGGESTED READINGS

Frankel, O.H. and M.E. Soulé. 1981. *Conservation and Evolution*. Cambridge University Press, Cambridge. Chapters 4 and 5 give some of the background of genetic thinking regarding minimum viable population sizes. For several reasons the "magic number" of 500 may be simplistic, but it is still considered to be about the right order of magnitude for many applications.

Shaffer, M.L. 1981. Minimum population sizes for species conservation. *Bio-Science 31:* 131–134.

Soulé, M.E. In press. *Viable Populations*. Cambridge University Press, Cambridge. The chapters in this book should fill in some of the theoretical details of the black boxes in Figures 3–6 of this chapter.

Soulé, M.E. and D. Simberloff. 1986. What do genetics and ecology tell us about the design of nature reserves? *Biological Conservation 35:* 19–40.

Wilcox, B.A. 1984. Concepts in conservation biology: Applications to the management of biological diversity. In *Natural Diversity in Forest Ecosystems: Proceedings of the Workshop*, J.L. Cooley and J.H. Cooley (eds.). Inst. of Ecology, Univ. of Georgia, Athens, Ga.

CHAPTER 3

INBREEDING IN NATURAL POPULATIONS OF BIRDS AND MAMMALS

Katherine Ralls, Paul H. Harvey and Anna Marie Lyles

Mating between close relatives increases the proportion of loci at which offspring are homozygous because such pairs are genetically more similar to each other than are pairs of individuals taken at random from the population. This increase in homozygosity can cause inbreeding depression. Although few data from natural populations are available, studies of captive and experimental animals consistently confirm the ubiquity and magnitude of its effects (Ralls and Ballou, 1983).

Packer (1979) lists three reasons why inbreeding depression should be expected to occur. First, "increasing homozygosity increases the chances of a detrimental recessive being expressed." Second, "the heterozygote may sometimes be fitter than either homozygote." And third, "an increase in homozygosity decreases the variability between offspring, with the effect that the chance of any one of an individual's progeny surviving a sudden environmental change may be reduced" (Williams, 1966).

Outbreeding can also entail costs. For example, outbreeding may require dispersal over an unfamiliar environment where the risks of mortality are high; and there may sometimes be genetic costs as well (Templeton, Chapter 6). The costs of inbreeding versus those of outbreeding will vary from species to species, depending on a range of variables (Bateson, 1983), the relative magnitudes of which have not been assessed in any particular case.

Inbreeding has been defined in a variety of ways, from the definition given in Webster's *New Collegiate Dictionary* of "mating between close relatives," to mating between relatives "that share a greater common ancestry than if they had been drawn at random from an entire species" (Shields, in preparation). The latter definition seems of little value when discussing inbreeding in birds and mammals, because almost all species of birds and mammals would then be described as inbreeding. If inbreeding avoidance exists, we would expect to find it where the costs are highest: among extremely close relatives.

This chapter assesses the frequency of mating between close relatives which we define as parents and their offspring or full siblings,

TABLE 1. Species names used in text.

Common name	Latin name
BIRDS	
Acorn woodpecker	*Melanerpes formicivorus*
Arabian babbler	*Turdoides squamiceps*
Bank swallow	*Riparia riparia*
Canada goose	*Branta canadensis*
Cliff swallow	*Hirundo pyrrhonota*
European mute swan	*Cygnus olor*
Florida scrub jay	*Aphelocoma c. coerulescens*
Great tit	*Parus major*
Japanese quail	*Coturnix coturnix japonica*
Laysan albatross	*Diomedea immutabilis*
Mexican jay	*Aphelocoma ultramarina*
Pied flycatcher	*Ficedula hypoleuca*
Pied kingfisher	*Ceryle rudis rudis*
Purple martin	*Progne subis*
Ring-billed gull	*Larus delawarensis*
Song sparrow	*Melospiza melodia*
Splendid wren	*Malurus splendens*
White-crowned sparrow	*Zonotrichia leucophrys*
White-fronted bee-eater	*Merops bullockoides*
Yellow-eyed penguin	*Megadyptes antipodes*
MAMMALS	
African wild dog	*Lycaon pictus*
Arctic ground squirrel	*Spermophilus parryii*
Bannertailed kangaroo rat	*Dipodomys spectabilis*
Barbary macaque	*Macaca sylvanus*
Beaver	*Castor canadensis*
Belding's ground squirrel	*Spermophilus beldingi*
Black-tailed prairie dog	*Cynomys ludovicianus*

using the available data from natural populations of birds and mammals. We then review the mechanisms that may reduce inbreeding between close relatives, irrespective of whether they evolved for that reason. Finally, we broaden our perspective to discuss the prevalence of matings between less closely related animals.

THE FREQUENCY OF CLOSE INBREEDING

There are two categories of evidence that allow us to estimate frequencies of close inbreeding in natural populations. The first is direct evidence where the frequency of matings between close relatives can be estimated on the basis of long-term studies of identifiable individ-

Common name	Latin name
Bonnet macaque	*Macaca radiata*
Brown antechinus	*Antechinus stuartii*
Cactus mouse	*Peromyscus eremicus*
California ground squirrel	*Spermophilus beecheyi*
Chimpanzee	*Pan troglodytes*
Deermouse	*Peromyscus maniculatus*
Dusky antechinus	*Antechinus swainsonii*
Dwarf mongoose	*Helogale parvula*
Fallow deer	*Dama dama*
Gray-tailed vole	*Microtus canicaudus*
Great gerbil	*Rhombomys opimus*
Horse	*Equus caballus*
Langur	*Presbytis entellus*
Lion	*Panthera leo*
Mantled howler	*Alouatta palliata*
Mongolian gerbil	*Meriones unguiculatus*
Mountain gorilla	*Gorilla gorilla beringei*
Old field mouse	*Peromyscus polionotus*
Olive baboon	*Papio anubis*
Pig-tailed macaque	*Macaca nemestrina*
Pika	*Ochotona princeps*
Prairie vole	*Microtus ochrogaster*
Red deer	*Cervus elaphus*
Rhesus macaque	*Macaca mulatta*
Spiny mouse	*Acomys cahirinus*
Stump-tailed macaque	*Macaca arctoides*
White-footed mouse	*Peromyscus leucopus*
White-tailed deer	*Odocoileus virginianus*
Wolf	*Canis lupus*
Yellow-bellied marmot	*Marmota flaviventris*

uals. The second category is indirect, when no data are available on the incidence of matings between various categories of relatives. Here inbreeding is inferred from a variety of sources, such as chromosomal or dispersal data, or electrophoretic surveys interpreted in the absence of pedigree information. We will defer discussion of the second category of evidence until the concluding section of the chapter.

Direct evidence of close inbreeding

There remain relatively few studies from which reliable rates of close inbreeding can be calculated for natural populations of birds and mammals. We have compiled a summary of the relative data in Table 2. A clear pattern seems to emerge. With two exceptions, parent–offspring and full-sib matings are between 0 and 6 percent of those observed. Over half of the studies put the figure at below 2 percent. One of the values over 6 percent is 9.8 percent from a rather unnatural population of European mute swans in the United States that was founded by five escaped individuals. However, the other high figure is 19.4 percent, from a natural Western Australian population of cooperatively breeding splendid wrens.

In contradiction to our review of the available data, Shields (1982) claims that "Relatively high frequences (10–100 percent of matings) of parent–offspring or full sibling incest have . . . been documented in natural populations of . . . various vertebrates usually considered *a priori* outbreeders." Among the vertebrates are a number of bird and mammal species: albatross (Fisher, 1967); swans (Reese, 1980); great tits (van Noordwijk and Scharloo, 1981); mice (Dice and Howard, 1951); wolves and other canids (Mech, 1977); and fallow deer (Smith, 1979). Shields (in preparation) adds pikas to this list on the basis of Smith and Ivins' (1984) study. We have checked through this list of case studies for examples of frequent parent–offspring and sibling matings and natural populations and we summarize our findings below. In short, the cited studies do not support Shields' claims. Matings between close relatives have either not been recorded in the studies (wolves, albatross, fallow deer, and pikas) or occur at a frequency below 10 percent (mice, swans, and great tits). Details on the data available for these species are given below.

Mech (1977) does not record any cases of close inbreeding in packs of wild wolves. In a recent paper Mech (in preparation) describes seven breeding pairs in a single pedigree, but none were known to be related. The only case of a mating between relatives that we have been able to find is a probable grandfather–granddaughter mating (Peterson et al., 1984). A computer model developed by Woolpy and Eckstrand (1979) is often cited as evidence that wolf packs must be highly inbred.

TABLE 2. Frequency of close inbreeding (between parents and offspring or full siblings) in natural populations of birds and mammals.

Species	Percentage[a]	Sample Size[b]	Source[c]
BIRDS			
White-fronted bee eater	0.0	81	S. Emlen
Pied flycatcher	0.4	276	P. H. Harvey & B. Campbell
Florida scrub jay	0.4	280	Woolfenden & Fitzpatrick, 1984
Arabian babbler	0.6	300	A. Zahavi
Yellow-eyed penguin	0.6	490	Richdale, 1957
Purple martin	0.7	140	E. Morton
Great tit (Hoge Veluwe)	1.5	460	van Noordwijk & Scharloo, 1981
Great tit (Wytham Wood)	1.5	1000	Greenwood, Harvey & Perrins 1979
Cliff swallow	1.7	59	P. J. Sikes & K. A. Arnold
Great tit (Vlieland)	3.0	834	A. van Noordwijk
Acorn woodpecker	3.2	218	W. Koenig
Mute swan	9.8	184	J. Reese, 1980 & unpublished
Splendid wren	19.4	211	I. Rowley
MAMMALS			
Brown antechinus	0.0	>150	A. Cockburn
Dusky antechinus	0.0	>100	A. Cockburn
Belding's ground squirrel	0.0	>300	P. Sherman
Lion (Serengeti)	0.0	>200Y	A. Pusey & C. Packer
African wild dog	0.0	18	Malcolm & Marten, 1982; J. Malcolm
Red deer	0.0	19	J. Pemberton
Olive baboon (Gombe—after 1976)	0.0	90Y	C. Packer
Yellow-bellied marmot	1.2	662Y	K. B. Armitage
Beaver	1.3	77	G. Svendsen
Chimpanzee	<1.7	1129C	A. Pusey
Black-tailed prairie dog	2.5	>300	J. Hoogland
Mountain gorilla	3.6	56	Fossey, 1983
Horse	3.9	129	J. Berger
Lion (Ngorongoro)	<4.4	121Y	A. Pusey & C. Packer
Dwarf mongoose	5.5	73	Rood, in preparation

[a]Percentages are rounded off to the nearest tenth.
[b]Sample sizes are number of pairs × number of breeding seasons except where indicated otherwise; Y = number of young, C = number of copulations.
[c]Data sources are personal communications from the investigators named unless otherwise indicated.

However, this conclusion is misleading. Their simulation model calculates the number of generations required for the fixation of one of a pair of alleles within a pack. It predicts that the average locus will become fixed within 20 years within any given pack. Unfortunately, the simulation stops once a pack becomes fixed for one allele. But in nature life goes on. Since different packs will be fixed for different alleles, dispersal will soon regenerate heterozygosity within many of the packs that were previously fixed for one allele.

Fisher (1967) does not mention any cases of close inbreeding in pairs of Laysan albatrosses. Several factors suggest that close inbreeding is uncommon. Population sizes are large, on the order of tens of thousands of pairs on Midway Atoll (Fisher, 1966; R.W. Schreiber and E.A. Schreiber, personal communication). Generation length must be long, as most birds do not breed until the age of eight to ten years (Fisher, 1976), and banded birds known to be at least 40 years of age are still breeding (R.W. Schreiber, personal communication). Some natal dispersal occurs between islands; for example, about 10 percent of the fledglings banded at Pearl and Hermes Reef and sighted again in subsequent years were found at Kure Atoll, about 275 km to the northwest (Amerson et al., 1974). Average natal dispersal distances (19 m for males and 25 m for females; Fisher and Fisher, 1969) are appreciably larger than territory diameters, which are frequently less than 1 m and only 60 to 80 cm in the most dense portions of the colonies (R.W. Schreiber, personal communication).

Smith (1979) points out that young female fallow deer may follow their mothers to rutting stands, and that rutting males may occupy the same stands in several years. Therefore, daughters could be mated by their fathers. However, Smith reports no examples of close inbreeding in this species (or other relevant data).

Smith and Ivins (1984) do not mention any instances of close inbreeding among pikas. Adult pikas live alone in territories that are adjacent to those of opposite-sex adults with whom they mate. As dispersal is limited and potential immigrants are repelled by residents, close inbreeding may well occur (Smith and Ivins, 1983). Although there are currently no data on the frequency of matings between close relatives, Smith (personal communication) is continuing his field studies of this species and hopes to be able to document the extent to which these occur.

Although Dice and Howard (1951) do not document examples of close inbreeding in their study of the deermouse, in a previous paper Howard (1949) specifically looked at the frequency of close inbreeding, which was estimated to lie between 4 and 10 percent. That study, as well as a more recent one by Spevak (in preparation), was not sufficiently detailed for us to include the data in Table 2, particularly in the light of Birdsall and Nash's (1973) work which suggests multiple paternity of litters in this species.

Three male and two female European mute swans escaped from captivity in the Chesapeake Bay area of the United States in 1962. From 1971 to 1979 Reese (1980) individually marked nesting pairs and young. Between 1972 and 1979, of 32 banded pairs recorded, two were siblings from the same brood, three were siblings from different broods, and one was a male paired with his daughter following the loss of his mate.[1]

Van Noordwijk and Scharloo (1981) reported on the incidence of inbreeding in an island and a nearby mainland population of great tits breeding in nestboxes in the Netherlands. Among mated pairs consisting of identified individuals in the island population, 19 percent were known to be related at any level (e.g., with "greatgrandparent in common"); however, only 3 percent of pairs were close relatives. Only 1 percent of pairs in the mainland population were known to be related at any level. In another mainland population, in Great Britain, Greenwood et al. (1979) also found that less than 2 percent of pairs were related. (In this study identified relatives were either parent–offspring or siblings.)

Inbreeding in human populations

Human populations can hardly be described as "natural," but the available data are of interest. Although mating between sexually mature siblings does seem to be avoided in human populations and there is a widespread incest taboo (van den Berghe, 1983), one recent study does demonstrate high rates of inbreeding between fairly closely related individuals, although not parents and offspring or full siblings (Devi et al., 1981, 1982). In Karnataka, southern India, 18 percent of Hindu marriages, 7.6 percent of Christian marriages and 4.7 percent of Muslim marriages are between uncles and their nieces (A.H. Bittles, personal communication). In the same religious groups 10.8 percent, 6.5 percent and 17.0 percent of marriages are between first cousins. Sample sizes in this study are large (43,112 pregnancies) and inbreeding depression has not been detected (Devi and Rao, 1981). Indeed, it seems that one widely cited study may have overestimated the levels of inbreeding depression in humans (Seemanova, 1971; Bittles, 1979).

[1] We are grateful to Dr. Reese for providing us with his full data set, including information on the pairs in which only one member was banded, from which we were able to calculate that the proportion of brother–sister and parent–offspring matings in this newly-established population was 9.8 percent.

Difficulty of statistical testing

Although matings between close relatives appear to be uncommon in most populations, the data in Table 2 do not demonstrate that these are less frequent than expected by chance. It is extremely difficult to test the hypothesis of random mating using such data because of the difficulty in determining the frequency of matings between close relatives expected by chance. A variety of "expected" frequencies can usually be calculated and the particular null hypothesis will determine the outcome of statistical testing. For example, we may assume that sex differences in the distribution of dispersal distances are as observed in the wild, or we may not wish to make this assumption. Expected frequencies of close inbreeding under the first null hypothesis will be lower than under the second null hypothesis. Other types of evidence, however, indicate that matings between close relatives do tend to occur less frequently than one might expect.

PROCESSES THAT REDUCE INBREEDING

Inbreeding will be reduced if close relatives rarely encounter each other during the breeding season (Process one; see below), or if close relatives that do encounter each other at this time mate less often than might be expected by chance (Process two). Of course, both might occur in a single species. Close relatives could usually be spatially isolated during the breeding season but tend not to mate if they did meet.

The first process involves temporal or spatial isolation and, for close relatives, it often implies that one or both sexes have dispersed from some common origin. It is useful to distinguish between two categories of such dispersal. Offspring can move from where they were born before breeding (natal dispersal), and adults can move between successive breeding seasons (breeding dispersal).

The second process requires some way for animals to identify those individuals likely to be close kin. Animals could then avoid breeding with these individuals in a variety of ways. Possibilities include failing to court, rejecting courtship advances, or even failing to come into breeding condition.

Process one: Reduction of opportunity

In several species of small carnivorous marsupials belonging to the genus *Antechinus,* all males die abruptly at the end of the short annual mating season (Lee and Cockburn, 1985), thereby eliminating any possibility of father–daughter inbreeding. Such extreme cases of temporal separation of close relatives appear to be uncommon, but high rates of population turnover due to environmental instability, short lifespans, and high juvenile mortality may reduce inbreeding in many small species (Horn and Rubenstein, 1984).

Spatial separation of close relatives is common in birds and mammals. Sex differences in natal dispersal are widespread in both groups (Greenwood, 1980). In gregarious species, one sex may tend to remain in the natal group, while the other moves to another group before breeding. In less gregarious or solitary species, one sex may tend to remain in the natal area or disperse a shorter distance than the other. The degree to which natal dispersal is sex biased varies across species from 0 to almost 100 percent. Reviews by Greenwood (1980), Dobson (1982), and Greenwood and Harvey (1982) indicate that juvenile females tend to be the predominant dispersers in birds, and juvenile males in mammals.

Sex differences in natal dispersal occur in a wide range of bird species, from colonial-breeding seabirds to many passerines (songbirds) (Greenwood and Harvey, 1982). In addition to sex-biased dispersal, inbreeding is probably further reduced among passerines by the combination of monogamous mating habits and high annual mortality rates. Sex-biased dispersal has not been found in nomadic species. Among those families where sex-biased dispersal does exist, the principal exception to the rule that juvenile females tend to be the predominant dispersers is the Anatidae (ducks and geese). Here pairs are usually formed on the overwintering ground and return to breed in the female's natal area.

Data from mammals are available mostly from those the size of ground squirrels and larger. Dobson (1982) found that juvenile males were predominant dispersers in 46 of 57 such species with promiscuous or polygynous mating systems. These mating systems prevail among mammals; relatively few species are monogamous (Kleiman, 1977). Juveniles of both sexes tend to disperse among 11 of 12 well-studied monogamous species (Dobson, 1982). Female-biased juvenile dispersal has been documented in only a few species (Greenwood, 1980; Dobson, 1982).

Primates have been particularly well studied. Dispersal in this order is comprehensively reviewed by Pusey and Packer (1986a) who conclude that "in many species characterized by male-biased dispersal, groups are composed almost entirely of female kin and immigrant males, while the converse is often true in species characterized by female dispersal. In these species, the like-sexed kin group may persist for generations (e.g. rhesus macaques: Chepko-Sade and Sade, 1979).

However, in the few species where both sexes transfer (e.g. mantled howlers), both males and females may be unrelated to like-sexed group members."

Although there have been many studies of dispersal in small mammals (Gaines and McCleneghan, 1980), particularly rodents, individual juveniles have generally not been followed until they breed and it is difficult to separate natal from breeding dispersal (Greenwood, 1980). Although dispersal is male-biased in deermice (Howard, 1949) and in many voles belonging to the genus *Microtus* (Lidicker, 1985), sex bias is lacking in the bannertailed kangaroo rat (Jones, 1984 and in preparation) and perhaps also in many other species. Male-biased, female-biased, and unbiased natal dispersal have all been documented in bats and all three types "are apparently sufficient to randomize adult population structure for the genetic markers examined" (McCracken, in press).

In both birds and mammals, breeding dispersal is almost invariably less frequent and over shorter distances than natal dispersal (Greenwood, 1980; Greenwood and Harvey, 1982). However, sex biases in breeding dispersal are common; in birds it is female-biased, whereas in mammals it is male-biased.

A high proportion of bird species is believed to be monogamous (Lack, 1968). Males and their mates generally breed in the same site in successive breeding seasons. Following the death of a mate or an unsuccessful breeding attempt, females are more likely to disperse than males (Greenwood and Harvey, 1982).

In contrast to birds, most mammal species are promiscuous or polygynous, and males of the larger gregarious species often survive long enough that mating with their daughters is a possibility (Pusey and Packer, 1986a). Nevertheless, it appears that father–daughter mating does not occur commonly among the relatively few well studied species (Berger, in press; Packer, 1979; Schwartz and Armitage, 1980; Hoogland, 1982).

Breeding dispersal in mammals has been well documented in primates (Pusey and Packer, 1986a), and in ground squirrels and marmots (Holekamp, 1984). The males of many primate species frequently transfer between breeding groups. For example, individual male olive baboons were observed in as many as five different groups (Packer, 1979). Return to the natal group has rarely been observed, and in several species the average length of time males stay in one group is known to be shorter than the time taken for their putative daughters to reach sexual maturity (Pusey and Packer, 1986a).

Hoogland's (1982) study of the black-tailed prairie dog demonstrates that sex-biased dispersal can effectively reduce the opportunity for mating between first degree relatives. The combination of male-biased natal and breeding dispersal eliminated the possibility of breeding with a close relative for 90 percent of females observed in estrus.

There has been considerable controversy over both the proximate (immediate) and ultimate (evolutionary) causes of sex-biased dispersal patterns (Moore and Ali, 1984; Packer, 1985; Dobson, in press). Dispersing animals are likely to face increased risks of mortality from starvation, predation, or aggression from unfamiliar conspecifics. Why then should any individuals disperse? Proximate causes are not easy to determine and may differ for males and females of the same species (Brody and Armitage, in press). The most common reason for dispersal in primates seems to be sexual attraction to members of other troops (Pusey and Packer, 1986a). Some primates may also be forced to disperse by the intense aggression of conspecifics (Pusey and Packer, 1986a) and dispersal appears to be correlated with dominance relationships in many birds (Gauthreaux, 1978). However, in many species of birds and mammals, natal dispersal occurs before sexual maturity (Greenwood, 1980) and individuals may disperse with no evidence of aggression by conspecifics (Bekoff, 1977). In yet other species, dispersal may sometimes be delayed (even beyond maturity) until vacant territories or other limiting resources become available (Waser and Jones, 1983; Emlen, 1984); a good example is provided by Jones' (1984 and in preparation) study of the bannertailed kangaroo rat.

Several evolutionary forces probably contribute to the patterns of natal and breeding dispersal found in birds and mammals. In summary, individuals will have been selected to disperse if they thereby enhance their reproductive success. For example, dispersing individuals may acquire access to better resources such as food, mates, or breeding sites protected from predation (Howard, 1960; Lidicker, 1962; Horn, 1978; Packer, 1979; Dobson, 1982; Waser and Jones, 1983; Moore and Ali, 1984).

Our focus in this section is dispersal that reduces inbreeding, but of course there is ample evidence that much dispersal among birds and mammals does not occur for this reason. For example, among birds we can find no evidence to indicate that breeding dispersal reduces inbreeding. Females tend to move after unsuccessful breeding attempts, even if this entails abandoning an unrelated mate (Harvey et al., 1979). Males, on the other hand, compete for good breeding territories and may move when better territories become vacant (Krebs, 1971). In a similar vein, male lions and langurs defend groups of females against competitors but are eventually forced to relinquish them to other males (Pusey and Packer, 1986b; Hrdy, 1977). Indeed, there is no direct evidence that breeding dispersal in male primates coincides with the sexual maturation of their daughters (Pusey and Packer, 1986a). Nevertheless, breeding dispersal in some mammals may be correlated with the avoidance of inbreeding. For example, Hoogland (1982) found that male black-tailed prairie dogs are more likely to move to a new breeding group if they have adult daughters present in their own group.

Sex-biased dispersal may act as an inbreeding avoidance mechanism. However, it is important to realize that there are many other sex differences among vertebrate species, for example in body size, foraging behavior, and patterns of parental investment. Such differences could select for sex-biased dispersal patterns, even in the absence of inbreeding avoidance. Once sex-biased dispersal has evolved, outbreeding and greater heterozygosity follow. Because heterozygous individuals are often more fit (Allendorf and Leary, Chapter 4), selection to avoid inbreeding will increase. Therefore, even if inbreeding avoidance currently contributes to the maintenance of sex-biased dispersal, it has not necessarily evolved for that reason.

How, then, can we identify cases of dispersal that are associated with inbreeding avoidance? One of the strongest pieces of evidence comes from studies of two species of *Antechinus,* small Australian marsupials (Cockburn et al., 1985). The unusual life history of these species makes it possible to eliminate alternative hypotheses. Male generations are discrete since all adult males die abruptly at the end of the short annual mating season. Females remain in their natal areas, but almost all juvenile males disperse, including those in litters that consist entirely of males and those in litters where there is only one male. Competition among males for mates can be excluded because there is no opportunity for interactions between adult and juvenile males, and interactions among sons would not explain the totality of male dispersal or male dispersal from litters with only one male. Mothers apparently cause the dispersal of their sons and recruit unrelated juvenile males to live with themselves and their daughters. As this exchange of males does not consistently lead to a reduction in the number of individuals in the nest, competition for resources can be excluded as a cause. Thus the only benefits to the mothers appear to be the exchange of sons for unrelated males to mate with their daughters.

More often the evidence is circumstantial but nevertheless strong. Based on his observations, Dobson (1979) believes that California ground squirrels avoid inbreeding because (1) all males left their natal areas, even those males maturing in colonies that had no resident adult males; (2) all young males recruited into colonies with unusually low numbers of adult males came from other colonies, and were not recruited from within; and (3) all males dispersed, even from colonies where supplemental food was provided. In fact, the probability of male dispersal does not appear to vary with population density or resource availability in a number of species of ground squirrels (Holekamp, 1984).

Pusey and Packer (1986a) use comparative evidence to support the view that sex-biased natal dispersal probably functions across the primates as a whole to reduce inbreeding. Only one sex needs to disperse in order to avoid inbreeding. If males leave, there is no reason why females should have been selected to pay the costs of dispersal and vice versa. The data, preliminary as they are, are in accord with this expectation (Pusey and Packer, 1986a). Among the primates there is negative interspecific correlation between the proportion of males and the proportion of females that disperse; this is a straightforward prediction of the inbreeding avoidance hypothesis.

Process two: Opportunity not realized

When sexually mature first-degree relatives occur within the same social group or local area, the potential for matings between them exists. However, long-term field studies of identifiable individuals indicate that such matings occur less often than one might expect. This implies the existence of the ability to recognize actual or probable kin. It also implies that, once a probable relative is recognized, there is some avoidance mechanism that operates to prevent inbreeding.

Kin recognition occurs in a variety of bird and mammal species (Colgan, 1983; Holmes and Sherman, 1983), and may or may not be based upon cues such as location or prior association (Beecher, 1982; Holmes and Sherman, 1983). Recognition of kin in the absence of prior association has been demonstrated in only a few species. Mammals believed to have this ability include white-footed mice (Grau, 1982), several species of ground squirrels (Davis, 1982, 1984; Holmes and Sherman, 1982), the pig-tailed macaque (Wu et al., 1980), and laboratory strains of rodents (Hepper, 1983; Kareem, 1983; Kareem and Barnard, 1982). Inbred Japanese quail prefer to associate with unfamiliar cousins over other classes of relatives as well as over nonrelatives (Bateson, 1982).

In many more species, kin recognition depends upon environmental cues. Adult birds commonly recognize their own eggs or altricial young on the basis of location in the nest. At the time fledglings become mobile or begin to mingle with nonrelatives, parents learn to discriminate their own young from others (Beecher et al., 1981a,b; Colgan, 1983; Holmes and Sherman, 1983; McArthur, 1982).

If kinship normally correlates highly with close association during early life, the correlation could be used to identify probable kin. This mechanism may provide the most common means of kin recognition (Bekoff, 1981; Holmes and Sherman, 1982). Sibling recognition generally seems to be based on familiarity during early life. The Bird species with such abilities include bank swallows (Beecher and Beecher, 1983), ring-billed gulls (Evans, 1970), and Canada geese (Radesäter, 1976). Many rodents also appear to recognize siblings on the basis of familiarity. These include prairie deermice (Hill, 1974), cactus mice (Dewsbury, 1982), prairie voles (Gavish et al., 1984), gray-tailed voles (Boyd and Blaustein, 1985), spiny mice (Porter et al., 1978; Porter and Wyrick, 1979), Mongolian gerbils (Ågren, 1981, 1984), great gerbils (Sokolov et al., 1981), and several species of ground squirrels (Davis, 1982, 1984; Holmes and Sherman, 1982; Holmes, 1984). Feral horses also tend to treat familiar animals in the same way as kin (Berger, in press; Duncan et al., 1984).

A single species may use more than one mechanism to identify kin or probable kin. Females of two of the best studied mammals, the Belding's ground squirrel and the Arctic ground squirrel, treat all individuals, related or not, that share their natal burrow as littermates. However, they can distinguish sisters they have not previously encountered from nonkin (Holmes and Sherman, 1982). Among Belding's ground squirrels, females can even distinguish between full and half sisters (Holmes and Sherman, 1982), which are often born in the same litter due to multiple paternity (Hanken and Sherman, 1981).

Some studies of kin recognition have addressed the question of whether discriminatory behavior towards kin wanes after a period of separation. Among several species of small rodents, siblings reared together will not normally mate, but after a period of separation of from one to four weeks, such siblings mate much more readily (Huck and Banks, 1979; Richmond and Stehn, 1976; Porter and Wyrick, 1979; McGuire and Getz, 1981; Dewsbury, 1982; Gavish et al., 1984). Berger (in press) found that feral horse stallions behave differently towards young animals born in their herd from those that were not, but after a separation of 18 months the discriminatory behavior ceases.

Although the exact way in which individuals avoid mating with close kin is often unknown, there are several possibilities. Females might express a preference for unrelated males by rejecting the courtship advances of male relatives or by actively seeking out unrelated males, or they might fail to come into estrus. Males might preferentially court unrelated females, fail to court related females or reject their sexual invitations, or fail to come into breeding condition. In theory, since females invest more than males in their joint offspring, we might expect these behavior patterns to be more commonly expressed by the female than the male of a related pair (Packer, 1977; Parker, 1979; Maynard Smith, 1982).

Fortunately, in recent years a number of long-term behavioral studies have begun to yield details of the behavior shown by sexually active males and females towards relatives versus nonrelatives. Studies on the black-tailed prairie dog (Hoogland, 1982), the chimpanzee (Pusey, 1980), and the acorn woodpecker (Koenig et al., 1984) are particularly informative.

Although the combination of male-biased natal and breeding dispersal usually eliminates the possibility of matings between close relatives in black-tailed prairie dogs (see above), Hoogland (1982) observed nine cases in which females had the opportunity to mate with close relatives. The chance of producing inbred offspring was either reduced or eliminated in a surprising variety of ways. In three cases, the female's group contained both a related and an unrelated male and she mated exclusively with the unrelated male. In another two cases, the female's group contained only a male relative, and she copulated with that relative and with another male who temporarily invaded the group on the day she was in estrus. The other four females behaved differently. One left her group, mated exclusively with an unrelated male in an adjacent group, and then returned. The second did not mate with her son since he had disappeared on the day she came into estrus (he was the only one of 89 adult males to disappear during the breeding season). The third came into estrus but failed to copulate. The fourth failed to come into estrus at all (the only other adult female who did not come into estrus was in extremely poor condition and died shortly thereafter).

The chimpanzee is one of the few mammals in which natal dispersal is female-biased; thus males often live in the same group as their mothers. In Pusey's (1980) study, males were never observed to court or copulate with their mothers, although many unrelated males did so. Most young females transferred to other groups either temporarily or permanently sometime during adolescence, but their behavior towards related and unrelated males in their natal group could be observed prior to their departure. Before their first estrus, young females tended to associate most often with close male relatives such as their mothers' sons. However, once these young females began cycling, their male relatives showed a lack of interest in them and rarely courted them, although they copulated readily with other females. During the rare courtships between related chimpanzees, the females were less receptive than they were to unrelated males.

The acorn woodpecker breeds communally in groups of up to 15 individuals (Koenig et al., 1984). Many birds within a group are close relatives, but mating between close relatives is uncommon. The males and females that join to form new groups tend to be unrelated, and sexually mature offspring rarely breed within their natal group as long as both parents are alive. Koenig et al. observed 15 cases where the young of one sex had the opportunity to mate with a parent because the other parent had died. No inbreeding occurred in 14 of these cases. In six cases, the offspring that were the same sex as the dead parent left the group and in another the remaining parent left. In a further three cases immigrants joined the group and bred, and the remaining offspring apparently did not breed. In the five remaining cases, no immigrant joined the group. In four of these the group did not breed for a season and, in the other, inbreeding probably occurred between a female and her putative father. Some of the woodpecker cases are consistent with the reproductive competition hypothesis as well as the inbreeding avoidance hypothesis (Shields, in preparation) but, as Dobson (in press) points out, these hypotheses are not necessarily mutually exclusive and both factors may well have influenced dispersal patterns in many species.

We believe the three studies discussed above provide particularly convincing evidence that individuals are avoiding inbreeding because of the variety of means by which this end appears to be achieved. However, similar evidence comes from a number of other studies. Among primates, male rhesus monkeys will rarely mate with their mothers under semi-natural conditions (Sade, 1968; Missakian, 1973; but see Sade et al., 1984), matings between captive closely-related stump-tailed macaques are less frequent than expected by chance (Murray and Smith, 1982), and none were observed in captive Barbary macaques (Paul and Kuester, 1985). In horses, males do not mate with young females that have matured in their group (Berger, in press). And among communally breeding bird species, in addition to the acorn woodpecker, Mexican jay relatives tend not to mate with each other (Brown and Brown, 1981).

Reduced or delayed breeding when the only available partners are close relatives has been demonstrated in a variety of rodents. Field observations showed that a young female black-tailed prairie dog is less likely to come into estrus if her father is still available as a mate (Hoogland, 1982). Similarly, young female yellow-bellied marmots are less likely to breed if the adult male in the group is their father (Armitage, 1984). However, most of the evidence comes from laboratory experiments comparing sibling and nonsibling pairs. For example, only 6 percent of prairie vole sibling pairs bred compared to 78 percent of nonsibling pairs (McGuire and Getz, 1981). Sibling pairs of deermice (Hill, 1974; Dewsbury, 1982), cactus mice (Dewsbury, 1982), Mongolian gerbils (Ågren, 1981, 1984), and great gerbils (Sokolov et al., 1981) also have a reduced probability of breeding. In the prairie vole and the great gerbil, the mechanism may be failure of the females to come into reproductive condition.

BROADENING THE PERSPECTIVE

It is noteworthy that the direct evidence from natural populations of birds and mammals usually reveals a rather low incidence of matings between close relatives (Table 2). However, frequent inbreeding has commonly been inferred from indirect evidence (e.g. mammals as a whole: Wilson et al., 1975; white-tailed deer: Nelson and Mech, 1984; black-tailed prairie dogs: Chesser, 1983). Although it may help to identify cases where close inbreeding could exist, such evidence is inconclusive. Some of the difficulties of drawing conclusions from indirect evidence will become apparent below.

So far, we have limited our discussion to matings between parents and offspring or between siblings. At this point, it is useful to broaden the discussion, but we immediately encounter a number of problems that must be addressed:

1. Different authors mean different things by the word "inbreeding."
2. Available measures of inbreeding are relative rather than absolute.
3. Not all phenomena which have been called "inbreeding" necessarily involve increased homozygosity.
4. Some authors unjustifiably conclude that matings between vaguely specified classes of fairly close relatives (not necessarily parents and offspring or siblings) must be common in their study population.

Uses of the word inbreeding

Inbreeding is clearly a continuous variable and it is not possible to adequately describe the continuum using the dichotomy "inbred" versus "noninbred." Since all members of a species have common ancestors and are therefore related, the question becomes, how close should the relatedness between mates be for their offspring to qualify as inbred? Different authors dichotomize the continuum at different (and often unspecified) points, which results in considerable confusion in the literature.

As we explained early in the chapter, there are two extreme uses of the word "inbreeding." We have been using it to refer to matings between close relatives, and it seems that such inbreeding is rare in birds and mammals. However, if we take the other extreme view that inbreeding occurs whenever mating is between "relatives that share greater common ancestry than if they had been drawn at random from the entire species" (Shields, in preparation), then inbreeding would be the rule in birds and mammals. Clearly, it would be useful to have a way to quantify different degrees of inbreeding in order to make meaningful comparisons.

Measuring inbreeding

One method of quantifying inbreeding is Wright's inbreeding coefficient F, the genetic correlation between uniting gametes. F can be calculated from pedigrees, but has been determined in only a few studies of natural populations of birds and mammals (for example Howard, 1949; Greenwood et al., 1978; Schwartz and Armitage, 1980; van Noordwijk and Scharloo, 1981) because of the difficulty of obtaining the data. More commonly, electrophoretic data are used to calculate various estimators of F (F_{ST}, F_{IS}, F_{IT}).

In practice, interpreting F (or any estimate of F) can be very misleading because F must be defined with respect to some reference generation or population (or both), and is thus a relative rather than absolute measure. Indeed, as Jacquard (1974) notes, "An inbreeding coefficient cannot be regarded as an estimate of any real quantity." For example, if we assume that all individuals of a particular species with an effectively infinite population size are descended from a single mated pair, and that pair is used as our reference population, then the species has an inbreeding coefficient of one, and is therefore totally inbred, irrespective of its mating system. However, if the reference generation for the same species is taken as one generation back from the present and mating is at random, the inbreeding coefficient is zero, and the species is therefore totally outbred.

Even at any single point in time we can measure inbreeding with reference to a variety of base populations. For example, consider a species that consists of only two small populations on two different isolated islands between which migration is almost impossible. Our measure of the degree of inbreeding will be much smaller if we use one of the populations instead of both as the base population.

A variety of base populations are available even within a continuously distributed species, since one could consider the base population to be those individuals contained within any geographic area from the local neighborhood to the range of the entire species. If the distance between the birth sites of a typical mated pair is short relative to the geographical range of the species, as seems to be the case in many vertebrates, the species can be portrayed as either inbreeding or outbreeding depending upon one's choice of base population (whether or not individuals avoid mating with close relatives).

Other difficulties in interpreting Wright's F statistics have been summarized by Foltz and Hoogland (1983) and Templeton (in preparation). Briefly, the values of these measures can be influenced by factors other than inbreeding. These include other forms of assortative mating, groups of siblings in the sample, differences in allele frequencies between male and female parents, selection acting on the animals before the sample is taken, differences in fertility among the parents, and sex-biased dispersal.

Nevertheless, F statistics are often used to infer the incidence of inbreeding in birds and mammals. For example, Melnick (1982) reviewed the available data from mammal species belonging to several orders and concluded that mammals generally do not inbreed. Because, as we have discussed, such measures are relative and can be influenced by factors other than inbreeding, we feel that this evidence is not conclusive. Indeed, Chesser (1983) concludes from such data that black-tailed prairie dogs are highly inbred, although as we have mentioned previously Hoogland's (1982) behavioral studies of the same species demonstrate that matings between close relatives are very rare.

Inbreeding and homozygosity

The problems of definition and measurement seem frequently to result in unjustified conclusions. Once an author has termed a phenomenon "inbreeding" there is often a temptation to conclude that homozygosity is therefore high. Alternatively, if homozygosity is high, it is often concluded that the frequency of inbreeding is also high. However, homozygosity can increase for reasons other than matings between very close relatives. For example, when populations are small and isolated from each other, genetic homozygosity within them can increase even when matings between close relatives are avoided.

The question then arises, Is dispersal between natural social groups, breeding groups, demes, or populations of birds and mammals often so low that individuals within these groups tend to be genetically homozygous? Population genetic models suggest that, in the absence of selection, very limited dispersal (on the order of one successful migrant between groups per generation) may often be sufficient to prevent such homozygosity building up. Relatively few bird or mammal groups or populations can be so isolated that levels of migration between adjacent ones are lower than this. Indeed, those who have attempted to evaluate the hypothesis that mammalian social structure impedes gene flow have rejected the idea (McCracken and Bradbury, 1977; Berry, 1978; Schwartz and Armitage, 1980; Baker, 1981a,b; Daly, 1981; Patton and Feder, 1981; Singleton, 1983; McCracken, 1984 and in press; Melnick et al., 1984; Lidicker and Patton, in preparation).

For example, in both yellow-bellied marmots and rhesus monkeys, permanent groups are centered around female matrilines, but natal dispersal by males is almost universal. The studies by Schwartz and Armitage (1980) and Melnick et al. (1984) demonstrate that considerable genetic heterozygosity is maintained within these groups, presumably by the extensive male migration each generation.

"Nonincestuous inbreeding"

A number of authors conclude that mating between close relatives (such as siblings or parents and offspring) may be rare in their study population, but that dispersal and mortality patterns are such that slightly more distant relatives (such as cousins) must frequently mate. Yet again, the evidence cited is always indirect.

Many authors view their results as supporting the "optimal outbreeding" or "optimal inbreeding" expectations of, respectively, Bateson (1978) and Shields (1982), who suggest that local "coadapted gene complexes" are preserved by matings between fairly close relatives. We know of no field evidence to support the idea that birds and mammals mate preferentially with fairly close relatives (say, cousins) rather than with more distantly related individuals. Indeed, coadapted gene complexes seem to be relatively unimportant in birds and mammals (Templeton, Chapter 6), and inbreeding is not the only way in which recombinational load can be reduced (McCracken, in press).

An example of this line of reasoning is provided by Nelson and Mech's study (in preparation) of white-tailed deer in northeastern Minnesota. No data on the degree of relatedness between mates are given, but the authors conclude that their results are consistent with the model of "Shields (1982) which considers dispersal as it occurs in most vertebrates to have evolved from selection for nonincestuous inbreeding. Adult home-range tenacity, female philopatry, relatively short male dispersal, short breeding movements, and little range overlap between deer from different yards suggests that deer populations composed of inbred groups are the rule." However, the movement patterns described by Nelson and Mech (in preparation) are not as restricted as the above quotation implies. Groups of deer have distinct winter ranges, but about 5 percent of deer have summer ranges that overlap with those of another group and it is on the summer ranges that about 80 percent of matings occur (L.D. Mech and M.E. Nelson, personal communication). In addition there appears to be considerable movement between groups by young animals; 5 of 13 marked yearling males moved outside the summer range of their natal group as, possibly, did one of 20 yearling females.

SUMMARY

The major conclusions that emerge from this review are that matings between close relatives are uncommon in most natural populations of birds and mammals (for unknown reasons, the splendid wren seems to be a notable exception), and that individuals often avoid mating with their closest relatives. To reach those conclusions, we have relied heavily on those relatively few studies that allow direct estimates of the frequency of matings between individuals of known relatedness, or provide details of the behavior shown by sexually active adults towards relatives and nonrelatives. The more indirect and circumstantial evidence provided by other studies largely supports this idea. However, there have been contrary claims in the literature. We have critically evaluated many of these and found them unjustified, mainly because the data do not warrant the conclusions drawn. In most cases, close inbreeding has not been recorded, although the potential seems to exist.

In sharp contrast to assertions made from weak evidence, a number of studies are beginning to provide data from which clear conclusions can be drawn. Observations of known individuals seem to us to provide some of the most useful data, but even these are open to criticism. It does not follow from the observation that a pair of animals is seen consorting or mating that the ensuing offspring have been fathered by the male involved. Recent electrophoretic studies which allow paternity exclusion show that assignments of paternity based on behavioral observation may often be incorrect (see Harvey, 1985 for a review). As a consequence, such studies are necessary in order to provide conclusive evidence in any particular case. In addition, even maternity is suspect among many bird species (Andersson, 1984). Females often manage to slip their eggs into other females' nests. As a consequence, both paternity and maternity studies are required (e.g. Gowaty and Karlin, 1984).

The strongest studies, therefore, are those that provide both behavioral and genetic data. These do not necessarily involve long-term field work over several generations. For example, Foltz (1981) recently used electrophoretic data on enzyme polymorphisms to test the conclusions of previous workers that the old field mouse is highly inbred. Because this species is often found living in monogamous pairs, he was able to estimate the genetic correlation between mates as well as between uniting gametes. He found no evidence of deviations from random mating.

There are very few studies that combine behavioral and genetic evidence to provide firm conclusions about the incidence of inbred young. And none of those provide the demographic data that are necessary to estimate the effects of inbreeding depression in the wild. Until such data are available, it is unjustified to conclude that close inbreeding, nonincestuous inbreeding, optimal inbreeding or, indeed, optimal outbreeding prevails in natural populations of birds or mammals.

SUGGESTED READINGS

Cockburn, A., M.P. Scott and D. Scott. 1985. Inbreeding avoidance and male-biased dispersal in *Antechinus* spp. (Marsupalia: Dasyuridae). *Animal Behaviour 33:* 908–915.

Greenwood, P.J. 1980. Mating systems, philopatry and dispersal in birds and mammals. *Animal Behaviour 20:* 1140–1162.

Hoogland, J.L. 1982. Prairie dogs avoid extreme inbreeding. *Science 215:* 1639–1641.

Pusey, A.E. and C.R. Packer. 1986. Dispersal and philopatry. In *Primate Societies,* B.B. Smuts, D.L. Cheney, R.M. Seyfarth, R.W. Wrangham and T. Struhsaker (eds.). University of Chicago Press, Chicago.

Pusey, A.E. 1980. Inbreeding avoidance in chimpanzees. *Animal Behaviour 28:* 543–552.

CHAPTER 4

HETEROZYGOSITY AND FITNESS IN NATURAL POPULATIONS OF ANIMALS

Fred W. Allendorf and Robb F. Leary

The loss of genetic variation caused by limited population size in captive populations is an important concern of conservation biology. It has long been known from studies with domestic animals that the loss of genetic variation has harmful effects on development, survival, and growth rate (Falconer, 1981). Recent studies by Ralls and her colleagues have shown that many mammals propagated in zoos show similar effects of the loss of genetic variation (reviewed in Ralls and Ballou, 1983). Thus, the effects of so-called "inbreeding depression" caused by reduced population size are well documented in a variety of animal species. The antithesis of inbreeding depression—heterosis—is an observed increase in general vigor (fertility, survival, growth, etc.) that is often associated with the increased heterozygosity produced by crossing different inbred lines or domestic strains.

The genetic mechanisms underlying inbreeding depression and heterosis are still not completely understood (Crow, 1948; Clarke, 1979; Frankel, 1983). Heterosis is sometimes incorrectly equated with heterozygous superiority. However, simple dominance alone can explain heterosis through the sheltering of deleterious recessive alleles. Inbred lines are likely to be homozygous for deleterious recessive alleles at different loci. Hybrids between lines will therefore be heterozygous at such loci and the deleterious recessive alleles will not be expressed in the phenotype. Thus, it is unclear whether it is heterozygosity *per se* or reduced homozygosity for deleterious recessive al-

57

leles that results in the superior performance associated with heterosis.

This chapter presents the empirical evidence relating heterozygosity to fitness in natural populations of animals. Two recent papers have reviewed the effects of heterozygosity on phenotypes of evolutionary interest (Mitton and Grant, 1984; Zouros and Foltz, in press). Our goal is to consider the implications of such results for conservation biology. It is notoriously difficult to estimate differences in fitness among genotypes in natural populations (Lewontin, 1974). We therefore first review the literature describing associations between heterozygosity and phenotypic effects that are likely to affect fitness: survival, growth rate, disease resistance, and developmental rate and stability. We then consider the implications of these results in the context of the relationship between heterozygosity and fitness. Finally, we discuss the broader issue of the relationship between heterozygosity and the evolutionary potential of populations, especially its relation to current issues in conservation biology.

PHENOTYPIC EFFECTS OF HETEROZYGOSITY

Survival

The best evidence for an association between survival and heterozygosity comes from studies of temporal changes in the level of heterozygosity in a single cohort. These data are most compelling when (1) immigration and emigration can reasonably be excluded as possible causes of the change; (2) a plausible mechanism for the association is demonstrated; and (3) it can be shown that selection is likely to be acting directly on the locus or loci that are monitored.

To our knowledge, only one investigation describing an association between heterozygosity and survival satisfies all of the above points. Watt (1977) found that heterozygosity at a locus coding for the enzyme glucosephosphate isomerase (GPI) significantly increased monotonically among four successive samples that spanned almost the entire adult lifetime of a single cohort of the sulfur butterfly, *Colias philodice eriphyle*. The dispersal capabilities of the insect and the geographic pattern of the allelic variation at the locus are not compatible with the alternative hypothesis that the increase in heterozygosity is the result of immigration or emigration (Watt, 1983).

Energy for flight in this insect is largely obtained by the biochemical degradation of glycogen and the sugar trehalose. GPI catalyzes one of the steps in these pathways. Analysis of in vitro catalytic properties indicated that the GPI extracted from heterozygotes catalyzed this reaction more rapidly than the GPI extracted from either homozygote (Watt, 1983). Watt expected that this catalytic superiority would enable heterozygous individuals to have greater endurance and to be able to fly over a broader range of environmental conditions than homozygotes. These predictions were subsequently verified with field studies (Watt et al., 1983).

The ability to fly for longer periods of time and over a broader range of environmental conditions is most likely responsible for the increased survival of the heterozygotes. This will afford them increased access to food, escape from predation, and escape from adverse weather (Watt, 1983). Furthermore, it allows males an increased access to receptive females (Watt et al., 1985). These studies provide compelling evidence of heterozygous advantage in which selection is acting directly upon the products of the investigated locus through differential survival and mating success of adults.

The Norway rat, *Rattus norvegicus,* often evolves resistance to warfarin in areas where warfarin commonly has been used as a rodenticide (Drummond, 1970). Resistance in some populations is due to a dominant allele at a single locus (Greaves and Ayers, 1967, 1969). The frequency of the resistant allele would be expected to rise rapidly in regions where warfarin has commonly been used to control population size for an extended period of time. Greaves et al. (1977), however, found that the frequency of rats resistant to warfarin remained stable at about 40 percent over nine years in such a region, suggesting that a stable equilibrium had been achieved. They also observed an excess of heterozygous individuals and a deficit of both homozygotes, suggesting some selection against resistant homozygotes. Individuals homozygous for resistance are much more susceptible to vitamin K deficiency than heterozygotes or warfarin susceptible individuals (Hermodson et al., 1969). If vitamin K deficiency is an important selective factor in populations, then this could easily account for the apparent superior survival of heterozygotes and the stability of the allele frequencies.

Temporal changes in heterozygosity within a cohort or population have been reported for a diversity of organisms (Table 1). In most of these studies, immigration and emigration can reasonably be excluded as possible causes for the observed change, suggesting that it is due to differential survival. Furthermore, the change in heterozygosity is often observed to occur during winter, a period with high mortality, adding circumstantial support to the conclusion that the change reflects differential survival. It appears from these data that a positive association between heterozygosity and survival exists during at least part of the life cycle in a variety of organisms. However, this is not universal; there are reports of negative associations (Table 1).

TABLE 1. Summary of studies reporting evidence for differential survival of homozygotes and heterozygotes at protein loci in natural populations.

Species	Observation	Heterozyg Status[a]
1. Blue grouse (*Dendragapus obscurus*)	Increased heterozygosity at *Ng* after severe winters	+
2. Wood louse (*Porcellio scaber*)	Increased heterozygosity at *Ldh* within a year class after winter	+
3. Western toad (*Bufo boreas*)	Increased heterozygosity at 9 loci after winter	+
4. Killifish (*Fundulus heteroclitus*)	Increased heterozygosity at 12 loci after winter	+
5. Field mouse (*Apodemus sylvaticus*)	Increased recapture rate of *Pgm* homozygote over two-year period	−
6. Hard clam (*Mercenaria mercenaria*)	Deficiency of heterozygotes at *Lap* in randomly mated experimental group	−
7. Eelpout (*Zoarces viviparous*)	Deficiency of heterozygotes at *EstIII* due to differential juvenile survival	−
8. White-tailed deer (*Odocoileus virginianus*)	Deficiency of heterozygotes at *Sdh* in fetuses	−
	Greater heterozygosity at β-hemoglobin in older animals	+
9. American oyster (*Crassostrea virginica*)	Heterozygosity at 4 loci increased with age	+
10. Marine bivalve (*Macoma balthica*)	Heterozygosity at 6 loci increased with age	+
11. Giant scallop (*Placopecten magellanicus*)	Heterozygosity at 6 loci increased with age	+
12. Sagebrush lizard (*Sceloporus graciosus*)	*Est* homozygotes monotonically decreased within cohorts	+
13. Japanese oyster (*Crassostrea gigas*)	Greater heterozygosity at *Mp-1* in older individuals	+

Less direct evidence for an association between heterozygosity and survival has been compiled by comparing levels of heterozygosity among cohorts, between juveniles and adults, or between size classes within a population (Table 1). Comparisons among cohorts have usually been conducted using marine bivalves, and they have invariably reported that heterozygosity increases from younger to older ages. Although suggestive of a positive association between heterozygosity and survival, the possibility that the observed differences are reflective of changes in allele frequency among cohorts cannot be excluded in many studies. In fact such changes were often observed. Samples of marine bivalves almost invariably have lower amounts of heterozy-

Species	Observation	Heterozygous Status[a]
14. Ribbed mussel (*Modiolus demissus*)	Greater heterozygosity at *Sod* in older individuals	+
15. California mussel (*Mytilus californianus*)	Greater heterozygosity at 2 loci in older individuals	+
16. Blue mussel (*Mytilus edulis*)	Greater heterozygosity at *Lap* in older individuals	+
17. Sand shiner (*Notropis stramineus*)	Increased frequency of heterozygotes at *Est* in larger individuals	+
18. Scallop (*Pecten maximus*)	Decreased heterozygosity at *Gpi* in larger individuals	−
19. Skipjack tuna (*Katsuwonus pelamis*)	Decreased heterozygosity at *Tfn* in larger individuals	−
20. Plaice (*Pleuronectes platessa*)	Heterozygosity at 5 protein loci decreased and then increased in three cohorts	+
21. Bleak (*Alburnus alburnus*)	Heterozygosity at *Est* increased with size	+

[a]A (+) indicates heterozygous advantage and a (−) indicates heterozygous disadvantage.

References: (1) Redfield, 1974; (2) Sassaman, 1978; (3) Samollow & Soulé, 1983; (4) Mitton & Koehn, 1975; (5) Leigh Brown, 1977; (6) Adamkewicz et al., 1984; (7) Christiansen et al., 1977; (8) Baccus et al., 1977; Chesser et al., 1982; (9) Singh 1982; Zouros et al., 1983; (10) Singh & Green, 1984; (11) Foltz & Zouros, 1984; (12) Tinkle & Selander, 1973; (13) Buroker, 1979; (14) Koehn et al., 1973; Chaisson et al., 1976; (15) Tracey et al., 1975; (16) Koehn et al., 1976; (17) Koehn et al., 1971; (18) Wilkins, 1978; (19) Fujino & Kang, 1968; (20) Beardmore & Ward, 1977; (21) Handford, 1983.

gosity than expected from random mating populations (reviewed by Burton, 1983 and by Zouros and Foltz, 1984). This suggests that they may contain individuals from genetically divergent populations because of the passive larval dispersal of these organisms (see Zouros et al., 1980; Adamkewicz et al., 1984; Singh and Green, 1984; Zouros and Foltz, 1984 for other possible explanations). Allele frequency and heterozygosity differences among cohorts could also be caused by changes in the proportion of larvae from different populations settling in an area. In addition, younger cohorts may have greater deficiencies of heterozygotes because of differential survival among individuals from different populations resulting in older cohorts having proportionally more individuals from fewer populations. Thus, heterozygosity deficiencies caused by population mixing would be decreased in older cohorts.

Some investigators have used size to infer the age of an individual and have then compared levels of heterozygosity among these "cohorts" (Table 1). The use of size to identify cohorts, however, is not precise because of overlapping size distributions between age classes. Many of these studies have used marine bivalves, and a positive association between growth and heterozygosity in marine bivalves has been reported (see section on growth). Thus, the positive associations between "age" and heterozygosity may be the consequence of the growth association. Larger, more heterozygous individuals of a cohort will tend to be classified as older and the smaller, more homozygous ones to be classified into a younger cohort. The report of increasing heterozygosity with size-determined age in the Japanese oyster, *Crassostrea gigas* (Buroker, 1979), is open to another interpretation. The trend was observed at only one of 11 loci, and at only one of three depths. Chance alone is expected to produce one statistically significant association out of 33 independent comparisons.

Baker and Fox (1978) used an experimental approach to deduce an association between heterozygosity at a peptidase locus and survival in the dark-eyed junco, *Junco hyemalis*. Under competitive conditions for food in aviaries the heterozygotes apparently had increased survival. The interpretation of these results, however, is confounded because males had higher survival, and 48 percent of the males in the study were heterozygotes versus only 24 percent of the females.

Studies examining temporal changes in heterozygosity within a cohort have provided good evidence for a positive association between heterozygosity and survival during part of the life cycle in a diversity of organisms. Other approaches have also revealed patterns of differential heterozygosity that are expected from a positive association between heterozygosity and survival. These results, however, are often open to alternative interpretations so that it is best to view them mainly as suggestive of the association.

Disease resistance

Transferrin is an iron-binding protein in the plasma of all vertebrates and in the eggs of birds. This protein has been shown to be capable of inhibiting or arresting the growth of a variety of iron-dependent microorganisms, many of which are pathogens (Schade and Caroline, 1944; Kochan et al., 1969). The locus coding for transferrin is invariably polymorphic with two alleles at nearly equal frequencies in populations of pigeons, *Columba livia* (Frelinger, 1972). Pigeon eggs, embryos, and young contain mainly maternal transferrin (Frelinger, 1971). The immune system of the offspring is incompetent and resistance to microbial infection is largely conferred by the presence of maternal proteins for a considerable time after hatching. Frelinger (1972) found that transferrin extracted from eggs of heterozygous females inhibited microbial growth more effectively than transferrin extracted from eggs of homozygous females. In addition, the proportion of hatching success of eggs from heterozygous females was 40 percent higher than eggs laid by homozygous females. This differential hatching success is likely a consequence of the increased ability of heterozygous eggs to resist microbial infection and is at least partially responsible for the maintenance of this widespread polymorphism (Frelinger, 1972).

Resistance to malignant falciparum malaria has been implicated to be responsible for the maintenance of many human blood protein polymorphisms (Wills, 1981). Certainly the most widely known and most conclusive case is that of sickle-cell anemia. The β-hemoglobin locus is polymorphic for the allele responsible for sickle-cell anemia *(S)* only in populations that inhabit regions where malaria is prevalent or have recently been founded from such populations (Allison, 1955). Because of this association, and the fact that without modern medicine *S* is often a lethal recessive, Allison proposed that the allele is present in high frequency because the heterozygotes are resistant to malaria. Wills (1981) has reviewed the evidence that conclusively substantiates this.

The alleles responsible for the α- and β-thalassemias, glucose-6-phosphate dehydrogenase deficiency, hemoglobin C, and hemoglobin E also exist in high frequency only in malarious regions. The first three "alleles" actually comprise many alleles. The evidence that these polymorphisms are maintained by heterozygous resistance to malaria, however, is largely circumstantial. Homozygotes for some of the α- and β-thalassemia alleles are conditional or unconditional lethals. It is difficult to account for the high frequency of these alleles unless the heterozygotes have an advantage. The other cases all involve instances in which selection against the homozygotes is uncertain and, as with the thalassemias, evidence for resistance to malaria is inferred by association only (Wills, 1981).

Most human populations are polymorphic at a number of loci coding for red and white blood cell antigens such as ABO, Rh, MN, and human leukocyte antigen-A (HLA). Many of the alleles at these loci have been found to be associated with a variety of diseases, suggesting that they may not be neutral (Mourant et al., 1978). Only at the HLA system, however, is there some evidence that the polymorphism may be being maintained because heterozygotes have increased resistance to disease (Degos et al., 1974; Wills, 1981; Black and Salzano, 1981).

In summary, most of the evidence indicating a positive association between disease resistance and heterozygosity suggests that the association exists. The transferrin polymorphism in the pigeon and sickle-cell anemia are the only examples that we feel conclusively demonstrate the existence of this association. The high frequency of the recessive lethal thalassemia alleles in some populations is most likely the consequence of heterozygous advantage. Whether or not this "advantage" involves disease resistance, however, is an open question at this time.

Growth and developmental rate

Associations between heterozygosity and growth have mainly been investigated in marine invertebrates, particularly bivalves. A number of different approaches have been used to demonstrate the existence of this phenomenon. In practically all cases, however, heterozygosity is measured as the number of heterozygous protein loci per individual.

The studies of Singh and Zouros (1978) and Zouros et al. (1980) exemplify one commonly used method. Planktonic larvae of the American oyster, *Crassostrea virginica,* were allowed to settle on experimentally placed scallop shells for one week in the natural environment. The shells were then removed to the laboratory until the natural settling period was over to minimize age differences. Subsequently, the shells with the growing oysters were placed in a tray and returned to the natural environment. The "population" was sampled after one year of growth, individuals were weighed, and their genotypes determined at seven enzyme loci. A significant positive correlation between heterozygosity and mean weight was observed in two year classes. This association is also present at an age of 2 and 3 years (Singh, 1982, 1984). Heterozygosity at each locus is apparently associated with weight gain independently of the other loci, suggesting that the association is due to the summation of small effects of individual loci (Foltz et al., 1983).

Using a similar experimental design, a positive association between heterozygosity and shell length was found in 3-month-old blue mussels, *Mytilus edulis* (Koehn and Gaffney, 1984), but the association was negative at an age of about 5 months (Diehl et al., 1985). No association between heterozygosity and growth rate was observed among half-sib families of blue mussels (Beaumont et al., 1983) or in a 1-year-old laboratory population of the clam *Mercenaria mercenaria* (Adamkewicz et al., 1984).

Other approaches have been used to deduce an association between heterozygosity and growth in marine bivalves. Garton et al. (1984) collected coot clams, *Mulina lateralis,* from a natural population and found that gain in length, weight, and height of individuals after 29 days in the laboratory was positively associated with heterozygosity.

Singh and Green (1984) determined the age of individuals in two natural populations of *Macoma balthica* from the number of annual winter rings on the shell. Growth rate was estimated from the distance between successive rings and was found to be positively associated with heterozygosity in both populations. In contrast, Foltz and Zouros (1984) used the same procedure to age individuals from a natural population of scallops, *Placopecten magellanicus,* and found no association between heterozygosity and the weight of individuals 4 to 11 years old.

Positive associations between heterozygosity and growth have been implicated for a marine polychaete, *Hyalinoecia tubicola* (Manwell and Baker, 1982), and a sea anemone, *Metridium senile* (Shick et al., 1979). Polychaetes that were more heterozygous at one or two alcohol dehydrogenase loci were significantly larger than individuals homozygous at both loci. The age of the individuals, however, was unknown; since the sample almost certainly contained individuals of different ages, the association may reflect differential survival rather than growth. Unknown age of individuals also confounds the interpretation of the anemone data. Furthermore, significantly more individuals heterozygous at the glucosephosphate isomerase locus were found in areas of low current velocity than in areas of moderate velocity. This nonrandom spatial distribution of genotypes may be solely responsible for the observed positive association between size and heterozygosity at this locus. Individuals in low velocity areas are larger than those in moderate velocity areas and within the areas there is no association between size and heterozygosity.

Associations between multiple locus heterozygosity and physiological parameters indicative of growth potential have been investigated in some marine invertebrates. Negative correlations between heterozygosity and oxygen consumption have been reported for the American oyster (Koehn and Shumway, 1982), coot clam (Garton et al., 1984), and blue mussels (Diehl et al., 1985). Under the experimental conditions it is presumed that oxygen consumption is directly related to the energetic costs of maintaining routine metabolic functions. Lower values reflect an energetically more efficient metabolism and such individuals are expected to be capable of channeling proportionally more energy into growth. Rodhouse and Gaffney (1984) found no association between heterozygosity and oxygen consumption in blue mussels but they did observe that more heterozygous individuals lost weight more slowly during starvation, indicating superior metabolic efficiency. Garton (1984) reports a positive association between heterozygosity and scope for growth in the oyster drill, *Thais haemastoma.* "Scope for growth" is the amount of energy consumed minus that necessary to maintain routine metabolic functions, and thus represents the energy available for growth and reproduction. The observed association, however, is apparently not due to better metabolic efficiency but to the larger size and greater feeding rate of more heterozygous individuals.

Associations between heterozygosity and growth also have been investigated in natural populations of vertebrates. A positive correlation between heterozygosity and length was found in five of six cohorts of larval tiger salamanders, *Ambystoma tigrinum* (Pierce and Mitton, 1982). This correlation was significant in four cases. When five of the cohorts were sampled again 4–6 weeks later, three of the correlations were negative with one being significant. In the laboratory, a full-sib family of larvae was divided into four replicate "populations." Each population was sampled in its entirety at one of three ages. No association was observed between heterozygosity and length in the 10-day sample, while a significant positive association was observed in the 15-day sample and a 22-day sample. Surprisingly, no association was found in another 22-day sample.

Heterozygosity was found to be positively associated with length in adult male and female as well as in juvenile mosquitofish, *Gambusia affinis* (Feder et al., 1984). The age of the individuals, however, is unknown so that this difference cannot be interpreted to be due to differences in growth rate with much certainty. Altukhov and Varnavskaya (1983) found that faster growing, earlier maturing male sockeye salmon, *Oncorhynchus nerka,* are more heterozygous at two enzyme loci than slower growing males. However, they found no association between heterozygosity, growth, and age at maturity in females.

Cothran et al. (1983) report a positive association between the number of heterozygous protein loci and fetal growth rate in the white-tailed deer, *Odocoileus virginianus*. They also report, without presenting the data, a positive association between heterozygosity and growth in adult females. In contrast, Smith et al. (1982) found no evidence for an association between heterozygosity and growth in adult males, from the same herd but sampled in different years than their study of females and fetuses.

In a study of two populations, human infants with low birth weights but normal gestation times were found to have reduced heterozygosity compared to babies of similar weight that were born prematurely (Bottini et al., 1979). In the Rome, Italy population there was no difference in heterozygosity between adults and premature babies, suggesting that fetal growth rate is positively associated with heterozygosity. A comparison between babies with normal gestation times would provide a more direct test of this association.

We feel that the comparisons among individuals of the same or known age and monitoring growth over specified periods in the laboratory provide convincing evidence for the existence of a positive association between heterozygosity and growth. The failure to find associations in comparable studies may indicate that the association is not universal, or that it is highly dependent on age, developmental stage, or environmental conditions. The data of Diehl et al. (1985), Pierce and Mitton (1982), and Smith et al. (1982) indicate that the association may be inconsistent.

To our knowledge, associations between heterozygosity and developmental rate have only been investigated in two species of fishes. DiMichele and Powers (1982a) report that killifish, *Fundulus heteroclitus,* that are heterozygous at a lactate dehydrogenase locus have a hatching time intermediate to the two homozygotes. It is likely that the observed differences are due to the locus examined, because the different genotypes are associated with differences in blood oxygen affinity (DiMichele and Powers, 1982b) and respiratory stress is known to trigger hatching in this species (DiMichele and Taylor, 1980, 1981). The developmental rate of the genotypes is inversely related to their association with blood oxygen affinity. In the rainbow trout, *Salmo gairdneri,* Danzmann et al. (1986) found that individuals more heterozygous at protein loci consistently hatched sooner than more homozygous individuals. The above studies provide direct evidence that heterozygosity either at single loci or multiple loci can influence developmental rate in fishes.

Developmental stability

"Developmental stability" is likely to be a rather nebulous concept to some readers, so some discussion of it may be necessary. The consideration that the phenotypes of individuals in natural populations tend to vary only within certain limits—despite presumed genetic and environmental variation—led Waddington (1942, 1948) to propose that developmental pathways are genetically adjusted by natural selection to have reduced sensitivity to this variation. This "buffering," "canalization," or "stabilization" of development enhanced the probability that the end result of development will be the phenotypic norm of the species, the norm being considered to be positively associated with fitness. Individuals with greater developmental stability will be favored by natural selection because they will have an increased tendency to express the phenotypic norm. A number of genetic mechanisms have been proposed to account for developmental stability, but the only one that concerns us here is Lerner's (1954) proposal that more heterozygous individuals have increased developmental stability.

A common way of testing for the presence of an association between heterozygosity and developmental stability has been to compare the amount of variation observed at particular characters between homozygotes and heterozygotes at protein loci. This approach assumes that increased variation within a genotypic class reflects a higher environmental component to the observed variability and thus reduced developmental stability. Homozygotes are predicted to be more variable than heterozygotes.

The study of Mitton (1978) is characteristic of this type of approach. Individuals were sampled from three populations of killifish. The counts of seven meristic characters and the genotypes at five protein loci were determined for each individual. The average amount of variation of the seven meristic characters was compared between heterozygotes and homozygotes at each locus in each sample. Data from the males and females were analyzed separately. A total of 30 pairwise comparisons were made, and in 22 cases the homozygotes had greater average variation (sign test $p < 0.05$). The tendency for homozygotes to have greater morphological variation than heterozygotes has been reported for a number of other species: monarch butterflies, *Danaus plexippus* (Eanes, 1978); guppies, *Poecilia reticulata* (Beardmore and Shami, 1979); American oysters (Zouros et al., 1980; Singh, 1982); house sparrows, *Passer domesticus* (Fleischer et al., 1983); blue mussels (Koehn and Gaffney, 1984); and humans (Livshits and Kobylianski, 1984). In contrast, there was no indication that homozygotes have greater morphological variation in rufous-collared sparrows, *Zonotrichia capensis* (Handford, 1980); plaice, *Pleuronectes platessa* (McAndrew et al., 1982); or the high cockscomb, *Anaplarchus purpurescens* (Yoshiyama and Sassaman, 1983).

The above studies suggest that a positive association between heterozygosity and developmental stability may be common in animals. Chakraborty and Ryman (1983), however, have shown that another mechanism other than increased developmental stability can result in lower morphological variation in heterozygotes. Assume that three loci control the count of a meristic character. Each locus has two alleles in equal frequency such that the substitution of a "high count" allele for a "low count" allele at any locus adds one more count. All of the variation of this character is due to additive variation (Falconer, 1981). The frequency distribution of the counts in the population and the number of heterozygous loci in individuals is portrayed in Figure 1. It is readily apparent that as the number of heterozygous loci increases, the number of possible counts and the maximum difference among them progressively decrease to zero. This results in decreasing variance with increasing heterozygosity. This association will be present even when allele frequencies vary among the loci and when there

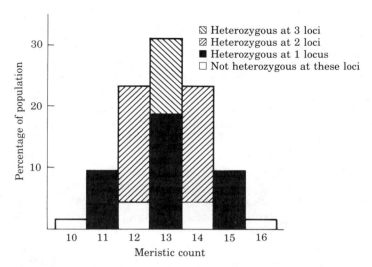

FIGURE 1. Frequency distribution of meristic counts for individuals heterozygous at 0, 1, 2, or 3 loci that additively affect the meristic count. Each "high count" allele increases the count by one. (See text for further explanation.)

is an environmental component to the variability of the character (Chakraborty and Ryman, 1983). Thus, the presence of additive genetic variation can account for the reduced morphological variation observed in heterozygotes. The failure to find this association in some studies may simply mean that the loci examined do not affect the characters in an additive fashion and are not marking portions of the genome that are.

In most of the studies cited, it is not possible to determine whether increased developmental stability or additive genetic variation is largely responsible for the observed association. The latter phenomenon is expected to result in alternate homozygotes at a locus having different means. If the locus examined, however, makes only a relatively small contribution to the variation, this difference is likely to be detected only in extremely large samples. Thus, the failure to find differences in means between homozygotes at individual loci (Eanes, 1978; Fleischer et al., 1983) does not provide strong support for rejecting the additive variance explanation. The observation that the morphology of individuals can be strongly influenced by additive variation lends credibility to this explanation (Falconer, 1981; Boag, 1983; Leary et al., 1985).

Even if the reduced variability of heterozygotes is largely due to additive variation, this does not necessarily mean that the association is irrelevant to conservation biology. From Figure 1 it is apparent that heterozygotes on the average will have a phenotype closer to the population mean. It has been shown in some studies that mean or modal values of certain characters can be positively associated with survival or other fitness components (Rendel, 1943; Karn and Penrose, 1951; Grant, 1972; Beardmore and Shami, 1976), although this is certainly not universal (Grant, 1972; Boag and Grant, 1981).

Phenotypic differences between the left and right sides of individuals for bilateral traits has been used as a measure of developmental stability. Bilateral characters of an organism are often asymmetric in size, shape, or number. Fluctuating asymmetry occurs when there is no tendency for one side to have a larger value than the other and symmetry is the normal phenotypic state (Van Valen, 1962). This occurs when the signed difference between a character on the left and right sides of individuals in a population is normally distributed around a mean of zero. This type of asymmetry reflects the inability of individuals to develop precisely along genetically determined pathways, because each side of an individual is genetically identical (Mather, 1953; Thoday, 1955, 1958). Fluctuating asymmetry, therefore, provides a measure of relative developmental stability. Increased asymmetry reflects reduced developmental stability.

A negative correlation between the average amount of heterozygosity at protein loci and the average amount of asymmetry among populations has been reported for the side-blotched lizard, *Uta stansburiana* (Soulé, 1979); two freshwater bivalves, *Elliptio complanata* and *Lampsilis radiata* (Kat, 1982); and the fish *Poeciliopsis monacha* (Vrijenhoek and Lerman, 1982). These data suggest that a positive association between heterozygosity and developmental stability exists among populations of these species. Whether this association exists among individuals within populations, however, cannot be concluded from these data. Fluctuating asymmetry has an environmental as well as a genetic component (Beardmore, 1960; Valentine and Soulé, 1973; Siegel and Doyle, 1975a,b,c; Bradley, 1980; Zakharov, 1981). It is usually not possible to determine which component is largely responsible for the different levels of asymmetry observed among populations. Thus, environmental differences among populations may result in a negative correlation between heterozygosity and fluctuating asymmetry among populations even if there is no association between these values among individuals within populations.

One can test more directly for an association between heterozygosity and developmental stability by comparing levels of heterozygosity and asymmetry among individuals from the same population. We have consistently observed a negative correlation between the number of heterozygous protein loci and asymmetric characters per individual in three species of salmonid fishes: rainbow trout, *Salmo gairdneri;* westslope cutthroat trout, *Salmo clarki lewisi;* and brook trout, *Salvelinus fontinalis* (Leary et al., 1983, 1984, 1985). Thus a positive association between heterozygosity and developmental stability appears to be a general phenomenon in populations of salmonid fishes. In contrast, Smith et al. (1982) found no association between heterozygosity and fluctuating asymmetry of antler characteristics in white-tailed deer.

Biémont (1983) sampled pregnant *Drosophila melanogaster* females from seven natural populations to create laboratory populations and then performed a series of brother–sister matings. He found that males and females with progeny survival greater than 90 percent also had significantly lower fluctuating asymmetry for wing length and higher heterozygosity at four enzyme loci than individuals with progeny survival less than 90 percent. These relationships were consistent in all seven populations, and suggest that heterozygosity at protein loci is positively associated with developmental stability. The association could be more directly tested by a correlation of individual heterozygosity and asymmetry.

Comparing levels of fluctuating asymmetry between individuals with extreme and central values of a morphological character has also been used to test for an association between heterozygosity and developmental stability. The rationale for this approach is evident on reconsideration of Figure 1. Individuals with extreme values for a character whose variation is largely due to additive genetic variation will tend to be homozygous at the loci that influence the trait. If the number of loci is large, then extremeness may provide a good estimate of genomic homozygosity. Individuals with extreme values are expected to be more asymmetric than those with central values.

This approach has provided highly variable results. Jantz and Webb (1980) found that humans with extreme palm ridge counts also had increased asymmetry. Similar results have been reported by Soulé and Cuzin-Roudy (1982) for wing length of houseflies *Musca domestica,* size of the external auditory meatus in kangaroo rats, *Dipodomys merriami,* and the weight of four wing bones in the English sparrow. In contrast, they found no evidence of a negative association between extremeness and asymmetry for other characters in the housefly, kangaroo rats, humans, and a butterfly, *Coenonympha tullia.* Similarly, we have found that rainbow trout with extreme numbers of vertebrae are significantly more heterozygous at protein loci and tend to be less asymmetric than individuals with central counts (Leary et al., 1985).

It is difficult to provide a simple interpretation of these results. The strength of the relationship between morphological extremity and asymmetry is likely to be highly dependent on how strongly the variation of a character is controlled by additive genetic variation and how many loci are involved. The association should be most easily detectable for those characters in which both these parameters have high values. Variation of these parameters among the characters examined in the studies could alone account for the varied results. This view suggests that finding that extreme individuals have high asymmetry indicates the existence of a positive association between homozygosity and asymmetry. The negative results are largely due to the selection of an inappropriate character.

It has also been suggested that extreme individuals will have increased asymmetry only for those characters for which extreme values are due to reduced developmental stability (Jantz and Webb, 1980; Soulé and Cuzin-Roudy, 1982). That is, the association will exist only for those characters in which extremeness and asymmetry are both measuring the same parameter, reduced developmental stability.

Direct evidence for a positive association between heterozygosity and developmental stability so far is available only for salmonid fishes and *Drosophila*. However, observations that relatively heterozygous individuals have reduced morphological variation, that populations with relatively high heterozygosity have reduced fluctuating asymmetry, and that morphologically extreme individuals have increased asymmetry all suggest that a relationship between heterozygosity and developmental stability exists in a diversity of species.

HETEROZYGOSITY AND CONSERVATION BIOLOGY

Heterozygosity and fitness

We have reviewed many studies demonstrating associations between heterozygosity and components of fitness (survival, disease resistance, growth and developmental rate, and developmental stability). Nevertheless, there are many problems in relating heterozygosity directly to fitness. The first problem is whether the observed effects are due to the loci examined or to loci that are linked to the loci examined (associative dominance or overdominance). There are thousands of loci throughout the genome that are likely to have deleterious recessive alleles at low frequencies because of the balance between deleterious mutations and natural selection. Any new mutation will initially be nonrandomly associated with allelic variation at linked marker loci (e.g., the protein loci examined by electrophoresis). This nonrandom association (linkage disequilibrium) is expected to decrease in each generation at a rate proportional to the amount of recombination between the locus at which the mutation occurred and the marker locus. Thus, homozygotes are expected to have a greater probability of also being homozygous for deleterious recessive alleles at linked loci.

The possibility of associative overdominance is an important concern in understanding the mechanisms underlying heterosis (Thomson, 1977). However, associative overdominance is not of critical concern from the perspective of conservation biology. Conservation biology is interested in the relationships between heterozygosity and fitness, irrespective of which individual loci are responsible for this relationship.

A second problem in relating heterozygosity to fitness is that heterosis may result either from dominance or heterozygous superiority at individual loci. Again, this issue is of importance for a basic understanding of heterosis but is not critical from the perspective of conservation biology when the concern is the effects of loss of genetic variation (see Ledig, Chapter 5, for a detailed discussion of this).

A third problem in relating heterozygosity to fitness is that the studies we have reviewed generally have measured one component of fitness, not fitness itself. Genotypes that have an advantage in some aspect of fitness may have disadvantages for other components of fitness. The pleiotropic effects of genes resulting in genetic "trade-offs" between components of fitness has been termed antagonistic pleiotropy (Rose, 1982). For example, heterozygous genotypes that seem to have an advantage because of greater growth rates may also have reduced longevity and thus have reduced overall fitness when both of these effects are considered.

We therefore cannot conclude for a particular species that an advantage of heterozygotes in some measured component of fitness demonstrates that heterozygotes have greater fitness. However, we can conclude from such a result that heterozygosity is affecting fitness; it is unlikely that an advantage in some fitness component is exactly balanced by some disadvantage. An examination of the relationship between heterozygosity and fitness components in many species provides information about the general relationship between heterozygosity and fitness. There are many well documented examples of advantages of heterozygotes in several components of fitness. There are fewer examples of disadvantages of heterozygotes for fitness traits. Thus, we conclude that in general there is a positive relationship between heterozygosity and fitness.

Frequency-dependent selection

Up to this point we have treated fitness as if a given genotype has a certain constant fitness. This assumption is not likely to be true under a wide variety of circumstances. Fitnesses are likely to change with environmental changes, as well as with changes in allele frequencies and population densities (Kojima, 1971). Frequency-dependent selection occurs when fitnesses change as allele frequencies change.

There is an increasing body of evidence that frequency-dependent selection plays a major role in the preservation of genetic variation in natural populations (Clarke, 1979). Frequency-dependent selection will act to maintain genetic variation whenever relatively rare alleles have some selective advantage. This may be true even if the heterozygous genotype has a lower fitness than the homozygous genotypes at equilibrium. Thus, frequency-dependent selection complicates the understanding of the action of selection in natural populations; fitness estimated in a population at or near equilibrium may not reflect the evolutionary dynamics of a locus.

Frequency-dependent selection appears to be of special importance in association with parasitism and disease resistance (Clarke, 1979). In turn, parasitism and disease resistance are especially important in conservation biology because of their potential role in extinction (Frankel and Soulé, 1981; Dobson and May, Chapter 16). It should be recognized that the full importance of genetic variation in populations cannot be determined by estimating the relative fitnesses of heterozygotes at a single point in time.

Heterozygosity and the preservation of threatened species

We have presented evidence from a wide variety of species that indicates there is a tendency for heterozygotes to have increased survival, increased disease resistance, increased growth rate, and increased developmental stability. We conclude, then, that the loss of heterozygosity is expected to make a particular species more vulnerable to extinction.

The recent detailed study of the cheetah, *Acinonyx jubatus,* by O'Brien and colleagues (1985) is an excellent case study that demonstrates most of these principles. There are an estimated 1500 to 25,000 cheetahs remaining in two wild populations in southern and eastern Africa. A genetic survey of over 200 structural loci indicates that the southern cheetah is almost completely lacking genetic variation. Reciprocal skin grafts between 14 cheetahs were accepted, which indicates that cheetahs have very little (if any) genetic variation at the major histocompatability complex (MHC). The MHC is extremely polymorphic in all other mammals that have been studied, and has been found to be associated with disease resistance in humans (Wills, 1981).

There is evidence that the extremely low heterozygosity in cheetahs has an effect on many aspects of survival and reproduction. Captive breeding of cheetahs is extremely difficult for a variety of reasons. Cheetah semen has low spermatozoal concentrations and a high proportion of morphologically abnormal spermatozoa. Cheetahs have relatively high juvenile mortality in captivity compared to other mammal species. Captive cheetah populations are also extremely sensitive to epizootics. Thus, the lack of genetic variation in cheetahs is apparently associated with a number of phenotypic effects that threaten the future of the species.

The genetic uniformity of cheetahs apparently is not the result of captive propagation—nearly all the animals studied were less than two generations removed from the wild. However, animals derived from the East African population have not been examined. These animals may provide a new source of genetic variation that could be introduced into captive breeding programs of cheetahs.

The cheetah is an important natural experiment for conservation biology. Reduction in the amount of genetic variation in endangered species through unwise genetic management is likely to have serious phenotypic effects that increase the probabilty of extinction of threatened species.

Heterozygosity and evolutionary potential

For conservation biology to be successful it must accomplish more than the successful propagation of species under captive conditions or in nature preserves. A conservation program is inadequate if it does not allow the future evolution of species. Therefore, in this last section we shift our focus away from the effects of heterozygosity on individual fitness to consider the importance of heterozygosity for the evolutionary change of populations and species.

There is a much clearer relationship between heterozygosity and the potential of a population to change genetically than there is between heterozygosity and individual fitness. A direct correlation between the amount of genetic variation in a population and the rate of change by natural selection was shown theoretically by R.A. Fisher (1930) in the Fundamental Theorem of Natural Selection: "The rate of increase in fitness of a population at any time is equal to its genetic variance in fitness at that time." The Fundamental Theorem applies strictly to allelic variation at a single locus, but a correlation between genetic variation throughout the genome and evolutionary potential is expected.

Experiments with *Drosophila* have demonstrated that the genetic response to selection is affected by increases or decreases in the amount of genetic variation present. Ayala (1965) found that the rate of evolutionary change in adapting to new environmental conditions was greater in mixed strains derived from two different natural populations than it was in a strain derived from a single natural population. Ayala (1969) also found that the the rate of evolutionary change could also be increased with mutation-inducing radiation. Frankham (1980) has shown that bottlenecks of two individuals reduces both the short- and long-term response to artificial selection by approximately 25 percent, as predicted by theory (James, 1971).

There are important differences between *Drosophila* and other species. *Drosophila* generally occur in natural populations that have population sizes measured by billions, and thus contain much more genetic variation than the species that are generally of concern in conservation biology (Soulé, 1976; Frankel and Soulé, 1981). The effects of increasing and decreasing genetic variation on evolutionary rates may be even more dramatic in species with less abundant genetic variation in natural populations. Unfortunately, we are not aware of similar experiments with other species.

Nevertheless, it is clear that the evolutionary potential of any species depends upon the amount of genetic variation it contains. In addition, the literature we have reviewed provides growing evidence that individual genetic variation (i.e., heterozygosity) is positively associated with components of fitness in animal species. Once genetic variation is lost it can only be replaced by the slow process of mutation, which takes many generations. Genetic variation must be preserved in order to increase the probability of both the short- and long-term survival of animal species.

SUGGESTED READINGS

Ayala, F.J. 1984. Molecular polymorphism: How much is there and why is there so much? *Developmental Genetics 4:* 379–391.

Clarke, B. 1979. The evolution of genetic diversity. *Proceedings of the Royal Society of London (B) 205:* 453–474.

Mitton, J.B. and M.C. Grant. 1984. Associations among protein heterozygosity, growth rate, and developmental homeostasis. *Annual Review of Ecology and Systematics 15:* 479–499.

O'Brien, S.J. et al. 1985. Genetic basis for species vulnerability in the cheetah. *Science 227:* 1428–1434.

Zouros, E. and D.W. Foltz. In press. The use of allelic isozyme variation for the study of heterosis. *Isozymes: Current Topics in Biological and Medical Research,* Vol. 13.

CHAPTER 5

HETEROZYGOSITY, HETEROSIS, AND FITNESS IN OUTBREEDING PLANTS

F. Thomas Ledig

This chapter explores the topics of heterosis and genic diversity, particularly with respect to tree species. A comparison of genic diversity for allozymes with the frequency of recessive deleterious genes indicates parallel patterns of variation. Consideration of mechanisms that could promote or maintain diversity leads to a hypothesis (1) that high genic diversity is a result of mutation; (2) that tree species are burdened with a greater mutational load than other life forms; (3) that inbreeding depression, resulting from homozygosity for deleterious recessive mutations, drives the breeding system of outcrossing plants; and (4) that selection promotes outbreeding mechanisms not to maintain diversity but to protect the individual from its excess. It turns out, however, that the lesson for conservation biology does not depend on whether genic diversity is maintained by mutation or by heterozygote advantage over homozygotes.

This chapter focuses largely on forest trees, particularly conifers, for several reasons. First, much of the recent literature on genic diversity and on heterozygosity and fitness in plants is from studies of forest trees. Second, individual heterozygosity is about twice as great in tree species as in annuals (Hamrick et al., 1979), and, being the extreme case, trees are a good place to look for the significance of heterozygosity. Third, their life history characteristics are also extreme: trees are the most long-lived plants, they reach reproductive age the latest, and their reproductive output exceeds that of any herbaceous perennial or annual. Finally, tree species pose the greatest problems for conservation biology because their large size means that a correspondingly large area is necessary to support reasonable populations, particularly in tropical rainforests, where many tree species are sparsely distributed (Hubbell and Foster, Chapter 10).

BALANCE AND NEUTRALIST INTERPRETATIONS OF GENIC DIVERSITY

Electrophoretic techniques have made it possible to assay enzymes of both wild and domesticated species (Lewontin and Hubby, 1966), and have revealed substantial stores of cryptic variation. Whether this variation is important to fitness is debated by two groups, the classic or neutralist school and the balance school.

The neutralists argue that the great store of variability that characterizes most organisms is merely a reflection of stochastic processes, mutation and random drift in gene frequencies (Kimura and Ohta, 1971). According to the opposing view, diversity is maintained by some form of natural selection that favors heterozygotes (for a full discussion, see Lewontin, 1974). To the neutralists, most genetic variants are of indifferent value in the short term, while to the balance school, heterozygosity is important to reproductive success and therefore to the maintenance of populations. Nonetheless, both groups would agree that change, or evolution, depends on the presence of genetic diversity, and that the long-term value of genetic diversity is probably proportional to its amount, which sets the limits to evolutionary change.

According to the classic viewpoint, each species must achieve a trade-off between short-term fitness and long-term flexibility (Mather, 1953). In the short term, progeny encounter an environment similar to the one to which their parents were adapted. Therefore, the production of genetically uniform progeny, identical to the parent, would increase fitness. Selfing should lead to increased resemblance of offspring to parents and to increased homozygosity, as well as doubling the contribution of the parental genes to the next generation for the same maternal investment (Fisher, 1941).

In light of the apparent disadvantage of producing genetically heterogeneous offspring, it is difficult to see why all species are not selfing. The usual explanation has been some form of the group selection argument: (1) selfing leads to loss of genetic variability; (2) environmental change is inevitable in the long term, and without variation, evolutionary response is impossible; (3) genetically depauperate lineages go extinct, and only those groups have persisted that, by chance, maintained diversity by outcrossing. This argument is not very satisfying and has led to a search for mechanisms that operate at the level of the individual to maintain genetic variation in populations.

Thus, several mechanisms of balancing selection, including heterozygote superiority, or overdominance, have been invoked to explain the widespread occurrence of genetic polymorphism (Ennos, 1983).

MECHANISMS AND MEASURES OF HETEROSIS

The term "heterosis" was coined around 1914 (Shull, 1952) to refer to the phenomenon of heterozygote superiority, or hybrid vigor. Dominance and overdominance are two major explanations for heterosis. Under the dominance hypothesis, heterosis counters inbreeding depression. Most individuals carry deleterious alleles in heterozygous or homozygous state, and unrelated individuals (referred to some arbitrary ancestral generation) are less likely to carry the same alleles than related individuals; in crosses of unrelated individuals, recessive deleterious alleles from one parent are masked by a dominant allele with normal function from the other parent. A homozygote for the normal allele at each locus would leave at least as many offspring as the heterozygote. With true overdominance, the heterozygote is actually superior to either homozygote.

Developmental homeostasis has been linked with heterozygosity and heterosis (Lerner, 1954). Bilateral symmetry in animals, an index of homeostasis, seems to be related to heterozygosity (Soulé, 1967, 1979; Kat, 1982; Leary et al., 1984). Variance in morphological traits is generally less among heterozygotes than among homozygotes (Eanes, 1978; Fleischer et al., 1983; Mitton, 1978; Zouros et al., 1980). In plants and ectotherms, heterosis is frequently expressed most strongly in fluctuating and stressful environments (McWilliam and Griffing, 1965; Parsons, 1971)—that is, heterozygous individuals maintain a more uniform phenotype than homozygotes across a range of environments, suggesting superior developmental homeostasis. Either the dominance or overdominance hypotheses could explain these results.

On a molecular level, heterozygotes for enzyme genes have two forms of the enzyme, or, if the functional enzyme is multimeric (a combination of two or more polypeptide chains), several forms. Each form, or isozyme, could have different optima for pH, temperature, etc., thus broadening the range of environments in which enzyme function was normal. Indeed, some heterodimeric enzymes have greater in vitro activity than the homodimeric forms (Scandalios et al., 1972; Schwartz and Laughner, 1969), but greater activity does not necessarily confer greater fitness in vivo.

Decades of biometric studies suggest that true overdominance either does not exist or is very rare (Jinks, 1983). Most cases of heterosis, when dissected into their component traits, turn out to be the product of multiplicative interactions between traits that themselves show additive or dominance inheritance, not overdominance (Sinha and Khanna, 1975; Williams, 1959). Where dominance occurs, the dominant allele is nearly always the one that confers a faster rate of development, which suggests that the recessive alleles have impaired function (see for examples Sinha and Khanna, 1975). "Pseudooverdominance" can occur in predominantly outbreeding organisms as a result of inbreeding depression (Ross, 1980).

Convincing examples of overdominance for fitness traits have been reported in some animals (Watt, 1985; Allendorf and Leary, Chapter 4). The relationship of vigor to heterozygosity in tree species seems best explained by the inbreeding or dominance hypothesis, although overdominance may play a role at some loci (Bush et al., in preparation; Strauss, 1985).

The great store of genetic diversity in plants could be maintained by several mechanisms other than true overdominance. One especially important mechanism in sessile organisms such as plants might be selection for homozygotes in multiple niches, followed by migration among niches. Second, density-dependent selection is a strong possibility because of the importance of true interference competition in plant communities and the environmental changes that accompany stand development; alleles that favor survival in a dense population may have a negative impact on fecundity and the potential for rapid multiplication after catastrophic events. Third, frequency-dependent selection could maintain genetic polymorphisms; a particularly well analyzed example is the sex polymorphism. And fourth, balanced polymorphisms might result from "hitchhiking"—a favorable mutant tightly linked with a deleterious recessive would increase in frequency until its selective advantage was balanced by the selective disadvantage of the linked deleterious gene (Hedrick, 1982).

Heterosis is important to conservation biology because of its relationship to fitness. However, fitness is a difficult concept to define and even more difficult to measure, especially in iteroparous perennial plants. Analysis is usually restricted to components of fitness; one of the most obvious and easy to measure is survival. In long-lived trees, both survival and reproductive output are strongly dependent on growth because of the importance of intra- and inter-specific competition. Heterosis in tree species has usually been evaluated by growth rate.

Growth is a good surrogate of fitness in trees because trees grow to tremendous size, which means their numbers must be greatly reduced with age. In loblolly pine *(Pinus taeda)* 2200 to 29,700 seedlings/ha reduces to 370 trees at maturity (Allen, 1961); in eucalypts *(Eucalyptus* spp.) competition may reduce cohorts from the level of 500,000 down to 25 per acre by the time stands are ready for regeneration (Barber, 1965); and in European beech *(Fagus silvestris)* 500,000 saplings/ha at age 10 may be reduced to 400 trees at age 100 (Assman, 1970). The most rapidly growing, either because of inherent factors or chance, have ample opportunity to overtop others in their cohort.

In addition to the effect of growth rate on survival, a wealth of literature indicates that the most rapidly growing individuals are most fecund and reach reproductive maturity soonest (for example, see Solbrig, 1981; Waller, 1984). Early reproductive episodes contribute in disproportionate measure to lifetime fitness. Also, the number of flowers or other reproductive structures increases with size in species of largely indeterminate growth, such as trees, and apical meristems may be considered demographically analogous to a population of individuals (White, 1980). Photoinsensitive plants must reach a certain size before flowering, so the correlation between earliness of flowering and vegetative growth is positive (Sinha and Khanna, 1975). After maturity is reached, reproductive output often reduces vegetative growth (Chalmers and van den Ende, 1975; Morris, 1951; Teich, 1975).

HETEROSIS IN PLANTS

Heterosis can occur at several taxonomic levels. Heterosis has been reported in crosses of different species, of different populations within species, and of unrelated individuals within populations.

Heterosis in interspecific hybrids

In forest trees, there is little evidence of heterosis in species hybrids (Brown, 1972; Schmitt, 1968), although many hybrids have proven valuable because they combine desirable features of both parents. Early reports of hybrid vigor (Righter, 1945) usually compared hybrids and parents in environments to which at least one of the parents was ill-adapted. One of the few convincing examples of a heterotic hybrid in conifers is the cross of Japanese and European larches *(Larix leptolepis × L. decidua*; Keiding, 1968). However, many hybrids between closely related species are inferior to either parent. Soulé (1967) predicted that hybridization of divergent genomes would lead to the breakdown of coadapted gene complexes, and that hybridization would be associated with developmental instability, which is apparently the case in organisms as diverse as sunfish (Graham and Felley, 1985) and spruce trees (Manley and Ledig, 1979). Black × red spruce hybrids *(Picea mariana × P. rubens)* are a well-documented case of negative heterosis. Photosynthesis, and therefore the whole seedling economy, was poorer in hybrids than in either parent. Even recurrent backcrosses had depressed rates of photosynthesis and growth. Apparently, divergence was too great to bridge for a complex process like photosynthesis, which requires a highly coordinated set of reactions matched with coadapted anatomical and morphological characteristics. Aberrations such as dwarfism are common in black and red spruce hybrids, and in hybrids of several other species within the pines and spruces (Langner, 1970; Manley, 1975; Schmitt, 1969).

Heterosis in interpopulation hybrids

In corn *(Zea mays),* the classic example of heterosis, there seems to be an optimum amount of divergence for hybrid vigor, beyond which hybrids prove inferior (Stuber, 1970). Hybridization of divergent genomes in wild plants rarely increases vigor or fitness. Interpopulation crosses between loblolly pine from widely separated areas were intermediate to their parents or were negatively heterotic (outbreeding depression) under normal conditions (Woessner, 1972). However, crosses between adjacent populations or subpopulations were often more vigorous than crosses within subpopulations. Inbreeding depression in loblolly pine was pronounced, ranging from 8.4 to 12.7 percent upon selfing (Woessner, 1975). In loblolly pine, spruces, larch, and in the insect-pollinated angiosperm, tulip poplar *(Liriodendron tulipifera),* crosses between trees from different subpopulations in the same geographic area grew more rapidly than progeny from uncontrolled open-pollination, which was interpreted as evidence that mating among relatives in natural stands led to homozygosity of recessive deleterious alleles (Carpenter and Guard, 1950; Coles and Fowler, 1976; Ledig, 1974; Park and Fowler, 1982, 1984; Park et al. 1984). Similar results were found in Scots pine *(Pinus sylvestris)* and Norway spruce *(Picea abies)*: open-pollinated progenies are inferior to progenies produced by crosses between unrelated individuals within the same population (Nilsson, 1973). The bulk of the evidence suggests that interpopulation crosses from widely separated areas do not result in heterosis, and the heterosis observed when crossing individuals from nearby neighborhoods or subpopulations rarely exceeds the level observed in crosses between unrelated individuals within the same neighborhood.

Heterosis within populations

The use of allozymes makes it possible to associate multiple-locus heterozygosity with the phenomenon of heterosis within populations. One of the first examples studied in plants was cylindric blazing star *(Liatris cylindracea),* a prairie perennial that sprouts from a subterranean stem, or corm. Heterozygosity was greater in older individuals than in younger, suggesting selection against homozygotes (Schaal and Levin, 1977). Although there were questions concerning the aging of plants in the field (Jones, 1977), greenhouse experiments demonstrated that the most heterozygous plants were larger and more precocious, and produced more flowers than the most homozygous individuals. In temperate tree species, where individuals can be aged by counting annual wood rings, several studies have found positive correlations between multi-locus heterozygosity and growth rate (Ledig et al., 1983; Mitton and Grant, 1980; Strauss, 1985), and for particular heterozygotes at single loci (El-Kassaby, 1983). For ponderosa pine *(Pinus ponderosa)* and lodgepole pine *(Pinus contorta)*, no relationship between heterozygosity and growth was apparent (Knowles and Grant, 1981; Mitton et al., 1981). However, only one stand was sampled for each species and heterozygote advantage is not always detectable, being more pronounced with age, population density, and climatic variability (Ledig et al., 1983). Positive relationships between growth and h?terozygosity may reflect either inbreeding depression—the result of homozygosity for recessive deleterious alleles—or overdominance at either the isozyme loci themselves or at closely linked loci.

Heterosis in natural populations of outcrossing plants is probably the converse result of inbreeding. Even outcrossing plants occasionally self, but the greatest source of inbreeding must be crosses among related individuals. The sessile nature of plants and the limited dispersal of seed and pollen is expected to result in "neighborhoods," or clusters, of related individuals (Levin, 1981; Levin and Kerster, 1974). In fact, isozyme genotypes often have clustered distributions, as in cylindric blazing star and in low-density stands of ponderosa pine (Schaal, 1975; Linhart et al., 1981). Clusters of related individuals would contribute to inbreeding. Seed set, a sensitive measure of inbreeding depression (Franklin, 1970), is reduced in crosses between adjacent plants, but tends to increase with distance between the parents (Coles and Fowler, 1976; Ledig, 1974; Levin, 1984; Park and Fowler, 1982; Price and Waser, 1979). In the insect-pollinated tulip poplar, seed yields were much improved when parents from different stands were crossed, compared to crosses among neighbors (Carpenter and Guard, 1950).

Heterosis: Summary

High levels of genetic variation and heterozygosity are common in outcrossing plant species, and particularly so in tree species. This variation can be maintained by several forms of selection, and selection for heterozygotes because of overdominance is a commonly invoked explanation. However, overdominance inheritance is seldom documented, and heterosis is rare in interspecies hybrids or between interpopulation hybrids. The bulk of the evidence suggests that heterosis in both natural and cultivated plant species arises merely from the avoidance of inbreeding. Because of limited seed dispersal, offspring of crosses between adjacent individuals in natural populations are likely to suffer from homozygosity of rare deleterious alleles, and crossing unrelated individuals restores vigor because dominance of the normal alleles masks the deleterious alleles. To investigate the alternative hypotheses (that genic diversity is maintained by overdominance or by avoidance of inbreeding), I will review the evidence for genetic diversity (particularly for conifers), including diversity for that unique class of loci, the recessive deleterious genes which constitute the true genetic load.

GENIC DIVERSITY IN TREES

Allozymes

It is convenient to use trees to discuss genic diversity; an extensive body of literature has accumulated during the last decade because of the ease of genetic analysis in the haploid tissue of conifer seeds (see Bergmann, 1973; Conkle, 1974). Genic diversity in conifers has been amply documented for seed isozymes (Table 1).

Tree species in general are highly outcrossing compared to herbaceous perennials and, especially, annuals (Bawa, 1974; Fowler, 1965a; Schemske and Lande, 1985). It is not a mere coincidence that they are also the most highly heterozygous plants (Hamrick et al., 1979). Although recent surveys of gene diversity have tended to lower earlier estimates of heterozygosity, Hamrick's conclusions are still justified, and can even be extended to tropical tree species (Loveless and Hamrick, 1985). Heterozygosities in trees generally exceed those in herbaceous plants by twofold. Nevertheless, even trees can be genetically depauperate. No variation was detectable within populations of Torrey pine *(Pinus torreyana)*, although populations of Coulter pine *(Pinus coulteri)*, a closely related species, are moderately heterozygous (Ledig and Conkle, 1983 and unpublished data).

Despite the generally great genic diversity in conifers and their several mechanisms to avoid selfing, slight deficiencies of heterozygotes are frequently observed among embryos (Dancik and Yeh, 1983; Danzmann and Buchert, 1983; Fins and Libby, 1982; Guries and Ledig, 1982; Knowles, 1984). Heterozygote deficiency has also been observed in eucalypt embryos, but heterozygote excess occurs in the parents (Moran and Brown, 1980). In natural populations of Monterey pine *(Pinus radiata)*, heterozygosity increases considerably between the embryo and the sapling stage and between saplings and mature trees (M. E. Plessas and S. H. Strauss, personal communication). Likewise,

TABLE 1. Genic diversity in tree species.[a]

Species	Expected heterozygosity[b]	Number of loci assayed	Percentage polymorphic loci		Sample size	Reference
			A[c]	B[d]		
PINES (*Pinus*)						
P. attenuata	0.125	22	73	27	10 pops., rangewide	Conkle, 1981
"	0.087	43	58	49	10 pops., rangewide	S. H. Strauss & M. T. Conkle, unpubl. data
P. banksiana	0.141	27	74	—	32 pops., Ontario	Danzmann & Buchert, 1983
"	0.115	21	81	52	3 pops., Alberta	Dancik & Yeh, 1983
P. brutia	0.13	29	45	38	10 pops., rangewide	M. T. Conkle, pers. commun.
P. caribaea	0.212	18	72	48	7 pops., Central America	P. D. Hodgskiss, pers. commun.
P. contorta	0.184	21	86	71	5 pops., Alberta	Dancik & Yeh, 1983
"	0.160	25	59	45	9 pops., British Columbia–Yukon	Yeh & Layton, 1979
"[e]	0.185	39	90	44	1 pop., California	Conkle, 1981
"[e]	0.116	42	68[f]	—	32 pops., rangewide	Wheeler & Guries, 1982
"[e]	0.135	9	44	44	4 pops., within 2 km in Colorado	Knowles, 1984
P. coulteri	0.148	33	49	45	8 pops., rangewide	F. T. Ledig, unpubl. data
P. halepensis	0.040	28	21	21	19 pops., rangewide	Schiller et al., in press
P. jeffreyi[e]	0.261	43	86	67	4 pops., central California	Conkle, 1981
"	0.255	20	90	90	14 pops., northern California–Oregon	Furnier, 1984
P. lambertiana[e]	0.275	19	79	58	58 individuals, California	Conkle, 1981
P. longaeva	0.327	14	79	—	5 pops., Nevada–Utah	Hiebert & Hamrick 1983
P. monticola	0.180	12	65	51	28 pops., rangewide	Steinhoff et al., 1983
P. muricata	0.084	46	67	56	18 pops., northern California	Millar, 1985

85

TABLE 1. (*Continued*)

Species	Expected heterozygosity[b]	Number of loci assayed	Percentage polymorphic loci		Sample size	Reference
			A[c]	B[d]		
P. nigra	0.272	4	100	75	28 pops., Yugoslavia–Mediterranean	Nicolic & Tucic, 1983
P. oocarpa	0.183	18	72	55	8 pops., Central America	P. D. Hodgskiss, pers. commun.
P. palustris	0.150	19	100	84	24 pops., rangewide	Duba, 1985
P. ponderosa	0.124	23	74	35	6 small, isolated pops., Montana	Woods et al., 1983
"	0.186	29	90	52	400 trees, Washington–Idaho–Montana	Allendorf et al., 1982
"	0.123	21	62	38	10 pops., pooled, Washington–Idaho–Montana	O'Malley et al., 1979
P. radiata	0.126	37	70	51	3 pops., California	M. E. Plessas & S. H. Strauss, pers. commun.
P. resinosa	0.007	27	15	4	2 pops., Minnesota	Allendorf et al., 1982
"	0.002	46	—	0	50 trees, Wisconsin	R. P. Guries, pers. commun.
P. rigida	0.146	21	100	76	11 pops., rangewide	Guries & Ledig, 1982
P. sabiniana	0.128	29	93	62	8 pops., rangewide	F. T. Ledig et al., unpubl. data
P. strobus	0.236	12[g]	83	66	27 pops., rangewide (grown in common garden)	Ryu & Eckert, 1983
"	0.330	17	53	—	35 selected clones	Eckert et al., 1981
"	0.309	16	100	81	14 pops., Scotland	Kinloch et al., in prep.
P. sylvestris	0.31[h]	11	100	91	9 pops., Sweden	Gullberg et al., in press
"	0.362	10	100	90	1 pop., North Carolina	Conkle, 1981
P. taeda	0.282	25	96	80	90 selected clones, southeastern U.S.	Conkle, 1981
P. torreyana	0	59	0	0	2 pops., rangewide	Ledig & Conkle, 1983

SPRUCES (*Picea*)

P. abies	0.22	34	82	65	9 pops., Poland	M. T. Conkle, pers. commun.
"	0.41[h]	7	—	—	21 pops., northern Europe	Bergmann & Gregorius, 1979
P. glauca	—	26	77	52	several pops., Alberta	King & Dancik, 1983
"	0.14	20	—	—	1 pop., Alberta	King et al., 1984
P. sitchensis	0.15	24	51	46	10 pops., Oregon–Alaska	Yeh & El-Kassaby, 1979

FIRS (*Abies*)

A. alba	0.50[h]	9	100	94	4 pops., Czechoslavakia	Kormutak et al., 1982
A. balsamea	—	14	57	—	4 pops. within 3 km, New Hampshire	Neale & Adams, 1981
A. balsamea, A. fraseri, and transitional pops.	0.13	20	65	60	12 pops., eastern U.S.	Jacobs et al., 1984
A. bracteata	0.052	30	37	23	5 pops., rangewide	F. T. Ledig & M. T. Conkle, unpubl. data

OTHER CONIFERS

Calocedrus decurrens	0.18	25	96	76	12 pops., California	Harry, 1984
Cupressus macrocarpa	0.16	28	61	61	2 pops., rangewide	M. T. Conkle, unpubl. data
Larix decidua	0.081	28	50	36	11 pops., rangewide	C. R. Niebling, pers. commun.
Larix occidentalis	0.074	23	39	26	19 pops., Washington–Idaho–Montana	Fins & Seeb, in prep.
Larix leptolepis	0.073	16	37	25	9 pops., rangewide	C. R. Niebling, pers. commun.
Pseudotsuga menziesii[e]	0.331	17	100	88	1 pop., Oregon	Conkle, 1981
"	0.155	21	86	67	11 pops., British Columbia	Yeh & O'Malley, 1980
Sequoiadendron giganteum	0.140	8	50	50	34 pops., rangewide	Fins & Libby, 1982
Thuja plicata	0.04					F. C. Yeh, pers. commun.

TABLE 1. (Continued)

Species	Expected heterozygosity[b]	Number of loci assayed	Percentage polymorphic loci A[c]	Percentage polymorphic loci B[d]	Sample size	Reference
		AUSTRALIAN EUCALYPTS (*Eucalyptus*)				
E. caesia	0.12	18	61	44	13 pops, rangewide	Moran & Hopper, 1983
E. cloeziana	0.24	—	—	—	?	Moran & Bell, 1983
E. delegatensis	0.27	—	—	—	?	"
E. grandis	0.18	—	—	—	?	"
E. saligna	0.26	—	—	—	?	"
		TROPICAL ANGIOSPERMS				
Acalypha diversifolia	0.273	29	68	—	3 pops., Barro Colorado Island	Loveless & Hamrick, 1985
Alseis blackiana	0.374	26	90	—	"	"
Hybanthus prunifolius	0.247	42	70	—	"	"
Psychotria horizontalis	0.152	20	50	—	"	"
Quararibea asterolepis	0.256	30	64	—	"	"
Rinorea sylvatica	0.106	35	35	—	"	"
Sorocea affinis	0.239	36	72	—	"	"
Swartzia simplex var. *ochnacea*	0.272	36	76	—	"	"

[a]Genic diversity calculations based on enzymes of the haploid female gametophyte for conifers and diploid leaf tissue for angiosperms, except as noted.

[b]Expected heterozygosity is calculated as the mean of within-population heterozygosities except as noted.

[c]Proportion of loci polymorphic under the criterion that a locus was polymorphic if any allelic variant was observed.

[d]Proportion of loci polymorphic under the criterion that the most common allele was present in a frequency < 0.95 in every population.

[e]Total heterozygosity and percentage polymorphic loci calculated from allele frequencies averaged across populations.

[f]Proportion of loci polymorphic under the criterion that the most common allele was present in a frequency < 0.99.

[g]Genic diversity calculations based on foliar enzymes.

[h]Only variable loci used.

heterozygosity in Scots pine was higher in a 300–350 year old forest overstory than in an 80–100 year old subcanopy (Tigerstedt, 1983), and heterozygosity in ponderosa pine seedlings was higher after some initial mortality than it was among embryos (Farris and Mitton, 1984). All of these observations suggest the occurrence of inbreeding, probably as a result of crosses among related individuals, followed by selection for outcrossed offspring.

Most of the variation (usually over 90 percent) in tree species is localized within populations. An exception is the knobcone pine *(Pinus attenuata)*, a fugitive species occupying a fragmented range, which has much greater diversity (19 percent) among populations than normally encountered in conifers (Strauss and Conkle, in preparation). Marginal populations tend toward lower heterozygosity than central populations (Bergmann and Gregorius, 1979; Guries and Ledig, 1982; Yeh and Layton, 1979), which might be expected because of isolation and restricted gene flow, and because marginal populations frequently owe their origin to colonizing events and suffer from the bottleneck of the founder effect. However, complete elimination of genic diversity is rarely expected even with a small number of founders. A population of pitcher plants *(Sarracenia purpurea)* known to be derived from a single founder, maintained a level of heterozygosity about half that found in ten other populations (Schwaegerle and Schaal, 1979). Although pollen dispersal around individual plants is limited, the total pollen output from populations of wind-pollinated trees is tremendous and high pollen counts may be found miles from the source (Silen, 1962; Wang et al., 1960; Lanner, 1966). Extensive pollen movement (Koski, 1970), and the small number of generations (perhaps less than 100) since expansion from glacial refugia 10,000 years ago, may explain the lack of enzyme differentiation usually found among populations of temperate zone trees.

Recessive deleterious genes

Many other lines of evidence suggest high genetic variation within species of outcrossing trees. Studies of quantitative traits such as growth rate, wood properties, and form usually show moderately high levels of additive genetic variance (see references in Zobel and Talbert, 1984). More pertinent for an understanding of the relationship between heterozygosity and fitness is the class of variants known as recessive deleterious genes. If the frequency of recessive deleterious genes was correlated with other measures of genetic variation, it would suggest that both were maintained by the same mechanisms or resulted from the same cause—namely, mutation.

Mutants that are obviously deleterious, recessive lethals such as pigment deficiencies, occur in high frequencies in conifers; 133 of 712 slash pine *(Pinus elliottii)* were carriers (Snyder et al., 1966). In loblolly pine, 22 different morphological mutants were found in the progeny of 30 of 119 selfed trees (Franklin, 1969). Conifers, the tree taxa for which the most data are available, have a high genetic load (Table 2), measured as embryonic lethal equivalents. Embryonic lethal equivalents are defined as true recessive lethal alleles or recessive deleterious genes of such number that if dispersed in different embryos would cause, on average, one selective death (Morton et al., 1956). They can be calculated from the reduction in seed yield following self-pollination. Douglas firs *(Pseudotsuga menziesii),* which have one of the highest average heterozygosities reported in conifers (Conkle, 1981), have about 10 embryonic lethal equivalents per tree up to a maximum of nearly 28 in a sample of 35 trees (Sorensen, 1969). These estimates are conservative because they fail to take into account polyembryony (Bramlett and Popham, 1971). Each ovule produces several egg cells in conifers, and if pollination is adequate, all will be fertilized. Therefore, not every embryonic death results in loss of reproductive capacity.

The true genetic loads for conifers are, in general, much higher than those reported for annual plants or animals, in agreement with the high level of allozyme heterozygosity in tree species. Embryonic lethal equivalents in conifers range from 0.3 to 10.8 (generally 5 to 9) per zygote (Table 2). In one review, Levin (1984) reported a range in lethal equivalents from 1.2 to 5.2 in agronomic plants and a similar spread, from 1 to 4, in humans, fruit flies, and flour beetles. In populations of phlox *(Phlox drummondii),* a normally self-incompatible annual, mean embryonic lethal equivalents ranged from 0.08 to 2.26, averaging 0.79 per zygote (Levin, 1984). In ferns, genetic load varies among species, increasing from zero in pioneering species to higher levels in species of more mature communities (Lloyd, 1974). Presumably, pioneer species are in low density during the colonization phase and are purged of deleterious genes by inbreeding.

The total genetic load is much higher than is indicated by embryonic lethal equivalents because additional mortality occurs after germination (Franklin, 1970). The difference in size between inbreds and outbreds increases with age (Cram, 1984; Libby et al., 1981), which should lead to further mortality of inbreds through competition. Following the initial purge of deleterious alleles during the embryonic stage, inbreeding depression in seedlings from selfing of mountain ash *(Eucalyptus regnans)* was nearly nil, but increased steadily with age (Eldridge and Griffin, 1983). By 12.5 years, inbreds had 57 percent poorer survival and the survivors were smaller than the outcrosses.

Obviously, the high levels of allozyme variation in forest tree species is paralleled by high values for the true genetic load.

Genic diversity and ecological success

Is genetic diversity related to success in any way? A broad geographic range might be considered a measure of success. On that basis, the endemic Torrey pine and the widespread Douglas fir represent opposite extremes of genic diversity; average individual heterozygosity is 0 in the two known populations of Torrey pine and 0.33 for Douglas fir, which ranges through Canada, the United States, and Mexico and from the Pacific Coast to the Rocky Mountains (Conkle, 1981; Ledig and Conkle, 1983). On the other hand, Monterey cypress *(Cupressus macrocarpa)*, which, like Torrey pine, is restricted to only two small populations, has an average heterozygosity of 0.16 (M.T. Conkle, personal communication) and red pine *(Pinus resinosa)*, with a latitudinal range of about 12 degrees and a longitudinal range of 42 degrees, has almost no variability (Fowler and Morris, 1977). Apparently, low heterozygosity does not preclude a wide range, although on average endemics do tend to have low variability compared to more widespread species (Hamrick et al., 1979).

The level of genic diversity is more likely a reflection of recent historical events, life history parameters, and the mating system than an adaptive mechanism. For example, it is likely that homozygosity in Torrey pine is the result of severe bottlenecks in the very recent past. Historical records from this century suggest that Torrey pine is in a recovery phase: at San Diego 200 trees were counted in 1916, 507 in 1930, and estimates in 1973 placed the population at 3401 (Fleming, 1916; H. Nicol, personal communication; California Department of Parks and Recreation, 1975). By comparison, fossil evidence suggests that the bristlecone pine *(Pinus longaeva)* was widely distributed in large continuous populations during the last full glacial, and is now at its nadir. Bristlecone pine is highly heterozygous (average heterozygosity 0.33) with little differentiation among isolated populations (Hiebert and Hamrick, 1983).

EXPLANATIONS FOR HIGH GENIC DIVERSITY IN TREE SPECIES

Reproductive system

Tree species have unusually high levels of genetic variation, as reviewed here for allozymes and recessive deleterious genes. Several factors may explain their high variation, including the breeding sys-

TABLE 2. Embryonic genetic load in conifers.

Species	Number of trees	Percentage self-fertility[a]		Embryonic lethal equivalents[b]		Reference
		Mean	Range	Mean	Range	
PINES (*Pinus*)						
P. banksiana	6	42	19–93	4.1[c]	0.3–6.6	Fowler, 1965b
P. contorta	27	—	—	8.3[c]	—	F. C. Sorensen, pers. commun.
P. monticola	14	61	9–96	2.5[c]	0.2–9.8	Bingham & Squillace, 1955
"	169	60	—	2.0[c,d]	—	Bingham & Rehfeldt, 1970
P. peuce	7	60	—	3.1	2–9	Koski, 1973
P. ponderosa	19	37	4–76	4.9[c]	1.1–13.1	Sorensen, 1970
P. resinosa		102	82–137[e]	0.3[c]	0–0.8[e]	Fowler, 1965b
P. sylvestris	80	—	—	9.4	2–20	Koski, 1971
"	100[f]	—	—	8.9	1–21[g]	Koski, 1973
P. taeda	116	—	—	8.5[c]	0–26[g]	Franklin, 1972

SPRUCES (Picea)

						Reference
P. abies	4	—	—	5.0	1–8	Koski, 1971
"	87	—	—	9.6	1–21[g]	Koski, 1973
P. glauca	18	16	3–49	8.7[c]	2.9–14.8	Coles & Fowler, 1976
"	20	13	2–45	9.8[c]	3.1–16.6	Fowler & Park, 1983
P. mariana	16	47	3–89	4.7[c]	0.5–14.1	Park & Fowler, 1984
P. omorica	12	—	—	4.7	2–12	Koski, 1973
P. pungens	72	29	—	5.0[c]	—	Cram, 1984

OTHER CONIFERS

Abies procera	10	69	29–118	1.8[c]	0–5.0	Sorensen et al., 1976
Larix laricina	20	7	0–47	10.8[c]	3.0–19.3	Park & Fowler, 1982
Pseudotsuga menziesii	35	11	0–46	11.2[c]	3.1–27.6	Sorensen, 1969

[a]Self-fertility percentage =

$$\left[\frac{\text{(sound seed} - \text{total seed) from selfing}}{\text{(sound seed} - \text{total seed) from outcrossing}} \right] \times 100$$

[b]Embryonic lethal equivalents = $-4 \log_e$(self-fertility). (After Morton et al., 1956.)

[c]Conservative estimate, uncorrected for polyembryony.
[d]Calculated from means.
[e]Range for population means.
[f]Trees were a selected sample with better than average growth and form.
[g]Estimated from histograms and frequency distributions.

tem, migration, spatial diversity of the habitat, temporal heterogeneity in conjunction with a long life cycle, and, perhaps most importantly, mutation rate.

Outcrossing is the rule in tree species, with a trend toward greater selfing in the progression trees → shrubs and herbaceous perennials → annuals. In a tropical flora, 76 percent of the tree species were obligate outbreeders by virtue of dioecy (male and female flowers on separate trees) or self-incompatibility systems (Bawa, 1974). Another 10 percent were monoecious (separate male and female flowers on the same trees). Monoecy is a mechanism to avoid self-pollination, and is widespread in temperate tree floras. Most wind-pollinated conifers are monoecious. Not only are male and female structures separate in monoecious conifers, but they are borne in different parts of the tree crown. Although there is overlap, female cones generally are produced near the top of the tree, and even when males and females occur on the same branch, females are distal. The location of female cones above males reduces self-pollination because pollen tends to fall rapidly in the absence of air turbulence (Colwell, 1951; Wright, 1952). In addition to physical separation, there is some degree of temporal separation, or dichogamy; the pollen sheds either before or after maximum female receptivity, depending upon the species and individual. Spatial and temporal separation and their role in promoting outcrossing in conifers were reviewed by Fowler (1965c).

Gametic competition and polyembryony are even more effective than spatial or temporal separation in promoting outcrossing. The period of pollen tube growth between pollination and fertilization is long in conifers (about one year in pines), and this provides ample opportunity for gametic competition. Within the ovule are several egg cells, up to six in pines and even more in other conifers (Lill, 1976; Willson and Burley, 1983). If pollen is not limited, several eggs will be fertilized. Some embryos abort, others seem to be outgrown in competition by more vigorous siblings. Very rarely will more than one embryo germinate. The process suggests ample opportunity for selective elimination of weak embryos—by implication, those homozygous for deleterious alleles (Sorensen, 1982).

Selective fertilization or pregermination selection are mechanisms that reduce the impact of self-pollination by the elimination of homozygotes without the loss of reproductive capacity (Fowler, 1964). Polyembryony, in particular, provides a way to reduce the cost of the genetic load by bearing it at the earliest possible stage (Haldane, 1957). Analogous phenomena are widespread in plants; for example, among the oaks (*Quercus* spp.) there are six ovules per ovary but only one matures an embryo (Mogensen, 1975). Pregermination selection is as effective in eliminating incompatible heterozygotic combinations as it is in eliminating homozygotes. For example, hybrids are produced with high efficiency when black spruce is control-pollinated with red spruce pollen or vice versa, but when a 50:50 pollen mixture is used, the proportion of hybrid seed set is 5 percent or less (Manley, 1975).

Even without competition, embryo mortality and reduced seed yields are common following self-pollination, as in wind-pollinated conifers (Franklin, 1970) and insect-pollinated eucalypts (Eldridge, 1970). In a survey of a Great Basin flora, Wiens (1984) found that the rate of ovule abortion was higher in woody perennials (67.3 percent) than in herbaceous perennials (42.8 percent) and was least in annuals (15 percent)—an important observation. Even more significant, the rate of abortion was not a function of the breeding system; it was as high in selfing as in outcrossing plants, suggesting that mutation rates may be critical, as discussed later.

Dispersal

Gene flow through seed and pollen dispersal is another factor that could contribute to the high genic diversity found in tree species. The ranges of most herbaceous plants are characterized by large gaps relative to their dispersal ability. By contrast, many tree species are relatively continuous in distribution over broad ranges. Furthermore, though most pollen falls near the parent, some can be carried long distances, and this facilitates gene flow—pollen has been known to be deposited in large quantities scores of miles from the source (reviewed by Lanner, 1966). Only a small amount of gene exchange is necessary to prevent drift and the decay of variation within populations. Long-distance pollen movement may explain the relative lack of differentiation for enzyme polymorphisms in tree species; usually over 90 percent of the total variation is found within local populations in temperate conifers (Guries and Ledig, 1982) or in tropical angiosperms (Loveless and Hamrick, 1985). Koski (1970) concluded that "the existence in the same subpopulation of neighborhoods that differ from one another cannot be regarded as possible" based on studies of pollen dispersal in conifers and birches (*Betula* spp.). Although an overstatement, Koski's results emphasized the vast potential of pollen to promote genetic coherence by gene exchange and to maintain diversity within populations (Soulé, 1973).

Spatial heterogeneity

It might be anticipated that plants are closely adapted to their habitat. Clones of clover *(Trifolium repens)* were restricted to narrow bands defined by elevational differences of less than 0.3 m, in this case marking a moisture gradient (Harberd, 1963). Patterns of variation in Douglas fir closely paralleled physiography within a single watershed (Campbell, 1979). And in balsam fir *(Abies balsamea)* the photosynthetic temperature optimum changed along an elevational gradient over distances of only a few hundred meters in a way that suggested close adaptation to local conditions (Fryer and Ledig, 1972). Edaphic differentiation can occur over even shorter distances; for example, heavy metal-tolerant and nontolerant races of bentgrass *(Agrostis tenuis;* Bradshaw, 1971) and the edaphic races on serpentine and nonserpentine soils in several annual plant species (Kruckeberg, 1951) and even in ponderosa pine (J. L. Jenkinson, personal communication). Such patterns must be maintained in spite of migration. Obviously, gene flow among different habitat types may help to maintain genic diversity, particularly in wind-pollinated plant species.

Temporal heterogeneity

Temporal variation has been suggested as another mechanism that maintains genetic diversity, but Hedrick et al. (1976) concluded that the conditions under which temporal heterogeneity would maintain genetic polymorphisms are restrictive. Temporal heterogeneity would more likely erode genetic variability than preserve it in annual plants; however, in long-lived iteroparous species with overlapping generations, temporal heterogeneity might promote some forms of polymorphism. Cohorts of sugar maple *(Acer saccharum)* on the same site were distinguishable on the basis of, presumably, heritable leaf proteins (Mulcahy, 1975). Apparently, conditions in the year of establishment selected for entirely different characteristics, depending on whether it was a wet or dry year, cold or warm. While the variability in each cohort might be eroded, different cohorts are similar enough in age to permit interbreeding over many decades or centuries, regenerating variability in the offspring. Even long-term climatic cycles of 50 years duration are not so long that they could completely eliminate variants in long-lived forest trees.

Mutation

The effect of life history characteristics on mutation rate, and the role of mutation in maintaining genetic variation, has been given little recognition. In animals, the germ line is cut off at an early stage of embryogeny. The germ line undergoes few divisions before the sexual division and production of the gametes, eggs or sperms (perhaps 50 cell generations between zygote and gamete in humans; Brewbaker, 1964). Plants have no germ line; spore mother cells derive from vegetative cell lines. Therefore, many more cell divisions and many more opportunities for DNA copying error intervene in plants than in animals.

Unless there are offsetting differences in the rate of mutation per cell division, more mutations would accumulate per generation in trees than in herbaceous perennials, and more in herbaceous perennials than in annuals. Mutation rate per year may be the same in annuals and trees, but the impact on individual fitness or reproductive capacity will be much greater in trees because each tree will have accumulated more mutations than an annual. The number of divisions between zygote and gamete can be estimated in conifers by assuming that four cells result from two divisions of each cell cut off by the apical meristem, and that the length of the wood cells, the primary tracheids, is 2 mm. Thus, a gamete produced in a cone at the top of a 12 m pine, represents 1500 divisions. Assuming a mutation rate of 10^{-8} per cell division, 1.5 new mutants would be expected in every 10^5 gene loci. For mammals, the same rate per cell division would result in only 5 mutations in 10^7 gene loci, or 30 times more in pine. For a 24 m pine, the rate should be double that in a 12 m tree. A sizeable proportion of new mutations will be deleterious, so long-lived perennial plants will have a tremendous genetic load compared to animals.

The hypothesis is testable by sampling gametes from divergent branches in large trees; genetic divergence should be proportional to the distance between the reproductive structures, measured along the branch and stem axes. A similar technique detected mutational divergence among dichotomously branched rhizomes in the royal fern (Osmunda regalis; Klekowski, 1976). Notably, genetic load tended to increase with clone diameter in sensitive fern (Onoclea sensibilis), although sample sizes were too small to demonstrate statistical significance (Saus and Lloyd, 1976).

Mutations may accumulate with age even in the absence of cell division; the viability load transmitted to progeny was greater in old fruit flies than in young ones (Andjelkovic et al., 1979). Mutations accumulate in stored seed even in the absence of division and at low rates of metabolic activity (Harrington, 1970). The great longevity of trees (centuries to millennia) provides the opportunity for the accumulation of many new mutations during their lifetime.

Several sets of circumstantial evidence suggest that mutational load may explain the high genic diversity in forest trees and may be an important force in shaping the breeding system. One line of evidence is a relationship recognized by M.T. Conkle (personal communication) between genic diversity and self-fertility. Conifers with the highest heterozygosity, such as Douglas fir, suffer the most depression in seed yield due to embryo abortion when selfed. At the opposite extreme is red pine, which has almost no detectable variation (Fowler and Morris, 1977), and can be selfed with no reduction in seed set (Fowler, 1965a). A second line of evidence, as cited above, is Wiens' (1984) observation that ovule abortion was highest in woody perennials, somewhat lower in herbaceous perennials, and lowest in annuals, which suggests that the genetic load increases with longevity or size.

Under extreme circumstances even conifers, such as Torrey pine and red pine, can be purged of variation; this facilitates inbreeding without the tremendous reduction in reproductive capacity that is normally attendant (Fowler, 1965d). Bannister (1965) noted that the breeding system of Monterey pine was flexible: because of the high reproductive output, inbreeding was possible in colonizing situations when competition was weak. Later, when normal population levels were reestablished, selection pressure brought a reversion to outbreeding.

No species of selfing trees were listed in a recent review on the evolution of self-fertilization (Schemske and Lande, 1985). Selection for outbreeding should not be as intense in short-lived plants, which have fewer divisions between zygote and gamete than tree species and should not accumulate as many mutations during their life cycle. In fact, annuals—monocarps that have only one opportunity to reproduce—are under pressure for facultative selfing to ensure the possiblity of leaving some offspring. Nevertheless, even annuals suffer from inbreeding depression (Waller, 1984).

Genic diversity: Summary

Genic diversity in tree species is generally high, as revealed by allozymes and other characteristics. However, the existence of genetically depauperate species suggests that genic diversity is not necessary for survival. Significantly, the true genetic load is high in conifers, in concert with other measures of diversity. The several mechanisms that promote outbreeding and favor gene flow in tree species might explain the maintenance of genic diversity within populations, but they are not the "cause" of the diversity. Selection in spatially and temporally heterogeneous habitats, followed by wide dispersal, may be an important contributing factor, particularly in long-lived perennials. However, for many loci the explanation may not be selection-dependent; due to the nature of plant growth, germ cells descend from vegetative cell lines, providing many more opportunities for accumulation of mutations in plants than in animals. Opportunities for the accumulation of mutations are greatest in large, long-lived tree species, and mutational load will drive the breeding system toward outcrossing, which will protect all forms of genetic variation, which in turn magnifies the effect of inbreeding—a truly vicious circle.

CONSERVATION OF OUTBREEDING PLANTS

The problem

The geographic distributions of most plants are being increasingly fragmented, except for weedy species that find their habitat along roads and railroads and in the wake of construction and tillage. In large part, the weeds include inbreeders or apomicts, often with intercontinental distributions, and they are in no need of conservation. Among outbreeders, we again take forest species as examples.

Fragmentation of the forest is not entirely new; in temperate climates and in the tropics, forests have been reduced to refugia many times (Critchfield, 1984; Prance, 1978). However, the impact of humans is new. Even where land use will allow the preservation of forest, the forest will be composed increasingly of planted trees of nonlocal origin, more or less modified by selection and breeding. Natural forest will exist as patches or islands in a matrix of production forests that conform to the agronomic model.

The consequences of forest domestication will be to decrease population sizes of native forest species and to reduce migration among patches. Based on our present understanding of genic diversity and its role, the ultimate result will be an increased expression of the genetic load as recessive deleterious alleles are exposed, reducing vigor and reproductive output. Populations will be harder to maintain and more susceptible to agents of stress, parasitic or climatic. Genetic contamination will be a problem because most gene flow will be from surrounding commercial forest, which may be of high vigor under domesticated conditions, but not necessarily adapted to natural conditions. If inbreeding within fragments of the native forest reduces reproductive output, the importance of gene flow in the form of pollen or seed from commercial forest will be accentuated. Even if natural populations can survive the short term while purging their genetic load, they could become cul-de-sacs with respect to long-term survival and evolution.

Management for conservation

Proposals for genetic conservation and the maintenance of viable communities have ranged from preservation of vast blocks of native vegetation (Yeatman, 1973) to direct intervention in the breeding system (Hagman, 1973). The protection of vast areas of commercially productive forest is probably not a politically practical solution. Where species are few, such as the extreme northern temperate forest, controlled pollination to ensure outcrossing, followed by direct seeding to allow natural selection its fullest scope, seems marginally feasible. Controlled outcrossing of plants in small, isolated populations is analogous to captive breeding in zoos, but permits some degree of evolution in natural habitats. For most species, such intensive management will be rejected because of the economic cost.

The most feasible course lies somewhere between "lock-up" of our remaining forestlands and intensive management of "captive" populations. As always, the question reduces to "how many individuals and how distributed?" The effect of a reduction in population size to below a certain level is an eventual increase in homozygosity, so it is important to return to the question of whether heterozygosity has fitness value.

The evidence suggests that heterosis results mostly but not exclusively from the dominance of normal alleles over recessive deleterious alleles—the genetic load of mutation. If so, then it is possible, in theory, to purge a population or species and result in a homozygous line that is as reproductively fit as any heterozygote. In practice, it is extremely difficult to obtain a vigorous homozygous line in outcrossing plants (for example, see page 215 in Allard, 1960; and Orr-Ewing, 1976). The optimum homozygote would virtually never occur in finite populations.

As an illustration, assume two genes affecting fitness in each of the 12 chromosomes of the haploid set in pines and 10 percent crossover between them. Given the number of embryonic lethal equivalents in forest trees, a load of 24 deleterious alleles is probably conservative. If the linkage is in repulsion phase (that is, a deleterious recessive linked with a dominant wild-type on one chromosome of a pair and the opposite combination on the other; Figure 1), to get a recombinant chromosome with both wild-type alleles is one-half the product of the crossover fraction. The probability of getting a gamete that carried only the dominant wild-type alleles on each of its 12 chromosomes is $[(\frac{1}{2})(0.1)]^{12} = 20^{-12}$. The probability that two such gametes, a sperm and an egg, will unite is $[20^{-12}]^2$, or 20^{-24}. In other words, for an even chance to get one plant of the desired genotype, 20^{24} progeny would have to be screened. For forest trees such as Douglas fir, this would require an area roughly 4.2×10^{15} times the size of the United States.

Although the best genotype can virtually never be found, selfing as a mating system has evolved repeatedly. Evolutionary events have resulted in purges of at least the most seriously deleterious alleles in many species. But even selfers show heterosis when outcrossed, suggesting that at some loci, less than the most-fit alleles were fixed

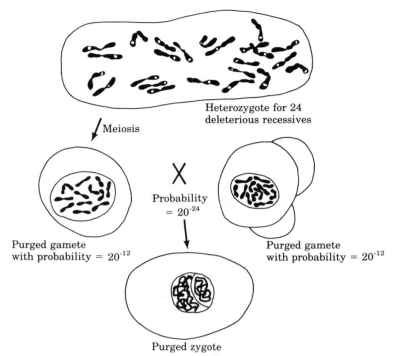

FIGURE 1. As an illustration, assume two genes affecting fitness in each of the 12 chromosomes of the haploid set in pines and 10 percent crossover between them. Given the number of embryonic lethal equivalents in forest trees, a load of 24 deleterious alleles is probably conservative. If the linkage is in repulsion phase (that is, a deleterious recessive linked with a dominant wild-type on one chromosome of a pair and the opposite combination on the other), to get a recombinant chromosome with both wild-type alleles is one-half the product of the crossover fraction.

(Lande and Schemske, 1985). Organisms that remain predominantly outcrossers may have also been purged; notable examples are red pine and Torrey pine among plant species, and the elephant seal (Bonnell and Selander, 1974) and cheetah (O'Brien et al., 1983, 1985) among animals. Despite the purges that must have occurred, it is likely that some deleterious alleles remain—certainly they do in the cheetah and the Torrey pine. The cheetah has a high incidence of sperm aberrations and, possibly, high infant mortality (O'Brien et al., 1985). In Torrey pine several characteristics seem maladaptive: for example, seed are retained in open cones by super-reflexive scales, seeds germinate in the cone where there is no chance for survival, and seed maturity apparently takes at least six months longer than is usual for pines.

Both the cheetah and Torrey pine have achieved near homozygosity, but at a loss of reproductive capacity.

In addition to the obvious role of dominance in the maintenance of vigor and fecundity in outbreeding plants, overdominance may operate at key loci. Homozygosity at these loci will further reduce reproductive success. Regardless of the relative importance of dominance and over-dominance to fitness, an increase in homozygosity will create a short-term crisis which may reduce reproductive output to such low levels that local extinction is a possibility. If the population survives such a purge, it will be at the expense of genetic diversity. The consequences of uniformity are not certain, but certainly the opportunities are greatest for chance extinction of genetically depauperate populations because all individuals are uniformly susceptible to stressful climatic events, new pests, or competitors (National Academy of Sciences, 1972).

The obvious management prescription is to reduce inbreeding *and* to promote heterozygosity by maintaining population numbers, encouraging outcrossing, and enabling migration among populations. Harris (1984) proposed a scenario to accomplish this in the Douglas fir forest of the Pacific Northwest. Fortunately, a system that provides for tree species would simultaneously promote the conservation of other plants and some animals as well. However, Harris' scenario assumed nuclei of old-growth forest buffered by forest plantations in various stages of development. The surrounding plantations of "domesticated" trees could result in pollen contamination of the central preserves, as shown for seed orchards (Friedman and Adams, 1985). Whether the effect of pollen contamination is positive, negative, or neutral is a matter of debate (Nienstaedt, 1980). The influx of non-native genes will reduce fitness according to one viewpoint, or promote evolution according to another. In the face of uncertainty, prudent planning dictates the avoidance of contamination.

Two ways exist to reduce contamination. One is to surround the nuclei of natural vegetation with buffers of forest managed for natural regeneration—forest in which harvesting can take place but be timed to coincide with abundant natural seeding, thus eliminating the need for reforestation with planted stock of off-site origin. Alternatively, plantation silviculture is acceptable if the surrounds are composed of different species than those preserved within the nuclei. Even better, the nuclei themselves should be interconnected to form the matrix and the commercially managed forest arranged as cells within the matrix.

The present system of natural areas, wilderness areas, and parks provide the nuclei for a conservation effort. However, in many areas—such as the southern United States—such nuclei are scarce, and the natural vegetation has been so greatly altered that it will be difficult to acquire sufficient examples. In the western United States, protected forest remnants are extensive, but do not constitute a random sample of habitat: low elevation forests (the most diverse for plant and animal species) have been cut over or subverted to more domestic uses (Harris, 1984). Noncommercial, high elevation forests are vastly overrepresented in the national system of reserves and parks (for documentation with respect to Douglas fir, see California Gene Resources Program, 1982). It is not possible to bring back what is lost, but we can work for a system of land use that will provide the best opportunity for the simultaneous preservation of genic diversity and adaptiveness.

SUMMARY

For at least 70 years, geneticists have debated whether heteroyzgotes are more fit than homozygotes. According to the overdominance hypothesis, heteroyzgosity confers stability in a broad range of environments and endows heterozygotes with properties not shared by homozygotes. But the correlation between genic diversity and ecological amplitude, at least as measured by geographic range, is not simple and direct. According to the inbreeding or dominance hypothesis, populations carry a genetic load of deleterious mutations which is expressed in homozygotes, but masked by the more frequent, dominant alleles in heterozygous combination. For several reasons, tree species are good material for investigating the relative importance of overdominance and inbreeding depression in the expression of heterosis. One reason is that genic diversity is much higher in tree species than in other plants.

Heterosis is not a general phenomenon in interspecific hybrids, nor is it common in wide interpopulation crosses. Complex physiological processes, conditioned by divergent, coadapted suites of genes, break down in hybrids. However, *within* natural populations of tree species and many other organisms, survival and growth—fitness surrogates— are related to heterozygosity. And in plants, reproductive output is correlated with size and vigor. The bulk of the evidence suggests that the relation between fitness, or its surrogates, and heterozygosity is the result of a correlation between homozygosity and inbreeding depression. Overdominance probably occurs at some loci, but is relatively less important.

If overdominance does not account for high levels of genic diversity, extensive pollen movement and a spatially and temporally heterogeneous habitat might provide an explanation. But mutation provides a better explanation than either selection or migration. Significantly, tree species have a higher genetic load than short-lived plants and animals. This load may result from the absence of a germ line in plants; reproductive structures originate from vegetative cell lines, providing many more cell divisions in the path from zygote to egg or sperm in long-lived plants than in animals. Each division provides opportunities for mutation, and the larger the plant, the greater the accumulated load. The load can be tolerated because of polyembryony, which pushes the cost of selection back into the earliest zygotic and embryonic stages without the wastage of seed. The genetic load is the force that drives the breeding system toward outcrossing, and the mechanisms that minimize inbreeding permit the accumulation of additional load. Heterozygosity upon outcrossing masks the load by the dominance of normal alleles over deleterious alleles.

The lesson for conservation biology is to maintain genic diversity. Because it would be virtually impossible to purge all recessive deleterious alleles, large populations are necessary to keep the level of inbreeding low and maintain high levels of heterozygosity for deleterious alleles. Many populations maintained as a few individuals over several generations would collapse, and most of the others would be fixed for mildly deleterious genes that would impair their reproductive capacity. Furthermore, populations with low diversity are vulnerable to new stresses such as pathogens and climatic change. And, finally, without variability, evolution is impossible.

SUGGESTED READINGS

Gowen, J.W. (ed.). 1952. *Heterosis.* Iowa State College Press, Ames. Still the best single source on heterosis.

Harris, L.D. 1984. *The fragmented forest: Island biogeography theory and the preservation of biotic diversity.* University of Chicago Press, Chicago. See particularly pages 94–104 on the importance of within-species diversity.

Mitton, J.B. and M.C. Grant. 1984. Associations among protein heterozygosity, growth rate, and developmental homeostasis. *Annu. Rev. Ecol. Syst. 15:* 479–499. Provides a contrasting opinion on the importance of overdominance and inbreeding to heterosis.

McDonald, J.F. 1983. The molecular basis of adaptation: A critical review of relevant ideas and observations. *Annu. Rev. Ecol. Syst. 14:* 77–102. A discussion of neutralist and neoselectionist arguments regarding molecular evolution.

Schonewald-Cox, C.M., S.M. Chambers, B. MacBryde and L. Thomas (eds.). 1983. *Genetics and conservation: A reference for managing wild animal and plant populations.* Benjamin-Cummings, London.

CHAPTER 6

COADAPTATION AND
OUTBREEDING DEPRESSION

Alan R. Templeton

Dobzhansky (1948) first used the word "coadaptation" to describe the phenomenon that sets of genes derived from different geographical populations of the fruit fly *Drosophila pseudoobscura* resulted in reduced fitness when brought together by hybridization. Wallace (1968) provided a more general definition: "Genes are said to be coadapted if high fitness depends upon specific interactions between them." Mayr (1963) further extended the concept, arguing that most of the genes in a species are coadapted because the integrated functioning of an individual, rather than the separate genes, is the "target of selection."

Mayr's concept of nearly all the genes in a species being coadapted is controversial. However, there is little or no controversy over the phenomenon of a fitness decline following hybridization; such declines can be and have been empirically demonstrated. In this regard, the word "outbreeding depression" is sometimes used to avoid the controversy associated with the word "coadaptation." An outbreeding depression simply refers to the phenomenon of a fitness reduction (usually in either fertility or viability) following hybridization (either in the immediate hybrids, or perhaps delayed until the backcross or later generations). The phrase "outbreeding depression" makes no judgment as to why hybridization leads to a fitness reduction; it merely refers to the fact that it occurs.

It is tempting to avoid the controversies over the genetic mechanisms of coadaptation by using the term outbreeding depression. However, ignoring the genetic mechanisms responsible for hybrid depression is not a good strategy, even for the narrow interests of applied conservation biology. The reason is simple: in implementing breeding

105

or release programs with endangered species, the applied biologist often has to solve the problems associated with outbreeding depression rather than merely to document its presence or absence. The effectiveness of a plan that attempts to eliminate or circumvent an outbreeding depression will in many cases depend critically upon the genetic details of the mechanisms responsible for that outbreeding depression.

This chapter addresses the issue of intraspecific outbreeding depression and its genetic basis, the implications of the genetic basis for solutions to the problem of outbreeding depression, and how this solution relates to the more general issue of just what we should be trying to save in endangered species programs. Many of the more general questions concerning outbreeding depression, including the very practical issues of detection and distinguishing an outbreeding depression from an inbreeding depression, will not be raised here, as they are discussed elsewhere (Templeton et al., 1986.)

OUTBREEDING DEPRESSION AND LOCAL ADAPTATION

By far the least controversial cause (in an ultimate, evolutionary sense) of outbreeding depression is local adaptation. Many species are distributed over a geographical range that encompasses a diversity of environmental conditions that influence the organisms' ability to survive or reproduce. Under these conditions, local populations of the species will often adapt to the local enviroment, particularly if dispersal is limited. Hybridization between different local populations can sometimes destroy the locally adapted gene complex. For example, when the Tatra Mountain ibex *(Capra ibex ibex)* in Czechoslovakia became extinct through overhunting, ibex were successfully transplanted from nearby Austria (Greig, 1979). However, some years later, bezoars *(C. ibex aegagrus)* from Turkey and the Nubian ibex *(C. ibex nubiana)* from Sinai were added to the Tatra herd. The resulting fertile hybrids rutted in early fall instead of the winter (as the native ibex did), and the kids of the hybrids were born in February—the coldest month of the year. As a consequence, the entire population went extinct (Grieg, 1979).

The phenomenon of local adaptation is well known and documented in a variety of plant and animal species and, as the ibex example illustrates, the presence of this phenomenon can play a critical role in determining the success or failure of an endangered species program. However, it is also well recognized today that local adaptation can occur without coadaptation. For example, one of the classic examples of local adaptation is industrial melanism in several different species of British moths. As the industrial revolution proceeded in

England, air pollution killed lichens on tree bark, resulting in a pronounced darkening of the general background color. Several species of moths normally rest on these trees and are subject to visual predation by birds. In unpolluted areas, these moths have light colored wings that are cryptic on the lichen-covered trees, but near industrial areas they are dark colored and match the color of the trees in those areas. This is an example of local adaptation, but in many cases where the genetic basis of this adaptation has been studied, it turns out to depend upon only one gene locus (Bishop and Cook, 1980). Obviously, a local adaptation accomplished through a single major gene does not involve fitness interactions between genes, and hence does not satisfy the definition of coadaptation. However, regardless of genetic mechanism, information about local adaptation should be obtained through studies on natural populations, and the information should be incorporated into endangered species programs whenever possible.

INTRINSIC COADAPTATION

Where genetic mechanisms do matter for the applied conservation biologist is in the case of intrinsic coadaptation (Templeton et al., 1986). Intrinsic coadaptation occurs when a genetic or karyotypic complex evolves in response to the state of other genes or chromosomes. For example, the process of gamete formation normally requires a matched set of chromosomes (with the exception of the sex chromosomes). As long as the set matches, meiosis is normal, but if the two sets of chromosomes differ in number or structure, abnormal meiosis and attendant fertility problems are possible. An example of this type of intrinsic karyotypic coadaptation is illustrated by fertility difficulties discovered in captive populations of the owl monkey (*Aotus trivirgatus;* de Boer, 1982). Several chromosomal races are known, and there is evidence that successful reproduction is enhanced when chromosomally similar forms are paired (Cicmanec and Campbell, 1977).

A more subtle type of intrinsic coadaptation depends upon interactions between specific gene loci rather than chromosome number or structure. This type of coadaptation will include those cases of local adaptation in which the genetic basis of the local adaptation depends upon interactions of two or more genes. However, in other cases, the intrinsic coadaptation can occur with no obvious local adaptation at all; instead, the genes seem more selected for the internal "genetic environment" defined by the other interacting genes than for an external environmental variable.

A good experimental example of this is provided by studies on the fruit fly *Drosophila mercatorum*. This species of fly is normally sexually reproducing, but if virgin females are isolated, they can reproduce parthenogenetically. Parthenogenesis is accomplished through a mechanism known as gamete duplication in which meiosis proceeds normally to produce a haploid egg nucleus. In the absence of fertilization, this haploid egg nucleus divides mitotically, and the mitotic cleavage nuclei fuse to form a diploid nucleus, which then initiates parthenogenetic development. Because diploidy is restored by the fusion of two genetically identical haploid nuclei, the resulting parthenogenetic fly is totally homozygous. Because the parthenogenetic flies are diploid and retain normal meiosis, they can be crossed to males and reproduce sexually (Templeton, 1983). In addition, because there is no crossing over in males of this species and visible genetic markers exist for all the major autosomes, it is possible to breed a sexually reproducing strain that is equivalent to a parthenogenetic strain (Templeton, 1983). In these sexual strains, the females are genetically identical to the parthenogenetic females, and the males have the same genotype as the parthenogenetic females except that a Y chromosome has been substituted for one of the duplicated X chromosomes. Using these sexual analogues, it is possible to hybridize two different parthenogenetic strains.

Templeton et al. (1976) hybridized two strains that had been reproducing parthenogenetically for several years prior to the hybridization. The hybrid females were then allowed to reproduce parthenogenetically, and a fitness decline was observed in the parthenogenetic progeny. This fitness decline could not be attributed to an inbreeding depression. All parental genes had been selected for their effects in a totally homozygous genetic background. Nevertheless, when these same genes were put into a recombinant homozygous genetic background, there was a substantial decline in fitness. Moreover, Templeton (1981) showed that different and incompatible coadapted complexes can arise between parthenogenetic strains isolated in the same laboratory environment from the same ancestral sexual population at the same time. Consequently, this is an intrinsic coadaptation that is not associated with local adaptation.

To test the hypothesis that this fitness decline was indeed an intrinsic coadaptation, the degree of hybridity was controlled by breeding females that were 100 percent, 60 percent, and 40 percent hybrid between the two parthenogenetic stocks (Templeton et al., 1976). As expected for an outbreeding depression, the absolute viability of the parthenogenetic progeny of the hybrid females declined with increasing levels of hybridity (Table 1). Moreover, it was possible with this design to actually estimate the fitness interactions between chromosome segments linked to the marker loci. A summary of these data for two locus systems is presented in Table 2.

TABLE 1. Number of parthenogenetic offspring per female *Drosophila mercatorum* surviving to adulthood as a function of the degree of hybridity of the female parent.

Percentage of parental genotype that is hybrid	Number of parthenogenetic offspring per female
0	14.54
40	10.25
60	5.36
100	1.63

Data from Templeton et al., 1976

Two important features are apparent. First, the fitness pattern is exactly what is expected under the coadaptation hypothesis. The positive fitness interaction terms mean that the parental genotypes had much higher fitness than the recombinant genotypes. However, these experiments also show that the whole genome of the organism is not necessarily welded into a single coadapted unit. At the highest level of hybridity (and hence the strongest level of selection induced by the outbreeding depression), significant fitness interactions occur between genes on the same chromosome arm, on the same chromosome but different arms, and on different chromosomes, as illustrated by the average fitness patterns given in Table 2 (see Templeton et al., 1976, for a detailed statistical analysis). However, at the lower levels of hybridity, significant fitness interactions only exist between genes on

TABLE 2. Average fitness interaction between pairs of marker loci in the surviving parthenogenetic offspring of females that display varying degrees of hybridity between two parthenogenetic strains of *Drosophila mercatorum*.

Percentage hybridity	Average fitness interaction for markers		
	Same chromosomal arm	Same chromosome, different arms	Different chromosomes
100	0.08	0.14	0.09
60	0.10	0.01	0.01
40	0.05	0.02	—

the same chromosome arm. This observation supports a theoretical prediction (Templeton et al, 1976) that the size of the genetic unit that can become coadapted through fitness interactions between loci is a compromise between selection building up multilocus complexes and the breakdown of multilocus associations by recombination in a broad sense (not only crossing over between homologous chromosomes, but independent assortment, sexual reproduction, extent of outbreeding, etc.). These experiments with parthenogenetic flies clearly document the interaction between the intensity of selection and the degree of recombination as predicted by this theory.

One implication of these results is that intrinsic coadaptation is most likely in those species with restricted recombination. Hence, intrinsic coadaptation is not expected to be found in all species, as has been confirmed by *Drosophila* studies (McFarquhar and Robertson, 1963; Richardson and Kojima, 1965; Singh, 1972). It is also significant that the phenomenon of coadaptation first described by Dobzhansky occurred in chromosome inversions, which act as crossover suppressors in *Drosophila*.

There are many ways of achieving restricted recombination other than by crossover suppression or other karyotypic structures. For example, population subdivision, small population size, and inbreeding will also cause an effective reduction in recombination, and hence facilitate the evolution of coadapted gene complexes. If a species is highly outcrossing and has large effective sizes, it is unlikely that an outbreeding depression will be encountered. The phenomenon of intrinsic coadaptation will only be likely in those species that are subdivided into inbred and/or small subpopulations. In the next section, a conservation program is outlined for a species that is a likely candidate for displaying intrinsic coadaptation.

COLLARED LIZARD POPULATIONS IN THE MISSOURI OZARKS

About 8000 years ago, the North American Midwest experienced a period of hot, dry weather known as the xerothermic maximum (King, 1981). During this period, the desert and dry-prairie adapted plants and animals of the Southwest were able to invade the Ozarks. The xerothermic maximum was short-lived, and by 4000 year ago the Ozarks were covered by an oak–hickory forest. However, certain rocky outcrops known as glades, particularly those with a southerly exposure, created xeric conditions that provided a refuge for some of these plants and animals from the Southwest (Ladd and Nelson, 1982). One of these relictual species is the collared lizard, *Crotaphytus collaris*.

The relictual populations on many Ozark glades have been destroyed through agriculture, residential–recreational development, and, perhaps most importantly, through fire protection practices (Sanders and Nightingale, 1982). Without periodic fires, the forest gradually encroaches upon the glade and destroys its unique biological character. As a consequence, the Missouri Conservation Commission has instituted a plan of glade recovery that involves cutting down the trees that have encroached upon glades, followed by periodic burnings. One area where this work is being done is on the Peck Ranch Wildlife Refuge in southern Missouri. This refuge encompasses about 23,000 acres, and contains a large number of glades based upon both dolomitic and rhyolitic outcrops. Although the glades vary in the amount of damage done by forest encroachment, none of the glades supported populations of collared lizards when the restoration program began. I was contracted by the Missouri Conservation Commission to establish collared lizards on these restored glades.

The first problem in such a release program is to obtain the lizards for release. The bulk of the species is still found in the American Southwest, so one option was to collect lizards from this area for release in Missouri. However, the lizards from the Ozarks show some life history differences from those in the central range (Sexton and Andrews, in preparation), so the problem of local adaptation (as in the ibex example discussed earlier) emerges as a real possibility. In this regard, lizards from other Ozark glades would be much better than lizards from the central range. Moreover, the Missouri Conservation Commission wanted to preserve the unique character of the Ozark populations of this species. Consequently, for both biological and political reasons, only lizards captured from other Missouri glades could be used in the release program.

Using Missouri lizards for the release project avoids the problem of local adaptation, but raises the problem of intrinsic coadaptation. As mentioned above, the Ozark populations have been isolated for about 4000 years, or 2000 lizard generations. Moreover, the population sizes per glade are very small, usually 20 to 50 lizards at the most (O.J. Sexton, personal communication). Finally, the lizards will not disperse through forested areas (O.J. Sexton, personal communication), so that glade systems that are surrounded by forests should have genetically isolated lizard populations. To see if this prediction is true, a genetic survey of collared lizards from several different glades was performed using protein electrophoresis for 36 loci and restriction site mapping of ribosomal and mitochondrial DNA using 27 different restriction enzymes. The genetic results are straightforward. Lizards from the same glade are genetically identical (with one exception that is associated with two natural glades having been connected by the construction of a parking lot and reservoir), but there are fixed genetic differences between glades, including glades that are on the same ridge system but with intervening forested areas (S. Davis, personal communication). This total lack of genetic variation within a glade and fixed differences between glades is readily explained by the accumulated effects of 2000 generations of genetic drift in populations with sizes smaller than 50 individuals.

Therefore, the Ozark lizards have a highly subdivided population structure in which each individual deme is so highly inbred that there is no detectable genetic variation within demes. Both population genetic theory and the experimental results obtained with *Drosophila mercatorum* indicate that this situation is ideal for allowing different glade populations to evolve different coadapted gene complexes.

One could in theory avoid the problem of coadaptation in the release populations by catching all the lizards for release on a particular restored glade from just one natural glade population. However, to insure that a glade population is established in light of the demographic variation observed in survivorship and fecundity on natural glades (Sexton and Andrews, in preparation), we decided that a minimum of ten reproductively mature lizards (five males and five females) were needed for release per glade. Unfortunately, capturing ten reproductively mature lizards from a single natural glade for release might well endanger the natural glade population, and indeed many glades do not even reliably have 10 reproductively mature lizards at the same time. In order not to endanger the natural glade populations, we decided that we should not collect any more than two lizards per glade.

The arithmetic of these constraints forces us into releasing together lizards caught from at least five different glades, each of which may have a uniquely coadapted gene complex. We will be able to detect such a coadaptation if it exists because we made sure that all five females released had different mitochondrial DNA markers (which are maternally inherited) and each male had a unique protein or ribosomal DNA marker. Replicate populations have already been established on two glades, and additional replicates will be established in the near future. By sampling the offspring over the years, we can directly monitor the reproductive performance of all the released lizards and their descendants and measure fitnesses of parental and recombinant genotypes. If coadaptation does exist, it should induce an outbreeding depression. The important question is, could this outbreeding depression endanger our release program?

An answer to this question is provided by experimental studies with *Drosophila mercatorum*. As mentioned earlier, the experiments of Templeton et al. (1976) indicated that the entire genome is not necessarily welded into a single coadapted unit, as envisioned by Mayr. Subsequent experiments (Templeton, 1979; Templeton et al., 1985) showed that even in the cases in which significant fitness interactions are detected among genes scattered throughout the whole *mercatorum* genome, the actual number of loci responsible for the relevant fitness interactions is quite small. This observation violates the predictions of Mayr that virtually all genes are held together in a coadapted complex. Indeed, a review of the literature on interspecific hybridization, where coadapted gene complexes are generally believed to be even more extensive genetically, reveals that coadapted gene complexes often involve very few genes (Templeton, 1981).

One implication of the number of genes involved in a coadapted complex being small is that selection can very efficiently reestablish either a parental genotype of high fitness, or occasionaly come across a particular recombinant that has even higher fitness. If coadaptation involved nearly all the genes of an organism, there would be virtually no chance of reestablishing a parental genotype, and the length of time needed to select a fit recombinant would be very long. Hence, the detailed genetic basis of coadaptation has important consequences for the long-term importance of an outbreeding depression.

The most important consequence of few genes being involved in coadaptation is that the outbreeding depression should be only a temporary phenomenon. This prediction was tested by additional experiments with *Drosophila mercatorum* (Annest and Templeton, 1978; Templeton, 1983). As in the previously discussed *mercatorum* experiments, hybrids between two different parthenogenetic strains were created and allowed to reproduce. As in the previous experiments, a severe outbreeding depression occurred, characterized by strong fitness interactions between loci. However, unlike the previously described experiments with this fruit fly, the parthenogenetic populations were allowed to continue to reproduce for a year, with approximately one generation every three weeks. The strong selective forces induced by the differently coadapted genomes resulted in rapid evolution in these populations (Annest and Templeton, 1978; Templeton, 1983). Not unexpectedly, a parental genotype was frequently favored and was fixed or nearly fixed by the end of the year. In these cases, the fitness traits (fecundity, survivorship, etc.) of the population derived from the hybrid cross gradually returned to parental levels in the generations subsequent to the initial outbreeding depression. Hence, by the end of the year, the populations of hybrid origin were just as fit as the initial parental populations. Selection in this case had reestablished one of the parental coadapted gene complexes and restored the fitness of the population. However, in one case (Templeton, 1983), a recombinant genotype with high fitness went to near fixation after one year. In this population, the fitness parameters were superior to those of either parental strain. Hence, selection had established a novel coadapted gene complex that was better than either of the original coadapted complexes.

Returning to our release program with collared lizards, these experimental results indicate that founding hybrid populations will have no deleterious long-term consequences even if an outbreeding depression is encountered. Selection should very rapidly eliminate the outbreeding depression, and indeed selection may favor a superior recombinant. Because the released lizard populations are marked genetically, we will be able to monitor the evolutionary changes that occur in these released populations. If our predictions are borne out, the release program can be greatly expanded in scope. At this moment, the best empirically supported expectation is that the hybrid nature of our release populations will have no deleterious long-term effects, but there may be beneficial effects even if an outbreeding depression occurs in the early generations.

One caveat, however, concerns the short-term effects of the outbreeding depression. Even if the outbreeding depression is temporary, it can greatly increase the chances of population extinction if it is severe during the early generations, when population size is small (Gilpin and Soulé, Chapter 2). Consequently, it is best to monitor the released populations carefully for the first few generations, as this is the time of greatest danger from oubreeding depression in both a genetic and demographic sense.

WHAT SORT OF "SPECIES" SHOULD WE SAVE?

My lack of concern about long-term outbreeding depression in collared lizards stems from my studies on the detailed genetic basis of coadaptation, a topic that may seem rather esoteric to most applied conservation biologists. Other esoteric issues in evolutionary biology also have important practical implications for conservation biology. One of these is the meaning of "species." Quite often "species" are treated as very discrete entities by conservation biologists. Thus, a species can be defined by having a particular kind of shape, size, color, etc. (or at least a restricted range of shapes, sizes or colors) that allows one to delineate it from other species. The purpose of many endangered species programs seems to be to preserve the unique type of shape, size, color, etc. represented by a particular "species."

In contrast, there is the "evolutionary species" concept, which defines a species as a lineage that shares a common evolutionary fate. This is not the most ideal definition of species for all purposes, but it does emphasize the fact that species are not static entities, and that the characteristics that "define" the present-day populations of a species are often characters that *evolve*. The question then becomes, what should we as conservation biologists be trying to preserve from extinction—a "species" that is a category defined by a currently-existing constellation of traits, or a "species" that represents a unique evolutionary lineage?

The definition of species one accepts has a profound impact on determining what one can and ought to do to save an endangered "species." One potential criticism of our collared lizard project is that a "new" type of lizard might emerge during the elimination of the outbreeding depression. The basic idea behind this criticism is that the endangered population must be preserved as is, and any genetic changes that arise are necessarily bad. This same criticism can also be raised against managed captive breeding programs. However, these changes are "bad" only under a rigid, categorical definition of species. Allowing a superior coadapted gene complex to evolve or eliminating an inbreeding depression are not at all "bad" under the evolutionary species concept.

For example, Templeton and Read (1983, 1984) deliberately used managed breeding to eliminate the genes responsible for inbreeding depression in a captive population of Speke's gazelle *(Gazella spekei)*. There has been a similar genetic reduction in inbreeding depression in a captive population of golden lion tamarins *(Leontidens rosalia)*, but in this case there was no deliberate attempt to eliminate the inbreeding depression (J. Ballou and K. Ralls, personal communication). It is almost inevitable that any managed breeding program will cause alterations in the gene pool, whether the alterations are deliberate or inadvertent.

I strongly feel that we should be preserving unique evolutionary lineages rather than unique present-day categories. This seemingly esoteric shift in definition of "species" would allow the conservation biologist to use evolutionary change as a beneficial management tool rather than to regard it as a deleterious situation that must be prevented at all costs. I am *not* advocating that humans should intervene extensively in the genomes of endangered species. Other than insuring that genetic variation is preserved, and thereby maintaining the potential for future evolutionary change, one should intervene as little as possible when managing an endangered species. However, situations do arise, as exemplified by the collared lizards and the Speke's gazelle, in which allowing natural selection to change the populations can be a very useful and beneficial component of a management program designed to preserve an endangered evolutionary lineage.

SUMMARY

An outbreeding depression occurs when fecundity and/or viability decline following an intraspecific hybridization. Coadaptation provides an explanation for outbreeding depression based upon the fitness interactions between genes. Two major causes of outbreeding depression are:

1. The phenomenon of *local adaptation,* in which different geographical populations of a species adapt to their local environments. Local adaptation may or may not involve coadaptation.
2. *Intrinsic coadaptation,* in which the genes in a local population primarily adapt to the genetic environment defined by other genes.

Drosophila experiments on intrinsic coadaptation lead to two major conclusions. First, intrinsic coadaptation is likely only in a species that is subdivided into inbred and/or small sized demes. Second, the outbreeding depression associated with intrinsic coadaptation is a temporary phenomenon that can be rapidly eliminated by natural selection. Indeed, sometimes a superior coadapted gene complex arises as the outbreeding depression is selectively eliminated. Thus, there are no long-term deleterious consequences of intraspecific intrinsic coadaptation. These conclusions are applied to the design of a release program for collared lizards in the Missouri Ozarks.

Finally, I believe that in saving endangered "species" we are really saving an endangered evolutionary lineage rather than an endangered constellation of present-day traits that defines a rigid category that we call a "species." Because species are evolutionary lineages, the conservation biologist should not ignore or try to suppress evolutionary change in all circumstances; rather, the conservation biologist should use evolutionary change as a beneficial and powerful management tool for the preservation of endangered evolving lineages (species).

PATTERNS OF DIVERSITY AND RARITY: THEIR IMPLICATIONS FOR CONSERVATION

DIFFERENT APPROACHES, DIFFERENT CONCLUSIONS

Diversity and rarity are synonyms for "everything" in ecology. If ecologists can explain and predict the patterns of diversity and rarity in communities, it means that they have understood the distribution and abundance patterns of their component species. Hence the diversity-rarity problem is one of the fundamental issues in biology.

It is therefore impressive if biologists working on different taxa and using different approaches arrive at similar descriptions of the patterns of rarity that they perceive. Though each chapter in this section uses a different perspective and employs a different typology, there are notable overlaps and commonalities. As shown in Table 1, three of the chapters recognize three components of rarity: a within-habitat component, a between-habitat component, and a geographic

TABLE 1. Some categories or components of rarity used in this section.

	"Within-habitat" component	"Between-habitat" component	"Geographic" component
Cody (Ch. 7)	α	β	γ
Rabinowitz et al. (Ch. 9)	Population	Habitat	Geographic distribution
Hubbell & Foster (Ch. 10)	Regeneration	Habitat	Immigration

component. Cody (Chapter 7) shows how the α, β, γ typology for diversity can be transformed into a parallel system for rarity.

By no means are these systems the same, however, either epistomologically or operationally, and it would be a serious mistake to try to collapse them into one. For example, the "immigration" component discussed by Hubbell and Foster (Chapter 10) is used to explain that some species will be rare in a particular patch because the center of abundance of that species is remote—the study site may be on the edge (ecologically or geographically) of the range of the species. In contrast, the geographic concept used by Cody (Chapter 7) is γ-rarity, which refers to the rate of geographic replacement (turnover) of similar species in a particular type of habitat. Still another meaning is employed by Rabinowitz, Cairns, and Dillon (Chapter 9), where "geographic distribution" refers to the amount of territory occupied by a species. Hence the similarities are more apparent than real.

One of the major sources of complexity in the analysis of rarity is the "sample" or "scale problem." This is the set of artifacts and biases that occur because of the inescapable arbitrariness and incompleteness of any reasonable choice of a sampling universe, whether it is a 50-ha patch of forest (Chapter 10) or all of the United Kingdom (Chapter 9). The smaller the sampling universe, the more species will be rare for a variety of "scale" reasons. In a very small habitat patch, for example, some species will be rare or absent because they are top predators with home ranges many times the size of the patch. Other species will be rare because their microhabitat is just beyond the border. Still others might be absent because they are "regeneration niche specialists," and occur only in a particular successional stage which by chance is absent at the time of sampling.

I have attempted to illustrate the relationship of this "sampling component of rarity" to the other components in Table 2. The totals at the bottom of the table suggest the obvious conclusion that scale problems decrease as the size and scale of the universe increase. Only one kind of rarity is detectable at the gross level of worlds: species whose absolute numbers of individuals are small, for whatever reason. ("Endemism" is meaningless at the level of worlds because all species will be endemic.) Implicitly, then, each level of scale contains information not available at other levels.

There is one approach to the problem of rarity that is missing. This is the taxonomic approach, where the sampling universe is not space—it is a relatively large, monophyletic taxon, such as a genus or family, in which the distributions, abundances, and evolutionary relations are known for most of the species. Such an approach could provide information not otherwise available, including the association of the dif-

TABLE 2. Occurrence matrix of relationships between some components of rarity and environmental scale.

Rarity type	Geographic scale				
	Habitat patches	Localities	Regions	Continents	Worlds
Rare within habitat	4[a]	1			
Habitat specialist		4	2	1	
Endemics			4	2	
Regeneration or successional specialist	4	1			
Trophic or life history rarity	3	2	2	2	1
Totals	11	8	8	5	1

[a]The numbers refer to the relative frequency of occurrence of the indicated interaction; "4" is the highest frequency.

ferent components of rarity with the putative age of the species. Old or primitive species, for example, might tend to be habitat specialists, or they might be low-density generalists with highly evolved mutualistic relationships. Hubbell and Foster (Chapter 10) make such comparisons among genera and families, but not within them.

It should also be noted that the chapters in this section all examine rarity within a trophic level or natural group. Such a choice is judicious. Sooner or later, though, someone is going to have to examine rarity at the community level—across trophic levels, and notwithstanding the conventional wisdom about the traditional energy and biomass pyramids.

As yet there appear to be no clear and predictable patterns that relate specialization and abundance. Rabinowitz et al. (Chapter 9) do not confirm for plants the generalization applied to animals (Brown, 1984) that generalist species are widespread and common. Hubbell and Foster (Chapter 10), on the other hand, find that generalists are more abundant. Here again, though, there are differences in scale between the two plant studies. In addition, edaphic and climatic conditions are more important for sedentary organisms such as plants than for animals. These differing patterns are important clues for further research.

A NONEQUILIBRIUM NEOTROPICS

Gentry (Chapter 8) and Hubbell and Foster (Chapter 10) both advance a dynamic view of species diversity in the New World tropics. Gentry points to the explosive speciation still occurring in Central America and the northern Andes, and to the continuing origination related to existing edaphic patchiness in the Amazon Basin. Hubbell and Foster find no evidence for a determininistic ecological ceiling on species diversity in the forests of Panama. Apparently there is still room in the tropics for more species.

The work of Gentry and other systematists is exploding the old myth of low levels of local endemism on continents in the tropics. Clearly, what is needed is a major campaign to document and understand this amazing diversity before it is too late (Wilson, 1985; Myers, Chapter 19).

These conclusions add new fuel to the old debate about why there are so many species in the tropics. They also provide evidence for the theory (similar to that from island biogeography) that the large number of species in some parts of the tropics is explicable in part by high rates of origination (speciation) and low rates of extinction. Geographic isolation in mountainous regions is strongly implicated as a source of species (and endemism) in the Neotropics. (In the northern temperate zones and the arctic, isolated mountains can't function well as refugia for the allopatric speciation of lowland species because their climate is so inhospitable.)

MANAGEMENT AND DESIGN IMPLICATIONS

Because sedentary organisms can't go looking for mates, the vulnerability of plant populations to extinction depends not only on some minimum overall population size (MVP, Chapter 2), but also on a minimum viable density (MVD). Hubbell and Foster point out that a tree species might have a hard time getting fertilized if its density falls below about 60 individuals per km^2. The notion of MVD is extremely important because a large reserve may have thousands of individuals of a particular species, but still be below the MVD threshold. This could happen if the source of recruits was historically from outside of the reserve area, or if there were an episode of unusually high mortality (natural or artificial)—harvesting and disease, for example.

If Gentry's theory (Chapter 8) of edaphic/geographic isolation and speciation is correct, and there are many more local endemics throughout lowland Amazonia than has hitherto been suspected, then the number of reserves will have to be vastly increased if we are to prevent the extinction of very large numbers of neotropical species. Cody (Chapter 7) also points to the need for manifold reserves in regions of high species turnover for birds. The importance of this cannot be overstated. It will require some very heroic proposals indeed if reserves of sufficient number and size are to be created to salvage these floras and their associated faunas. Hubbell and Foster (Chapter 10) argue that each such reserve will have to be hundreds of km^2 in size in order to provide sufficient habitat diversity, MVPs for mutualist animals, and buffers against large, stochastic environmental changes. The chapters in Section III further document this conclusion.

M.S.

CHAPTER 7

DIVERSITY, RARITY, AND CONSERVATION IN MEDITERRANEAN-CLIMATE REGIONS

Martin L. Cody

Rarity may seem at first glance a straightforward notion. To conservationists it signals species at risk and in need of protection; it supplies a sufficient description and points to worthwhile goals. Yet the only easy generalization about rare species is that, inevitably, the list of rare and endangered species will become longer, and the category of "safe" species will diminish as more and more of the Earth's surface is converted from natural habitat to use by humans for agriculture, industry, recreation, or other activities. To the evolutionary ecologist or biogeographer, rare species are a diverse assemblage, including both species that were rare before humans influenced their numbers and distribution and those whose rarity is caused by humans.

Main (1982) introduced several different and useful categories of rarity, and his theme is developed in this chapter (see also Rabinowitz et al., Chapter 3, and Gentry, Chapter 8). In this chapter I link different sorts of rarity to different components of species diversity. I define three such components, discuss the sorts of rarity they are each associated with, and attempt to isolate the environmental factors that control each diversity component and thus the production of rare species (which I see as a direct consequence of these environmental factors). This theme is developed within one well-defined and well-studied set of habitats—those produced by the Mediterranean-type climate in five regions of the world.

The Mediterranean climate regions of the world are defined as regions of modest annual precipitation (275–900 mm) with at least two-thirds falling in winter, and with warm summers (20°–25°C monthly means) and cool winters (< 15°C) in which frosts are rare. Such conditions are found in five widely-separated geographic areas on five continents: California, Chile, around the Mediterranean basin in southern Europe and North Africa, the Cape region of southern Africa, and southern and southwestern Australia. Because these regions are relatively small in area and are widely separated, with generally dissimilar floras, faunas, and evolutionary histories, they have been the subject of many studies on parallel and convergent evolution (reviewed in Cody and Mooney, 1978).

These regions are of interest to conservation biologists for a number of reasons. Most are heavily populated, with large cities and the disruption of natural landscapes that this implies. Moreover, areas around the Mediterranean Sea have been major population centers for thousands of years. Mediterranean-type climates are ideal for a wide diversity of agricultural endeavors in the valleys and for a variety of fruit and nut crops on hillsides; together with grazing, clearing, and burning, these effects have removed or drastically altered the natural vegetation in some regions and at least restricted or modified it in others. Competing demands on land use from urbanization, agriculture and forestry, recreation, and conservation have produced new and complex landscapes in Mediterranean climates.

Along with broad climatic variation within these regions, topographic and substrate differences modify local environments to produce a variety of natural habitats. In the lower, flatter, and drier sites these range from grasslands and low, semideciduous shrublands to microphyllous and broadleaved sclerophyllous shrublands (variously known as chaparral, matorral, macchia, maquis, fynbos, heath, or kwongan). In wetter sites, woodlands of broadleaved evergreen trees predominate, and at higher elevations, especially in the northern hemisphere, narrowleaved evergreens become dominant. Besides natural habitats, some regions support vegetation that appears to be largely anthropogenic. The garrigue of southern Europe, the goat-resistant *Euphorbia* scrublands in Morocco, and the *Acacia caven* valley savannas of Chile are some examples.

In this chapter I will concentrate on the diversity patterns of birds and plants in Mediterranean-climate regions, developing the viewpoint that rarity and the conservation problems it entails are often a natural consequence of high diversity, and can best be understood by describing the patterns in different components of diversity and by determining the factors that produce and modify these patterns.

RARITY AND SPECIES DIVERSITY

Anthropogenic rarity

In some instances rare species are a direct result of human modification of natural environments, activity that aggravates already restricted distributions, reduces already low population densities, and/ or decimates and isolates populations of otherwise common or widely distributed species by habitat destruction. Examples of these influences are the effects of road cuts on the manzanitas *Arctostaphylos hookeri* and *A. hispidula* and perhaps of changing land use on the condor *Gymnogyps californianus* in California; of urban development of coastal flats near Cape Town, South Africa on local geophytic plants; and of clearing in wandoo and salmongum woodlands *(Eucalyptus wandoo, E. salmonophloia)* for wheat farming in southwestern Australia on mammals such as numbats *(Myrmecobius fasciatus)* and rat kangaroos *(Bettongia)*. In the case of anthropogenic rarity, the causes (usually habitat destruction) and their obviation (habitat restoration) are often equally obvious, although the latter may not be possible, practical, or likely.

Natural rarity as a product of high diversity

Natural rarity can be produced independently of Man's activities. Some natural rarities are the products of endemism in limited areas, such as islands. Thus more than 50 percent of the Canary Islands plant species, which are about 95 percent endemic, and about two-thirds of the 155 endemic plant species on Crete are considered endangered (Lucas and Synge, 1978). Habitat destruction in these cases has added to the perils of the extremely restricted natural ranges in these island species (Bramwell and Bramwell, 1974).

Island species apart, to what extent are natural rarities in continental areas the products of high species diversities? That is, to what extent will adaptive radiation within a taxon such as the genus produce rare species, species with small geographic ranges, narrow habitat limits, or species with but a small share of the community pie? The floras of the California Floristic Province, the Cape Floral Kingdom, and the Southwestern Province of Western Australia are all relatively richer than their adjacent continental areas, and about a quarter of these floras is considered endangered relative to the 10 percent norm in larger regions (Table 1). This trend to proportionally more rare species in richer floras is continued in the relatively species-rich genera within these floras. For example, the threatened species of Australian *Acacia* (113 of approximately 850 species), *Eucalyptus* (111 of 600) and *Grevillea* (57 of 250) are at least twice the proportion of the representation of these genera in the overall flora. The four largest genera of shrubs or subshrubs in California chaparral are *Arctostaphylos, Ceanothus, Eriogonum,* and *Ribes;* their 309 species constitute 6.12 percent of the Californian flora. However, the representation of these genera in the list of 1136 threatened plant taxa— expected to be $0.0612 \times 1136 = 70$ spp., is significantly larger than this figure, with 112 species recorded as threatened or endangered (Chi-square $= 26, p < 0.001$).

Components of diversity

There are three basic components of species diversity, which can be considered with little modification to apply to any taxonomic group

TABLE 1. Areas, number of plant species, and numbers of extinct (X), endangered (E), and rare and endangered (R&E) species in Mediterranean-climate regions relative to the continents and the world as a whole.

Region	Area (10^6 km^2)	Number of plant species	Species richness[a]	Number plants X, E, R&E[b]	% of flora X, E, R&E
CONTINENTAL					
World	148	ca. 250,000	1.7	ca. 25,000	10
U.S.A.	9.36	ca. 20,000	2.1	1,641	8
South Africa	2.67	ca. 20,000	7.5	2,373	12
Australia	7.58	ca. 22,000	2.9	2,206	10
Europe	5.68	ca. 14,000	2.5	1,490	11
MEDITERRANEAN CLIMATE					
California	0.41	5,050	12.3	1,136	23
Cape flora	0.09[c]	6,000	67	1,621	27
S.W. Australia	0.31	3,630	11.7	853	24

(Data from various sources, including Leigh et al., 1982; Powell, 1974, 1980; and Hall et al., 1984.)
[a]Calculated as 10^3 plant species per 10^6 km^2 of area.
[b]Half of these species have distributional ranges of less than 100 km^2.
[c]Excluding the Little Karroo.

(Cody, 1975, 1983a; Whittaker, 1972). The number of species coexisting within a uniform habitat is termed α-diversity: an inventory or level of species packing characteristic of the habitat in question. As habitats change along topographical or climatic gradients, new species are encountered as other species drop out, and this species turnover rate is termed β-diversity: a function of changing habitat (Figure 1).

The third diversity component is γ-diversity: the rate at which additional species are encountered as geographic replacements within a habitat type in different localities. Thus γ-diversity is a species turnover rate with distance between sites of similar habitat, or with expanding geographic areas. All three components of diversity are potentially independent of one another, although in practice correlations among them often exist.

Natural rarity is a product of high diversity in one or another of its components. For example, species in communities of high α-diversity may each receive a small share of the total available resources and therefore persist only at low densities. The lower densities of species at higher trophic levels might be included as a subset of these sorts of rarities. Such rare species, which I term high α-rarities, are at risk from a variety of consequences of their low population sizes or densities, including susceptibility to chance environmental fluctuations in their resources, impairment of density-dependent effects in competition or reproduction, inbreeding, or reduced social facilitation. Especially vulnerable are predators, extreme resource specialists, social vertebrates, and plants with insect-mediated pollination or dispersal systems. Examples are Eleonora's falcon *(Falco eleonora),* a resource specialist dependent for food on migratory land birds high over the Mediterranean Sea (Walter, 1979); and the mistletoebird *(Dicaeum hirundaceum),* which ranges throughout Australia and feeds exclusively on mistletoe berries (it was seen only twice during my five months of field work there).

High β-diversity results from a high habitat specificity. High-β habitat specialists may be particularly rare and restricted if their

FIGURE 1. The three major components of species diversity. Species are ▶ ranked on the vertical axis S; as the habitat gradient H is traversed from the origin, "new" species are gained at a rate $g(H)$ and "old" species lost at a rate of $l(H)$. The difference between these gain and loss curves at any point on H, $(g - l)$, is α-diversity: the number of species coexisting in that habitat. The slope of a line midway between the gain and loss rates, $d/dH[(g + l)/2]$, measures β-diversity as a function of H: the rate of species turnover with respect to habitat change. Species turnover with distance D, independent of habitat, calculated as $d/dD[(g + l)/2]$, is γ-diversity: the rate of accumulation of ecological counterparts within a habitat among different geographical regions.

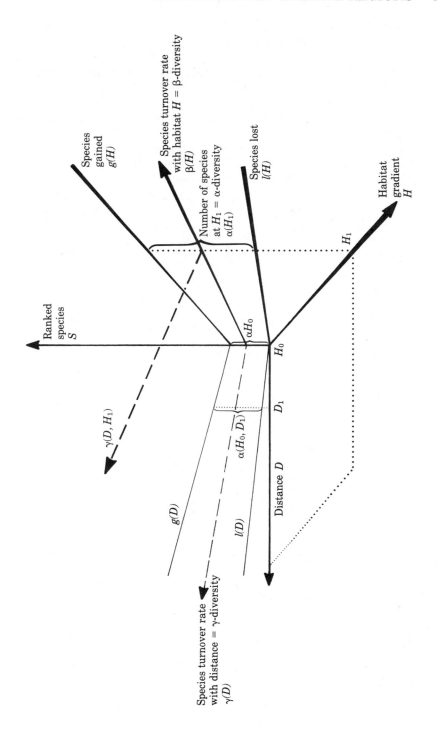

preferred habitats are scarce. This is the case in many plant taxa restricted to specific substrates, such as patches of serpentine soil in California. A southwestern Australian example is *Eucalyptus caesia,* a mallee comprising some 2000 individuals restricted to outcroppings of granitic boulders (Hopper et al., 1982). The extremely localized Eyrean grasswren *(Amytornis goderi)* is in fact reasonably common in its preferred tussock grassland habitat in the Simpson Desert of central Australia (Ovington, 1978), but such habitats are themselves rare and restricted, coinciding with the most unpredictable of Australian climates (Gentilli, 1971). In southern Africa, where sylviine warblers are diverse throughout scrub habitats, three habitat specialists (Karoo green warbler, *Eremomela gregalis;* Victorin's scrub warbler, *Bradypterus victorinii;* and rufous-eared warbler, *Malcorus pectoralis;* see Figure 2) were proposed as threatened species in the first South African Red Data Book (Siegfried et al., 1976), but were excluded along with several other warblers in the category from the revised list (Brooke, 1984) when it was realized that they are relatively common within their particular and restricted habitats.

Many rare species are local endemics, but are members of species-rich genera which collectively occupy broadly distributed habitats.

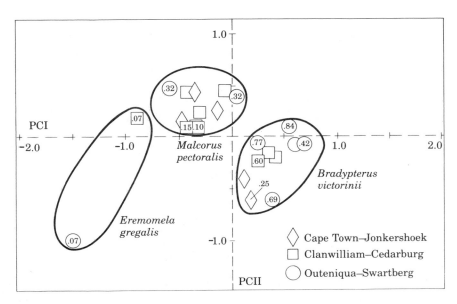

FIGURE 2. Some South African sylviine warblers can be considered "β-rarities," common only in their preferred habitats. The axes are principal components of vegetation structure (PCI, PCII); enclosed numbers are pairs/ha at the indicated site. (From Cody, 1983a.)

Such species constitute ecological equivalents in different parts of the habitat's range, and are the products of high γ-diversity. Such local rarities are especially typical of habitats that occur as variously isolated patches. Island endemics are a special subset within this category of rarities, and many continental examples come from geographically restricted plant species which form local isolates by dint of their low vagility (see examples in Gentry, Chapter 8), but where ecologically equivalent and often closely related taxa exist in other parts of the habitat range (see next section). An avian example is the helmeted honeyeater *(Lichenostomus melanops cassidix),* restricted to a few square kilometers of wet, shrubby woodland in South Australia, but with subspecies and other species of its diverse (18 species) genus in similar habitats elsewhere.

Determinants and regulation of diversity components

The upper limit of α-diversity may be constrained in many vertebrate communities by resource diversity, although ecologists differ in the extent to which they believe resources to be limiting and the consumer communities to be at competitively determined equilibria (Pimm, Chapter 14). However, equilibrium communities are expected to reflect their resource base in the absence of major disturbances, influences from other trophic levels (such as predators), and resource superabundance. On the other hand, β-diversity appears to have little to do with resource allocation and community theory, and this is not unexpected. Empirically, β-diversity appears to be correlated with the relative and absolute areal extents of different habitat types, and to the contiguity and spatial arrangement of the various habitats. The theory of these relationships is just developing.

Even less developed is our understanding of what controls γ-diversity. We know that γ-diversity tends to be highest in taxonomic groups with the greatest potential for isolation and speciation, and that the evolution of ecological counterparts has much to do with habitat fragmentation, biotic and abiotic barriers to dispersal, and species vagility relative to invasion opportunities—the very components of speciation itself. As such, stochastic and historical factors will play a large role in the determination of γ-diversity, with minimal influence from more deterministic processes such as resource competition. The opposite may apply more closely to α-diversity, with β-diversity likely intermediate. However, with long-lived organisms such as tropical trees, in which population processes such as germination and establishment are strongly subject to the vagaries of chance, stochastic factors may more directly influence α-diversity (Hubbell and Foster, Chapter 10).

Rarity as a product of high diversity in any of its components is in a sense self-regulating. That is, rare species tend to go extinct, and such negative feedback should reduce and regulate diversity components to more stable levels in line with the extant environment conditions of resource availability, habitat extent, and habitat configuration. The extinction of high α-rarities may free up resources which are then utilized by other species in the community. The surviving species, by compensatory increases in density in the absence of a competitor or by density compensation, may become more common and less likely to follow the same route to extinction. If a habitat specialist or high-β rarity becomes extinct, α-diversity may decrease if resource use is adjusted within the community by density compensation. Alternatively, α-diversity may remain at the same level but β-diversity might decrease if resources are reallocated via habitat extension by species from adjacent habitats into that vacated by the defunct population. The demise of ecological counterparts that contribute to γ-diversity likewise may lower all diversity components through resource reallocation and adjusted habitat use, to produce more stable diversity patterns and a reduced likelihood of further extinctions. Some of these aspects of diversity components and their environmental correlates are amplified in the next section.

COMPONENTS OF PLANT SPECIES DIVERSITY

The determinants of plant species diversity have been the subject of a great deal of research recently, to which the late Robert Whittaker contributed much of his effort and insight. He was one of the first to separate the different components of diversity (Whittaker, 1972), but he later concluded that a number of attributes of plant assemblages made both the description of plant diversity patterns and the determination of their controlling factors a much more difficult task than it is in, for example, bird communities (Whittaker, 1977). This is because plants have nondeterministic sizes and reproductive efforts and are much more subject than animals to the vagaries of chance in dispersal and habitat selection. In addition, plants have patterns of resource use and allocation that are much less apparent than are, for example, territory or prey choices in birds (in part because all plants share similar requirements of light, water, and nutrients, and in part because subterranean resource uptake and above-ground light absorption and gas exchange require more subtle measurement).

Species packing in plant communities

A variety of factors is thought to contribute to within-community or α-diversity in plant communities. In the annual plants of Mediterra-

nean-climate regions, natural or anthropogenic disturbance may be a
prime factor which enhances α-diversity, at least at moderate levels
(Naveh and Whittaker, 1980). Even in tropical forests, events such as
cyclonic storms or local treefalls may preclude a stable and deterministic community structure and permit at best a long-term dynamic
stability (Hubbell and Foster, Chapter 10).

Nevertheless, a number of promising avenues to understanding
plant α-diversity are opening. Grubb (1977) has suggested that plant
species may segregate along four sorts of niche axes: habitat, time
(phenology), growth form, and regeneration tactics. The last category
is perhaps the most complex, involving seed production, dispersal, and
seedling establishment. Harper (1977) gives many of the specific ways
in which coexisting plants may differ in these respects, and Grime
(1979) presents a useful overview. Seeds and their dispersal agents
may interact in a mutualistic fashion to maintain α-diversity, as in
the fynbos shrub *Mimetes cucculatus* and the native ants which eat
the oily bodies (elaiosomes) off the seeds in the process of dispersing
them. The delicate balance in this system is illustrated by the fact
that seed dispersal, and therefore the survival of the plant, is severely
impaired in fynbos where the introduced ant *Iridomyrmex humilis*
displaces the native ants (Bond and Slingsby, 1984).

The patterns of α-diversity in Mediterranean-climate chaparral
(fynbos, macchia) were recently reviewed by Bond (1983), who pointed
up the impressive diversity of the South African fynbos and the southwestern Australian kwongan (or heaths; see Pate and Beard, 1984),
both of which are many times richer than the equivalent vegetation
on other continents. This high diversity correlates with the extreme
impoverishment of the soils in these two regions. Tilman (1982) has
produced models that predict this relation of higher plant diversity
where resource levels are low by showing that a wider range of resource levels and combinations may be available and utilized by plants
in such habitats.

Plant segregation by growth form may result from the evolution
of interspecific differences in root systems below ground (see, for example, Dodd et al., 1984, for kwongan plants) or in aboveground
branching patterns and leaf morphology. Differences in such structural niches may be a major feature of some plant communities, such
as deserts (Cody, 1985b). In particular, the leaf morphologies of protead
shrubs, a major component of fynbos, show a distinct segregation of
the 6–9 species within sites (Figure 3), although the specific morphological niche of any one species may shift among sites, and a specific
niche may be occupied by different but morphologically equivalent
species from one site to another. In the southwest Australian kwongan,
where 5-ha α-diversity in *Acacia* reaches about a dozen species, the

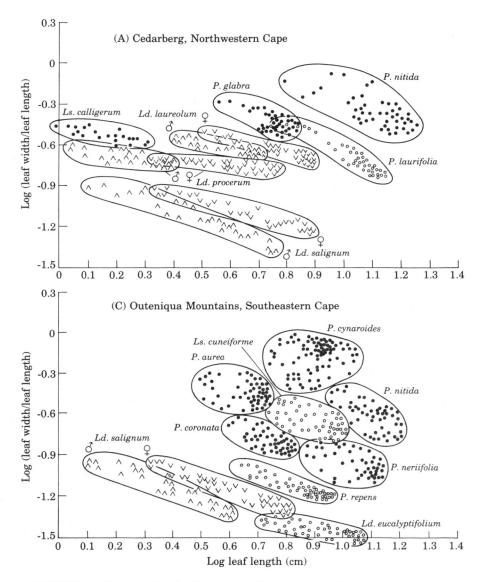

FIGURE 3. Segregation by leaf morphology of protead species of the genera *Protea, Leucadendron,* and *Leucospermum* at four sites in South African fynbos. Each species (and in *Leucadendron* each sex) displays a distinct leaf morphology with little interspecific overlap, although species composition changes among sites and, in some cases, a specific leaf morphology also changes among sites. (From Cody, 1985b.)

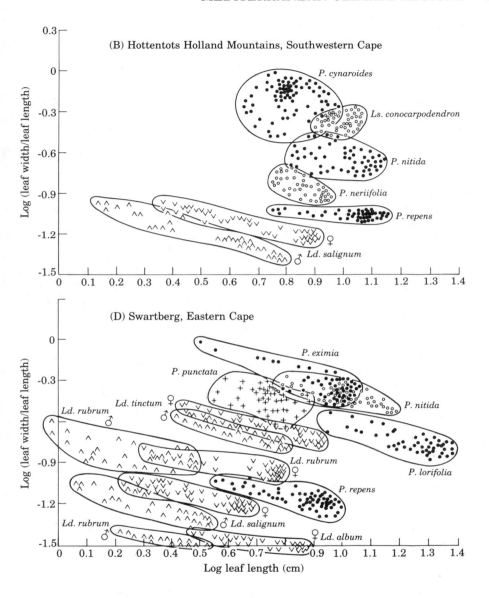

leaf-like phyllodes of these plants show a similar morphological segregation, but with rather greater interspecific overlaps (Cody, in preparation). It appears that these interspecific differences in leaf or phyllode morphology are associated with different carbon-fixing and gas exchange schedules. In *Acacia*, at least, the schedules covary strongly with the phenology of growth and reproduction (Cody, 1985b and in preparation). Leaf morphology also varies with overall growth form and probably with root systems, and illustrates the importance of structural niches to the α-diversity of the vegetation.

Other diversity components in plants

It appears that a major part of the extraordinary species richness of South African fynbos and southwest Australian kwongan is in the β-component of diversity—species turnover with such habitat variables as substrate and climate, and topographic variables such as slope and aspect, and (especially in γ-diversity) species turnover among sites independently of habitat change. There are few critical studies of these diversity patterns, and even fewer studies of the environmental variables which might help to explain them, but some descriptive data exist (Kruger and Taylor, 1979; Pate and Beard, 1984). A striking feature of fynbos is the difference in species composition between similar sites, as shown by the turnover in protead species between the Outeniqua and Swartberg Mountains, just 150 km or so apart (Figure 3C,D). Hopkins and Griffin (1984) comment on the local floristic turnover among kwongan sites, with approximately a 10 percent species turnover per 25-km shift. Maslin and Hopper (1982) show similarly high species turnover in *Acacia* species among different sites in central Australia.

Figure 4 plots species turnover as a function of distance in each of two important and species-rich genera of shrubs of the Mediterranean-climate regions of California, South Africa, and southwestern Australia. The genera *Ceanothus* and *Arctostaphylos* each have 43 species in California, and although about a quarter of the species in the former genus are restricted to one or two counties, fully 40 percent of the latter's species are local endemics (with ranges less than 50,000 km^2). In both genera the percentage of species turnover among ½° × ½° latitude–longitude quadrats (about 50 km × 50 km) along the coast is nearly linear (Figure 4A), with a 4–5 percent turnover per unit shift in distance over a 500-km interval (abscissa). But species turnover from coastal points (Santa Cruz, Santa Barbara) inland, over a steeper climatic gradient together with topographical changes, is much steeper—about 20 percent per unit shift in distance.

In South Africa the genera *Protea* and *Leucadendron,* with 69 and 80 species respectively in the southwestern Cape, have higher turn-over rates than the California genera over short distance (50–100 km) along the steep environmental gradients running south to north from the Indian Ocean coast (Figure 4B). Along the shallow environmental gradients running west to east and in parallel with gradual climatic change and few if any topographical barriers, turnover rates are lower, and in line with those measured in California. In southwestern Australia similar patterns are illustrated with *Banksia* (58 species) and *Acacia* (approximately 250 species in southwest Australia). In this area topographical diversity is low, and climatic gradient shallow; in Figure 4C the steepest environmental gradients cross climatic contours from south to north, and the shallowest run along climatic contours northwest to southeast. In consequence, species turnovers over short distances are modest, but over longer distances reach very high levels, largely reflecting the γ-diversity component of the flora.

The modal range size of species in these South African and southwestern Australian genera is in the range 10^4–10^5 km^2, with around half the species in each case qualifying as local endemics. But the data needed to clearly separate the β and γ components of diversity in these plants are lacking; consequently turnover rates cannot be plotted independently against habitat factors and against distance apart of census sites. When this becomes possible, the interactions among diversity components and the adaptive radiations within genera can be better explored.

COMPONENTS OF BIRD SPECIES DIVERSITY

In this section I will extend the notion that rarity and high species diversity (in one or another of its components) are necessarily related, using data on bird diversity in various Mediterranean-climate regions. The components of bird diversity have been more accurately quantified and their determinants better studied than in plants. For birds, the patterns of diversity and the processes affecting the production of rare species are somewhat better understood.

Vegetation structure and α-diversity

A quarter of a century ago Robert MacArthur's fieldwork showed that bird species α-diversity in temperate habitats is directly related to vegetation structure (MacArthur and MacArthur, 1961; MacArthur, 1965). While other factors, such as productivity and its predictability, also play a role (Cody, 1974), it appears to be a minor one, with

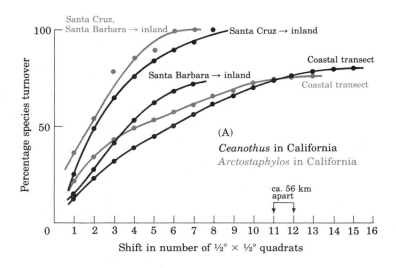

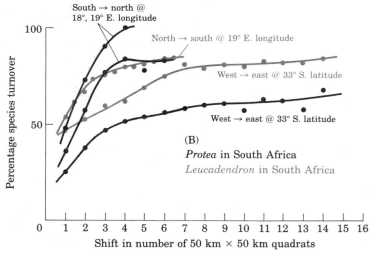

generally at least four-fifths of the variation in α-diversity attributable to variations in such habitat variables as the density and distribution of vegetation over height. This relation presumably holds because there is a greater variety of food resources and, especially, a greater variety of ways of exploiting and subdividing these food resources in taller, denser vegetation. The linear relation MacArthur

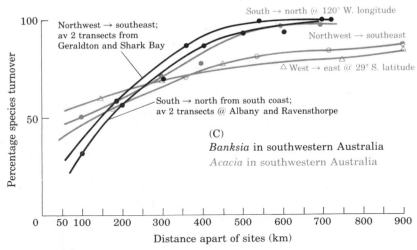

FIGURE 4. Species turnover rates in two large plant genera each from California (A), the South African fynbos (B), and the southwest Australian kwongan (C). Turnover is calculated as the mean percentage of the species that the sites have in common, subtracted from 100 percent (ordinate), computed for sites at different distances apart (abscissa). Transects cross steep or shallow climatic gradients, with or without topographic relief (see text). The abscissas are drawn approximately to the same scale. [Distributional data are taken from (A) Munz, 1968; (B) Rourke, 1980, and Williams, 1972; (C) George, 1984, and Maslin and Pedley, 1982.]

described between α-diversity and vegetation structure appears to hold well in continuous canopy habitats occupied mainly by generalized sorts of birds, most of them insectivores. It begins to break down, however, in more open vegetation such as heaths and deserts, in habitats occupied by more specialized birds with specific resource requirements (such as frugivores in tropical areas and nectarivores in Australian and South African heaths and fynbos; see below), and particularly where habitats are fragmented into isolated patches, such as relict woodlots in largely cleared farmland (Willson, 1974; also see below).

Alpha-diversity in Mediterranean areas: Habitat patchiness and isolation

Bird α-diversity is related to habitat structure in Mediterranean habitats as shown in Figure 5A, where sigmoidal relations are compared among different continental areas (Cody, 1975, 1983a). While there

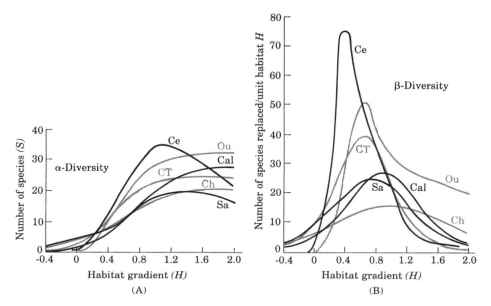

FIGURE 5. Parallel habitat gradients for birds in Chile (Ch), California (Cal), Sardinia-Corsica (Sa), and southern Africa (CT, Ce, Ou) show similar α-diversities (A), but quite different β-diversities (B). In (A) and (B) abscissas are the first principal components of several vegetation structure measurements (see Figure 6). The curves are fitted. These data are replotted in the α–β plane in (C), and the value of habitat H corresponding to the α–β values is indicated on each region-specific curve. The difference between the diversity curves of different regions is largely attributable to the difference in available habitat among regions, as shown in (D). Here lines connect a given habitat type in different regions, and circled figures give the cumulative area (in 10^5 km^2) reached by that point on the habitat gradient H. Note that diversity increases with increasing habitat areas, which affect largely β-diversity at low H (low, medium scrub), but increasingly enhance α-diversity at high H (woodlands). Sardinia (dashed circles), which is an island, shows discordantly lower diversity values relative to continental areas. (After Cody, 1983a.)

are general similarities among these curves, the apparent differences from one continent to another are related to differences in the availability of habitat of various types. This point is more clearly made in Figure 6, which contrasts α-diversity and vegetation relations in four areas: California, Sardinia, Morocco, and the Canary Islands. Over increasing vegetation height and density on the left half of the vegetation gradient (abscissa), α-diversity increases in a similar fashion

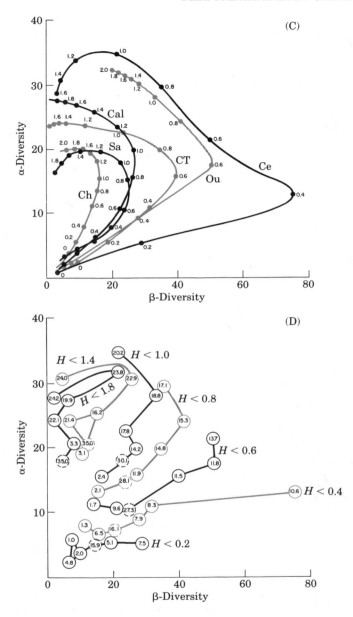

in all four areas, with the single exception of the higher bird diversities in the open scrub habitats in Morocco (points 1 on the figure). These habitats are far more extensive in this region than in the other three, largely because of increasing desertization.

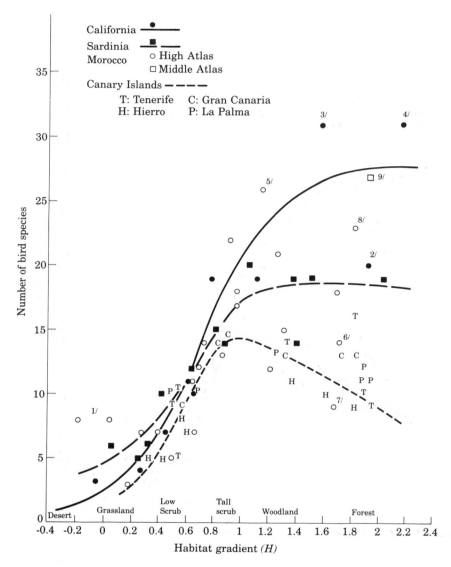

FIGURE 6. Numbers of bird species censused in a range of Mediterranean-type habitats, from semi-desert and grassland at left, through chaparral (center), to woodland and forest at right. The number of bird species (α-diversity) increases smoothly and is similar among regions over the first part of the habitat gradient, but shows a great deal of variability over the second half, where habitat availability varies due to clearing and topographic effects. Californian sites are generally richer and Canary Islands sites generally the poorest in species. Labeled points are: 1/ Moroccan semi-desert; 2/ Californian closed-cone pines; 3/ oak woodlands; 4/montane coniferous forest; 5/ Moroccan maquis; 6/ isolated woodland patches of oaks; 7/ isolated woodland patches of pines; 8/ High Atlas cedars; and 9/ Middle Atlas cedars.

The right half of the habitat gradient represents taller woodlands and forests. These are the habitats that are naturally rarer in these regions, being restricted to more mesic sites or higher elevations, an effect that is minimized in mountainous California but is particularly pronounced in the Canary Islands. Also, habitat clearing, for charcoal production in Morocco and for crop plants in all areas, affects these taller and more productive habitats in particular, and such anthropogenic influences are concentrated in this part of the habitat gradient. In the taller habitats, bird species are most diverse in California where woodland forest is still plentiful; intermediate in Morocco and Sardinia-Corsica where this habitat is much reduced; and lower still in the Canary Islands, where most of the mid-elevation laurel forest has been cleared and the less productive pines at higher elevation, though still largely extant, are more restricted in area.

Figure 6 also shows that the closed-cone pine forests of coastal California (point 2 on the figure), which are patchily distributed and of limited extent, are much less diverse than the interior oak woodlands and montane forests (points 3 and 4). In Morocco the highest bird diversities are reached in *Juniperus–Cistus* dominated maquis (point 5); diversity is much lower in the isolated and relictual woodland patches (points 6 and 7). A census of the few remaining hectares of high-elevation cedar forest on the Jbel Ayachi (point 8) in the High Atlas failed to produce nine bird species typical of this habitat (and still common in the Middle Atlas at Azrou, point 9, where cedars remain plentiful).

Beta-diversity and the effects of habitat area

In contrast to the similarities in patterns of α-diversity (Figure 5A), comparisons among continents in β-diversity show striking disparities (Figure 5B; Cody, 1975, 1983a). It appears the reason for this is that continents are dissimilar in the areas occupied by structurally analogous habitats. Habitat extent is largely a function of topography, which varies widely between different regions. In Figure 5C each of the six curves in Figures 5A and 5B are plotted in the α–β diversity plane, and on each curve the position on the habitat gradient at which the corresponding diversity values are reached is indicated. But each habitat type is differentially common in each of the six regions, and these data can be shown as frequency distributions over the habitat gradient, H. In Figure 5D the cumulative areas under such frequency distributions, in 10^5 km^2, are shown for each region at each of several values of habitat H (see also Cody, 1983a). An increased availability of habitat area results in increased diversity levels; commoner habitats produce data points located radially further out into the diversity plane, especially enhancing β-diversity in lower-H habitats and α-diversity in higher-H habitats. The data are internally consistent, with the exception of the Sardinian data, which show lower diversity levels than those expected from habitat areas. But this data set is the only one collected from an island, and so appears to reflect once again that diversity is reduced in isolated habitats. Clearly habitat availability has a strong effect on β-diversity patterns, and a lesser but still considerable influence on α-diversity.

Ecological consequences of reduced α- and β-diversity

Competitive release is expected to occur in the low-diversity sites, resulting in density compensation and expanded habitat use. Density compensation produces higher densities per species in habitats with lower α-diversities. This is illustrated in a comparison between Morocco and the Canary Islands, where habitats are well matched; in general the Canary Islands avifauna is a subset of the Moroccan. I have compared species numbers and densities in three habitats: pine forests, maquis, and euphorbia scrub, with several censuses in each habitat in each area (Cody, 1985a). In pines, where species numbers are nearly equal, densities per species are higher in Morocco, but in maquis and euphorbia scrub, where species numbers are two-thirds higher in Morocco, densities per species are around twice as high in the Canary Islands.

A second illustration of density compensation comes from the evergreen woodlands that occupy the south-facing slopes of the mountain ranges along the Indian Ocean coast of southern Africa (Cody, 1983b). These woodlands occur in smaller and more isolated patches from east to west, and bird species numbers follow patch size and decrease with patch isolation. A census in the more extensive and contiguous woodland to the east produced 43 bird species, and numbers decline to just 15 species in the smallest and most isolated patch on Table Mountain in the west. But although species numbers varied almost 300 percent, total bird density varied only 10 percent (Figure 7), with density compensation being complete in the small insectivorous birds and partial in other bird groups such as generalized omnivores, although absent in such bird guilds as pigeons and sunbirds.

In addition to density compensation, habitats with low α-diversity are expected to permit expanded habitat use and to be correlated to low β-diversity. This effect is seen in the habitat utilization by sylviine warblers in the Canary Islands as compared to Morocco (Cody, 1985a). The distributions over elevation and vegetation of 14 Moroccan sylviine species are shown in Figure 8. While a few species—especially *Sylvia cantillans* and to a lesser extent *S. melanocephala* and *Phyl-*

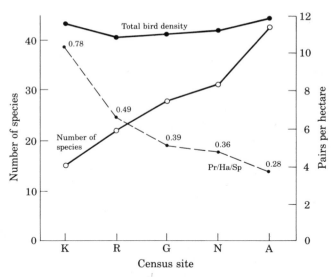

FIGURE 7. Density compensation in the bird species of Afromontane wood-land patches in southern Africa. Five census sites are ranked from the smallest and most isolated patch at left through increasingly larger and less isolated patches to the right. Patch size and isolation correlates with bird α-diversity, which varies 300 percent, but total bird density in the patches remains almost constant, varying only about 10 percent among census sites. (From Cody, 1983b.)

loscopus bonelli—are quite broadly distributed, over half the species are restricted to specific habitats at one or two sites. The mean "habitat range" for these 14 species is 9.3, calculated in the plane of this figure in units of 100 m of elevation and of vegetation structure intervals. On the Canary Islands sylviine warblers number only five species (three conspecifics and two congenerics with Morocco), and each oc-cupies a much broader range of habitats, averaging 27.0 units. This is especially dramatic in *Phylloscopus bonelli,* which is found in vir-tually every habitat in the Canary Islands, from lowland semidesert through maquis, woodland, and pine forest, to high-elevation alpine scrub. It seems reasonable to attribute such examples of extended habitat utilization on islands to the one outstanding attribute of island communities—their lower overall species diversity.

Habitat contiguity and γ-diversity

Gamma-diversity may be measured by censusing similar habitats at different geographic locations, and relating the species turnover be-

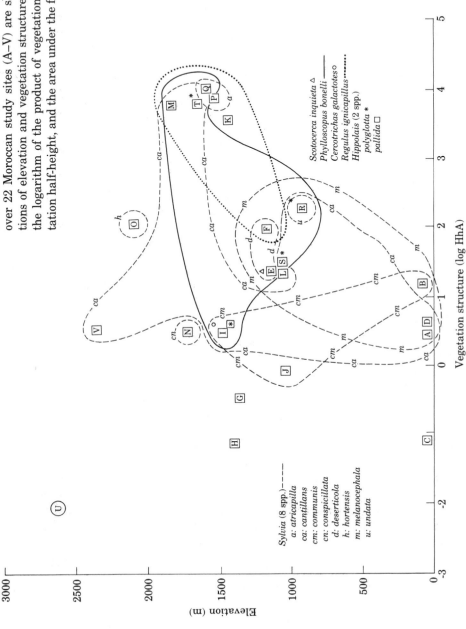

FIGURE 8. The ranges of 14 sylviine warbler species over 22 Moroccan study sites (A–V) are shown as functions of elevation and vegetation structure (computed as the logarithm of the product of vegetation height, vegetation half-height, and the area under the foliage profile).

Sylvia (8 spp.) —————
  a: atricapilla
  ca: cantillans
  cm: communis
  cn: conspicillata
  d: deserticola
  h: hortensis
  m.: melanocephala
  u: undata

Scotocerca inquieta △
Phylloscopus bonelli —————
Cercotrichas galactotes ○
Regulus ignicapillus ••••••••••
Hippolais (2 spp.) *
  polyglotta *
  pallida □

144

tween census sites to such measures of site separation or relative isolation as distance, differences in intervening climate or habitat, or topographical barriers. If patches of a given habitat type are small and isolated, then local extinction rates will be high, and each may reach low α-diversity with different types and numbers of species, and permit the influx of various species from the mix of surrounding habitats. This process allows for the evolution of ecological counterparts in the different habitat patches, and contributes to γ-diversity.

Three habitat types—renosterveld, fynbos, and kloof woodland—were censused in different parts of Cape Province, South Africa, to provide measures of γ-diversity (Cody, 1983a). In renosterveld four censuses averaged 14.0 ± 0.8 bird species, and γ-diversities among sites, calculated as (number of species gained + number of species lost)/2 and expressed as a percentage of α-diversity, averaged 46.6% ± 8.9. The figures for fynbos are: α-diversity = 29.0 ± 6.7 species and γ-diversity = 35.4% ± 10.4; for kloof woodland, α-diversity = 14.7 ± 2.9 species and γ-diversity = 41.9% ± 10.6. A proportion of the variation in γ-diversity, up to 43 percent, is attributable to the distance apart of the sites compared; these distances varied from 110 km to 450 km, and γ-diversity increases significantly with site separation. However, other factors seem at least as important in the determination of γ-diversity, and particularly high figures were obtained when one of the sites compared was locally rare or surrounded by very different habitats. This permits an influx of species that are more typical of the surrounding habitat type, and elevates γ-diversity.

But γ-diversity figures of 35–47 percent give a rather misleading impression of high species turnover. Actually the commoner bird species are constant among census sites within a habitat type, and it is the rare bird species that comprise the bulk of the turnover. Thus each habitat type supports a community of "core species" which are predictably present and common wherever the habitat occurs, and an additional group of "peripheral species" which vary among sites with the variety and extent of different adjacent habitats. This point is made in Figure 9, in which core species are ranked in each habitat type from commonest (highest density in all censuses combined) to rarest on the abscissa, and the cumulative percentage of the total bird density at a census site is scored on the ordinate. Each separate curve shows how total bird density is accumulated at a particular census site over the (same) ranked species. (Most of the curves do not reach 100 percent because the peripheral species were excluded.) Note that the core species set in renosterveld, a constant *(Elytropappus rhinocerotis-*dominated) and widely distributed habitat type, are most similar in their contributions to total bird density (e.g., 63–73% at ½ mean α-diversity), while in kloof woodland, which is very patchy in occur-

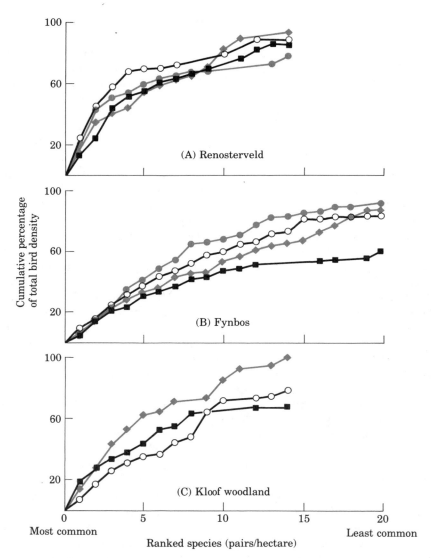

FIGURE 9. Three South African habitats (renosterveld, fynbos, and kloof woodland) were censused for birds in four different areas of Cape Province. Although there are differences, often substantial, between censuses in the same habitat type, the "core species" of the habitat, ranked from most common (at the left) to least common (in pairs/ha), are accumulated and contribute to total bird density in similar ways among censuses. However, the composition and density of the core species is most predictable in renosterveld (A), nearly as constant in fynbos (B), and least consistent in kloof woodland (C). This corresponds to greater structural replicability and a more continuous distribution in renosterveld, and a more discontinuous distribution with variable structure in kloof woodland. Symbols represent censuses in regions of Cape Town (◆), Cedarburg (■), Outeniquas (○), and Swartberg (●). (After Cody, 1983a.)

rence and variable in plant species composition, cumulative density curves in core species are least similar (45–73% at ½ α-diversity). Fynbos, which is widely distributed but variable in structure, is intermediate in these respects. These data illustrate that (1) α-diversity and species identities are predictable and constant in common habitats, but less so in patchy habitats, and (2) species that are rare in a habitat are the major contributors to γ-diversity, and these become the predominant class of species in more patchily distributed habitats, which have fewer common core species.

Habitat fragmentation also contributes to γ-diversity in southwest Australian bird communities (Cody, unpublished). The results from a total of 39 bird censuses from five habitat types are analyzed in Figure 10. Six censuses were made in mulga (*Acacia aneura* dominant), an extensive and little-disturbed habitat continuously distributed over arid inland areas; the six sites produced a total $S_T = 48$ species, with α-diversities (S_α) averaging 23.8 ± 2.4 species, and γ-diversities 36%. Thus $S_T/S_\alpha = 2.0$, so on average 50 percent of the mulga species can be seen at a single site. The upper curve in Figure 10 shows that nearly 20 percent of all mulga species occur at all census sites, and illustrates the way in which S_T is accumulated by including more census sites.

Eleven censuses were made in structurally similar sclerophyll woodlands, which are variously dominated by different *Eucalyptus* species in different sites. This woodland occurs on the richer soils of southwestern Australia and has been extensively cleared for wheat farming, leaving much of the woodland as relictual patches. Here $S_T = 70$ species, with α-diversity averaging 26.1 ± 6.9 species, γ-diversity = 48%, and 37 percent of the woodland species are seen at a single site. The distribution of S_T among the census sites is shown by the second highest curve in Figure 10, illustrating that a smaller proportion of the total of 70 censused species occurs at any one site than in the mulga. And lastly in Figure 10 are the three lower curves, each derived from censuses in heath or kwongan, protead-, *Acacia-*, or mallee-dominated, respectively. These habitats are the most patchily distributed of those censused in southwestern Australia, being mainly restricted to local sites of nutrient-poor sandy or laterite soils. In these habitats α-diversities are lower (averaging 12.4 ± 3.3, 18.0 ± 7.6, and 16.1 ± 9.0 species respectively), γ-diversities higher (65, 71, and 71 percent respectively), and a smaller proportion (27–30 percent) of the species totals of 41–58 species can be seen at any one site.

The high species turnovers among census sites, labeled γ-diversity, occurs despite the existence of a set of core species for each habitat type. In fact it is the various categories of rare species that are re-

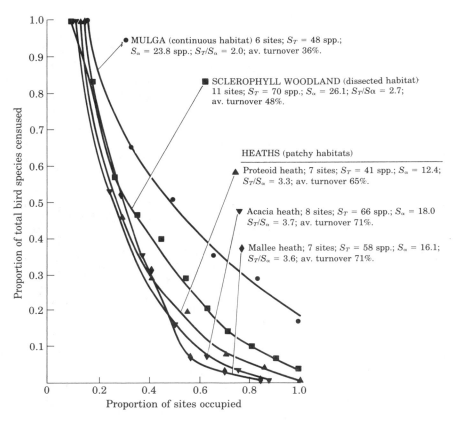

FIGURE 10. Bird censuses taken in the same habitat but in different areas of southwestern Australia show varying amounts of γ-diversity. Gamma-diversity is lowest (36 percent) in mulga, a habitat of wide and continuous extent. One-sixth of all birds censused were found at all census sites, and the top curve shows how the proportion of total bird species encountered in mulga increases as birds are included which occur at progressively fewer mulga census sites. In sclerophyll woodland, which is much dissected by clearing and agriculture, γ-diversity is higher and total bird species are accumulated less rapidly by adding in increasingly site-specific birds. In the three types of heath, which are very patchily but widely distributed, γ-diversities are higher still, and a much higher proportion of the total bird species censused is found in only a few sites.

sponsible for the turnover, and I illustrate this by looking at 12 (out of 70; 17 percent) species that occurred in only one of the sclerophyll woodland census sites. These 12 can be divided into several categories. First there are species which are common in other habitat types but are rare, local, or accidental in the habitat type in question; these are species which really do not "belong" in the censuses. In the woodlands this category includes three species that are common and regular in mulga (redthroat, *Sericornis brunneus;* spiny-cheeked honeyeater, *Acanthagenys rufogularis;* and crested pigeon, *Ocyphaps lophotes);* three species common in taller and more mesic forest (white-naped honeyeater, *Melithreptus lunatus;* New Holland honeyeater, *Phylidonyris novaehollandiae;* and red-capped parrot, *Purpureicephalus spurius);* one species common in *Acacia* heaths (singing honeyeater, *Lichenostomus virescens);* and one species more typical of grasslands and low scrub (stubble quail, *Coturnix pectoralis).*

In the second category are species which are locally common in the habitat (in which they reach their highest densities), but with restricted geographic ranges; they have ecological counterparts elsewhere in the habitat's range. An example from the woodlands is the yellow-rumped pardalote *(Pardalotus xanthopygus),* which occurs in woodlands of southeastern Western Australia in which the broadly distributed striated pardalote *(P. striatus),* ubiquitous in woodlands elsewhere, was not recorded.

The third group comprises species which are broadly distributed and most common within the habitat type, but in fact are of patchy occurrence and nowhere common. In the woodlands this includes the restless flycatcher *(Myiagra inquieta)* and crested shriketit *(Falcunculus frontatus).* This leaves one more species recorded at a single site: the flightless emu *(Dromaius novaehollandiae),* which must have been common in woodlands before hunting and fencing restricted its distribution. Thus the rarities contributing to species turnover between census sites within habitats are chiefly of three types: α-rarities that are patchily distributed; true γ-rarities, which have narrow geographic distributions; and species common in other habitats (β-restricted) and found in the habitat in question only sporadically, chiefly as a function of the type and preponderance of the adjacent habitats.

As might be expected from the patchy nature of heaths, in these habitats the percentage of species that occurred at just one site increases to 47–54 percent. It is in the heaths that most of the rare and endangered local endemics in Western Australia occur, species such as the noisy and rufous scrubbirds *(Atrichornis clamosus, A. rufescens),* and western whipbird *(Psophodes nigrogularis).*

In Figure 11 the determinants and rates of bird species turnover are compared among habitats and related to structural differences in vegetation within a particular habitat type (woodland, heath), and distance apart of the census sites. The directions of the habitat-specific arrows show the relative influences of these two variables in determining species turnover (or which axis to move along to obtain maximum species turnover). The width of the arrows is proportional to

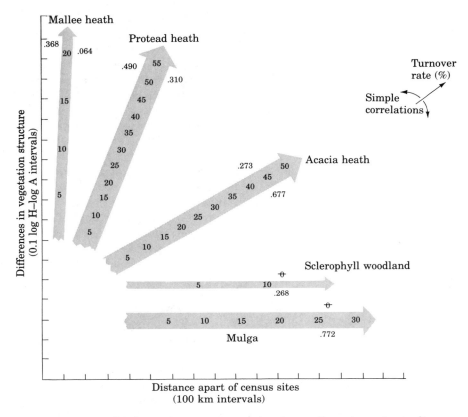

FIGURE 11. Bird species turnover rates in southwestern Australia are shown relative to structural differences in the vegetation (β-diversity; ordinate) and to distance apart of the census sites (γ-diversity; abscissa). Arrows indicate the relative extent to which vegetational differences or distance produces species turnover, and simple correlations are included for species turnover versus either axis. Figures within the arrows show the turnover rates, all to the same scale. Thus β- and γ-diversity have different values and rates in different habitat types, apparently related largely to habitat contiguity.

the amount of the variance in turnover that is explained by the two factors. The numbers within the arrows are turnover percentage rates and show how fast the turnover of species occurs with shifts along the axes. The correlation coefficients beside the arrows show the relative importance of the two axes in contribution to species turnover. Turnover in mulga and woodland is uncorrelated with any structural differences in the vegetation (and therefore is pure γ-diversity), but is correlated with site separation, especially for mulga (see correlation coefficients beside arrows). Species turnover is highest in protead- and *Acacia*-dominated heaths, and is correlated both to differences in vegetation structure and to site separation. In mallee heaths there is little distance effect, and species turnover is very largely β-diversity with changing vegetation structure. This corresponds with the biogeography of the heaths, for mallee is limited to the south-central part of the region, whereas the other heaths extend from both west and south coasts inland to central Australia. Structurally different heaths support rather different bird species, a result in accord with the natural cycling of vegetation structure following bushfires. Fires presumably initiate a postfire production sequence of different nectar, fruit, and insect resources. These resources are found and utilized by different bird species for as long as they last, after which they are sought in the appropriate habitat elsewhere. Thus there is little distance effect for such site-opportunistic but resource-specialized species. In general, series of replicated censuses within habitats not only identify the rare species and classify them, but reveal the environmental parameters that contribute to their production.

CONSERVATION HINTS FROM DIVERSITY PATTERNS

In this chapter we have seen how patterns of diversity and its components are related to patterns of rarity within Mediterranean-climate regions, and in particular how the effects of habitat area, isolation, and fragmentation contribute to diversity and rarity patterns. This final section attempts to draw conclusions from the available information, incomplete as it is, that might be helpful in predicting the most critical effects of habitat destruction and in identifying the taxa most sensitive to extinction.

The first message is that biotic diversity varies substantially among the five Mediterranean-climate regions; this variability is least apparent in the α-diversity component, more conspicuous in β-diversity differences, and most obvious in γ-diversity. In relatively small taxonomic groups such as birds, α-diversities are similar among regions even where the total number of species differs greatly (fewest bird species in Chile, most in South Africa). In contrast, in large taxonomic groups such as plants (largest floras in South Africa and Australia), α-diversities differ substantially among regions. As habitat clearing and fragmentation "progresses," it is first the γ-rarities (local endemics) that become endangered and are lost, and this happens first in the regions with the richest biotas.

But different taxonomic groups vary in their persistence times in relictual patches of habitat; plants and lizards are generally good persistors, while birds and mammals (with their high metabolic rates) are not. Further, there are differences among taxa in colonization potential of habitat patches. Many plants (except those with specific pollination and dispersal requirements) and most birds (except weak-flying species such as Australian malurid wrens and terrestrial parrots) are good colonizers, whereas most mammals and lizards are not. While these generalizations are crude, they suggest that habitat fragmentation and isolation will be most critical for mammal populations (which neither survive nor recolonize well), somewhat less so for birds (which can recolonize well) and lizards (which can survive well), and least critical for plants (which both survive and recolonize well).

Where more specific data are available, more specific conservation rules might be established. Figure 11, for example, suggests that mulga birds might be protected with several well-separated reserves in different areas, regardless of structural differences in vegetation among the sites, whereas fewer sclerophyll woodland reserves with a smaller geographic scope would accomplish the same result (since there is lower species turnover with site separation in the woodlands). On the other hand, diversity in heath birds can only be maintained by conserving a range of structural variability in these habitats, because sites with small structural differences support rather different species. Thus a wide range in the ages of postburn vegetation is necessary to maintain the full suite of heath bird species.

Diversity and rarity patterns will surely be found to be dependent on both taxon and habitat type, and further studies of these patterns should produce additional conservation guidelines, from the most general (such as the ways in which reduced habitat availability affects taxonomic groups or diversity components) to more specific algorithms for maintaining high levels of all diversity components in a specific habitat, given its geographical distribution and the recolonization and persistence abilities of the taxa which live in it.

CHAPTER 8

ENDEMISM IN TROPICAL VERSUS TEMPERATE PLANT COMMUNITIES

Alwyn H. Gentry

One important criterion for determining the optimal design for conservation units is the degree of biotic endemism. If most of a region's species are wide-ranging, one or two large and well chosen reserves might effectively conserve all or nearly all of that region's species. On the other hand, if most of a region's species are local endemics, several reserves scattered throughout the area will be necessary, even at the expense of losing some of the well known advantages of very large contiguous reserves (Terborgh, 1974; Soulé and Simberloff, in press; Wilcove et al., Chapter 11). Even if local endemism is high, however, concentration of locally distributed species in one or two centers of endemism might make possible preservation of most of a region's species within a few reserves chosen to include the high-endemism areas.

PATTERNS OF ENDEMISM

Patterns of endemism[1] are different in different taxa and in different regions. This chapter will focus on endemism in plants, which typically have more localized distribution patterns than do more widely known groups such as birds and mammals, and may call for different conservation strategies. Endemism occurs at different scales and may arise

[1] In this chapter, the general term "endemism" refers to any localized distribution.

from many different causes. On a worldwide scale, taxa may be "endemic" to a continent or subcontinent; at the opposite extreme, many plant species appear to be naturally restricted to areas as small as a few square kilometers. Some endemics, called paleoendemics, were formerly more widespread. Many such species may now have restricted distributions because of climatic changes and the reduction of once more widespread favorable habitats; other paleoendemics are archaic taxa with primitive traits and are on the verge of natural extinction. In contrast neoendemics are recently evolved species that have restricted distributions because they have not yet had time to spread over a large geographic area.

"Endemic" does not necessarily equate with "rare." Some species with restricted distributions are very common locally, while others are extremely rare. Among the latter are a rapidly increasing number of species whose endemism is an artifact of habitat destruction by humans. Terborgh and Winter (1982) concluded that birds with ranges of less than 50,000 km^2 may be conveniently regarded as having localized distributions. Except as otherwise noted, in this chapter "local endemics" are considered to be plant species with distributions of 50,000 km^2 or less. At this scale, distribution areas often coincide with geographic units for which floristic data are available (for example, Panama, Choco, or western Ecuador). I have also tried to compare endemism at a much finer scale, on the order of single ridgetops or mountains, although the very limited data sets available are rarely strictly comparable.

In some cases, an endemic plant species may not be very distinctive, whereas in others an endemic may be strikingly differentiated. Some genera have given rise to numerous poorly differentiated geographic isolates whose status as separate species is open to question. For the purposes of this chapter, I will assume that preservation of "species" is an appropriate focus of conservation and that all taxa recognized as species are equal from the viewpoint of conservation biologists.

Nevertheless, it must be noted that the taxonomic data base for tropical plants is woefully weak. Many of the "species" recognized here are recognized on the basis of superficial herbarium comparison of gross morphology rather than on the basis of thorough taxonomic revisions. Moreover, many of these taxa are undescribed and very many more await discovery. We estimate that as many as 10,000 species of vascular plants probably remain to be described in the Neotropics alone (P. Raven, personal communication; Gentry, 1982a). Almost certainly a disproportionate number of these taxa will be local endemics, which potentially alters the analysis of endemism patterns presented here. However, the addition to the data base of 10,000 still-to-be-discovered neotropical endemics can only strengthen my basic conclusion. Not only do tropical forest communities have a much greater diversity of species than do temperate ones, but they also have many more locally endemic species and greater habitat specialization (β-diversity; see Cody, Chapter 7) as well. The first part of this conclusion has long been accepted, but the high degree of local endemism in tropical forests is only now beginning to become apparent.

Temperate zone endemism patterns

Endemism is unevenly distributed in the temperate zone. Therefore, the comparison of "temperate" versus "tropical" endemism is exceedingly complex. In order to set the stage, I will first review some of the temperate zone patterns. Figure 1 summarizes endemism in the United States by state. Although many endemic species whose distri-

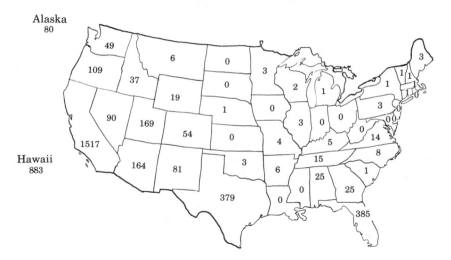

FIGURE 1. Number of endemic plant species by state.[2]

[2] Data mostly from a computer list of state endemics extracted by L. Morse from the Federal Register of December 15, 1980 and November 28, 1983, "Endangered and threatened wildlife and plants: Review of plant taxa for listing as endangered or threatened species." The original list includes only those species proposed for endangered or threatened status and thus underestimates the number of endemics in many states; in other cases, taxonomically questionable decisions are involved and too many species are listed. The following states' endemics were taken from the indicated sources rather than from the list prepared by Morse. Florida: James, 1961; Georgia: Harper, 1946; Alabama, Mississippi, Louisiana: S. McDaniel, personal communication (Alabama may have as many as 30); California: Raven and Axelrod, 1978; Texas: Correll and Johnston, 1970; New Mexico: New Mexico Native Plant Protection Advisory Committee, 1984; Missouri: Steyermark, 1963; Wyoming: Dorn, 1977; Maryland: L. Morse, personal communication; Alaska: Hulten, 1968; Hawaii: Wagner and Gagné, in press.

bution area happens to span a state boundary are not indicated in this figure, the general pattern is clear. By far the greatest concentration of locally endemic species is in the Mediterranean habitats of California, a well known and much-discussed pattern (Cody, Chapter 7; Stebbins and Major, 1965; Raven and Axelrod, 1978). In contrast local endemism is virtually nonexistent in most of north central and northeastern United States. Even an area like the Ozarks, noted for its relict and endemic species, has only 38 endemic species plus a few varieties and subspecies (Steyermark, 1963), all but three found in more than one state. The even more famous Appalachian endemics constitute at most a mere 188 species (Harper, 1947), and that number should be considerably reduced (Harper, 1948). Only 87 of the species on White's (1982) list of the plants of Great Smoky Mountains National Park were considered to be Appalachian endemics, and only 59 are strictly endemic to the southern Appalachians.

There is a conspicuous but often overlooked second center of endemism in temperate North America, in the southeastern coastal plain. This area of high local endemism is best developed in Florida but extends across the border into Georgia and Alabama. It is remarkable that Florida, only 152,000 km^2 and with virtually no topographic relief, should rank second only to California (and Hawaii) in number of endemic species; it is even more remarkable when we consider that the endemic plant species are concentrated in northern and central Florida, not the subtropical southern part. There are only about 120 endemic species in the subtropical southern tip of Florida and the Florida Keys (James, 1961; Long and Lakela, 1971: 9 percent endemism in a notably depauperate tropical Florida flora of approximately 1280 native species) compared with 189 in the Lake District of central Florida and 73 in northern Florida (R. Howard, unpublished data, cited by James, 1961). It is probable that these endemics represent a combination of relicts surviving from the Pleistocene glaciations (for example, *Illicium parviflorum* Michx. ex Vent.; *Taxus floridana* Nutt.; *Torreya taxifolia* Arn.) and of recent speciation on various types of specialized sandy habitats (James, 1961).

In addition to the concentrations of endemic species in the Mediterranean-climate California floristic province and in the southeast coastal plain, there are significant numbers of locally endemic species in Texas and scattered throughout the Great Basin and the Southwestern deserts. The Intermountain region as a whole (a Texas-size area of 694,000 km^2) has at least 215 locally endemic species plus 58 additional varieties and subspecies (Cronquist et al., 1972). Farther north there are progressively fewer local endemics; for example, the giant state of Alaska has only 80 endemic plant species (Hulten, 1968).

European floristic diversity (and endemism) is also concentrated in the Mediterranean countries and is very low in the north (Webb, 1978). The vascular flora of the entire British Isles is only 1443 species, and only 17 of these are endemic (excluding apomictic microspecies: Dandy, 1958, quoted by Raven and Axelrod, 1978). The Mediterranean-climate South African Cape region is extremely species-rich and has an incredible concentration of locally endemic plants (Cody, Chapter 7; Goldblatt, 1978; Bond and Goldblatt, 1984; Oliver et al., 1983). More southerly Tierra del Fuego, which does not have a Mediterranean climate, includes only 15 endemics in its flora of 417 native species (Moore, 1983), but the nearby area of Chile, with a Mediterranean climate, has a rich flora full of localized endemics.

In an important survey of endemism in California, Stebbins and Major (1965) emphasized that the state's high endemism results from several different causes. Not only are there many paleoendemics, left behind by extinction of their relatives and deterioration of formerly more widespread favorable habitats, but also many neoendemics, recently evolved species that have yet to attain their maximum potential distribution. Raven and Axelrod (1978) emphasize the importance of edaphic specialization, e.g. on serpentine soils, in giving rise to the latter. Stebbins and Major (1965) emphasize the roles of climatic change, hybridization, and polyploidy in speciation. They conclude that climatic and edaphic diversity, occurring on ecotones or border regions between different biotic provinces are factors which most actively promote the speciation of higher plants.

TROPICAL ENDEMISM PATTERNS

Tropical community structure and diversity

Although tropical forests are widely thought to be much more species-rich at the community level than most temperate zone ecosystems, there are very few comparative data available. To set the stage for a comparison of tropical and temperate endemism, a brief summary of some relevent aspects of tropical plant community composition and diversity is appropriate.

Tropical plant communities are extremely diverse. In Amazonian Peru a single hectare sample of forest with 600 individuals $\geq$ 10 cm diameter may include as many as 300 species (Gentry, 1986); in western Ecuador, 365 species of vascular plants (including herbs) in 0.1 ha (Gentry and Dodson, 1985). Moreover, wet tropical forests, at least in the New World, are massively richer than any other vegetation in the world in herbs and shrubs, as well as in lianas and trees (Gentry and Dodson, 1986).

Despite the difficulties engendered by so many species and inadequate taxonomic resolution, some clear patterns are beginning to emerge. Strong circumstantial evidence is accumulating that, unlike temperate zone forests, tropical forests, at least in the Neotropics, have predictable floristic compositions and diversities, probably maintained at environmentally determined equilibria (reviewed in Gentry, 1982b, 1986). Some of the predictable trends are clearly relevent to endemism. For example, mammal dispersal of fruit, which is generally correlated with local endemism (Gentry, 1983b), is more prevalent in wetter forests, suggesting that the protection of wet forests is a good strategy to preserve more endemics.

Of course, even better conservation planning would result from actual data on the distribution of endemism. Perhaps even more than in the temperate zone, generalizations about "the tropics" are dangerous, and patterns of endemism are no exception. It is now clear that several contrasting patterns of endemism predominate in different parts of the tropics.

Island endemism

Tropical islands constitute a special case. Many tropical islands are famous for their endemism. Hawaii's high endemism has often been noted (e.g. Carlquist, 1974), even though the revised Hawaiian Flora now in preparation will substantially reduce the number of species recognized. Other large, isolated islands also tend to have high percentages of endemic species (Table 1). On the other hand, the Galapagos and Puerto Rico—smaller, relatively drier, and with little altitudinal relief—have relatively low endemism.

Since islands are isolated, it is hardly surprising that they have high rates of endemism. Moreover, their small sizes almost automatically give them a high concentration of endemics per unit area. But it is perhaps less appreciated how exceedingly localized on any given island many of its endemic species are. Thus many island endemics are restricted to single mountain peaks, or unusual rock outcrops. For example, 19 of the 22 endemic Hispaniolan species of Bignoniaceae are known only from one small area or from a few small outcrops of serpentine.

Puerto Rico provides numerous well-documented examples of endemism (R. Woodbury, personal communication; Woodbury and Figueroa, 1985). One of the best is *Crescentia portoricensis,* a taxonomically isolated shrubby tree endemic to serpentine outcrops in the Maricao-Sosua area and known only from four individuals plus two

TABLE 1. Endemism on some islands.

Island	Area (km^2)	Genera	Endemic genera	Species	Endemic species	Percentage endemism
Cuba[a]	114,914	1308	62	5900[b]	2700	46
Hispaniola[a]	77,914	1281	35	5000[b]	1800	36
Jamaica[a]	10,991	1150	4	3247[b]	735	23
Puerto Rico[a]	8,897	885	2	2809[b]	332	12
Galapagos[c]	7,900	250	7	701	175	25
Hawaii[d]	16,600	253	31	970[e]	883	91
New Zealand[c]	268,000	393	39	1996	1618	81
New Caledonia[f]	17,000	787	108	3256	2474	76

[a] From Liogier, 1981.
[b] Includes adventives.
[c] From Raven and Axelrod, 1978.
[d] W. Wagner, personal communication; Wagner and Gagné, in press.
[e] Note that this value is significantly lower than older estimates of the number of species in Hawaii.
[f] From Morat et al., 1984.

small, recently discovered prereproductive populations. R. Woodbury (personal communication) estimates that a new species of *Calyptranthes* from El Yunque also has a population of about four trees, while *Calyptranthes luquillensis* Alain from the Luquillo Mountains and *Auerodendron pauciflorum* Alain from Quebradillas have about five individuals each. There are 37 species known only from the Maricao Forest Reserve, over 42 from the Luquillo Mountains, 10 from the Guanica Reserve, and 9 from Quebradillas—all areas of a few hundred km^2 or less. Examples from many other genera could be cited.

Hawaii is especially famous for its endemic species. Although modern taxonomic concepts considerably reduce the number of Hawaiian flowering plant species, perhaps to 970 species, the uniquely high (91 percent) endemism rate remains almost the same (W. Wagner, personal communication; Table 1). In Hawaii, unlike Puerto Rico, many plant species are known to have gone extinct; in fact, over 10 percent of the native flora is thought to be extinct (Wagner et al., 1985). Many other species hover on the brink of extinction or are reduced to very few individuals. The whole genus *Hibiscadelphus* may soon be extinct: of its six species, only a grand total of 14 individuals are known to survive in the wild; one of these species, only recently described, is known from a single individual (Hobdy, 1984). Two other species are already extinct and another, extinct in nature, persists as two cultivated seedlings (Wagner et al., 1985). For Hawaii in general, the older and more dissected an island, the more locally endemic are (or were) its species (Wagner et al., 1985).

Almost certainly such examples stem in part from the near extinction of formerly more populous species in notoriously fragile island ecosystems. However, other factors are surely operating as well. Even on a single island, plants of different habitats may show very different patterns of endemism. For example, the upland forests of Sri Lanka are rich in endemics, while that island's lowland forests have little endemism due to its recent land connection with India. In Puerto Rico habitat destruction, which has been largely reversed with the virtual abandonment of agricultural enterprise on the heavily subsidized island, may not be entirely responsible for the reduced numbers of individuals in many of the endemic species. Almost all of the local endemics listed above are currently in well-protected forests, yet they show no signs of increasing their populations. Some obvious relict species, such as *Crescentia portoricensis* (Figure 2D) and *Goetzea elegans,* are virtual living fossils, and we may actually be witnessing natural extinction in such cases.

At the other extreme, some of the locally endemic species on islands are extremely abundant. A good example is *Tabebuia revoluta* Urb., one of the commonest understory species of the pine forests of the Jarabacoa region of La Vega Province, Dominican Republic, the only region in which it occurs.

Relict endemics

There are perhaps three main patterns of local endemism in continental tropical America. The first pattern is that of relict species. Unlike the parallel temperate zone situation, the paleobotanic data base generally does not document that such species were once more widespread, but this is probably a safe assumption in many cases. Examples include *Styloceras brokawii* (Gentry and Foster, 1981), a "missing link" species of a putatively old and archaic genus that is locally exceedingly common in the Manu Park area of Madre de Dios, Peru; and *Freziera forerorum* A. Gentry (Gentry, 1978), an isolated species of an ancient family that may have the most asymmetric leaf base of any angiosperm and is restricted to the extreme summit of Cerro Tacarcuna on the Panama–Colombia border. One endemic neotropical species whose relict nature is established by a long fossil record is *Pelliciera rhizophorae* (Figure 2A; Gentry, 1982c; Jimenez, 1984).

The Guayana Highlands are especially famous for relict endemics (Maguire, 1970; Steyermark, 1979). There are 79 plant genera endemic to the upper parts of this region's tepuis, with 13 genera endemic

FIGURE 2. Endemic neotropical plant species. (A) A paleoendemic species, *Pelliciera rhizophorae,* widespread during the Cenozoic but reduced today to a few small enclaves. (B) An anthropogenic endemic, *Dicliptera dodsonii,* known only from this single plant in the last remnant of Ecuadorian coastal wet forest at Rio Palenque. (C–D) Naturally restricted island endemics. (C) *Tabebuia rigida,* the most common tree on the highest peaks of the Luquillo Mountains in Puerto Rico, but found nowhere else. (D) *Crescentia portoricensis,* a relict species endemic to serpentine outcrops near Maricao, Puerto Rico, and known from only four mature plants. (E–F) Neoendemics of cloud forest habitat islands. (E) *Palicourea tuminodosa,* endemic to Cerro Tacarcuna on the Panama–Colombia border. (F) *Gasteranthus atratus,* endemic to the now-destroyed Centinela forest of western Ecuador. The black leaves are the natural color—an unusual trait shared with several other now apparently extinct species endemic to the same ridgetop.

to Cerro Neblina alone (Steyermark, 1979; Gentry, 1985b). There is even an endemic family, Saccifoliaceae (Maguire and Pires, 1978).

There are many similar locally endemic representatives of putatively Old World genera in Central America, many of them recently discovered (Gentry 1982c, 1983a; Robbrecht, 1982). Even more striking is *Meliosma alba,* the only neotropical representative of its otherwise Asian subgenus (Gentry, 1980). Most such examples surely represent relicts of a once more widespread tropical Laurasian flora (Raven and Axelrod, 1974). Stebbins (1974) emphasized the importance of such relict taxa in tropical forests, which he regarded as museums more than active centers of speciation.

Anthropogenic endemics

Another kind of relict endemic is becoming increasingly prevalent in the Neotropics. These are the species whose highly restricted distributions are largely due to human intervention and habitat destruction. In extreme cases, such as *Dicliptera dodsonii* Wasshausen (Figure 2B) of the Rio Palenque Biological Station in western Ecuador (see Figure 3), they are reduced to single individuals. Many of the island endemics mentioned above probably fall into this category. Western Ecuador provides many especially well documented examples of this kind of now extremely localized species "pseudoendemic" to the last small patches of appropriate habitat (Gentry, 1977, 1979). For example, almost 100 new plant species have been described from the Rio Palenque Field Station in Los Rios Province in the process of preparing a local florula (Dodson and Gentry, 1978). Most of these are now known only from the less than 1 km^2 field station forest, the last remnant of the narrow strip of wet forest that once paralleled the base of the Andes in coastal Ecuador. However, we assume that such species, though now mostly endemic to western Ecuador, were much more widespread before their habitat was destroyed. An especially striking example is *Persea theobromifolia* A. Gentry (Figure 4), formerly the most important timber tree of the region, which is now reduced to a population of less than a dozen mature trees at Rio Palenque (Gentry, 1979).

Equally striking examples come from Jauneche (see Figure 3), the last intact patch of coastal moist forest in central or southern Ecuador (Dodson et al., 1985), where such highly distinctive species as *Aspidosperma jaunechense* A. Gentry, *Annona hystricoides* A. Gentry, and *Inga jaunechensis* A. Gentry are now restricted to remnant populations of perhaps less than a dozen individuals each. Other species, described long ago but collected in recent years only at Jauneche, prove the artifactual nature of these locally endemic taxa. For exam-

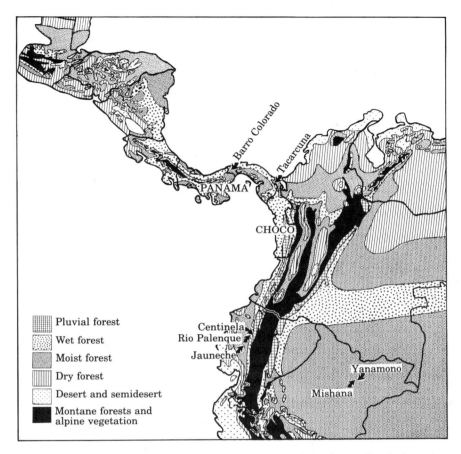

FIGURE 3. Ecological map of Central America and northwest South America showing localities emphasized in the text. Choco Department, Colombia, is outlined in black, as are international borders. (From Gentry, 1978.)

ple, *Duguetia peruviana* R. E. Fr., the second most common species in the Jauneche forest (36 trees larger than 2.5 cm dbh in 0.1 ha) was described from an eighteenth century collection made by Ruiz and Pavon in "Perou" and never recollected in the next century and a half, though it must once have been fairly widespread in the Ecuadorian coastal moist forest.

Neoendemics in the tropics

Contrary to Stebbins (1974), very active local speciation appears to have taken place in many parts of the tropics. Two very different

FIGURE 4. An economically important endemic. *Persea theobromifolia,* related to the cultivated avocado and endemic to the coastal wet forest of western Ecuador, was formerly the most important timber tree of the region but was not scientifically "discovered" until 1977, by which time the only extant population was reduced to less than a dozen mature trees in a forest remnant of less than 1 km². Details of floral structure (lower right) indicate that this species belongs to the segregate Asian genus *Caryodaphnopsis*, previously unknown from the Neotropics. (From Van der Werff and Richter, 1985.)

models of speciation seem to have been involved, but the result in both cases is a high degree of local neoendemism. In Amazonia as well as in tropical Asia (P. Ashton, personal communication) local endemism in plants typically involves specialization for marginal habitats. In montane regions, such as Central America, the Afromontane region (J. Lovett, personal communication), New Guinea, and especially the Andes, local endemism seems to result mostly from a veritable explosion of speciation in relatively few taxa, mostly of shrubs, herbs, and epiphytes. These groups constitute almost half of the neotropical flora, and they account for most of the excess floristic diversity of the Neotropics as compared with the Paleotropics (Gentry, 1982a). Both of these models (habitat specialization and explosive, recent speciation in montane areas) contrast greatly with the Pleistocene refugia speciation model now in vogue. The implications for conservation also differ.

Local endemism in Amazonia

According to the conventional wisdom, local endemism in Amazonia is largely derived from the fragmentation of the flora during arid periods in the Pleistocene (see Prance, 1982 and references therein). Here I will argue for an alternative model of Amazonian speciation, one based on edaphic specialization.

Geographic restriction to habitat islands is prevalent in Amazonia. Most of the taxa involved are canopy trees and lianas, with derivative species in specialized habitats such as white sand catingas and campinas or seasonally inundated varzea or swamp forest. In some cases, such as *Phryganocydia* (Gentry, 1983b), it is clear that the locally endemic habitat specialists are derived from wide-ranging ancestors of the terra firme forest.

Habitat specialization has clearly been the prevalent evolutionary theme in giving rise to local endemics in some Amazonian families. For example, there are 67 species of Bignoniaceae in Amazonia that are restricted to only one of the phytogeographical regions proposed by Prance (1977). Of these, only 24 species occur in the widespread terra firme forest, whereas 37 occur only in specialized habitats (Table 2). If only the 48 most locally endemic Amazonian bignons are considered—i.e., those species with total ranges of only a few hundred to a few thousand km^2—habitat endemics are even more prevalent. Only a quarter of these locally endemic species of Amazonian Bignoniaceae with known habitat preference occur in terra firme forest. For Latin America as a whole, 65 percent of the locally endemic species of Bignoniaceae are habitat specialists (excluding *Amphitecna*—see below—and the poorly known Cuban species of *Tabebuia*). Prance

TABLE 2. Habitats of neotropical Bignoniaceae with restricted distributions.[a]

Habitat type	All species	Amazonian species	Locally endemic Amazonian species
Terra firme forest	79	24	12
Cloud forest	13	2	2
White sand savannas and campinarana	19	18	17
Riverine forest	5	5	2
Limestone outcrops	6	0	0
Swamps	9	1	0
Dry forest	70	4	4
Other	24	7	6
Not known	24	6	5
Totals	247	67	48

[a] Includes all species with ranges restricted to a single phytogeographical region or equivalent except Cuban species and *Amphitecna*. For the West Indies, species occurring on two or more islands are excluded.

(1982) reports a similar situation for Chrysobalanaceae, especially in the Guiana–Venezuela region where only 8 of 31 locally endemic species occur in terra firme forest.

Even when local endemism is not involved, habitat specialization is one of the main reasons there are so many species of plants in Amazonia. The *Passiflora vitifolia* complex provides a good example of how habitat specialization increases species diversity (Gentry, 1981a). In the Iquitos area there are four species, one restricted to terra firme forest, one to seasonally inundated forest, one to white sand substrates, and one to forests on richer, noninundated alluvial soils; outside of Amazonia, only a single ecological generalist occurs in any given area.

The taxonomic data base is generally inadequate to determine the overall correlation of habitat specialization and local endemism in Amazonia; few families are as well known taxonomically as Bignoniaceae and Chrysobalanaceae. However, new data do document the extent to which habitat specialization contributes to high Amazonian species diversity (Gentry, 1981b, 1985, and in preparation). In a series of samples 0.1 ha in size from forests on different substrates near Iquitos (Table 3), the diversity of plants more than 2.5 cm dbh ranged from 163 species to 249. Overlap in species between these different plant communities was extremely low, with about 20 or fewer shared species between adjacent but different habitats. The only two habitats to share a significant number of species were the two adjacent sites on white sand soil at Mishana on the Rio Nanay. Even here, the hilltop site and lowland floodplain site shared only 55 species.

I believe that local endemism in Amazonian plants generally results from edaphic specialization, as it does in Bignoniaceae and Chrysobalanaceae. This cannot be proved until a much better taxonomic data base is developed. We do know that many new, and presumably locally endemic, species have been discovered on the specialized substrates near Iquitos, especially in the white sand forests. Many of the unidentified species from these areas that cannot be matched in the herbarium are also likely to be undescribed local endemics. Moreover, habitat specialization is not specific to plants and is almost certainly important in determining many localized distributions of animal species as well (Kinzey and Gentry, 1979; Remson and Parker, 1983).

The habitat specialization model proposed here has very different implications for conservation than does an alternative explanation— that local endemism in Amazonia largely derives from the effects of Pleistocene forest refugia (Haffer, 1969; Prance, 1973, 1982 and included papers). According to the Pleistocene refugia model, local endemics should be concentrated in the high rainfall areas that putatively constituted isolated forest islands during the dry periods of glacial advance. If so, as many of the refugial centers as possible should be protected by reserves (Wetterberg, 1976). As I have shown,

TABLE 3. Number of species shared by 1000 m^2 samples of Iquitos area forest types.

	Yanamono No. 1	Yanamono No. 2	Yanamono tahuampa	Mishana lowland	Mishana campinarana	Mishana tahuampa
Yanamono:						
Terra firme No. 1	212	91	20	24	12	14
Terra firme No. 2		230	20–21	19	9	8
White water tahuampa			163	9	5	ca.19
Mishana:						
Lowland noninundated				249	55	17
Campinarana (white sand)					196	3
Black water tahuampa						168

however, many endemic Amazonian plant species are edaphic specialists, widely scattered in pockets of suitable substrate, not concentrated in high rainfall areas. Nor is it clear to what degree the origin of endemics in terra firme forest relates to the postulated refugial phenomena. The hypothetical refugia, therefore, may be of less conservational significance than has commonly been supposed. Much of the evidence for the refugial hypothesis can be interpreted equally well as support for alternate models based on present day ecology and barriers to gene flow (Endler, 1982; Beven et al., 1984).

The implications for conservation of the Pleistocene refugia model and the habitat diversity model proposed here are very different. We are apparently denied the easy solution of automatically conserving endemic species of many unrelated taxa by conserving a single carefully chosen refugial area. I conclude from the above that, given the serious limitations of the current taxonomic data base, a better conservational approach for preserving locally endemic plants, especially in Amazonia, would be to focus on preserving areas of high habitat diversity, perhaps as indicated by such potential indicator groups as Chrysobalanaceae or Bignoniaceae, instead of concentrating on preservation of the hypothesized refugial centers. Even if other nonbotanical criteria dictate that few large reserves be given priority over many small ones, a series of supplementary reserves in different habitat types would be an extremely high priority from the botanical viewpoint.

Endemism in the northern Andes and Central America

Endemism in the northwestern South American Andean region presents a very different picture. An analysis based on tabulating distributions from a data set of 8117 recently monographed neotropical plant species revealed that the great majority (71 percent) fall into two well-defined habit-phytogeographical categories (Gentry, 1982a): (1) taxa that are predominantly canopy trees and lianas and that, without exception, have Amazonian distributional and diversity centers; or (2) taxa that are predominantly epiphytes, shrubs, or herbs and have fundamentally extra-Amazonian diversity centers, with species concentrated in the northern Andean region and often also in southern Central America. Local endemism is very high in the latter "Andean-centered" groups, and the individual genera tend to be larger than in the Amazonia-centered canopy taxa.

I interpret this concentration of species, mostly along the moist lower slopes of the Andes and to a lesser extent in the cloud forests of southern Central America, as due to very active speciation, apparently somehow related to the peculiarities of the broken terrain and/or the complex juxtaposition of different vegetation types. Quite probably the exceedingly dynamic, even explosive, evolution of these taxa is an accident of the Andean orogeny. In such groups there is no evidence for any kind of ecological equilibrium or limits on species diversity (Hubbell and Foster, Chapter 10). Additional discussion of these points and of other neotropical phytogeographical patterns may be found in Gentry (1982a).

Most of the data documenting this extraordinary endemism comes from three western Ecuador sites (Figure 3): a wet forest remnant at the Rio Palenque field station (Dodson and Gentry, 1978); the moist forest Jauneche Reserve (Dodson et al., 1985); and the now destroyed cloud forest of the 400–600 m high Centinela ridge east of Rio Palenque (Dodson and Gentry, in preparation). Other localities for which there are data include Colombia's Choco Department (Forero and Gentry, in preparation; see also Gentry, 1982c), and three Panamanian localities, including the moist forest on Barro Colorado Island (Croat, 1978; Hubbell and Foster, Chapter 10); Cerro Tacarcuna and the adjacent upper part of the Serrania del Darien on the Panama–Colombia border (Gentry, 1985a and in preparation); and Panama as a whole (W. D'Arcy, personal communication).

One way to compare endemism is to compare distributional patterns of similarly selected subsets of the species of different regions. Local florulas provide a convenient data set for this type of analysis. Only three local florulas are available: Barro Colorado Island, Rio Palenque, and Jauneche, all written in the last few years. The distributional patterns of the individual plant species of these florulas are tabulated in Table 4. At all three sites, widespread species predominate in the floras. Nevertheless there are many species with geographically restricted ranges at each site. Endemism is much higher at the wet forest Rio Palenque site. The other two sites, both moist forest, have similar numbers of endemic species, but the more geographically isolated Ecuadorian site has more endemism than does the Panamanian one. All three tropical values for regional endemism (7–20 percent) are in the same range as those in the regions of North America where endemism is greatest (12 percent for the Santa Cruz Mountains, probably well over 10 percent for the Santa Barbara region), but much lower than in the Mediterranean climate area of South Africa (compare the first line of Table 4 with the "endemic to area" column of Table 9.).

At a different scale, exteme local endemism, here meaning distributions naturally restricted to a few hundred km^2, is definitely present in some tropical areas. This has never been well documented, however. Although many species with restricted ranges are known, the collection data base and limits of taxonomic resolution are so inadequate

TABLE 4. Distributional patterns of naturally occurring plant species in local florulas.

Distributional pattern	Barro Colorado[a] Panama No. spp.	%	Jauneche[b] Ecuador No. spp.	%	Rio Palenque[c] Ecuador No. spp.	%
Locally endemic (may have ranges up to 75,000 km^2)	92[d]	7	85[e]	14	172[e]	20
Wide endemic (distributional area ca. 100,000–200,000 km^2)	122[f]	9	8[g]	1	51[g]	6
Regional	180[h]	14	63[i]	11	154[i]	18
Widespread on continent (Central America to Amazonia but not West Indies)	473[j]	36	111	19	167	19
Pan-American (Central America and West Indies to Amazonia)	436	33	279	47	271	31
Trans-Andean disjuncts (Western Ecuador and Amazonia but not reaching Panama)	—	—	16	3	35	4
Central America and West Indies to coastal Ecuador (but not in Amazonia)	—	—	10	2	20	2
More or less widespread in South America and/or West Indies but not reaching Central America	—	—	5[k]	1	—	—
Disjunct between northern Colombia/ Venezuela and coastal Ecuador	—	—	7	1	—	—
Other	—	—	4[l]	1	—	—
Total species analyzed	1316		589		870	

[a] From Croat, 1978.
[b] From Dodson et al., 1985.
[c] From Gentry, 1982c.
[d] Endemic to Panama (52,000 km^2).
[e] Endemic to western Ecuador (ca. 64,000 km^2).
[f] Endemic to Panama plus Costa Rica, or to Panama plus northern Colombia.
[g] Endemic to coastal Ecuador and adjacent Colombia.
[h] Central America; mostly Mexico to Panama.

[i] Central America to western Ecuador (but not Amazonia).
[j] Croat's Type IV plus non-West Indian portion of Type V.
[k] *Trixis antimenorrhea, Cayaponia cruegeri, Gurania spinulosa, Guapira olfersiana, Cuphea racemosa.*
[l] Three more or less Andean species (*Cuphea strigulosa, Stenomeria pentalepis, Phyodina gracilis*) and one species disjunct from Paraguay and Bolivia (*Cydista decora*).

that it is rarely possible to be sure that a species does not also occur elsewhere. Moreover, with habitat destruction so rampant, most distributions restricted to a habitat island like Rio Palenque are surely artifactual. Two of the data sets available for analysis do represent naturally delimited small areas and are appropriate for comparison with the much more thorough data base for temperate zone areas with high local endemism. These two data sets are from Centinela, Ecuador, and from Cerro Tacarcuna, on the Panama–Colombia border, both of which are isolated cloud forests (Table 5).

Cerro Tacarcuna is the highest mountain (1900 m) in the 600 km of lowlands that separate the Central American Costa Rica-Chiriqui

TABLE 5. Endemism in some isolated cloud forests.

Distributional pattern	Centinela[a]		Tacarcuna[b]	
	No. spp.	%	No. spp.[c]	%
Extremely endemic (range mostly 5–10 km^2)	38+ (possibly up to 90)[d]	10	71	24
Locally endemic (includes above)	65+ (possibly up to 114)[e]	12	71	24
Species shared with Rio Palenque	465	50	—	—
Disjunct from Chiriqui–Costa Rica uplands	—	—	44	15
Shared with montane South America	—	—	29	10
Shared with montane South and Central America	—	—	17	6
Also in lowlands	—	—	127	43
Disjunct from Cerro Jefe	—	—	4	1
Totals	925[f]		292[g]	

[a] Centinela cloud forest in western Ecuador (400–600 m).
[b] Cerro Tacarcuna montane forest, Panama–Colombia border (1400–1900 m); Gentry 1985a and in preparation.
[c] Excluding ferns.
[d] The first value is for species described or ready for description; second value includes all unidentified taxa that have not been matched in the Missouri Botanical Garden Herbarium.
[e] Includes species also known from Tenafuerte (35 km south of Centinela), from Rio Palenque (8 km west), and from around Santo Domingo.
[f] Includes indeterminate morphospecies.
[g] Excludes 26 species with inadequate identification.

mountains from the western Cordillera of the Andes. The vegetation above 1500 m on the main massif is a montane oak forest, sharply differentiated floristically from the surrounding lowland forest. Although floristic knowledge of this inaccessible region is incomplete, enough of the plants we collected have been identified to compile the distributional data presented in Table 5. No less than 24 percent of the montane forest angiosperms seem to be strictly endemic to the top of Cerro Tacarcuna and the adjacent ridge. I think that this is a typical value for local endemism in isolated cloud forests in southern Central America and western South America. For example, Cerro Pirre, south of Cerro Tacarcuna across the Tuira Valley, also has very many endemic species in the cloud forest on its summit (e.g., 15 endemic aroids; T. Croat, personal communication), but almost none of these are shared with Cerro Tacarcuna.

My second example of high local endemism in neotropical cloud forests comes from western Ecuador. Centinela, the first Andean foothill ridge, about 8 km east of Rio Palenque, is isolated from the main Andean range farther east by a broad valley about 15 km wide. At its north end it merges with the flat but higher altitude terrain formed by the less eroded outwash from the Andes of the Santo Domingo area. The Centinela ridgetop, now deforested, is only 600 m high at its highest point and thus only 300 m higher than the valley that separates it from the Andean foothills proper. Nevertheless it is high enough to engender a local cloud forest effect and at one time had a dramatically different flora than adjacent regions. There were at least 38 species of plants endemic to this unprepossessing ridge and, depending on the outcome of taxonomic decisions in taxonomically difficult groups, possibly as many as 90 species, or 10 percent of the entire known flora of the ridge. The entire ridge is only about 20 km long and 1 km wide, an area of some 20 km^2. However, all but a few of the endemic species were concentrated at the south end of the ridge, in an area of only 5 to 10 km^2. Even in this small area, there were tremendous differences from place to place along the ridgetop—in some patches the trailside trees were full of endemic species of epiphytes, whereas a similar-looking adjacent patch may have had only common widespread species. Furthermore, different patches of high endemism had different sets of species.

Nor were the species of this ridgetop obscure taxonomic segregates. Many of them were both strikingly beautiful and quite unlike any known congener. One noteworthy feature shared by many species of quite unrelated families is that the upper surfaces of their leaves are black, and usually intricately bullate as well. How did so many distinctive species come to exist at Centinela? *Gasteranthus* of the Gesneriaceae (Figure 2F) provides an instructive example. This small genus of about 25 species ranges from Guatemala to Peru, mostly in cloud forests, generally with a few endemic species in each country in which it occurs. In Panama, for example, there are five species—two endemic and three shared with Costa Rica (Skog, 1978). Many *Gasteranthus* species have rather similar-looking orange flowers. On Centinela there were no fewer than six species—a quarter of the world total! Four of these species were restricted to the high-endemism area near the south tip of the ridge, a fifth is also known from a single collection from north of Santo Domingo, and the sixth occurs also at Rio Palenque. Moreover, most of these species were dramatically different from each other and from their congeners. Indeed, *Gasteranthus* showed as much morphological diversity in the few square kilometers of Centinela as in the entire remainder of its range. It seems inescapable that the endemic species of Centinela represent an amazing level of in situ evolution.

Nor do Centinela and Cerro Tacarcuna seem to represent unusual situations. *Anthurium* is a good example of a genus with extensive very local endemism. Almost every isolated cloud forest in Panama has locally endemic species. There are 8 *Anthurium* species endemic to Cerro Jefe, 8 to Cerro Pirre, 3 to Cerro Sapo, 4 to Cerro Tacarcuna, and 7 endemic to the Cocle Province forests (Croat, 1985). The Bignoniaceae genus *Amphitecna* is another that shows an extremely high rate of local speciation and endemism, in this case mostly confined to Central America (Gentry, 1983b). Most of the 18 species are found only in one or two adjacent countries; one-third of the species of its whole mammal-dispersed tribe Crescentieae are known only from single localities, mostly isolated cloud forests. We can safely conclude that extremely local endemism is a prominent characteristic of many neotropical cloud forests.

For larger tropical areas (around the order of magnitude of 50,000 km^2) endemism also tends to be high. Table 6 tabulates the data for such locally endemic species in the largest genera on a list of 3553 plant species currently known from Choco Department, Colombia (Forero and Gentry, in preparation), mostly from the lowlands below 500 m. For the Choco as a whole, endemism is currently tabulated as 20 percent, and for some genera (such as *Anthurium)* it reaches more than 70 percent. It should be noted that many of the included species are undescribed; for example, only a dozen of the 90 *Anthurium* morphospecies have names. Since every collecting trip into this area brings back many new species, it is obvious that these figures are very tentative.

Because of the incompleteness of the Choco data base, I have compiled some similar figures for Panama, where the long history of plant collecting and identification associated with the Flora of Panama proj-

TABLE 6. Endemism in the largest genera in Choco Department, Colombia[a]

Genus	Number of species	Number of Choco endemics[b]	Percentage endemism
Anthurium	90	65	72
Psychotria (incl. *Cephaelis*)	90	15	17
Piper	87	37	43
Miconia	65	25	38
Peperomia	44	13	30
Solanum	43	4	9
Thelypteris	41 (+2 var.)	2	5
Columnea	37	27	73
Cavendishia	33	17	52
Clidemia	32	12	38

[a] From Forero and Gentry, in preparation. Data recorded for 3553 species.
[b] A Choco endemic is defined as occurring only in coastal Colombia and the adjacent wet forest of Ecuador, neither crossing the Panamanian border nor reaching Magdalena or Cauca Valleys.

ect makes the flora perhaps the best known in the continental Neotropics. The Panamanian data (Table 7) come from a computerized data base currently being compiled by W. D'Arcy (personal communication). Overall, Panamanian endemism is slightly lower than the Choco value, but this value would be much higher if phytogeographically similar Costa Rica and Panama were combined as a single geographic entity. The available Panamanian data seriously underestimate endemism in many taxa, since only those taxa specifically noted by an author as endemic are so recorded. I have corrected some of these, but only for species I happened to know to be endemic. It is likely that the final figure will be closer to 20 percent. In Panama, as in Choco, endemism at the species level in large genera like *Anthurium* can exceed 50 percent. In *Ardisia* no less than 82 percent of the 105 Panamanian species are endemic.

One attempt has been made to compare neotropical endemism and species diversity at an even larger scale, by phytogeographic region. Figure 5 summarizes the distribution of species and regional endemism in the Neotropics, based on a sample of 8117 recently monographed species. The data on endemism are separately tabulated for canopy taxa (trees and lianas) and for epiphytes and palmettos. Tree

TABLE 7. Endemism in the largest genera in Panama.[a]

Genus	Number of species	Number of Panama endemics	Percentage endemism
Anthurium	148	82	55
Pleurothallis	135	ca.43	32
Piper	121	39	32
Psychotria (incl. *Cephaelis*)	119	52	44
Ardisia	105	86	82
Epidendrum	91	3	3
Peperomia	77	25	32
Miconia	67	7	10
Solanum	59	6	10
Maxillaria	54	2	4
Columnea	52	24	46

[a] From Flora of Panama computer bank (W. D'Arcy, personal communication). Ferns are not included in the data base.

and liana taxa are predominantly distributed in Amazonia (44 percent of all their species, compared to 16 percent or less in any other region). The Amazonian species are also overwhelmingly endemic to that region (80 percent).

The distributional pattern of epiphyte and palmetto taxa, on the other hand, is almost the mirror image of that of predominantly canopy taxa. Most of these taxa are concentrated in the northern Andes (27 percent) and Central America, especially southern Central America (22 percent). Only 11 percent of these species are found in Amazonia. The rate of regional endemism in these taxa is very high almost everywhere (up to 86 percent for coastal Brazil) but the endemic species are concentrated in the Andes and Central America because there are so many more species there. Note also the surprisingly low endemism in the West Indies as compared to mainland phytogeographical regions.

It is apparent from these data that an adequate system of preserves must span the entire Neotropics. A more controversial implication might be that on a unit area basis much greater focus should be placed on the smaller Andean and coastal Brazilian phytogeographical areas. Although these regional data do not necessarily reflect local endemism, they do show that herbs and shrubs, more prone to local ende-

FIGURE 5. Neotropical phytogeographic regions and their floras. T = percentage of total sample of all 8117 recently monographed species that occur in that region. C = percentage of species in that region that are canopy trees and lianas; the values in parentheses are the percentages of the canopy trees and lianas that are endemic to that region. E = percentage of species in that region that are epiphytes and palmettos; the values in parentheses are the percentages of epiphytes and palmettos that are endemic to that region. (From Gentry, 1982a.)

mism than trees and lianas, are concentrated in areas like the Andean foothills. In turn, this suggests that the establishment of more protected areas in this region would be expected to preserve more locally endemic plant species than would a similar focus in epiphyte-poor central Amazonia.

TEMPERATE VERSUS TROPICAL ENDEMISM

The preceding sections document that very high local endemism exists in many parts of the Neotropics and at many different scales. How does this endemism compare with that in temperate regions? Table 8 compares the Panama and Choco data with data for some temperate zone regions. The tropical endemism figures (15–20 percent) are intermediate between those for Mediterranean climate areas, where known regional endemism is higher (30–68 percent) and other temperate areas, where it is much lower (1–9 percent). There also tend to be more species per genus in Mediterranean climate areas than in the known tropical floras. The percentage of a flora constituted by its ten largest genera remains practically constant for all these areas and may not be a useful indicator except in extreme cases such as Hawaii.

From such an analysis one might conclude that local endemism is as high or higher in Mediterranean regions than in the tropics. However, the tropical data base is very incomplete. Especially poorly known are the local endemics, because so much of the area remains botanically unexplored. How many more Centinelas and Rio Palenques are scattered along the base of the Andes? Probably many hundreds. If the botanically unexplored areas are as rich in local endemics as the areas explored to date, an area like the Choco may approach or equal the Mediterranean climate regions in percentage of endemism. More data are urgently needed.

One way to circumvent the data limitations is to compare field station florulas, where intensive collecting in a single area has led to a relatively well known local flora. (Note, however, that almost 100 additional species have been found at Rio Palenque since the flora was published, and at least 15 on Barro Colorado Island; R. Foster, personal communication). Table 9 summarizes endemism data for some temperate zone florulas which may be compared with the tropical data in Tables 4 and 5. All three of the tropical florulas are most similar to the Mediterranean climate ones in degree of local endemism. All three have many more endemic species than do any of the high endemism areas delimited by Stebbins and Major (1965) for the Central Coast range of California. The California endemism centers have between 16 and 44 species endemic to the Central Coast range, plus a few endemic relict species in some cases. The Central Coast range area includes 18,500 km^2, so is probably quite equivalent to the distribution areas of most of the tropical florula species tabulated as endemic in Table 4, because the latter's actual distributions are ecologically limited to a small part of the area of "Panama" or "western Ecuador."

The tropical florulas are clearly much more strongly endemic than are those of any of the California high-endemism areas. Moreover, unlike the California study areas, the tropical florula areas were selected for study because they happened to still have patches of natural forest, not because they were known to have high endemism. Although many Rio Palenque species are now known only from the field station, this is clearly an artifact of habitat destruction, not an indication of

TABLE 8. Number of species and genera and potential speciation indicators for some continental temperate and tropical regions.[a]

Region	Area × 10³ km²	Number of species	Number of genera	Endemic species No.	Endemic species %	Species/ genus ratio	Percentage species in 10 largest genera
Panama[b]	75	ca.6800	ca.1500	ca.1034	15	4.5	14
Choco Department[c]	47	3553[d]	1069	ca.711	20[e]	3.3	16
California	411	5046	878	1517	30	5.7	16
California Floristic Province	324	4452	795	2125	48	5.6	15
Cape Region, S. Africa	90	8578	989	5850	68	5.9	20
Southwest Australia	320	3600	287	ca.2450	68[f]	12.5	?
Texas	751	4196	1075	379	9	3.9	10
Carolinas	217	2995	819	23	1	3.5	15
British Isles	308	1443	545	17	1	2.7	18
Tierra del Fuego[g]	66	417	228	15	3	1.8	24
Europe	10,000	10,500	1340	ca.3500	33	7.8	14
Southern Africa	2573	18,550	1930	ca.14,800	80	7.7	15
Gray's Manual area	3238	4425	849	599	14	5.2	22
Central America and Mexico		25,000–30,000	?	?	60[h]	?	?

[a] Data from Raven and Axelrod, 1978, or Bond and Goldblatt, 1984 except as otherwise noted.
[b] Tentative data from Flora of Panama computer data base, still undergoing revision. (W. D'Arcy, personal communication; additional endemics indicated in some groups.)
[c] From Forero and Gentry, in preparation.
[d] We estimate that this may be approximately half the actual flora.

[e] Estimate from Gentry, 1982c.
[f] Possibly higher; Hopper, 1979, quoted by Bond and Goldblatt, 1984.
[g] From Moore, 1983.
[h] From Gentry, 1982a.

TABLE 9. Endemism among some temperate zone local florulas.

Site	Area (km²)	Number of native species	Species endemic to study area[a] No.	%	Species endemic to region[a] (≤ 50,000 km²) No.	%
Santa Cruz Mts., CA[b]	3600	1246 (incl. var.)	27[c]	1.5	152[d]	12
Santa Barbara region, CA[e]	7300	1390	145	10	?	?
Cape Peninsula, So. Africa[f]	470	2256	157	7	?	ca.90
Southwestern Georgia[g]	13,000	1540	33[h]	2	38	2
Florida Caverns State Park[i]	5	485[j]	7[k]	1	17	4
Pensacola Region (FL/AL)[l]	16,000	1601	10	1	33[m]	3
Baldwin County, AL[l]	4000	1148	0	0	6	1
Santa Rosa County, FL[l]	2700	956	2	0.2	18	2
Great Smoky Mts. Nat. Park[n]	2000	1150	6[o]	1	59[p]	5
Tyson Reserve, MO[q]	8	465	0	0	0	0
Konza Prairie, KS[r]	35	369	0	0	0	0

[a] Endemics of larger geographic area include those of smaller.
[b] From Thomas, 1961.
[c] There are six species endemic to a 700 km² central ellipse from Año Nuevo to Mt. Hermon. (Stebbins and Major, 1965.)
[d] Coast ranges from Humboldt to San Luis Obispo County, thus including two counties from Stebbins and Major's (1965) North Coastal Region.
[e] From Smith, 1976; includes Channel Islands.
[f] From Raven and Axelrod, 1978.
[g] From Thorne, 1974; an area of 17 counties.
[h] Includes localized taxa that cross the border into adjacent Alabama and Florida.
[i] From Mitchell, 1963.

[j] Species treated minus 29 introduced taxa.
[k] No species are endemic to the park, but seven species are highly localized in the vicinity.
[l] From Wilhelm, 1984; area of six counties, four at the tip of the Florida Panhandle and two in adjacent Alabama.
[m] "Narrowly endemic to north-central Gulf coast area." The 30 mapped endemics, plus three species listed as equally endemic.
[n] From White, 1982.
[o] Three species limited to park plus three just crossing border.
[p] Strict southern Appalachian endemics.
[q] R. Coles, personal communication.
[r] From Freeman, 1980.

extremely local endemism. I conclude that there is generally more local endemism in the species of a typical small tropical area than in even the most species-rich small temperate ones, with the possible exception of the South African Cape. Data from genuinely highly endemic tropical areas prove this point. Centinela, Cerro Tacarcuna, and presumably many other cloud forests probably reflect biological reality. Extremely local endemism in these areas (compare Tables 5 and 8) would seem to exceed anything else known in the world, including the Cape region.

I have concluded (Gentry, 1982a) from patterns such as those indicated above that extremely active, even explosive, speciation has occurred and is still occurring in the Andean foothills region of western South America and in adjacent southern Central America. The rate of speciation (and resultant local endemism) in this area is similar to that in Mediterranean climate regions and much higher than that in other temperate zone areas (compare Tables 4 and 5 with Table 8). This picture of highly dynamic speciation in the tropics contrasts dramatically with, and would appear to falsify, Stebbins' (1974) conclusion that "Comparisons between tropical and temperate regions . . . do not support the hypothesis that the well-known richness of tropical floras is due chiefly to extensive speciation in progress at the present time. . . . In spite of the much greater richness of tropical flora in total numbers of species, genera, and families, there is no evidence that speciation is more active in them than it is in the varied communities of the marginal temperate regions and it may even be less active."

It seems highly probable that there is very much more endemism and many more locally endemic species in the tropics, at least in the Neotropics, than in any part of the temperate zone. Moreover, the percentage of endemics on individual habitat islands like Cerro Tacarcuna is as great as on such real islands as Jamaica, Puerto Rico, and the Galapagos (compare Table 5 with Table 1), perhaps justifying an equal conservation effort on habitat islands as on oceanic ones.

Even though speciation (and the resultant local endemism) in many tropical forests is as great as in Mediterranean climate regions, the underlying causes of this speciation seem very different from those proposed for California by Stebbins and Major (1965). Rather than strong selection in rigorous and changing environments, I see accidental and often suboptimal genetic transilience (Templeton, 1980), associated with founder effect phenomena in a kaleidoscopically changing landscape, as a dominant evolutionary mode in the epiphyte, understory shrub, and palmetto genera that constitute the bulk of the Andean foothill forests' species (Gentry, 1982a). If this interpretation is correct, conservationists may be faced with the philosophical question of whether preservation of such randomly generated and often rather nondescript species should be accorded the same priority as preservation of a *Sequoia,* or for that matter a *Persea theobromifolia.* In any event, it now seems very clear that from a world perspective tropical forests deserve more attention than temperate zone ecosystems, not only because of their greater species richness but also because of the greater concentration of local endemism in many of them.

It is already too late for many tropical forests. The last patch of forest on the Centinela ridge was cleared a few months ago. Its prospective 90 new species have aleady passed into the realm of botanical history, most of them before they were described. We must find a way to mobilize the general public so that they realize that their world includes these wonderfully dynamic living evolutionary laboratories of the tropics just as much as it does a Furbish lousewort or a California condor. Otherwise, multitudes of other species will soon follow their counterparts from Centinela into extinction.

CHAPTER 9

SEVEN FORMS OF RARITY AND THEIR FREQUENCY IN THE FLORA OF THE BRITISH ISLES

Deborah Rabinowitz, Sara Cairns and Theresa Dillon

A central concern of conservationists is the protection of rare species. Our colloquial notion of "rarity" brings to mind organisms that are uncommon or unusual, and these rare creatures sometimes take on connotations of being particularly fragile, precious, or valuable. Indeed, many biologists contend that rare species are more susceptible to extinction than common ones for a variety of reasons (Terborgh and Winter, 1980). The most pressing reason for concern is the presumed impending extinction of many species due to the destruction of tropical forests (Myers, 1984 and Chapter 19), but various taxa throughout the world are threatened as well.

However astute their natural history may be, biologists and conservationists lump together many kinds of organisms under the term "rare," possibly because English does not have a rich lexicon of words to describe rarity. This lack of precision obscures much interesting biology for a very heterogeneous group of organisms. Clearly, before we can effectively protect rare species, we must understand what kinds of rarity exist and in what important ways rare species differ from one another. Let us illustrate with two examples.

Argyroxiphium macrocephalum A. Gray (Compositae), the Haleakala silversword (Figure 1), is an extremely restricted endemic, found
only in the "crater" of Haleakala Volcano on the Hawaiian island of
Maui (Degener et al., 1976). Approximately 47,000 individuals exist,
and within its tiny range, it is surprisingly common. Nonetheless, this
is a plant that clearly merits the judgment of being rare on the basis
of its restricted geographic range.

But some plants with very broad distributions are also justifiably
termed rare. *Setaria geniculata* (Lam.) Beauv. (Gramineae)—knotroot
bristlegrass—has a huge geographic distribution, from Massachusetts
to California in North America, and southward through tropical America to Argentina and Chile (Hitchcock and Chase, 1950). This grass is
not selective about the habitats in which it grows but it is never found
in large local populations. That is, *S. geniculata* is chronically sparse,
and is rare in the sense that it is never common anywhere.

The issue we are addressing is the kinds of rarity that plants can
exhibit and we focus on three questions:

1. Can we clarify the concept of rarity? Are there distinct kinds?
2. If so, how frequent or unusual are the different forms of rarity?
3. What are the consequences of such a classification for conservation
   biology?

Thomas and Mallorie (1985) address similar issues with reference to
the conservation of Moroccan butterflies, and Hubbell and Foster
(Chapter 10) consider forest dynamics and conservation of tropical
tree species.

KINDS OF RARITY

Naturalists have noted that plants may be rare in several ways. They
may occur only in rare habitats, they may be very localized in a small
area, or they may have few individuals. We begin by distinguishing
three traits that all species possess.

1. *Geographic range:* Whether a species occurs over a broad area or
   whether it is endemic to a particular small area.
2. *Habitat specificity:* The degree to which a species occurs in a variety
   of habitats or is restricted to one or a few specialized sites.
3. *Local population size:* Whether a species is found in large populations somewhere within its range or has small populations wherever it is found.

FIGURE 1. *Argyroxiphium macrocephalum,* the Haleakala silversword, on Maui in the Hawaiian Islands.

Of course, these three traits are really continuous variables, but for convenience, we can dichotomize them into an eight-celled $2 \times 2 \times 2$ table (Table 1; see also Rabinowitz, 1981). Only one of these eight categories contains species which are really common in the ordinary sense—plants with wide ranges, many habitats, and large population sizes. *Chenopodium album*, lamb's quarters or fat hen, is an example.

Endemics such as *Argyroxiphium macrocephalum* are species with small ranges and restricted habitats. *Lloydia serotina* provides a striking example (Woodhead, 1951). In Britain, it is found only in Snowdonia, in tiny vertical fissures on almost sheer cliff faces. The plants are often solitary and never common, making the alpine lily rare in each of the three dimensions. These are the classic rare species that attract biologists' attentions (Stebbins and Major, 1965; Kruckeberg and Rabinowitz, 1985). In contrast, sparse species such as *Setaria geniculata* are particularly inconspicuous; they illustrate the distributional form opposite to endemics: wide range in numerous habitats, but small populations.

Plants restricted to specific habitats but with broad geographic range, such as mangroves *(Rhizophora mangle)* and salt marsh species, usually can easily be found since these plants are generally as predictable as their habitat. An interesting example in the British flora is *Draba muralis,* wall whitlow grass, a small herb in the mustard family. Native only to shallow immature soils of carboniferous limestone, it is widely scattered over England in populations where it is never common (Ratcliffe, 1960). Curiously, as an introduced species, *D. muralis* is capable of spreading as a garden weed. The remaining

TABLE 1. Seven forms of rarity, based on three traits.

Geographic distribution		Wide		Narrow	
Habitat specificity		**Broad**	**Restricted**	**Broad**	**Restricted**
Local population size	Somewhere large	*Chenopodium album,* lamb's quarters	*Rhizophora mangle,* red mangrove	*Primula scotica,* Scottish bird's-eye primrose	*Argyroxiphium macrocephalum,* Haleakala silversword
	Everywhere small	*Setaria geniculata,* knotroot bristlegrass		*Draba muralis,* wall whitlow grass	*Lloydia serotina,* alpine lily

(Adapted from Rabinowitz, 1981)

two categories have narrow geographic range but broad habitat specificity. Botanists have difficulty citing examples of these types of rarity, but one interesting case is *Primula scotica,* Scottish bird's-eye primrose, confined to the extreme northeast coast of Scotland and Orkney. Within its tiny range, it is found in populations of hundreds of individuals in a variety of habitats as variable as maritime cliffs, dune slacks, and pastures (Ritchie, 1954). The geographic limitation of such species is particularly puzzling.

Armed with this system of classification, we can determine the frequencies of different forms of rarity in a flora. To do so, we need both a data base (a flora with sufficient information to give us data on the three traits independently) and a method to assign species to the eight categories in an unbiased manner.

We ideally wish to consider commonness and rarity of a large number of species from data collected over their entire ranges, but flora inventories are not extensive enough to permit us to do so. The two studies in this volume that examine rarity from large data sets, ours and the study on the trees of Barro Colorado Island in Panama by Hubbell and Foster (Chapter 10), are perforce geographic samples of larger "universes." Hubbell and Foster census a 50-ha plot and find that many trees which are rare within their area of concern are common in nearby portions of the forest. We use the flora of the British Isles as our data base and find similarly that species with geographically restricted distributions in Britain may have much larger distributions when viewed on a global scale. *Thymus serpyllum,* for example, is confined in Britain to a tiny area of dry sandy soil on the breckland of East Anglia, but its range extends through European Russia to Siberia and China (Pigott, 1955). This change in status for rarities as geographic area increases is not really a drawback of these two studies, but it emphasizes that rarity must be considered on a variety of spatial scales, and in particular in delimited samples versus "global" censuses. Conservationists face similar decisions when they deal with whether to protect geographic outliers of species with vast populations in other areas, for instance prairie plants which extend into the eastern United States.

THE BIOLOGICAL FLORA OF THE BRITISH ISLES AS A DATA BASE

The British flora, one of the best known in the world, is our data base. The *Biological Flora of the British Isles,* published in the *Journal of Ecology,* is a series of articles each containing detailed information on habitat distribution and abundance for particular species. A total of 177 native species or 9.7 percent of the native flora have been described in the *Biological Flora* as of early 1985. The *Atlas of the Flora of the British Isles* (Perring and Walters, 1976; Perring and Sell, 1978) was the source for a dot map of geographic range for each species on a 10-km grid (Figure 2). In addition to providing sufficient information to document the three traits of interest, there are two other important advantages of this data set. The flora of the British Isles corresponds to a natural geographic area, and the *Atlas'* dot maps are based on an aerial grid of equal-sized entries (Stott, 1981). There are many excellent floras containing dot maps which do not lend themselves to quan-

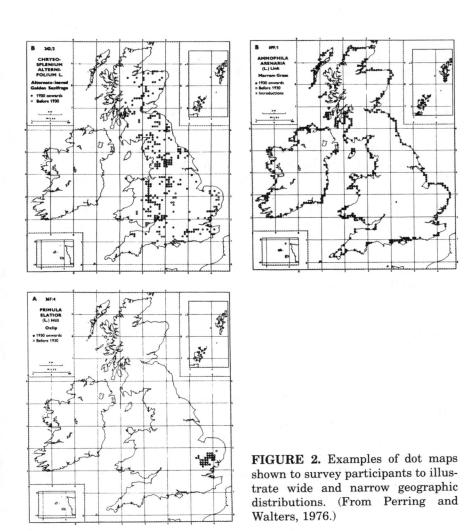

FIGURE 2. Examples of dot maps shown to survey participants to illustrate wide and narrow geographic distributions. (From Perring and Walters, 1976.)

titative analyses because they are either based on arbitrary political boundaries (for example, Pennsylvania; Wherry et al., 1979) or the aerial units are unequal (such as counties in the United States; Great Plains Flora Association, 1977).

What historical forces have shaped the flora of the British Isles and how does their island nature influence our study? Even for islands, the British Isles have a depauperate flora and a low level of endemism. The former is because of repeated glaciation and the latter because the British Isles have only recently become separated from Europe, about 5500 years ago (Pennington, 1974). Because they are islands, they possess a long coastline and consequently well-developed maritime habitats. Members of these communities are habitat specialists to a large degree—for instance, *Ammophila arenaria,* marram grass (Figure 2). The British flora, of course, is in a dynamic state and is pervasively influenced by human activity (Perring, 1974).

BIAS IN SPECIES SELECTION FOR THE BIOLOGICAL FLORA

Since the *Biological Flora* is a subset of the entire British flora, we need to know (1) how its entries were selected, and (2) whether they are a biased set from our perspective. The entries to the *Biological Flora* are contributed on a voluntary basis by taxonomists and ecologists (British Ecological Society, 1975). If these biologists have a tendency to select rarer species (because they attract attention) or common species (because they are easier to work with), then our frequencies of the seven forms of rarity will be influenced. Probably both kinds of bias have operated during different periods (P. Grubb, editor of the *Biological Flora,* personal communication). Earlier contributions (1940s and 1950s) were from naturalists and taxonomists who were interested in "unusual" species. In the 1960s and 1970s, as demographic analyses became popular under J.L. Harper's influence, entries in the *Biological Flora* were selected primarily for their experimental tractability. Thus they tended to be common species.

The net effect of these biases can be quantified in two ways with the dot maps. In the *Atlas,* each species in the entire flora is marked with an A, B, or C, depending on the number of vice-counties in which it is found (Figure 3). We can compare these frequencies in the flora as a whole with the frequencies in the subset in the *Biological Flora.* The species occurring in the fewest counties (A) are in the same proportions on both lists, but widespread species (C) are overrepresented in the *Biological Flora* and intermediate species (B) are underrepresented (G-test; $G = 7.10$, $0.025 < p < 0.05$; Sokal and Rohlf, 1981).

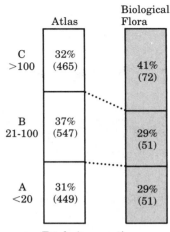

	Atlas	Biological Flora
C >100	32% (465)	41% (72)
B 21-100	37% (547)	29% (51)
A <20	31% (449)	29% (51)

Total vice-counties
in which a species occurs

FIGURE 3. The total number of vice-counties in which a species is found. Sample sizes are 1461 for the *Atlas of the British Flora* (not including the *Supplement*) and 174 for the *Biological Flora of the British Isles* (3 of the 177 total native species in the survey do not have an ABC designation in the *Atlas*).

A second means of assessing bias in species selection is to compare the distribution of number of dots on the maps for species in the *Biological Flora* with those from the *Atlas*. Each dot in the maps represents the occurrence of the species in a 10 km^2 area. Total dot counts per species thus measure the total geographical area in which the species is found.

Of the 177 species in the *Biological Flora,* 80 were selected at random and their total dots counted from the maps printed in the *Atlas.* Similarly, total dot counts were obtained for 80 species chosen at random from the 1822 species in the *Atlas* (introduced species and subspecies were not included). The distribution of the total dot counts for the sample from the *Atlas,* representing all the British flora, was highly non-normal, being skewed to the right (Figure 4). The square root transformation commonly used for count data did not normalize the distribution, but did make it less skewed. The species in the *Biological Flora* show a more uniform distribution.

The random samples from the *Atlas* and the *Biological Flora* were compared using the Kolmogorov-Smirnov Two-Sample test (Sokal and Rohlf, 1981), and were significantly different ($D = 0.30$, $p < 0.01$). Inspection of Figure 4 confirms the finding that moderately widespread species are underrepresented, widespread species are overrepresented, but infrequent species are present in their actual proportion. The *Biological Flora* is in fact a biased subset of the British flora, at least in terms of geographic range.

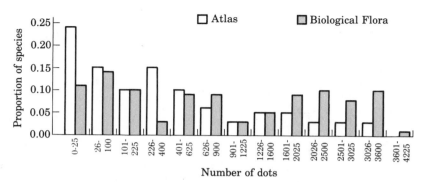

FIGURE 4. Histogram of the frequency of total dot number on distribution maps for species in the *Atlas of the British Flora* and in the *Biological Flora of the British Isles*. The ABC analysis (Figure 3) was based on the number of vice-counties in which a species occurred. Although this measure is correlated with total dot counts, the two are not identical measures; total dot counts for species classified as "A" in the two samples ranged from 1 to 182, those classified "B" from 71 to 2094, and "C" from 392 to 3745. Moreover, the ABC classification necessarily uses arbitrary cutoff values in dividing a continuous variable into discrete categories.

ASSIGNMENT OF SPECIES TO RARITY CATEGORIES

Our second requirement for estimating the frequencies of the forms of rarity in a flora was a method of assigning each species to one of the eight cells in an unbiased fashion. If one person were to classify all the species in the *Biological Flora,* the results would reflect that person's biases and preconceptions about rarity, albeit based on much field experience with rare plants. Since there were no objective boundaries between, for instance, wide and narrow geographic distributions, our judgments about assigning species to categories were necessarily subjective, but we hope not arbitrary. Our approach was to use a survey format and ask a number of biologists to classify each species. We specifically did not tell our participants what criteria to use in making their choices. Our classification of species therefore rests on educated opinion rather than on objective, arbitrary criteria. We are not advocates of the idea that science is an objective and impersonal march toward truth and reality. Science is a human, socially contingent activity (Gould, 1981), and there is a large, generally accepted role for informed opinion in scientific inquiry. We are simply utilizing it more systematically here.

Dot maps from the *Atlas* were used to indicate geographic range, and quotes were extracted from the *Biological Flora* to describe habitat specificity and local abundance. For geographic distribution, our judges were asked to decide whether the distribution was "Wide" or "Narrow." For habitat specificity, the choices were "Broad" or "Restricted," and for local abundance "Somewhere Large" or "Everywhere Small." Before filling out the survey, participants were shown unambiguous examples of species belonging to each of the two choices available for the three traits. Maps (Figure 2) were used to demonstrate two types of wide geographic distribution, and we emphasized that only total range, not density, of dots should be considered. For example, *Chrysoplenium alternifolium,* golden saxifrage, shows wide geographic distribution over most of the interior of Britain, but *Ammophila arenaria,* marram grass, has a wide geographic distribution along the coast. The map of *Primula elatior,* oxlip, was given as an example of a species with narrow geographic distribution. Examples given for the other two traits are shown in Table 2. For habitat specificity, the first quotation is an example of broad specificity, the next two quotes are examples of narrow restriction. The second quote illustrates different descriptions of a single habitat. For local population size, the first two quotes are examples of species that should be scored as having populations that are somewhere large, the last as a species which has populations that are everywhere small.

In the survey itself the three traits were in separate sections, each in a different random order. The dot maps for the geographic distribution section were presented four to a page with the species names removed. Tables 3 and 4 give examples from the sections on habitat specificity and local abundance. Each of the three traits was thus judged independently and without reference to species identity. Each choice always included a category of "can't tell," which participants

TABLE 2. Examples shown to survey participants to illustrate the kinds of habitat specificity and local abundance.

HABITAT SPECIFICITY

1. "Marshes, fens, heaths, woodlands, and waste ground. A common weed of arable land."
2. "Edges of roads, along paths and cart-ruts, and along ridges in woods."
3. "Restricted to soil-filled crevices in scree slopes."

LOCAL POPULATION SIZE

1. "In favorable situations, it forms large clumps or pure stands which may cover considerable areas."
2. "Locally abundant."
3. "Scarce when present. Occurs as widely scattered individuals."

TABLE 3. Sample page from survey administered to participants, describing habitats of individual species from the *Biological Flora of the British Isles*.

Wide	Narrow	Can't tell	472.1[a]	Open grassland where the herbage is short. Gateways, lanes and roadsides. A nuisance in lawns.
Wide	Narrow	Can't tell	669.4	Banks of slow-flowing rivers and canals with inorganic substratum. Almost or completely stagnant ponds and dykes, ungrazed marshland in river floodplains, peat basins with strongly tidal water.
Wide	Narrow	Can't tell	166.3	Chalk and limestone grassland, occasionally found on alkaline sandy soils. Restricted to exposed, unshaded sites.
Wide	Narrow	Can't tell	678.1	Hedges, dry banks, sides of dykes, sea cliff tops, pastures, hedgerows, meadows, rough grassland.
Wide	Narrow	Can't tell	123.7	Rock ledges, in crevices, on screes and mountaintop detritus, and on sea cliffs.
Wide	Narrow	Can't tell	55.1	A member of the floating leaf community of still, or slowly flowing, fresh waters. Usually occurs in water 2–10 feet deep over mud, silt, or peat, rarely over shingle.

[a]Numbers are codes from the *Atlas of the British Flora* for identification of the species.

were encouraged to use in instances of ambiguous data. This procedure helped to eliminate the effects of preconceptions of how species "ought" to behave, and so the process was desirably "blind."

Consistency of judges' decisions

The 15 judges were ecologists and systematists, all members of the Cornell academic community. Fourteen were North Americans with little direct experience with the British flora; the single Briton was a lepidopterist. How closely do the judgments among these scientists agree? Consistency among judges was examined in two ways. The first was to look at scores for each judge over all three traits and to compare between judges. Judges were quite consistent in deciding the proportion of all species considered to have a "wide" geographic range (Table 5). They were less consistent about habitat specificity and local abundance, and their written comments on the survey indicated that these were the more difficult decisions to make.

To quantify the degree of agreement between judges, we calculated a species-by-species Spearman rank correlation (corrected for ties as in Siegel, 1956) betweeen all possible pairs of judges by coding "wide" decisions as 1, "narrow" as 3, and "can't tell" as 2. Thus a disagreement between "wide" and "narrow" would have the largest magnitude $(3 - 1 = 2)$. With this coding, judges' choices were highly correlated for geographic distribution (average correlation = 0.80, range 0.63–0.94), less so for habitat (average correlation = 0.58, range 0.45–0.73) and local abundance (average correlation = 0.54, range 0.31–0.75). An oblique principal components analysis (VARCLUS procedure in SAS; SAS Institute, 1982) showed that there were no grounds for splitting the judges into discrete clusters. That is, there are no distinct "schools of thought" among the participants.

The second analysis of the judges' consistency was based on the fact that the dot maps of four species selected as being particularly ambiguous were put into the survey twice (mixed randomly with the other maps). Within-judge consistency was then tested by examining how frequently judges changed their minds. Seventy percent of the 60

TABLE 4. Sample page from survey administered to participants, describing the local abundance of individual species from the *Biological Flora of the British Isles*.

Somewhere large	Everywhere small	Can't tell	600.1[a]	Many form pure societies on patches of sand in woodland. Frequent and often locally dominant constituent of oak and beechwood.
Somewhere large	Everywhere small	Can't tell	715.1	Dominance in many moorlands, generally abundant in the north and west of British Isles.
Somewhere large	Everywhere small	Can't tell	58.4	Never abundant anywhere, usually only a few scattered individuals are found. Seldom, if ever, found in large populations.
Somewhere large	Everywhere small	Can't tell	61.1	Moderately common and locally extremely abundant.
Somewhere large	Everywhere small	Can't tell	176.1	In the island scrub it may attain subdominance. Sometimes locally abundant as a subordinate tree layer in mature beechwood.
Somewhere large	Everywhere small	Can't tell	158.1	Occurs as scattered individuals or more frequently in colonies.
Somewhere large	Everywhere small	Can't tell	685.3	Tends to form pure stands, particularly if left undisturbed. A common component of lightly grazed grassland communities.

[a]Numbers are codes from the *Atlas of the British Flora* for identification of the species.

TABLE 5. Decisions of judges from the survey.

Judge	Geographic range			Habitat specificity			Local population size		
	Wide	Narrow	Can't tell	Broad	Restricted	Can't tell	Somewhere large	Everywhere small	Can't tell
1	146	26	5	28	110	39	163	1	13
2	154	16	7	85	78	14	144	18	15
3	142	24	11	26	77	74	149	4	24
4	156	15	6	77	58	42	137	5	35
5	124	31	22	69	74	34	128	9	40
6	143	21	13	59	81	37	109	18	50
7	134	36	7	65	96	16	145	14	18
8	154	17	6	75	94	8	137	10	30
9	151	16	10	83	90	4	161	9	7
10	121	24	32	44	85	48	153	4	20
11	150	26	1	101	72	4	158	7	12
12	152	14	11	24	104	49	154	3	20
13	145	26	6	56	86	35	135	3	39
14	150	23	4	79	83	15	149	13	15
15	124	31	22	74	69	34	121	20	36

judgments were consistent. Of the 18 inconsistent decisions, 83 percent (15 of 18) changed from or to the "can't tell" category. Only 18 percent (3 of 18) reversed a judgment from "wide" to "narrow" or vice versa.

In summary, the judgments of participants were moderately correlated with one another for habitat specificity and local abundance and highly correlated for geographic range; judges were remarkably consistent when asked to make the same difficult decision twice.

Data analysis: assignment of species to categories of rarity

After collecting data from the participants, each species was assigned to one cell of the eight-celled table. We performed two analyses, one using conservative criteria, the other using liberal criteria. For the conservative analysis, if three people disagreed or if there was a consensus that the data were ambiguous for any one of the three traits, a species was excluded from the tabulation. This resulted in 128 exclusions, leaving a sample of 49 species. For the liberal analysis, a simple plurality (six respondents in agreement) was considered adequate to classify a species. This technique produced a sample size of 160, with only 17 species excluded. Species excluded from consideration had 6:6 ties or a plurality of "can't tells."

Some examples will clarify this procedure (Table 6). *Deschampsia cespitosa,* tufted hair-grass, was included in the tabulations according to both the conservative and liberal criteria because there was very close agreement on all three traits. For the conservative analysis, *Agrostis setacea,* bristle-leaved bent, was omitted because five judges disagreed on geographic range, but it was included in the liberal analysis because more than six agreed for each trait. Similarly, *Draba muralis,* wall whitlow grass, was included in the liberal but not the conservative analysis. *Myosotis alpestris,* alpine forget-me-not, was excluded from both analyses because there was no consensus on any trait and because a plurality agreed that the geographic and local population data were ambiguous.

The final step in classifying the species is to place each species in the appropriate cell and tally up the total frequencies; Table 6B shows the categories of the species in the example. *Deschampsia cespitosa* falls into the common category in the upper left cell—wide geographic range, broad habitat specificity, and large local populations. *Agrostis setacea* falls into one of the endemic categories, with narrow geographic range, restricted habitat specificity, but large local populations. *Draba muralis* falls in a third cell; it is a species with wide geographic distribution but restricted to particular habitats, within which it is never common.

TABLE 6. Assignment of species to categories of rarity.

A. Survey results for four species.

Species	Geographic range			Habitat specificity			Local population size		
	Wide	Narrow	Can't tell	Broad	Restricted	Can't tell	Somewhere large	Everywhere small	Can't tell
Deschampsia cespitosa	15	0	0	15	0	0	14	0	1
Agrostis setacea	2	10	3	0	15	0	15	0	0
Draba muralis	14	0	1	1	9	5	2	9	4
Myosotis alpestris	1	6	8	1	8	6	4	5	6

B. Placement of species in categories.

Geographic distribution	Wide		Narrow	
Habitat specificity	Broad	Restricted	Broad	Restricted
Local population size				
Somewhere large	*Deschampsia cespitosa*, tufted hair-grass			*Agrostis setacea*, bristle-leaved bent
Everywhere small		*Draba muralis*, wall whitlow grass		

Results of the conservative and liberal analyses were similar (Table 7). The proportion of species which had broad versus restricted habitat specificity was compared for those species which had both large local abundance and wide geographic distribution (the two categories with the most species to test whether the two analyses gave different results). Of 129 such species, 58 (45 percent) had wide habitat specificity under the liberal criterion, while 6 out of 20 (30 percent) had wide habitat specificity under the conservative criterion. The 95 percent confidence intervals are 34–54 percent and 12–54 percent respectively. Since the two intervals completely overlap, the different percentages (45 percent versus 30 percent) under the two different criteria can be accounted for by the variability expected in smaller samples, and there is no evidence that the different criteria led to different proportions of species being assigned to these two categories. Thus, for the remainder of the discussion, we restrict our attention to the larger data set of the liberal analysis.

Independence in the liberal data

The test for independence of the three traits was performed using the general loglinear model described in Fienberg (1981). The results indicate that the model of complete independence of the three traits adequately fits the data ($G^2 = 5.9047$, $p < 0.10$). This is an interesting result because it tells us that geographic range, habitat specificity, and local population size are three independent variables in the British flora. Thus each variable supplies information on rarity not provided by the other two.

This conclusion might seem to contradict the views held by many ecologists that generalists are widespread and that abundance is positively correlated with range (Brown, 1984). This is not the case, however. Even though *local* abundance is independent of geographic range and habitat specificity in our contingency table, we still might find a correlation for a collection of species between *total* number of individuals for each species and the total area it occupies.

Naturally, we have no idea whether the independence of the three traits is universal, since this is the first test of its existence (but see Thomas and Mallorie, 1985). Again, the issue of scale emerges as critical (Hubbell and Foster, Chapter 10). Were the global ranges of the species considered, rather than simply their occurrence in the British Isles, particular species would shift among the eight categories, but perhaps the proportions in each would remain similar. As yet, we do not know.

TABLE 7. Results of the liberal (boldface) and conservative (parentheses) analyses.[a]

Geographic distribution		Wide		Narrow	
Habitat specificity		Broad	Restricted	Broad	Restricted
Local population size	Somewhere large	58 (21)	71 (23)	6 (0)	14 (4)
	Everywhere small	2 (0)	6 (0)	0 (0)	3 (1)

[a]Numbers are species counts out of a total of 160 for the liberal and 49 for the conservative analysis.

Frequencies of the marginal totals and cells

Most species are common somewhere (Table 7), as Brown (1984) on regional and continental scales and Shmida and Ellner (1984) on a within-habitat scale have recently argued. In the British Isles, 149 species out of 160 have large populations somewhere within their ranges, but only 11 have chronically small populations. Many more species (137) have wide geographic ranges than narrow ones (23). A slight majority of the species have restricted habitat specificity (94), but many (66) occur in a variety of habitats.

Again we have an apparent but not actual contradiction with the ecological generalization of the lognormal distribution of species abundance (May, 1975). Within a single site, very few species have large populations, and most species are represented by only a few individuals. That is, *locally* most species are rare. Our results say that most species are common *somewhere* within their ranges, but it may be that most species have locally small populations throughout the majority of the sites where they occur.

Species of wide geographic distribution, restricted habitat, but high local population size are the most numerous category of plants in our sample of the British Isles (71 species; Table 7), even more numerous than ordinary common species (58 species). These plants are predictable habitat specialists—for example, sand dune, marsh, bog, and fen species, or forest floor species under particular canopies. Plants in this category may be especially numerous in Britain due to its long coast and predominance of maritime species. Nearly one-third (30 percent) of the 71 species have exclusively coastal distributions.

These plants will be as frequent in their occurrence as is their habitat. *Ammophila arenaria,* marram grass (Figure 2), is a good example from the British flora (Huiskes, 1979). Often the only species present on semi-fixed and mobile sand dunes along most of the coastline, this maritime species is abundant in a frequently encountered habitat. *Limonium humile,* lax-flowered sea lavender, a resident of muddy substrates in salt meadows, also has a wide geographic distribution and occurs in locally large populations (Boorman, 1967). But in contrast to *A. arenaria,* the habitat appropriate to *L. humile* is curiously infrequent, and as a result, the lavender is quite a rare plant.

The second most numerous form of rarity (14 species) is the classic restricted endemic: habitat specialist, small range, but common when you find it (Table 7). Such rarities have been the object of study by biogeographers and evolutionists for decades (Cain, 1944; Kruckeberg, 1954; Stebbins and Major, 1965). The British flora contains numerous interesting cases of this form of rarity, but especially striking is *Orchis purpurea,* lady orchid. Confined to chalk coppice and beechwoods in extreme southeast England, the lady orchid is "still locally plentiful" (Rose, 1948).

As both Robert Whittaker and Ledyard Stebbins have suggested, there seems to be a cell in this table that is impossible or nearly so. Plants with narrow range, small populations, but broad habitat specificity seem not to exist in the British flora, at least when one samples from 177 species. It is interesting to note that Thomas and Mallorie (1985) found a single butterfly species (out of 39) in this category in the Atlas Mountains, *Gonepteryx cleopatra.* Here we return to the issue of bias in selection of species for inclusion in the *Biological Flora.* For this and the following forms of rarity, inconspicuous or pedestrian species of intermediate abundance or range are underrepresented because they do not attract the notice of biologists. We hope that our classification will bring these species to the greater attention of researchers.

Truly sparse species (the opposite of endemics), with large ranges, many habitats, but chronically low local population sizes, are infrequent. We sampled only two from a population of 177 species. This result surprised us, but perhaps sparse species are especially underrepresented in Britain or in the *Biological Flora.* An interesting case is *Hypochoeris maculata,* spotted cat's ear, a composite with a scattered distribution in Britain and the Continent, found in habitats as varied as "chalk grassland, ancient earthworks, old quarry workings, blown sand and maritime cliffs on calcareous and serpentine-derived soils" (Wells, 1976). Yet it is not particularly common anywhere.

The remaining three categories are infrequent as well, for reasons we don't understand. We have already discussed instances of these rarities (Table 1). *Draba muralis* has a wide geographic distribution but is habitat-restricted in small populations. *Primula scotica* grows in several habitats in considerable populations, but its range is tiny. And *Lloydia serotina* is rare on all three fronts. It is important to recognize the existence of such species and to determine in the future whether they are truly unusual distributional forms in other floras as well and if so, why this might be true.

IMPLICATIONS FOR CONSERVATION BIOLOGY

The seven forms of rarity give us some recommendations and justifications for conservation practice. How one goes about protecting a species will hinge on the dimension in which it is rare, and it is useful, perhaps necessary, to specify precisely what kind of rarity one is dealing with. For example, for a rare plant such as *Primula scotica*, with a tiny range in northeastern Scotland, its large populations in several habitats suggest that the species possesses substantial ecological plasticity and tolerance for a variety of environmental conditions. This form of rarity indicates that there probably is available uncolonized but suitable habitat where the species does not occur and that transplant experiments stand a good chance of success. In contrast, the large range and patchy restriction of *Limonium humile* to large populations in a narrow habitat suggest that in the past, dispersal has "sampled" a wide variety of maritime habitats, and colonization has already occurred on the patches that are suitable in some subtle way we cannot detect. Transplants as a method of conservation have less likelihood of being successful in such a species, and protection of reserves against environmental perturbation is probably critical. The primrose and lavender are thus likely to be endangered by very different forces. On the other hand, there are some rare species such as *Hypochoeris maculata* which, despite small local populations, occur over a wide geographic range in a variety of habitats. Such sparse species are unlikely to be endangered except by massive habitat destruction. In a triage system for rare species, they can be left to look after themselves and may even provide us with clues on how to protect close relatives in more precarious states.

Rarity thus comes in a variety of forms, and our results are a first attempt at determining how frequent these are, based on a method applicable to any well-studied flora on a local, regional, or continental scale. An interesting second step would be to compare the distribution of endangered species or extinct species with species as a whole in a flora. This examination would tell us whether endangered species were a random subset of floras or whether they were drawn from particular sorts of rarity. The exercise might suggest the causes of the endangered state.

The British flora shows that the preponderant attention which conservationists pay to endemic species is well justified (Kruckeberg and Rabinowitz, 1985). We already knew that these plants of small range and narrow habitat specificity, such as *Orchis purpurea,* are easily threatened or extirpated by habitat destruction. Our results show that they are a numerically important part of the flora, not just isolated cases.

Findings for both endemics and species of wide geographic range underscore that most species occur in restricted habitats. Conserving and studying habitats is probably the most effective means of conserving such species (Thomas and Mallorie, 1985), in much the way that the reserve in Snowdonia protects *Lloydia serotina.*

These data do give us a hopeful indication that we can move from our colloquial notions of rarity to a concept based on a solid quantitative footing. The independence of geographic range, habitat specificity, and local abundance tell us that conservationists concerned with rare species need to pay attention to all three traits. The consistency of our judges' decisions is an encouraging sign that observers converge in their notions of rarity. So conservationists can "objectify" what has previously been the exclusive province of "natural history." Our results bode well for the possibility of advancing the conservation of rare species to a discipline based on sound ecological principles.

SUGGESTED READINGS

Brown, J.H. 1984. On the relationship between abundance and distribution of species. *American Naturalist 124:* 255–279.

Rabinowitz, D. 1981. Seven forms of rarity. In *The Biological Aspects of Rare Plant Conservation,* H. Synge (ed.). Wiley, Chichester.

Thomas, C.D. and H.C. Mallorie. 1985. Rarity, species richness and conservation: Butterflies of the Atlas Mountains in Morocco. *Biological Conservation 33:* 95–117.

APPENDIX

Species from the *Biological Flora of the British Isles* included in the survey and the category into which each fell.

A note about error. Since our approach is statistical, misclassification of particular species are inevitable, especially considering the absence of a standard list of habitats in the *Biological Flora of the British Isles.* For example, *Senecio jacobaea* occurs in large populations (C. Thomas and H. Robertson, personal communication), yet it fell into the category of "everywhere small" for local population size based on quotes from the *Biological Flora*. There are undoubtedly other errors of this sort. If they are random with respect to category, the tallies for the eight cells will not be biased, but we have no way to assess whether this is the case.

WIDE Geographic Distribution
BROAD Habitat Specificity
SOMEWHERE LARGE Local Population Size

Acer campestre L.
Agropyron repens (L.) Beauv.
Arrhenatherum elatius (L.) J. & C. Presl
Arum maculatum L.
Calluna vulgaris (L.) Hull
Catapodium rigidum (L.) C.E. Hubbard
Chenopodium album L.
Chenopodium rubrum L.
Dactylis glomerata L.
Deschampsia cespitosa (L.) Beauv.
Deschampsia flexuosa (L.) Trin.
Eleocharis uniglumis (Link) Schult.
Endymion non-scriptus (L.) Garcke
Epilobium angustifolium (L.) Scop.
Erica cinerea L.
Eriophorum angustifolium Honck.
Galeobdolon luteum Huds.
Holcus lanatus L.
Holcus mollis L.
Hypochoeris radicata L.
Ilex aquifolium L.
Juncus conglomeratus L.
Juncus effusus L.
Juncus inflexus L.
Lolium perenne L.
Menyanthes trifoliata L.
Narcissus pseudonarcissus L.
Nardus stricta L.
Papaver argemone L.

Phragmites communis Trin.
Plantago lanceolata L.
Polygonum lapathifolium L.
Polygonum nodosum Pers.
Polygonum persicaria L.
Potentilla fruticosa L.
Puccinellia maritima (Huds.) Parl.
Quercus robur L.
Ranunculus acris L.
Ranunculus ficaria L.
Ranunculus repens L.
Rhamnus catharticus L.
Rumex crispus L.
Rumex obtusifolius L.
Sesleria albicans Kit. ex Schult.
Silene alba (Mill.) Krause
Silene dioica (L.) Clairv.
Silene nutans L.
Sonchus asper (L.) Hill
Spergula arvensis L.
Stellaria media (L.) Vill.
Succisa pratensis Moench
Teucrium scorodonia L.
Thymus drucei Ronn.
Trifolium repens L.
Urtica dioica L.
Urtica urens L.
Vaccinium myrtillus L.

WIDE Geographic Distribution
RESTRICTED Habitat Specificity
SOMEWHERE LARGE Local Population Size

Allium ursinum L.
Alnus glutinosa (L.) Gaertn.
Alopecurus myosuroides Huds.
Ammophila arenaria (L.) Link
Anthemis cotula L.
Aster tripolium L.
Atropa bella-donna L.
Carex arenaria L.
Carex flacca Schreb.
Chrysanthemum segetum L.
Cirsium acaulon (L.) Scop.
Cladium mariscus (L.) Pohl
Colchicum automnale L.
Corynephorus canescens (L.) Beauv.

Crambe maritima L.
Cuscuta europaea L.
Cynosurus cristatus L.
Dryas octopetala L.
Eleocharis palustris (L.) Roem. & Schult.
Elymus arenarius L.
Empetrum nigrum L.
Erica tetralix L.
Eriophorum vaginatum L.
Frangula alnus Mill.
Fraxinus excelsior L.
Gentiana pneumonanthe L.
Glaucium flavum Crantz
Glyceria maxima (Hartm.) Holmberg

Halimione portulacoides (L.) Aell.
Helianthemum canum (L.) Baumg.
Helianthemum chamaecistus Mill.
Hippocrepis comosa L.
Hippophae rhamnoides L.
Hornungia petraea (L.) Rchb.
Hymenophyllum tunbrigense (L.) Sm.
Hymenophyllum wilsonii Hook.
Juncus acutus L.
Juncus squarrosus L.
Juncus subnodulosus Schrank
Lathyrus japonicus Willd.
Limonium humile Mill.
Limonium vulgare Mill.
Lobelia dortmanna L.
Mertensia maritima (L.) S.F. Gray
Narthecium ossifragum (L.) Huds.
Nuphar lutea (L.) Sm.
Nuphar pumila (Timm) DC.
Nymphaea alba L.
Papaver rhoeas L.
Plantago coronopus L.
Plantago major L.

Plantago media L.
Poa annua L.
Polygonum hydropiper L.
Quercus petraea (Mattuschka) Liebl.
Rorippa microphylla (Boenn.) Hyland
Rorippa nasturtium-aquaticum (L.) Hayek
Rubus chamaemosrus L.
Saxifraga hypnoides L.
Saxifraga oppositifolia L.
Schoenus nigricans L.
Sinapis arvensis L.
Sonchus oleraceus L.
Sparganium erectum L.
Spartina x townsendii H. & J. Groves
Suaeda maritima (L.) Dum.
Tamus communis L.
Tuberaria guttata (L.) Fourreau
Viola lactea Sm.
Vulpia membranacea (L.) Dum.
Zostera angustifolia (Hornem.) Rchb.
Zostera marina L.

WIDE Geographic Distribution
BROAD Habitat Specificity
EVERYWHERE SMALL Local Population Size

Hypochoeris maculata L.   *Senecio jacobaea* L.

WIDE Geographic Distribution
RESTRICTED Habitat Specificity
EVERYWHERE SMALL Local Population Size

Draba muralis L.   *Saxifraga cespitosa* L.
Papaver hybridum L.   *Silene acaulis* (L.) Jacq.
Papaver lecogii Lamotte   *Subularia aquatica* L.

NARROW Geographic Distribution
BROAD Habitat Specificity
SOMEWHERE LARGE Local Population Size

Arbutus unedo L.   *Pinus sylvestris* L.
Lobelia urens L.   *Polemonium caeruleum* L.
Melampyrum cristatum L.   *Primula scotica* Hook.

NARROW Geographic Distribution
RESTRICTED Habitat Specificity
SOMEWHERE LARGE Local Population Size

Agrostis setacea Curt.   *Helianthemum apenninum* (L.) Mill.
Corrigiola litoralis L.   *Linum anglicum* Mill.
Daboecia cantabrica (Huds.) C. Koch   *Orchis purpurea* Huds.
Draba aizoides L.   *Saxifraga rosacea* Moench
Dryopteris villarii (Bell) Woynar   *Spartina maritima* (Curt.) Fernald
Erica mackaiana Bab.   *Suaeda fruticosa* Forsk.
Frankenia laevis L.   *Thymus serpyllum* L.

NARROW Geographic Distribution
BROAD Habitat Specificity
EVERYWHERE SMALL Local Population Size
No species in this category

NARROW Geographic Distribution
RESTRICTED Habitat Specificity
EVERYWHERE SMALL Local Population Size

Arabis stricta Huds.   *Saxifraga hartii* D.A. Webb
Lloydia serotina (L.) Rchb.

CAN'T TELL   for at least one of the three categories
   (Geographic Distribution, Habitat Specificity, and/or Local
   Population Size)

Allium vineale L.   *Oxalis acetosella* L.
Anemone pulsatilla L.   *Phyllodoce caerulea* (L.) Bab.
Anthemis arvensis L.   *Ranunculus bulbosus* L.
Arum italicum Mill.   *Senecio integrifolius* (L.) Clairv.
Chrysanthemum leucanthemum L.   *Sibbaldia procumbens* L.
Gentiana verna L.   *Thymus pulegioides* L.
Hypericum linarifolum Vahl   *Vaccinium vitis-idaea* L.
Juncus filiformis L.   *Viola lutea* Huds.
Myosotis alpestris F.W. Schmidt

COMMONNESS AND RARITY IN A NEOTROPICAL FOREST: IMPLICATIONS FOR TROPICAL TREE CONSERVATION

Stephen P. Hubbell and Robin B. Foster

This chapter addresses the implications of a large plot study of species-rich tropical forest for strategies for preserving tropical tree diversity. We outline some of the reasons for commonness and rarity in tropical trees in the forest on Barro Colorado Island, Panama, and make some general recommendations for conservation based on these abundance patterns. Before giving these recommendations, however, we present data on the species composition and spatial structure of the Barro Colorado forest. We ask managers to consider the data thoughtfully because our current evidence regarding temporal dynamics and long-term changes is circumstantial rather than direct. We base our conclusions about forest dynamics and about strategies for tree conservation on the static distribution and abundance of tree species in the Barro Colorado forest at a single census—not on dynamic data from recensuses of the forest through time.

QUESTIONS AND CONSERVATION ISSUES

The puzzle of tree species richness in tropical forests is often reduced to the question of how so many rare tree species are maintained, ignoring the question of species origin. The related conservation issue is whether to attempt to conserve rare tropical tree species, and if so, how. Conservation strategies and tactics are complicated by the many biological causes of plant rarity (Rabinowitz, 1981; Rabinowitz et al., Chapter 9). In the case of tropical forests, we also are hampered by a lack of basic knowledge about the local population biology and regional biogeography of tropical tree species. We often don't even know which tree species are present, let alone which ones are rare or why they are rare.

To answer the question of tropical tree species maintenance, we must describe the patterns and explain the causes of both tree commonness and rarity:

1. What is the distribution of species abundances?
2. Are there density-dependent biotic factors which prevent common species from assuming complete dominance?
3. Are the rare species self-maintaining (reproducing) in the forest, or sustained by continual immigration?
4. Are there minimum critical population sizes for tropical tree species?
5. Do rare species exhibit greater specialization to habitat or conditions of regeneration than common species?
6. Are rare species more often found in certain genera or families than others, suggesting that rarity occurs for reasons of life history or physiology, rather than for reasons of chance or circumstance?
7. How do the answers to these questions depend on scale (e.g., the size of the forested area or reserve under consideration)?

Many of these questions relate to a larger issue of general concern to tropical forest conservation, namely the extent to which these forests represent stabilized communities of particular taxonomic assemblages of tree species in or near equilibrium (see Pimm, Chapter 14). If biotic mechanisms tending to stabilize existing assemblages of tree species in tropical forests are weak or absent, then the preservation of tree diversity will be a different and more difficult task than if these forests have strong stabilizing forces (Hubbell, 1984; Hubbell and Foster, 1985b). We will be attempting to assess the strength of various potentially stabilizing forces for the forest on Barro Colorado Island over the next several decades.

THE STUDY SITE

Barro Colorado Island is a 15 km^2 former hilltop in artificial Gatun Lake in the zone of the Panama Canal (Figure 1). The vegetation is classified as tropical moist forest, and the climate is seasonal, with a

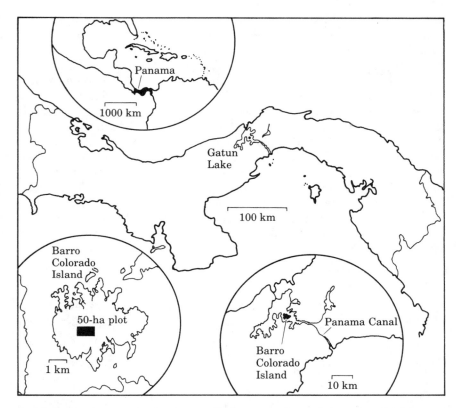

FIGURE 1. Geographic locations of Panama in Central America, the Panama Canal and Gatun Lake within Panama, Barro Colorado Island within the lake, and the 50-ha plot within the island.

pronounced dry season from late December until mid April (Leigh et al., 1982). Half of the island is covered with "old-growth" forest; the other half is mostly second-growth forest dating from precanal days (70–150 years old). In 1980 a permanent plot of 50 ha was established in the old-growth forest on the island (Hubbell and Foster, 1983).

By 1982 a census and detailed map had been completed of all free-standing woody plants in the plot with a stem diameter of 1 cm dbh— a total of over 238,000 individuals. The data were subjected to an elaborate system of checks and rechecks, so we are confident of the accuracy of the census and species identifications, including the abundances of the rarest species. Canopy structure and canopy gaps have also been mapped annually since 1983 (Hubbell and Foster, 1985a) with sample points every 5 m, and spatial patterns of sapling regeneration in each species are correlated with these canopy maps (Hubbell and Foster, 1985c). Growth, survival, and new sapling recruitment of each species will be recorded in a complete plot recensus every 5 years.

The site is relatively flat, with about 25 ha of 0–3 percent slopes, 13 ha of 3–10 percent slopes, and 10 ha in moderate (10–21 percent) slopes to the east and south. There is also a small (approximately 2 ha) seasonal swamp in the middle of the plot (Figure 2). Second-growth forest borders the plot along its northeastern edge, and includes a section of about 1 ha inside the plot. The remainder of the forest in the plot is 500–600 years old. Before this time, beginning around 500 A.D., occasional small clearings were made. These clearings were probably the work of hunters and gatherers since there is no carbon or phytolith fossil evidence of slash-and-burn maize agriculture anywhere in the plot (D. Piperno, personal communication; Hubbell and Foster, 1985a).

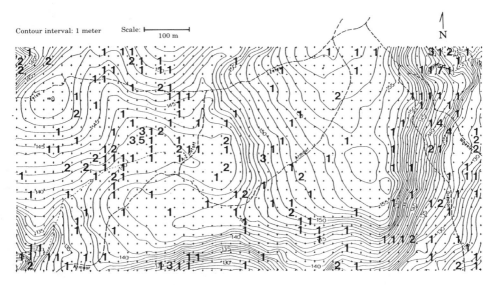

FIGURE 2. Topographic map of the 50-ha plot on Barro Colorado Island, showing the aggregate distribution of individuals of the rarest species (< 10 individuals per species). Contour lines are drawn for each meter of elevational change. Dotted lines mark permanent trails through the plot. Grid marks indicate the corners of 1250 20 × 20 m quadrats. A creek (lower left corner) drains a seasonal swamp (left center). Grades on steeper slopes are from 10 to 21 percent. Note the patchy distribution of rare species. A major patch occurs in the swamp, and lesser patches (such as that in the northeast corner) correspond to large treefall gaps.

Fifty hectares is a lot of ground, and it is reasonable to ask: Why study so much? The answer is that many tropical tree species are rare by temperate zone standards. A large plot is necessary to obtain adequate population samples for determining dispersion pattern, demography, and regeneration requirements of rare species (Hubbell, 1984). In fact such a large sample allows us to be more precise about the meaning, causes, and consequences of tropical tree rarity.

In this chapter we are concerned only with local tree species abundance and the relationship of local abundance to habitat specificity and gap regeneration requirements. Although the flora of Barro Colorado Island is well known (Croat, 1978), we do not consider geographic range as a parameter because the regional taxonomic affinities of some tree taxa are still unclear. Indeed, until the genetic architecture of tropical tree populations and of geographic races are better understood, the concepts of "species" and "geographic range" in tropical trees will remain uncertain. Our conclusions appear to differ from those of Rabinowitz et al. (Chapter 9), but as will be explained, our seemingly contrasting results can be largely explained by the difference in spatial scale of the two studies.

RELATIVE SPECIES ABUNDANCE

At the completion of the 1982 census, the Barro Colorado plot contained 303 species of free-standing woody plants greater than 1 cm dbh, from shrubs to overstory trees. Based on the size of the mature plant, we classified 58 species as shrubs (adults less than 4 m in height), 60 species as understory treelets (adults 4–10 m), 71 species as midstory trees (adults 10–20 m), and 114 species as overstory (canopy plus emergent) trees (adults taller than 20 m). The overstory category also includes four strangler figs which begin life as hemiepiphytes.

Species abundances range over 4.6 orders of magnitude, from 25 extremely rare species represented by single individuals, to 39,877 shrubs of *Hybanthus prunifolius* (Violaceae). Species abundance patterns in a number of temperate zone plant communities have been successfully fitted by the lognormal distribution (Whittaker, 1975); but in the case of Barro Colorado, the pattern of total species abundance deviates from the lognormal in having too many rare species (Figure 3). Pooling counts of species having different growth form could be the cause of an apparent excess of rare species if, for example, large overstory trees are relatively rare when grouped with smaller and potentially more numerous shrubs. However, when stems of all size classes of species are enumerated and species are grouped into four abundance classes by orders of magnitude, shrubs as a group are not overrepresented among common species (Table 1); in fact, there is a tendency for too many rare shrub species. Although the commonest species is a shrub, the second most common is an understory tree, *Faramea occidentalis* (Rubiaceae), with 23,452 individuals, and the third most abundant species is an overstory tree, *Trichilia tuberculata* (Meliaceae), with 12,932 individuals. As will be discussed later, a relative abundance distribution with too many rare species might also arise if occasional propagules from many new species arrive from a species-rich source area and survive, but do not establish reproductively viable and growing local populations.

For present purposes we define a rare species as one having an average density of less than one individual per ha, or less than 50

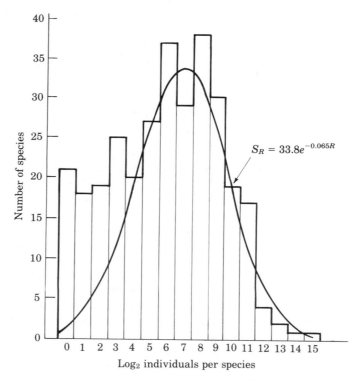

FIGURE 3. Species abundance curve for all free-standing woody plants of 1 cm dbh or larger in the 50-ha plot. Abundance categories are in octaves of abundance (individuals per species), shown in units of log base 2. The lognormal distribution given was fitted using the octaves to the right of the mode and two to the left of the modal octave. Note the excess of species in the rarer abundance categories.

TABLE 1. Number of species by abundance class and growth form in the Barro Colorado Island 50-ha plot.

Growth form[a]	Abundance class (individuals per species)				Total species
	1–9	10–99	100–999	1000+	
Shrub	21 [12.1][b]	12 [16.1]	16 [21.4]	9 [8.4]	58
Understory	9 [12.5]	17 [16.6]	26 [22.2]	8 [8.7]	60
Midstory	8 [14.8]	17 [19.7]	34 [26.2]	12 [10.3]	71
Overstory	25 [23.7]	38 [31.6]	36 [42.1]	15 [16.6]	114
Abundance class totals	63	84	112	44	303[c]

[a] Growth forms are defined as follows: Shrubs are plants whose adults are less than 4 m; understory treelets, adults 4–10 m; midstory trees, adults 10–20 m; overstory trees, adults above 20 m (includes hemiepiphytic fig species).
[b] The first entry is the observed number of species; the entry in brackets is the expected number based on the proportion of species of the given growth form in the plot.
[c] A contingency analysis indicates no significant association of growth form and abundance class (Chi-square = 19.90, 12 df, $p > 0.05$); but the two most deviant cells are the observed excess of rare shrubs (21 versus 12.1) and a deficit of rare midstory trees (8 versus 14.8). However, the number of common species (more than 1000 individuals per species) of each growth form is almost exactly what would be expected from the proportion of species of each growth form.

individuals in the entire plot. Slightly more than one third (111) of all species fall into this category, and more than one fifth (67) of all species have fewer than 10 individuals in the ½ km^2 plot. The extreme rarity of these species is best understood by walking through the Barro Colorado forest, but it can also be appreciated by noting that each of these species on average represents less than one plant out of the mean total number per ha, 4711. Collectively, the rarest third of all species add up to only 1419 individuals, or just over 0.6 percent of the plants in the plot.

COMMONNESS AND RARITY

One can think of many potential reasons for rarity in tropical tree species. Of these, three are perhaps of greatest interest to conservation biologists because they have a direct impact on management policy and practice. A species can be rare because:

1. Its required habitat is spatially a small proportion of the total habitat available.

2. Conditions for its successful regeneration (such as required light gap conditions or escape from enemies) occur infrequently and have not recurred for some time or are no longer expressed. (In the latter case the species is on its way to local extinction.)
3. The species is a recent immigrant from population centers outside the study area (whose local population may or may not increase subsequently).

We will consider first the combined evidence for the first two reasons—habitat and regeneration niche specialization—and then the evidence for immigration of rare species into the Barro Colorado plot.

Habitat and regeneration niche specialization

Evidence for habitat or regeneration niche specialization may be sought in spatial correlations between microsite and tree species distributions. By "habitat niches" we refer to microsite conditions which are relatively fixed in space and time, such as conditions of drainage controlled by topography and edaphic characteristics. "Regeneration niches" refer here to microsite conditions which are relatively ephemeral, particularly regenerative conditions in canopy gaps created by falling branches or by single or multiple treefalls. In referring to the spectrum of conditions from gap to shaded understory of the mature forest, the term "regeneration niche" is used here in a somewhat more restricted sense than its use by Grubb (1977).

In what follows, we devise a simple two-way classification scheme for species, with gap regeneration requirements on one axis, and habitat specialization (slopes, swamps, etc.) on the other. This classification is obviously simplistic but it is useful; the dominant variables influenced by topographic and edaphic site characteristics are moisture and nutrients, whereas the dominant variable influenced by treefall gaps is light quality and quantity.

It is relatively straightforward to classify species by their distributions among habitats defined by fixed topographic or edaphic features of the plot (Hubbell and Foster, 1983). Thus, there is a group of species which is found in much greater abundance on the steeper slopes than on flatter terrain (Figure 4), whereas other species are more common on the plateau and less so on the slopes (Figure 5). Some species are narrow specialists, found mainly, for example, in the seasonal swamp (Figure 6), whereas others appear to be generalists, occurring in virtually all the major habitats in the plot (Figure 7). The habitat categories we distinguish for present purposes are "Slope," "Indifferent," "Plateau," "Streamside/Ravine," and "Swamp." Species were classified as habitat specialists if they had a significant positive

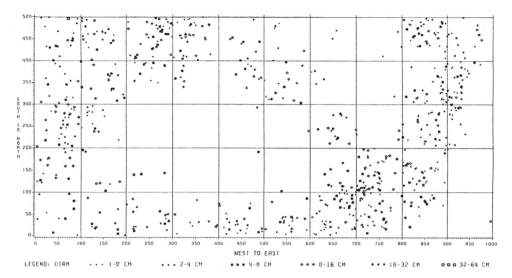

FIGURE 4. Distribution of *Unonopsis pittieri* (Annonaceae), a midcanopy tree which is a slope specialist. Grid lines mark hectare (100 m) boundaries. Map symbols refer to different diameter classes. Note that even though *U. pittieri* is more common on the steeper slopes, it occurs at low density throughout most of the rest of the plot. This is a general pattern for most of the habitat specialists.

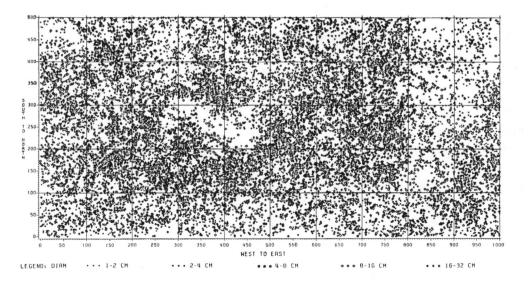

FIGURE 5. Distribution of a flat terrain specialist, *Faramea occidentalis* (Rubiaceae), an understory tree.

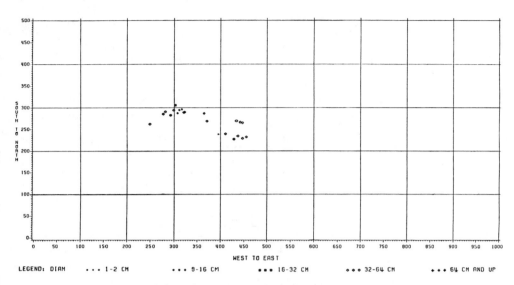

FIGURE 6. Distribution of a rare specialist, the oil palm *Eleais oleifera* (Palmae), found only in the 2 ha seasonal swamp in the center of the plot. This is a case of an extreme specialist, strictly limited to a rare habitat.

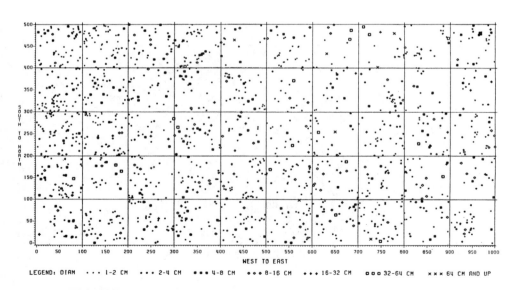

FIGURE 7. Distribution of the generalist overstory tree *Pterocarpus rohrii* (Leguminosae, Papilionoideae), which is largely indifferent to the habitat variation occurring in the plot.

association of individuals with the spatially delimited habitat area at a critical p value of 0.05.

We are beginning to understand the mechanistic basis of these habitat differences in terms of the seasonal patterns of soil water potentials in each site (P. Rundel and P. Becker, personal communication). The plateau of Barro Colorado, for example, becomes seasonally dry much sooner than the slopes and ravines around the edge of the plateau. Groundwater beneath the plateau drains out along the slopes in many seepage areas and springs, so that the dry season is shortened and ameliorated. Consequently, many of the slope specialists are tree species characteristic of wetter forest than those common on the plateau.

The detailed maps of canopy structure over the Barro Colorado plot have also allowed us to classify species into "regeneration guilds" based on the distribution of their small saplings into recent gaps, older regenerating gaps, and the understory of mature-phase forest (Hubbell and Foster, 1985c). At present we are able to recognize four guilds: "sun," "partial sun," "shade," and "indifferent" species, examples of which are shown in Figure 8. Species were classified to regeneration guilds (in all but the "indifferent" guild) on the basis of the canopy height categories with the largest Chi-square sampling deviations from expected.

Saplings in the "sun" guild of "pioneer species" are strongly skewed into recently opened gaps (0 to 2 m canopy). Saplings of species in the "partial sun" guild appear in gaps after the regenerating canopy in the gap has increased to a height of 5 to 10 m. Generally these saplings were present earlier, but they reach a censusable size (1 cm dbh) later because they grow slightly more slowly than the pioneers. Saplings of "shade" species are skewed away from gaps into the understory beneath mature-phase forest canopy (20 to 30 m), and are clearly highly shade tolerant. The surprise result was the discovery of a large guild of "indifferent" species whose saplings were distributed indistinguishably from random across all canopy height categories. Many of these species are sufficiently common to accept the null hypothesis of random distribution with high statistical power (Hubbell and Foster, 1985c).

When all woody plant species with 10 or more individuals in the census are classified with regard to both habitat and regeneration niche, we discover that the categories are very unequal in their number of species (Table 2). The category with the most species represents species indifferent to both habitat and regeneration microsite distinctions in the plot—generalist species able to regenerate equally well (or equally poorly) over the range of environmental circumstances

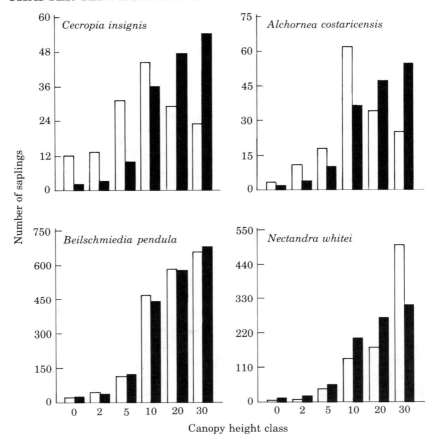

FIGURE 8. Distribution of 1–4 cm dbh saplings in six canopy height classes for four tree species representing four regeneration guilds of Barro Colorado Island trees. Canopy height measurements were taken at the corner of 20,000 5 × 5 m quadrats over the plot, and scored in the following height classes: "0" (0–2 m); "2" (2–5 m); "5" (5–10 m); "10" (10–20 m); "20" (20–30 m); and "30" (30 m and up). The black bars give the expected distribution of saplings in each category if the saplings were distributed at random with respect to canopy height class. The open bars give the observed number of saplings in each canopy height class. The species, from top left to bottom right, represent the guilds "sun" *(Cecropia insignis,* Moraceae), "partial sun" *(Alchornea costaricensis,* Euphorbiaceae), "indifferent" *(Beilschmiedia pendula,* Lauraceae), and "shade" *(Nectandra whitei,* Lauraceae). Note that the saplings of *Cecropia* are significantly skewed into low canopy height categories (recent gaps) ($p < 0.0001$), and that the saplings of *Alchornea* are concentrated in the intermediate canopy height categories (regenerating gaps) ($p < 0.001$). Saplings of the highly shade tolerant species *Nectandra* are strongly skewed away from gaps into the understory mature forest ($p < 0.001$). Saplings of the indifferent generalist *Beilschmiedia* are nearly randomly distributed in all canopy height categories *(p > 0.8).*

TABLE 2. Number of woody plant species, and patterns of species abundance, in relation to habitat and regeneration niche on Barro Colorado Island.

Habitat	Sun			Partial sun			Indifferent			Shade			Subtotals			Total species
	1000[a]	100	10	1000	100	10	1000	100	10	1000	100	10	1000	100	10	
Slope	0	0	0	0	6	4	0	15	4	5	4	2	5	25	10	40
Indifferent	0	9	18	3	17	8	14	21	18	9	2	0	26	49	44	119
Plateau	0	3	6	1	15	15	4	17	3	5	1	0	10	36	24	70
Streamside/ravine	0	0	2	0	1	0	0	0	2	0	0	1	0	1	5	6
Swamp	0	0	1	0	1	2	0	0	0	0	0	0	0	1	3	4
Subtotals	0	12	27	4	40	29	18	53	27	19	7	3	41	112	86	239
Total species		39			73			98			29			239		

[a] Column headings 10, 100, and 1000 refer to abundance classes 10–99, 100–999, and 1000+

encountered in the 50 ha. Summing across regeneration guilds, half (119) of the species were indifferent to slopes, flat ground, ravines or swamps. There was greater sensitivity to regeneration niche differences, with fewer species (98; 41 percent) exhibiting indifference to canopy height categories (i.e., gaps or forest interior).

Not only are there more generalist species than any other category, but these generalists are also more common. Dividing species into abundance classes by orders of magnitude, 10–99, 100–999, and 1000+ individuals respectively, we find that the largest group of common species occurs in the category of species indifferent to both habitat and regeneration microsite distinctions (Table 2). Moving away from this category to increasing specialization, we find that the number of occasional and rare species increases, and the number of common species declines. For example, only 4 (5.5 percent) of the species of the "partial sun" guild are common (more than 1000 individuals), and none of the extremely shade intolerant species ("sun" guild) is common. Finally, certain combinations of specialization do not occur among the Barro Colorado species. We find no species simultaneously exhibiting extreme shade intolerance and preference for slopes, for example; nor are there species with high shade tolerance ("shade" guild) that are specialized for swamps.

Specialization of rare species

We may now return to one of the questions posed about rare species, namely whether they are more specialized as a general rule than the common species. It is difficult to say much about specializations in species with fewer than 10 individuals; but for the 47 species with between 10 and 49 individuals, it is possible to construct another table like Table 2. We can then ask: Are rare species distributed in like manner to common species having an abundance of 50 or more individuals?

The answer is, No (Table 3). Rare species in aggregate show significantly greater degrees of specialization for both habitat and regeneration niches, as Chi-square tests of the marginal totals reveal. With regard to regeneration niches, there are too many extreme heliophiles ("sun" guild) among rare species, and too few "indifferent" generalists. With regard to habitats, there are too many rare species specialized to streamsides, wet ravines, and swamps—each a rare habitat in the plot—and too few species specialized for the more abundant habitats such as flat terrain ("plateau") or moderate grades ("slopes"). The picture for habitat specialization is less clear than for regeneration

TABLE 3. Distribution of species among habitat and regeneration niches on Barro Colorado Island.

A. Distribution of rare species (10–49 individuals/species) among habitat and regeneration niches.

Habitat	Regeneration niche				
	Sun	Partial sun	Indifferent	Shade	Totals
Slope	0	1	1	1	3
Indifferent	11	10	5	1	27
Plateau	3	6	0	0	9
Other (streamside, ravine, swamp)	3	2	2	1	8
Totals	17	19	8	3	47

B. Chi-square test over habitat categories.[a]

	Slope	Indifferent	Plateau	Other	Totals
Common species	37	92	61	2	192
Rare species	3	27	9	8	47
Totals	40	119	70	10	239

Chi-square over habitat categories: 29.1, 3 df, $p < 0.0001$

C. Chi-square test over gap regeneration niche categories.

	Sun	Partial sun	Indifferent	Shade	Totals
Common species	22	54	90	26	192
Rare species	17	19	8	3	47
Totals	39	73	98	29	239

Chi-square over regeneration niche categories: 25.6, 3 df, $p < 0.0001$[a]

[a] Chi-square tests of the marginal distributions are based on comparisons of the distributions of rare (10–49 individuals per species) with common (50 or more individuals per species) species. The marginal sums were tested in preference to a full two-way contingency analysis because of low frequencies of species in some categories.

niches because there are slightly more habitat generalists than expected among the rare species. This may be due in part to a weakened ability to statistically reject the generalist hypothesis for rare species. In spite of this, however, the percentage of generalists for both habitat and regeneration niche among rare species is only half (5; 10 percent) of the percentage found among all species (53; 22 percent) (Tables 2 and 3).

Thus, although many common species are generalists and many rare species specialists, rare generalists do exist. In these cases, the explanation for rarity must obviously lie elsewhere. The results suggest that being a generalist is a necessary but not a sufficient condition for becoming common in the Barro Colorado forest. For example, a species could be a generalist but be limited in abundance by an inability to regenerate near itself, by intense predation or fungal attack on seeds and seedlings, or by a host of other factors. Whatever the effects of these factors may be, however, they do not obscure the community-wide correlation of commonness with generalist species and rarity with specialist species, particularly with regard to gap regeneration niches.

Nevertheless, we still need to exercise caution in interpreting these results because such conclusions about specialization, especially habitat specialization, are dependent on spatial scale. The definition of a habitat specialist obviously depends on the habitat diversity included in the study, which increases monotonically with area considered. Thus, Rabinowitz et al. (Chapter 9) conclude that on geographic scales of distribution, species of high local abundance but restricted habitat are the most abundant class of species in the British flora. On local spatial scales such as the Barro Colorado Island forest plot, however, one might expect a strong correlation between total abundance and relative ability to grow in all available habitats.

Conclusions about specialization for regeneration niches are probably less sensitive to spatial scale, particularly in the case of specialization to gap-phase regeneration conditions. For example, we have found that rare species with populations everywhere sparse are more often extreme heliophiles that require large gaps to regenerate. Heliophilic species are expected seldom to become common in the Barro Colorado forest because the normal disturbance regime is one which mostly produces small gaps (Brokaw, 1982, 1985; Hubbell and Foster, 1985a,c). These species could increase in abundance, at least temporarily, following catastrophic disturbances that might occur on time scales of centuries. They may also become generally more common in the tropics as more and more tropical forest is converted to second growth by man.

Immigration of rare species

The preceding discussion makes the tacit assumption that the rare species are the result of reproductive events occurring within the plot; but this may not be true, especially in the case of the heliophilic rare species. Rarity in these species may be a combination of long-distance immigration from population centers outside the plot, coupled with an inability to increase once established because of rarity of required habitat or an unfavorable regime for regeneration. We know that many of the rare heliophiles are more common in second-growth forest. The Barro Colorado plot is bounded by 70–150 year old secondary forest on its northeastern edge; we believe that many of the 111 rare species were established as immigrants from this second-growth forest, and are now in decline. We can identify nearly two-fifths of the rare species in the plot as common (or commoner) species in second-growth forest elsewhere on the island (Table 4). This group includes nearly all of the rare heliophiles ("sun" and "partial sun" species) in Table 3.

TABLE 4. Some causes of rarity among species with fewer than 50 individuals in the Barro Colorado Island 50-ha plot.

Number of species[a]	Percentage of rare species	Apparent cause of rarity
9	8.1	Most plants too small[b]
12	10.8	Habitat restriction[c]
42	37.8	Common in second-growth forest[d]
5	4.6	Other (hybrid, selectively cut)[e]
43	38.7	Too rare to determine
111 (total)		

[a] The 111 species include all species in Table 3 and all additional rare species.
[b] A few species are artificially rare because most individuals are too small for our census.
[c] Excludes second-growth species, but includes species found along streamsides, in wet ravines, or in swamps that are not found in second-growth forest. Also includes species which are common in much drier (Pacific coast) or much wetter (Caribbean coast) forests.
[d] Includes all species that are habitat specialists within second-growth forest as well.
[e] A few species in this category may be rare because they were selectively cut. Although there is no present evidence for such cutting, these species are present in greater abundance in virgin forests of similar type elsewhere. The last of the other "species" is a hybrid between two species in the plot.

There are a number of rare species in the plot which are not second-growth heliophiles, but which are rare apparently because the plot catches only the edge of their true range in the old forest (Figure 9). As a group, rare species exhibit a strong "edge" effect, tending to occur significantly close to the perimeter of the plot. We divided the plot into the 26 border hectares and the 24 interior hectares and tested for significant concentration of individuals in the edge hectares, using the 68 rare species having 5 or more individuals. Nearly one-third (22) of these 68 rare species were significantly concentrated in edge hectares, versus only 5 percent of the common species.

Other reasons for rarity

Table 4 also indicates that some species are rare for spurious reasons: some species are more common than the census indicates because most individuals are too small. It is also possible that a few species may have been selectively logged out some time prior to the 1920s when the reserve was created. However, eyewitness accounts do not support the idea of logging on the Barro Colorado plateau, at least during the decade or so before the canal. Twelve species are rare because of restriction to streamsides, wet ravines, and swamps.

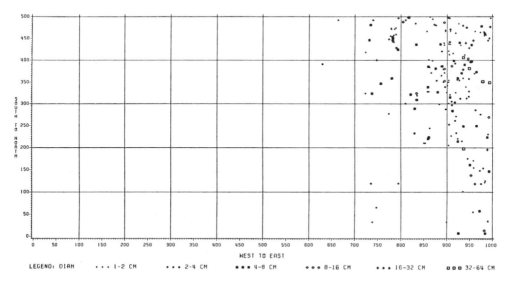

FIGURE 9. Distribution of *Inga pezizifera* (Leguminosae, Mimosoideae) in the Barro Colorado plot. This species is common to the east of the plot, but only the western edge of its range makes it into the 50-ha plot.

Forty-three species are so rare that little can be said about the causes of their rarity. For many species in this group we know next to nothing about their distribution elsewhere on the island or in central Panama. Nevertheless, we suspect that a number of these very rare species are probably also habitat or regeneration niche specialists. This conclusion is suggested by the pattern made by the joint distribution of all species with fewer than 10 individuals in the plot (Figure 2). This pattern reveals a concentration of individuals in the swamp and in several large light gaps. Tests of the light requirements of these rare species should reveal which ones are heliophiles.

FOREST IN NONEQUILIBRIUM

We tentatively conclude that many—quite possibly more than half—of the rare species (fewer than 50 individuals) in the Barro Colorado Island plot are second-growth species whose populations were probably established by immigration from the 70–150 year old secondary forest. Many may have immigrated when the second-growth forest was younger and their populations were at higher density. If this were the case, and if they are not currently self-replacing in the plot, then we might expect to find many "senescent" populations among the rare second-growth species—populations dominated by large individuals with few juveniles. An analysis of senescence in relation to abundance for 169 tree species confirms this expectation. Trees were divided into three classes of adult abundance: rare (1–9 adults); occasional (10–99), and common (100 and up) (Hubbell and Foster, 1985b). The percentage of species with more than 20 percent adults declined from 54.0 percent in rare species to 42.5 percent in occasional species, and further to 35.9 percent in common species. Because many of the rare Barro Colorado species are moderate or extreme heliophiles (Hubbell and Foster, 1985a), the pattern is as expected.

The conclusion that all rare species in the plot are in decline is unwarranted, however. Some species are probably rare because they are recent invaders of the plot. An example of such a species is *Psidium anglohondurense* (Myrtaceae), which appears to be expanding rapidly from a single original point of colonization (Figure 10). Sixteen rare species are represented solely by juveniles. In some of these cases, one or more adults may once have been present and subsequently died, but in most cases the widely scattered pattern of juveniles strongly suggests that long-distance dispersal from multiple parents was involved in their establishment. In a smaller number of cases already mentioned, rare species may be more persistent components of the tree flora of the plot, but are rare because their habitats are rare and local.

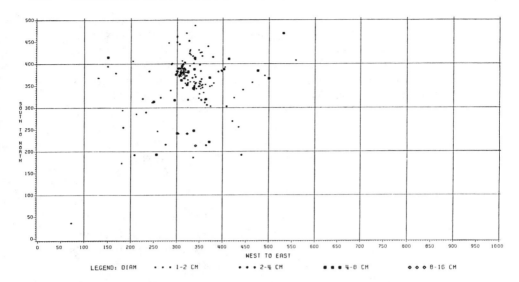

FIGURE 10. Distribution of the invading species *Psidium anglohondurense* (Myrtaceae) in the Barro Colorado plot. Most of the adults of this species are concentrated in the central nucleus, believed to be the single point of original colonization. (See caption to Figure 4 for additional explanation.)

Nevertheless, many of the rare species in the Barro Colorado plot do appear to be rather ephemeral components of the forest, waxing and waning in abundance with their changing abundance in forests surrounding the plot. It might therefore seem that inconstancy would characterize rare as opposed to common species, but our evidence suggests that many of the most common tree species in the island forest are themselves subject to major changes in abundance through time.

This nonequilibrium view of the forest is reviewed elsewhere (Hubbell and Foster, 1985b), but three observations are worth mentioning here. First, with the exception of the one or two most abundant species, most Barro Colorado tree species do not seem to exhibit negative density dependence sufficiently strongly to limit population sizes at current abundances. The most common canopy tree species, *Trichilia tuberculata* (Meliaceae), does exhibit strong density dependence. This density dependence is manifest in a statistically significant scarcity of *Trichilia* saplings in the vicinity of *Trichilia* adults (Hubbell and Foster, 1985d) and in a strong negative relationship between number of adult and juvenile trees per ha (Hubbell and Foster, 1985b). However, the remaining species do not show these patterns. Thus, although density-dependent factors may prevent extremely common species from increasing to total dominance in the forest, most tree species in the Barro Colorado forest appear to be at intraspecific densities far below those at which strong density dependence would be expressed.

Second, the size class distributions of populations of common species often show considerable spatial heterogeneity from one part of the plot to the next, suggesting that the populations are far from stationary (Hubbell and Foster, 1985c). Finally, there is little evidence that rare species in the Barro Colorado forest have any frequency-dependent per capita reproductive advantage over common species, as might be expected from some equilibrium community models. In fact, extreme rarity appears to present a major reproductive problem for dioecious species in the forest, which may suffer reduced fertilization rates at low densities (an Allee effect). There are very few dioecious species below an absolute density of 32 trees in the 50 ha plot, whereas many hermaphroditic species are this rare or rarer (Hubbell and Foster, 1985b).

TAXONOMIC PATTERNS OF RARITY

We left the question of the taxonomic status of rare species for last. Are they a random draw from genera and families, or are some taxa more likely to be rare than others? Given the suggestion that many rare species are habitat or regeneration niche specialists, it might seem that certain taxa would be more consistently rare than others. It is difficult to do statistics at the generic level except for the few Barro Colorado genera which are sufficiently rich in species; but for these genera, we can use the frequencies of very rare, rare, occasional, and common species in Table 1 to generate a null distribution for a test. From this we find, for example, that it is extremely unlikely by chance that all eight *Ficus* species would be as rare (66, 15, 12, 11, 9, 9, 8, and 6 individuals, respectively; $p < 0.004$), or that all four *Eugenia* species would be as common (2076, 952, 809, and 484 individuals, respectively; $p < 0.046$). However, species in some genera exhibit a considerable range in species abundances (for example, from one to 912 individuals in the case of the 14 species of *Inga*).

At the familial level, there are few clearcut differences of commonness and rarity, specialization or generalization. For example, most of the Barro Colorado genera of Bombacaceae are represented by occasional or rare shade intolerant heliophilic species (for example, *Bombacopsis, Cavanillesia, Ceiba, Pseudobombax,* and *Ochroma),* but *Quararibea* is a very common and quite shade tolerant species. For

Barro Colorado tree species in general, habitat and regeneration niche requirements are often quite similar among species of the same genus, yet quite different between genera. This reveals that these ecological traits have had multiple origins within families, developing independently at generic or specific levels many times.

Given diversification of habitat and regeneration niche requirements within families, we wondered whether families represented by many species in the plot might also have more abundant species than families represented by few species—or vice versa. The answer is potentially interesting in either case. Since common species are often generalists in the Barro Colorado forest, a "yes" answer suggests a possible link between ecological plasticity, abundance, and adaptive radiation (speciation). A "no" answer or no correlation suggests that niche differentiation, adaptive radiation, and speciation may be largely uncoupled from the local population and community ecology of tropical tree species.

No significant patterns between familial species richness and mean species abundance were found (Table 5). This result supports the conclusion that the principal control of tree species richness in most local tropical forests may not depend heavily on the community ecology, biotic interactions, and niche character displacement of the current species residents (Hubbell and Foster, 1985b). There appear to be many generalist tree species all behaving very similarly ecologically in a few broad adaptive zones. We suspect that the maintenance of high tree species diversity has more to do with regional tree species richness and availability of potential immigrants, which in turn are dictated by the interaction of climate, the historical dynamic biogeography of particular tree taxa, and speciation processes on a regional and subcontinental scale.

TABLE 5. Relationship between number of rare species and species richness of plant families of Barro Colorado tree species.

Species per family	Number of families	Observed number of species	Expected number of rare species	Observed number of rare species
1	14	14	5.1	7
2–3	19	44	15.9	13
4–7	14	74	26.8	22
8 and up	11	175	63.3	69

Chi-square = 2.64, 3 df, $p > 0.4$

IMPLICATIONS FOR TROPICAL TREE CONSERVATION

Our discussion presupposes that trees are the objects of conservation, not endangered primates, birds, or other organisms. However, given the diffuse mutualisms which exist between many tropical trees and their animal pollinators and dispersal agents, conservation efforts which ignore the needs of these animals, including provision for minimum viable populations (Gilpin and Soulé, Chapter 2), may fail in their objectives of conserving given tree species, or the animals that depend on them for food (Terborgh, Chapter 15).

Reserve size is a critical factor. There are probably few tropical forest reserves as small as ½ km², so our conclusions from the Barro Colorado plot would need to be scaled up to larger areas. Nevertheless we discuss 1 km² as a convenient minimum area for analysis, and we try to make some educated guesses at "hard" numbers whenever possible. Although we are in as good a position to make such estimates as anyone, the hardness of our numbers is difficult to gauge.

Diversity in one square kilometer of forest

Based on the species–area curve for the Barro Colorado plot, we can anticipate that between 30 and 50 more species would be added by doubling the plot's size to 1 km². This would bring the total to about 350 shrub, understory, midstory, and overstory tree species. This number of species is only about half the number which could be found in a similar space in the richest tropical forests of South America and the Far East. Of these 350 species, 130 would have fewer than 100 individuals (less than 1 plant per ha). Fifty to 75 of these species would be rare because of regeneration niche specialization (mainly large-gap requiring heliophiles), and an additional 15–20 species would be rare because of specialization to fixed habitats of limited area, tied to special topographic or edaphic site factors within the square kilometer. The remaining rare species are probably rare for similar but more extreme reasons, but are too rare to tell.

Many of the rare species would be present as a result of infrequent long-distance dispersal events and accidental establishment away from population centers of the species. The rare habitat specialists are expected to be common where their habitat area is extensive. However, extreme heliophiles are probably never very common in the closed-canopy forest, and become common only in secondary forests (which can occur naturally after large disturbances such as river course changes and forest destruction following major tropical storms, land-slides, and earthquakes).

Because most tropical forest reserves of the future will be surrounded by second growth, we can anticipate that the reserves will be invaded by many second-growth species. Thus, the diversity and perhaps the population density of these second-growth species will be artificially high in the reserve forest. This will be due to continual reimmigration from surrounding unprotected areas rather than to continual self-regeneration. A similar conclusion has been reached for the invasion of second-growth species into relict stands of mature dry forest in Guanacaste, Costa Rica (Janzen, 1983).

Tropical tree conservation strategies

The following recommendations can be made for strategies of tropical tree conservation, based in part on the present study. These recommendations are intentionally very general, but if more specific information should be required, the Barro Colorado results are sufficiently detailed to define preservation strategies for a large number of individual tropical tree species. The results are of more general signficance because many of the species in the study are widely distributed geographically, or are representative of genera found throughout the Neotropics, and many are of commercial importance.

1. Reserves of mature tropical forest should be large—minimally tens of square kilometers in size, preferably hundreds of square kilometers or larger. Among overstory trees in the Barro Colorado forest, only 22 species had over 100 adults per km^2. Given the patchy nature of the distributions of most tropical tree species, a 10 km^2 reserve would not guarantee a tenfold increase in adult population size of these species over what it would be in a 1 km^2 reserve (Hubbell and Foster, 1983).

2. Minimum critical population sizes are unknown for tropical trees. However, the nearly complete absence of rare dioecious tree species from the Barro Colorado forest suggests that low densities do exist, below which obligately outcrossed tropical trees suffer catastrophic decline to local extinction. For Barro Colorado dioecious overstory and midstory trees, we might guess these critical densities to be below 60 adults per km^2. Therefore, preservation strategies for obligately outcrossed trees must take into account local population sizes.

3. Larger reserves are required than might seem necessary based solely on current tree population densities. Adequate provision must be made to maintain the integrity of the pollinator and dis-

perser animal communities on which many of the trees depend for successful reproduction. Some of these animals (for example, birds, bats, and possibly insects) may depend on critical resources provided by certain "keystone" tree species (Gilbert, 1980; Terborgh, Chapter 15; Jordan, Chapter 20) and the animals may have to range widely to find these resources in seasons of scarcity. In the absence of detailed information about the resource needs of the pollinator and disperser communities in particular tropical forests, we should err on the side of too large rather than too small. This argues for reserves which are at least hundreds of square kilometers in size.

4. Large reserves are also required if tropical forests are nonequilibrium communities. Major fluctuations in tree species abundance and spatial migrations of populations can be expected to occur, and large areas and population sizes can better accomodate and buffer these fluctuations (Wright and Hubbell, 1983). With climatic change on time scales of centuries or millenia, the species composition of forests in reserves is certain to change. Moreover, change may occur on considerably shorter time scales if modifications of local climate (such as rainfall patterns) result from massive tropical deforestation. (The possibility of climate change was one rationale behind Brazil's attempt to locate reserves in areas thought to have been Pleistocene refugia.)

5. If possible, reserves should encompass diverse habitats to conserve as many habitat specialists as possible. The present study has shown that tropical tree species can respond to subtle differences in topography and soils. In addition, much of the niche differentiation of tropical trees is expressed in terms of gap regeneration requirements. Therefore, some attention to the natural disturbance regime would be desirable. For example, meandering streams, which cut off oxbow lakes and cause continual long-term forest succession, are an excellent complex natural disturbance regime, used to maintain high diversity in Manu National Park, Peru.

6. Gentry (Chapter 8) has suggested that many tropical tree species may be local endemics. If this is true, it may be difficult to save many of these species with the destruction of 90 to 95 percent of the world's tropical forests. Perhaps a more realistic goal in this case would be to design tropical forest reserves which effectively conserve populations of the many common mature forest tree species. Our data do not reveal how many rare Barro Colorado tropical tree species are local endemics, but the data do suggest that many rare species are present locally only because of continual long-distance recolonization. If this is so, efforts to preserve rare species will fail unless the main population centers of dispersal of these

species are preserved. In either case, we have another argument for large reserves.

7. Reserves should have minimal perimeter-to-area ratios (Lovejoy et al., Chapter 12; Janzen, Chapter 13). They should be round or square rather than linear to minimize the edge across which second-growth species can invade. This study has shown that second-growth species tend to be more concentrated around the perimeter of the plot, many species invading less than 100 m. If a forest buffer zone of, say, ten times this distance (1 km) could be established around the main block of the reserve, the invasion of second-growth species into the reserve interior might be kept to a minimum for very long time periods.

SUMMARY

Our current view of the structure and dynamics of the Barro Colorado Island forest can be summarized as follows:

1. The largest group of common species are generalists with respect to both gap regeneration and habitat requirements. Species richness and abundance per species both decline with increasing specialization in habitat and regeneration niche.
2. A greater fraction of rare species are apparently specialized by regeneration niche (moderate or extreme heliophiles) than are specialized by habitat, although both types of specialization are more prevalent among rare than among common species.
3. Many (at least one third) of the rare species (fewer than 50 total individuals) do not appear to have self-maintaining populations in the plot. Their presence appears to be the result of immigration from population centers outside the plot, and their numbers are probably kept low by a combination of unfavorable regeneration conditions, lack of appropriate habitat, or both, in the plot.
4. A smaller number of rare species are probably self-maintaining, especially the habitat specialists; and some rare species may be invading and increasing in abundance.
5. Rare species have no detectable per capita reproductive advantage over common species, based on current evidence, so we suspect that they are not maintained in the community by frequency-dependent competitive advantages.
6. Common species do not seem to exhibit strong density-dependence, unless they become extremely common—at which point density effects do prevent them from becoming complete dominants. Most species, however, appear to be at population densities far below intraspecific saturation of the habitat.

7. The forest is probably not in species equilibrium, and we predict continual changes in the relative abundance of tree species. Common species will not be immune from these changes, but rare species will suffer higher local extinction and turnover rates.

SUGGESTED READINGS

Foster, R.B. 1980. Heterogeneity and disturbance in tropical vegetation. In *Conservation Biology,* M.E. Soulé and B.A. Wilcox (eds.). Sinauer Associates, Sunderland, MA, pp. 75-90. Discusses tropical forest patchiness and the scale of natural disturbances as critical factors in the design of tropical forest reserves.

Hubbell, S.P. and R.B. Foster. 1983. Diversity of canopy trees in a neotropical forest and implications for conservation. In *Tropical Rain Forest: Ecology and Management,* S.L. Sutton, T.C. Whitmore and A.C. Chadwick (eds.). Blackwell Scientific, Oxford, pp. 24-41. Discusses how the diversity of tree species in the Barro Colorado forest increases with area, and how forest patchiness affects the species–area relationship—both issues of importance to tropical tree conservation.

Hubbell, S.P. and R.B. Foster. 1985a. Canopy gaps and the dynamics of a neotropical forest. In *Plant Ecology,* M.J. Crawley (ed.). Blackwell Scientific, Oxford, Chapter 3. Goes into greater detail on Barro Colorado forest regeneration and the dynamics of canopy gap formation and closure.

Hubbell, S.P. and R.B. Foster. 1985b. Biology, chance, and history, and the structure of tropical rain forest tree communities. In *Community Ecology,* T.J. Case and J. Diamond (eds.). Harper & Row, New York, pp. 314-329. Outlines various theories for the maintenance of tree species richness in tropical forests, with special reference to Barro Colorado Island.

Leigh, E.G. Jr., A.S. Rand and D.M. Windsor (eds.). 1982. *The Ecology of a Tropical Forest: Seasonal Rhythms and Long-Term Changes.* Smithsonian Institution Press, Washington, DC. An excellent collection of papers about the natural history of the Barro Colorado Island forest, including the ecology of the animals as well as plants.

Whitmore, T.C. 1984. *Tropical Rain Forests of the Far East,* Second Edition. Oxford University Press, Oxford. The best text on the dynamics and variability of tropical forests.

THE EFFECTS OF
FRAGMENTATION

THE DIMENSIONS OF THE EXTERNAL THREAT

One of the oldest and most powerful concepts in modern biology is the surface area-to-volume relationship. Depending on the scale, surface area-to-volume relationships determine, for example, the maximum sizes of cells, the minimum sizes of endothermic animals such as birds and mammals, and the thicknesses of leaves. Wherever the passive exchange of matter or energy is involved, two of the most important factors are the nature of the surface and the distance between the center and the outside. Analogously, nature reserves are like patches of epithelial cells on the skin of the earth. The health of these patches depends on, among other things, what happens at their edges and the distance between the periphery and the core. In general, the smaller the ratio of the periphery to the area, the better protected the patches will be from external threats.

Not all surface-acting effects, however, are from the sides or edges. As seen from space, a terrestrial nature reserve is a flat bit of biological film. Therefore, most of the exposure is "dorsal" to agents *in* the atmosphere (e.g., pollutants), agents *of* the atmosphere (e.g., greenhouse effects, and regional drying as a result of deforestation or desertization), or agents *through* the atmosphere (e.g., radiation, which is increasing as a result of ozone depletion). The spatial scale of these effects varies. For example, pesticide pollution is mostly local; drying and acid precipitation are mostly regional; and greenhouse warming is global.

Granted, these dorsal surface effects are not all caused by fragmentation (at least, not directly) and are therefore not immediately germane. Nevertheless, the degree of these changes and perturbations is proportional to the amount of habitat destruction and to the size of the human population. In other words, fragmentation and atmospheric impacts have the same ultimate causes, namely people. In the long run, I believe that general surface effects will come to be of even greater concern than edge effects.

233

Besides lateral (edge) and dorsal (surface) effects, fragmentation produces a third kind of impact. This is the "distance effect." As employed in island biogeography, this term refers to the inverse relationship of distance and the rate of colonization. However, the authors in this section point to another distance effect: the effect of fragmentation and isolation on seasonal migrations. As patches become more isolated, there are distance thresholds beyond which essential seasonal migrations become virtually impossible. Janzen (Chapter 13) provides the example of sphingid moths that must migrate between increasingly isolated patches of wet and dry forest. Wilcove, McLellan, and Dobson (Chapter 11) also give examples, and Terborgh (Chapter 15) notes with concern the threats from fragmentation to many tropical avian and mammalian frugivores, whose obligatory nomadic or migratory wanderings in search of episodic fruiting events make them vulnerable to starvation in a fragmented landscape.

THE "EDGE WEDGE"

In recent years, the variety of defined area and edge effects has increased by an order of magnitude. As indicated in Chapter 12 (Lovejoy et al.), it is also becoming difficult to distinguish between the effects of edge and the effects of size. In some cases, this is because the two types of effects are inherently inextricable. The rate of fire encroachment, for example depends on the humidity of the reserve and on its topographic relief, both of which are often correlated with size. Another example is the rate of penetration of livestock and the impact of native herbivores on a reserve; these rates may depend on the viability of large carnivores—also an effect, though an indirect one (Chapter 2), of area.

The effects of edge, size, and "sedge" (*size-edge*) can be categorized using a number of systems. Lovejoy et al. (Chapter 12) mention primary and secondary effects. An instance of a secondary effect is the increase in the populations of omnivores and small predators following the local extirpation of large predators, the probability of which is inversely related to reserve size. A tertiary effect would be the extinction of ground-nesting birds as a result of the above-mentioned secondary increase in small nest predators (Wilcove et al., Chapter 11).

It is important to recognize that some of the edge and sedge effects feed upon themselves autocatalytically. The result is a creeping edge that can eventually reach the core of even a relatively large reserve.

Janzen (Chapter 13) describes two such positive feedback processes: fire and the rain of seeds from secondary edge species.

"Edge creep" can also occur as a direct or indirect result of human activities that open up a closed-canopy forest or other kinds of mature habitat such as chaparral and coral reefs (Chapter 17, Johannes and Hatcher). The grazing of livestock and the cutting of vegetation for fuelwood or for building materials both have a tendency to facilitate access to deeper and deeper parts of a reserve. Another kind of edge creep is the result of poaching in the periphery of the reserve, assuming this eventually changes the density of hunted animals in the center.

Edges benefit some plants and animals because of the luxuriant growth of secondary species. Many successional plants are the hosts of insects, and many provide shelter and palatable food for a variety of vertebrates. For example, Lovejoy et al. (Chapter 12) point out that tamarins and marmosets flourish in such habitats, especially in small reserves free of competitors and predators. Of course, the populations of such secondary species in small reserves are subject to the demographic, genetic, and environmental stochasticities described in Chapter 2.

Such observations naturally lead to the proposal that edge creep in small reserves might be exploited to establish habitats of exceptional quality for species that depend on resources in successional stages. That is, rather that adopting a garrison mentality in the face of edge effects in small reserves, managers might consider sacrificing the primary vegetation and the deep forest animals of such reserves in order to benefit secondary species, especially in light of the deep penetration of some edge effects (Janzen, Chapter 13) and the expense of combating them.

There are, however, hazards attending such a passive strategy, depending on the objective of the particular reserve. For example, this approach would be counterproductive if the reserve were established to protect examples of vegetation in regions of high local endemism (as suggested by Gentry in Chapter 8). In addition, small reserves are unlikely to retain, for long, populations of many animal species that have critical roles in many ecological interactions, including plant dispersal and reproduction (e.g., pollinators such as bats; Culver, Chapter 21). Unfortunately, only long-term studies will inform us about the utility of small reserves in the tropics. Projects of the type initiated by Lovejoy et al. (Chapter 12), designed especially to describe such processes and to test hypotheses, should be established throughout the tropics, particularly in Africa and Southeast Asia, not to mention marine habitats.

MANAGEMENT IMPLICATIONS

This entire overview has emphasized management implications, so only a few more words need be expended. First, fragmentation and its consequences is a theme that runs throughout this book, though it is emphasized in the following three chapters.

Second, the majority of the challenges (including many social, economic, and political ones) to be faced by managers will be a consequence of fragmentation and its attendant effects, some of which will be surface effects. At first, many of the effects of fragmentation will be subtle and will only be detected by field work, censuses, and monitoring. The health of reserves, therefore, will depend on communication and cooperation between field workers and managers. Though their ranking of priorities will never be identical, researchers and managers must eventually acknowledge that their long-range interest—the preservation of biological diversity—is a shared concern of the highest ethical weight.

Third, as pointed out by Janzen (Chapter 13), the check-list of possible edge effects is virtually infinite, and every reserve will be subject to its own unique problems. Therefore, there can be no substitute for constant vigilance on the part of managers and field workers.

M.S.

CHAPTER 11

HABITAT FRAGMENTATION IN
THE TEMPERATE ZONE

David S. Wilcove, Charles H. McLellan and
Andrew P. Dobson

In this chapter we examine three questions relating to habitat fragmentation in the temperate zone: (1) What is the effect of fragmentation on the species originally present in the intact habitat? (2) How does fragmentation lead to the loss of species? (3) For an already fragmented landscape, are there any guidelines for the selection and management of nature reserves? Here we shall set as our goal the long-term preservation of those species whose continued existence is jeopardized by habitat destruction. At the outset we note that this chapter is slanted towards vertebrate communities (especially birds) and forested habitats. Our bias reflects, in part, a bias in the existing literature. On the other hand, by virtue of their low population densities, birds and mammals are among the taxa most likely to disappear from isolated fragments (Wilcox, 1980).

INTRODUCTION

Fragmentation occurs when a large expanse of habitat is transformed into a number of smaller patches of smaller total area, isolated from each other by a matrix of habitats unlike the original. When the landscape surrounding the fragments is inhospitable to species of the original habitat, and when dispersal is low, remnant patches can be considered true "habitat islands," and local communities will be "isolates" (*sensu* Preston, 1962). If the matrix can support populations of many of the species from the original habitat, or if dispersal between patches is high, communities in the fragments will effectively be "samples" from the regional faunal or floral "universe" (Preston, 1962; Connor and McCoy, 1979). The process of isolate formation through fragmentation has been termed "insularization" by Wilcox (1980). The challenge to conservationists is to preserve as much of the species pool as possible within these fragments, in the face of continual habitat destruction.

Habitat fragmentation has two components, both of which cause extinctions: (1) reduction in total habitat area (which primarily affects population sizes and thus extinction rates); and (2) redistribution of the remaining area into disjunct fragments (which primarily affects dispersal and thus immigration rates). Both Lovejoy et al. (1984) and Haila and Hanski (1984) stress the need to partition extinctions into those caused purely by habitat destruction and those in which insularization is an important additional component.

Temperate communities are widely believed to be more resistant to the effects of habitat fragmentation than are tropical communities. Temperate species tend to occur in higher densities, be more widely distributed, and have better dispersal powers than their tropical counterparts. These attributes should allow populations to persist in smaller patches of suitable habitat. Although local extinction rates may be high (due to high levels of population fluctuation and shorter individual lifespans; see Diamond, 1984a), high vagility can facilitate rapid recolonization from other fragments following extinction (Brown and Kodric-Brown, 1977).

On the other hand, one of the main reasons why habitat fragmentation seems less severe in the temperate zone is that most of the damage was done long before most people were aware of it. For example, in Great Britain reduction and fragmentation of the original forest cover began some 5000 years ago with permanent clearances by Neolithic farmers, and was well advanced by the time of the Norman Conquest in 1066 (Figure 1). Species whose extinctions in Great Britain were certainly related to the destruction of the original forest (as well as other causes, especially hunting) include: brown bear (extinct by the time of the Norman Conquest), wild boar (18th century), wolf (18th century), goshawk (19th century; now reestablished in new conifer plantations), and capercaille (18th century; now reintroduced in new conifer plantations).

Much the same story can be told for the fauna of the deciduous forest of the eastern United States, where widespread forest destruction began with the arrival of European settlers (about 300 years ago) and reached a peak about the time of the Civil War. Here, too, a number of species vanished from the east as a result of habitat destruction combined with hunting. These include wolf (19th century), mountain lion (20th century, although a few persist in Florida), elk

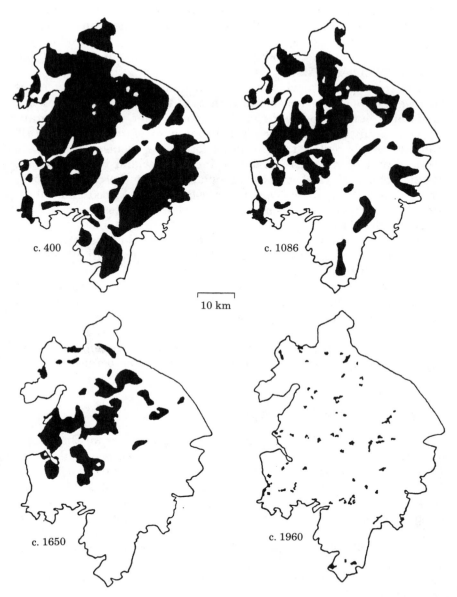

FIGURE 1. Forest fragmentation in Warwickshire, England, from 400–1960 A.D. Forested areas are shown in black. (Redrawn from Thorpe 1978.)

(19th century), passenger pigeon (20th century), and ivory-billed wood-pecker (20th century). For both the British and American species, it is probably correct to say that even if they could be reintroduced to their former haunts, the outcome would be disappointing: suitable habitat no longer exists.

The species that survived this initial round of habitat fragmentation were the ones better able to withstand the human impact on the landscape, but by no means is the problem over. In Great Britain today, pressure on the land is so great that many of the "semi-natural" habitats which replaced the original forest are themselves severely reduced and fragmented. Examples include lowland heaths (Moore, 1962; Webb and Haskins, 1980), upland moors (Porchester, 1977; Parry et al., 1981) and calcareous grasslands (Blackwood and Tubbs, 1970; Jones, 1973). There is now a growing list of species characteristic of, or restricted to, these habitats whose declines or extinctions can be attributed at least in part to fragmentation (see Hawkesworth, 1974 for a general survey). In the United States, the continuing fragmentation of such habitats as old-growth Douglas fir forests in the Pacific northwest (Harris, 1984), deciduous forests in the east (Burgess and Sharpe, 1981; Whitcomb et al., 1981), and grasslands in the midwest has prompted concern for the continued survival of the species that inhabit them (including small whorled pogonia, greater prairie chicken, spotted owl, and Delmarva fox squirrel). Fragmentation remains the principal threat to most species in the temperate zone.

A MODEL OF FRAGMENTATION

Analyses of the effect of fragmentation (and guidelines for the design of nature reserves) have generally been based on the conceptual framework of island biogeography (Preston, 1962; MacArthur and Wilson, 1967; Soulé and Wilcox, 1980; Burgess and Sharpe, 1981). This theory suggests that the number of species on an oceanic island represents a balance, or dynamic equilibrium, between processes of immigration and extinction. The equilibrium number of species on an island depends upon the characteristics of the island—in particular, its size and isolation from potential sources of colonists—and the characteristics of the species themselves—in particular, their dispersal abilities and population densities.

We have recently (McLellan et al., 1986) developed a computer model that simulates the effects of habitat fragmentation on two pools of species with different minimum area requirements and dispersal abilities. This model has led to a number of insights regarding the extent of fragmentation that different species can tolerate.

TABLE 1. Scientific names of plants and animals mentioned in the chapter.

Plants	Dog's mercury: *Mercurialis perennis*
	Small whorled pogonia: *Isotria medeoloides*
Insects	Large blue butterfly: *Maculinea arion*
Amphibians	Red-spotted newt: *Notophthalmus viridescens*
Birds	American woodcock: *Scolopax minor*
	Blue-gray gnatcatcher: *Polioptila caerulea*
	Blue jay: *Cyanocitta cristata*
	Brown-headed cowbird: *Molothrus ater*
	Capercaille: *Tetrao urogallus*
	Common grackle: *Quiscalus quiscula*
	Goshawk: *Accipiter gentilis*
	Greater prairie chicken: *Tympanuchus cupido*
	Great spotted woodpecker: *Picoides major*
	Ivory-billed woodpecker: *Campephilus principalis*
	Kirtland's warbler: *Dendroica kirtlandii*
	Louisiana waterthrush: *Seiurus motacilla*
	Passenger pigeon: *Ectopistes migratorius*
	Spotted owl: *Strix occidentalis*
Mammals	Bobcat: *Lynx rufus*
	Brown bear: *Ursus arctos*
	Delmarva fox squirrel: *Sciurus niger cinereus*
	Eastern chipmunk: *Tamias striatus*
	Elk: *Cervus elaphus*
	Gray squirrel: *Sciurus carolinensis*
	Mountain lion: *Felis concolor*
	Opossum: *Didelphis virginiana*
	Raccoon: *Procyon lotor*
	Short-tailed weasel: *Mustela erminea*
	White-tailed deer: *Odocoileus virginianus*
	Wild boar: *Sus scrofa*
	Wolf: *Canis lupus*

The pattern of fragmentation of our hypothetical habitat is based largely on that of heathland in Dorset, England, as reported by Moore (1962) and Webb and Haskins (1980). In our model, the original habitat is reduced from five extremely large tracts to an archipelago of over 450 fragments totalling 5 percent of the original area. For simplicity we have shown the total area of habitat decreasing linearly over time; in reality, the rate of destruction usually increases with time. The total number of fragments increases exponentially over time, reflecting a distribution increasingly skewed towards a large number of very small fragments (Figure 2). At each stage of fragmen-

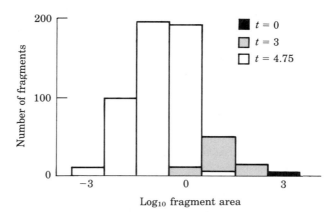

FIGURE 2. The frequency distributions of fragment sizes at various stages of the fragmentation sequence. The top diagram corresponds to the original ($t = 0$), largely contiguous habitat. Fragmentation results in a larger number of smaller patches (lower diagrams) with a total area less than that of the original habitat. (From McLellan et al., 1986.)

tation, the remaining area of habitat is distributed among the growing number of fragments in a roughly lognormal fashion. There is no information on interfragment distances for the heathlands, so we have borrowed a pattern from another system, namely woodland in Cadiz township, Wisconsin. Sharpe et al. (1981) have shown that mean nearest-neighbor distances increased until about 10 percent of the original area was left, and then remained constant despite further losses.

Details of the model

The two species pools were chosen to represent the extremes of susceptibility to fragmentation: one (resistant) pool consisted of species with good dispersal abilities and a low average proneness to local extinction. The other (susceptible) pool had species with poor dispersal abilities and a high average proneness to local extinction. The results therefore define qualitatively the range of patterns of species loss to be expected in habitats undergoing fragmentation. The fates of the P species in each pool were modeled using a species-by-species ("molecular"; see Gilpin and Diamond, 1981) formulation of the equilibrium model. The basic variable is the probability J_i that a given species i ($i = 1,2 \ldots P$) occurs as a breeding population in a fragment. This probability, termed the "incidence" (Diamond, 1975a) of the species, increases with fragment area (A), due to increasing population size and thus decreasing chance of stochastic extinction, and decreases with distance (D) from a source of colonists due to decreasing frequency of immigration. Observed values of J_i plotted against area for a given distance, or vice versa, are called "incidence functions" (Diamond, 1975a; Figure 3). Elaborations such as dependence of extinction rates on distance (due to the "rescue effect;" see Brown and Kodric-Brown, 1977) and dependence of immigration rates on area (due, for example, to "passive sampling;" see Connor and McCoy, 1979) are not incorporated in this model.

This model has the general form

$$\frac{dJ_i}{dt} = I_i - E_i \tag{1}$$

where I_i is the net rate at which unoccupied fragments are colonized by species i and E_i is the net rate at which the species becomes locally extinct. We chose to use a version of this general model first presented by Levins and Culver (1971):

$$\frac{dJ_i}{dt} = a_i J_i (1 - J_i) - b_i J_i \tag{2}$$

where a_i is an instantaneous colonization rate per occupied fragment and e_i is an instantaneous local extinction rate. This model differs from the version described by Gilpin and Diamond (1981) in that the instantaneous rate of colonization of unoccupied fragments ($a_i J_i$) is a

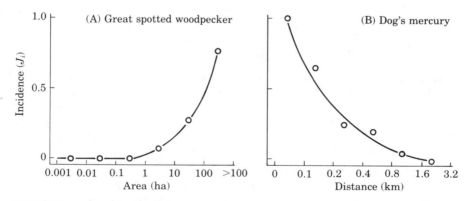

FIGURE 3. Incidence functions for two species in forest fragments in Britain. (A) Incidence (J_i, the proportion of woods in each class occupied by a population) of the great spotted woodpecker plotted against woodland area. (From Moore and Hooper, 1975.) (B) Incidence of dog's mercury, a woodland herb, in secondary forest in Lincolnshire plotted against the distance to the nearest older forest containing the species. (From Peterken and Game, 1981). Curves were drawn by eye.

function of J_i rather than a constant. This is more appropriate for fragmented systems, where colonists must come from other (occupied) fragments rather than from some large, inviolate, "mainland" area. At equilibrium I_i and E_i are equal, and J_i is constant at the equilibrium incidence level J_i^*. For equation (2) this is

$$J_i^* = 1 - \frac{b_i}{a_i} \tag{3}$$

Thus incidence is positive only when the colonization rate (a_i) exceeds the extinction rate (b_i).

Incidence functions were generated by specifying the dependence of a_i on distance and b_i on area. We used an exponential function for $a_i = f(D)$

$$a_i = c_i \cdot \exp \frac{-D}{D_i} \tag{4}$$

where c_i is a colonization coefficient and $1/D_i$ the death rate per unit distance of migrants (Gilpin and Diamond, 1976). An inverse hyperbolic function was used for $b_i = f(A)$

$$b_i = e_i/A \tag{5}$$

where e_i is an extinction coefficient (Gilpin and Diamond, 1976). Substituting equations (4) and (5) into equation (3) gives the expression for J_i^* as a function of area and distance

$$J_i^* = 1 - \left[\frac{e_i}{c_i} \cdot \exp \frac{D/D_i}{A}\right] \tag{6}$$

When $D/D_i \simeq 0$ (i.e., in nonisolated fragments), equation 6 describes a hyperbolic function with increasing area, with the parameter e_i/c_i corresponding to the area at which incidence is zero. We distributed this "minimum area" lognormally with variance 1.0 in each of the species pools, with the susceptible pool having a mean value an order of magnitude greater than the resistant one. (For theoretical justification of the lognormal in such models see Gilpin and Armstrong, 1981, and for empirical evidence see Gilpin and Diamond, 1981.) For a given area, equation (6) describes an exponential function with the parameter D_i (the "mean dispersal distance;" see Gilpin and Diamond, 1976) corresponding to the distance required to reduce J_i to $1/e$ (36.8%) of its values at $D = 0$. We treated D_i unrealistically as constant among species in a given pool, assigning the susceptible pool a value half that given to the resistant pool.

Results of the model

The results of this exercise are illustrated in Figure 4. Initially, when a large amount of habitat remains, mostly in large fragments, few or no species are lost from either pool. As fragmentation proceeds we eventually reach some critical level of reduction and fragmentation where species begin to die out. The susceptible pool loses species earlier and loses more species in total than does the resistant pool. When the resistant pool begins to lose species, it loses them very rapidly, because by this time the fragments are small and there is little habitat left.

Insularization causes extinctions over and above those expected through reduction in the total area of habitat. More species persist at equilibrium if the remaining habitat is concentrated into a single large patch rather than distributed over many small fragments (Figure 4). We stress that the results in Figure 4 are equilibrium patterns; depending on the relative time scales of habitat destruction and species'

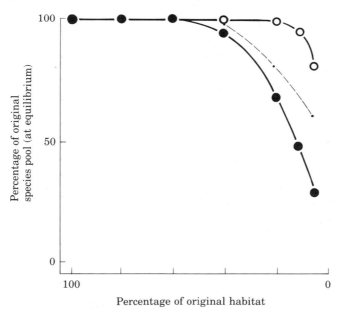

FIGURE 4. The number of species remaining in each species pool as fragmentation proceeds. Closed circles show the pool of species with large area requirements and low vagility. Open circles show the species with less stringent area requirements. The small dots connected by the dashed line depict the proportion of the first pool that would be present when the habitat is minimally fragmented. (From McLellan et al., 1986.)

population dynamics, extinctions may either closely track the changing landscape patterns or lag behind them.

An important omission from the model is the explicit inclusion of population size as a variable (see Schoener, 1976; Williamson, 1981). Species' carrying capacities are assumed to be directly proportional to fragment size, and extinctions simply the result of demographic stochasticity. It seems more likely both that habitat heterogeneity means that the carrying capacity is not a simple function of area, and that factors like environmental stochasticity and population structure are important in determining extinction rates (see MacArthur and Wilson, 1967; Richter-Dyn and Goel, 1972; Leigh, 1975, 1981 for details; Diamond, 1984a; and Gilpin and Soulé, Chapter 2 for a summary). Colonization will also be affected by habitat heterogeneity and population dynamics. If a fragment does not contain enough suitable habitat for a given species, establishment of a breeding population will not occur no matter how high the immigration rate. Similarly, abundant species will produce more colonists than scarce ones. Such factors will need to be included in more realistic (albeit more complex) models.

Despite these limitations, we believe the model provides a clear message: Even where most of the habitat has already been destroyed, subsequent fragmentation should be minimized, lest a rapid loss of species occur. Furthermore, insularization can cause extinctions independent of habitat reduction.

MECHANISMS OF EXTINCTION

The above theoretical discussion has taken something of a black box approach to fragmentation. In this section, we shall focus on the proximate mechanisms of extinction. These include home range size, loss of habitat heterogeneity, effects of habitats surrounding fragments, edge effects, and secondary extinctions.

Home range

Some fragments will be smaller than the minimum home ranges or territories of certain species. This is often the case for large animals. For example, a single pair of ivory-billed woodpeckers may require 6.5–7.6 km^2 of undisturbed bottomland forest (Tanner, 1942). The European goshawk has a home range of approximately 30–50 km^2 (Cramp and Simmons, 1979). Male mountain lions in the western United States may have home ranges in excess of 400 km^2 (Seidensticher et al., 1973). However, species often disappear from habitat fragments that far exceed their minimum home range sizes; mechanisms other than home range limitations must be operating in such cases.

Loss of habitat heterogeneity

One common consequence of fragmentation is a loss of habitat heterogeneity. Even a seemingly uniform expanse of habitat such as forest or grassland is, at some level of discrimination, really a mosaic of different habitats. Individual fragments may lack the full range of habitats found in the original block. Patchily distributed species or species that utilize a range of microhabitats are especially vulnerable to extirpation under these circumstances. An example of such a patchily distributed species is the Louisiana waterthrush of eastern North America. It nests and forages near open water, especially fast-moving streams (Chapman, 1907). Woodlots without open water do not provide suitable habitat for the waterthrush, and, not surprisingly, this bird is one of a number of forest songbirds that are rarely encountered in small woodlots (Robbins, 1980).

While the habitat requirements for most songbirds are far less obvious, they may nonetheless play a major role in the response of these birds to fragmentation. Lynch and Whigham (1984) studied the bird communities and vegetation of 270 woodlots in Maryland. They discovered that structural or floristic characteristics of the vegetation strongly influenced the local abundance of each bird species. These vegetation characteristics, in turn, varied with the forest size. A qualitatively similar result was noted by Bond (1957) in forest fragments in southern Wisconsin. Although generalizations are risky, we might expect this problem of patchiness to be most acute for plants, and for insects that depend upon specific host plants.

When species require two or more habitat types, fragmentation may make it impossible for them to move between habitats. Karr (1982a) has attributed many of the extinctions of landbirds of Barro Colorado Island, Panama to just this mechanism. Within the temperate zone, this problem is likely to befall many kinds of organisms. The red-spotted newt is typical of a number of amphibians in having both a terrestrial and an aquatic stage. The terrestrial efts may remain ashore for up to three years, but eventually must return to the water to breed. Among birds, the blue-gray gnatcatcher in California moves from deciduous oak woodlands to chaparral and live oaks over the course of the breeding season (Root, 1967). Other temperate zone birds are also believed to make seasonal shifts in their ranging behavior (MacClintock et al., 1977). Unfortunately, too little is known about the behavior of most temperate zone animals to say with certainty what their habitat requirements are, or how these requirements may change seasonally. This lack of knowledge is always an obstacle to predicting the effects of fragmentation on individual species. Detailed information on habitat usage will be crucial to devising successful conservation programs for many species.

Effects of habitat between fragments

In the case of a true island, the ocean is an impassive barrier, and potential colonists will either traverse it successfully or perish in the attempt. In the case of a habitat fragment, the ocean has been replaced by a landscape of human dwellings or agricultural land. This landscape can also be a formidable barrier to colonists from the fragments. Unlike an ocean, however, a human-created landscape can contribute directly to the extinction of species within fragments. It does so by building up populations of animals that are harmful to species within the fragments. A good example of this problem comes from studies of forest-dwelling songbirds in forest fragments in the eastern United States.

Breeding populations of songbirds have been declining in small woodlots throughout the eastern United States since the late 1940s (Robbins, 1979; Whitcomb et al., 1981; Wilcove, 1985a). A number of factors have contributed to this decline, two of the most important being high rates of nest predation (Wilcove, 1985b) and brood parasitism by the brown-headed cowbird (Mayfield, 1977; Brittingham and Temple, 1983). In recent decades, the numbers of nest predators and cowbirds have increased greatly as a result of human-induced changes in the landscape.

Among the nest predators, blue jays, raccoons, and gray squirrels all occur in higher densities in suburban communities than in more natural habitats like forests (Flyger, 1970; Fretwell, 1972; Hoffman and Gottschang, 1977). Prior to the arrival of European settlers, the cowbird was largely confined to the grasslands of the midcontinent, where it followed the grazing mammals and ate the insects they stirred up. With the disruption of the eastern deciduous forest and the introduction of livestock, the cowbird spread throughout the eastern United States and Canada (Mayfield, 1977). More recently, the cowbird population in eastern North America has increased tremendously due to an increase in their winter food supply—waste grain in southern rice fields (Brittingham and Temple, 1983). The advent of mechanical harvesters has simultaneously increased the amount of land under rice cultivation and the amount of waste grain. This range expansion and population increase has brought the cowbird in contact with populations of forest-dwelling songbirds, most of which lack behavioral defenses against cowbird parasitism (Rothstein, 1975; May and Robinson, 1985).

No habitat preserve is immune to the effects of human activity outside its borders, and wildlife managers must concern themselves with the ecological effects of land development outside the boundaries of protected areas. To quote Janzen (1983), "No park is an island."

Edge effects

Wildlife managers have long extolled the virtues of forest edge (see for example Dasmann, 1964, 1971; Yoakum and Dasmann, 1969; Burger, 1973), in a tradition dating back to the writings of Aldo Leopold (1933). Certainly a variety of game animals, including white-tailed deer and American woodcock, do well in edge habitats. But it is becoming increasingly clear that the forest edge has a strong negative impact on other members of the woodland flora and fauna (Hubbell and Foster, Chapter 10; Lovejoy et al., Chapter 12; Janzen, Chapter 13).

Ranney et al. (1981) believe that the seed rain into the cores of small woodlots is dominated by the seeds of the edge species. This may ultimately change the species composition of the woodlots, as the shade tolerant plants of the interior are replaced by shade intolerant forms from the edge. Such an effect would require the number of plants germinating to vary with the number of seeds set in the interior. Ranney et al. note that very small or irregularly shaped forest reserves may be unable to sustain populations of forest interior plants.

Field studies by Gates and Gysel (1978), Chasko and Gates (1982), and Brittingham and Temple (1983) have shown that the nesting success of songbirds is lower near the forest edges than in the interior (Figure 5). This is because many nest predators (blue jay, American crow, common grackle, eastern chipmunk, short-tailed weasel, raccoon) and brood parasites (brown-headed cowbird) occur in higher densities around forest edges (Bider, 1968; Robbins, 1980; Whitcomb et al., 1981; Brittingham and Temple, 1983).

For management purposes, it is important to know how far into the forest the influence of the edge is felt. Studies by Ranney (1977) and Wales (1972) show that the major vegetational changes caused by the edge extend only 10–30 m inside the forest, depending on whether the edge has a northerly or a southerly exposure. However, by placing artificial nests at varying distances from the edge, Wilcove (1985a) has shown that the edge-related increase in predation may extend from 300–600 m inside the forest (Figure 6). It should not be surprising

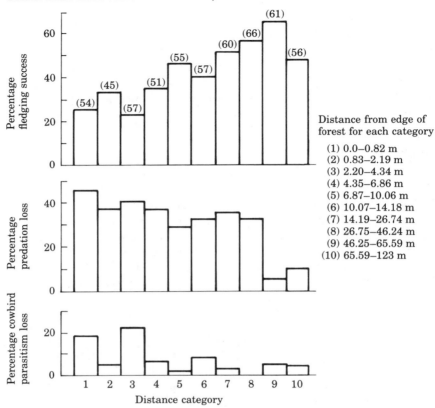

Distance from edge of
forest for each category

(1) 0.0–0.82 m
(2) 0.83–2.19 m
(3) 2.20–4.34 m
(4) 4.35–6.86 m
(5) 6.87–10.06 m
(6) 10.07–14.18 m
(7) 14.19–26.74 m
(8) 26.75–46.24 m
(9) 46.25–65.59 m
(10) 65.59–123 m

FIGURE 5. Nesting success of songbirds as a function of distance from the forest edge, based on a study in Michigan. Distance categories 1–10 are delimited at right, sample sizes are in parentheses. (From Gates and Gysel, 1978.)

that the faunal effects of a forest edge would exceed the floral effects. Birds like crows, grackles, and cowbirds are not intolerant of the forest interior. Similarly, mammals like the raccoon, weasel, and chipmunk, while concentrating their activities near the forest edge, will also frequent the forest interior (Whitaker, 1980).

One consequence of these observations deserves special emphasis: if 600 m is taken as a liberal estimate of the faunal edge effect, then circular reserves smaller than 100 ha will contain no true forest interior. In the case of forest songbirds, this finding suggests that reserves should contain *at least* several hundred ha of uninterrupted forest. In fact, far larger areas may be needed to ensure the long term survival of these birds (Whitcomb et al., 1976).

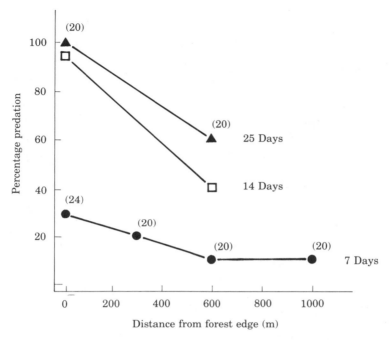

FIGURE 6. Percentage of experimental nests preyed upon as a function of distance from the forest edge. Nests are wicker, open-cup baskets containing fresh quail eggs. Solid circles are the predation rates after 7 days, open squares are after 14 days, and closed triangles are after 25 days. The numbers in parentheses are the numbers of experimental nests. The data suggest that the edge-related increase in predation may extend from 300–600 m inside the forest. (From Wilcove, 1985a.)

Secondary extinctions

Fragmentation often disrupts many of the important ecological interactions of a community, including predator–prey, parasite–host, and plant–pollinator relations, and mutualisms (Gilbert, 1980; Terborgh and Winter, 1980). The disruption of these interactions may lead to additional extinctions, sometimes referred to as "secondary extinctions." Typically, these secondary extinctions are associated with the decay of complex tropical communities, but they are certainly not unknown in the temperate zone. For example, small woodlots in the eastern United States support few, if any, large predators like mountain lions, bobcats, and large hawks or owls that may regulate populations of smaller, omnivorous species like raccoons, opossums, squirrels, and blue jays (Matthiae and Stearns, 1981; Whitcomb et al.,

1981). These omnivores, in turn, prey upon the eggs and nestlings of the forest songbirds. As noted earlier, the rate of nest predation in small woodlots is very high, and this may be one reason why songbird populations have declined. A similar explanation has been invoked to explain some of the avian extinctions on Barro Colorado Island, Panama (Terborgh, 1974).

A more complicated example involves the extinction of the large blue butterfly in Britain (Thomas, 1976; Ratcliffe, 1979). This butterfly has a remarkable life history in that the larvae must develop within the nests of the red ant *Myrmica sabuleti*. The large blue was brought to the brink of extinction when land development and reduced grazing by livestock eliminated the open areas it required. The remaining populations vanished through a complex chain of events. An epidemic of myxomatosis in the mid-1950s depressed rabbit populations, and as a result many of the sites became overgrown with scrub. The *Myrmica sabuleti* ants were unable to survive in the overgrown areas, and their decline meant the end of the large blue.

We suspect that as more data on fragmentation are gathered, secondary extinctions will prove a common occurrence in temperate communities. The prevention of these extinctions will require synecological studies involving threatened species, coupled with active management of preserves.

GUIDELINES FOR TEMPERATE ZONE RESERVES

Although blanket prescriptions for the design of nature reserves (Wilson and Willis, 1975; Diamond, 1975b) have come under criticism in recent years (examples include Simberloff and Abele, 1976; Abele and Connor, 1979; Higgs and Usher, 1980; Game, 1980; Margules et al., 1982; Boeklen and Gotelli, 1984), we believe that the theory of island biogeography provides a useful framework within which more detailed studies of particular cases can be planned. In this final section, we focus on three questions:

1. How much of the available habitat must be set aside as reserves, and in what distribution of sizes?
2. Should reserves be clustered together in close proximity to each other, or spread out over a broad area?
3. What is the optimum shape for reserves?

It is important to realize that the "correct" answers to these questions may depend very much on the scale of the conservation effort, because local, regional, and national conservation operations usually operate under very different budget constraints and spatial scales. Once again, we note that our perspective on these questions is rather ornithocentric.

How much and how large?

A characteristic pattern in habitats undergoing fragmentation is an increasingly skewed distribution of fragment sizes as the total area of habitat declines. In general, a large proportion of the remaining area of highly fragmented habitats should be targeted for protection in order to avert (or at least minimize) the biotic collapse which models suggest can occur in such systems. All other things being equal, priority should go to the largest remaining fragments, for several reasons. First, as emphasized in our model, different species have different area requirements, and the large fragments will often be the only refuge for species which exist at low densities (such as top predators and large herbivores) or who are habitat specialists whose requirements are only satisfied in large areas. Second, the large fragments may well serve as sources of immigrants for marginal populations in neighboring small fragments. If many species are maintained in these small fragments by the "rescue effect" (Brown and Kodric-Brown, 1977), then the small fragments do not represent a viable reserve strategy on their own (although they may be useful in an integrated regional strategy; see below). Third, the trend will always be for large fragments to be eroded unless protected. Because of the cost involved, the responsibility for acquiring and managing these large reserves must rest primarily with national conservation organizations.

The foregoing discussion is not meant to denigrate the value of small reserves. Indeed, their selection emerges as a logical strategy when one considers the different levels of organization and scale at which conservation policy is determined (McLellan et al., 1986). We have argued above that the largest fragments of threatened habitats should generally be obtained as reserves by *national* conservation organizations. However, in a heterogeneous environment these reserves may not encompass all of the habitat variation (and thus all of the characteristic biota) present in the ecosystems concerned. Thus, we suggest the primary task for conservation organizations operating on a *regional* scale should be to distribute their funds for land acquisition among a series of medium sized reserves designed to capture this variation. The optimal trade-off between capturing more habitat heterogeneity (by purchasing several smaller reserves) and maintaining viable populations of area-sensitive species (by purchasing fewer larger reserves) will have to be determined by detailed studies of each particular system (Simberloff and Abele, 1982). Conservation on a *local* scale, as in a township, operates under the tightest budgetary constraints and thus the most restricted size range for possible reserves (all of which will be small in absolute terms). However, it is possible to state with some confidence that the best strategy at this scale is to go for single "large" reserves rather than several (very) small ones. There are two main reasons: (1) The slope (z) of the species–area relationship is normally steep (>0.35) at small areas where fragments contain a small proportion (<0.25) of the species pool (Martin, 1981); (2) The similarity in species composition among the small local reserves will usually be very high (>>0.5) because of their physical proximity and likely similarity in habitat, and because the effective species pool which can colonize them may be considerably less than the regional pool due to minimum area effects. Higgs and Usher (1980) have shown that under these circumstances more species will be contained in single large reserves (Figure 7).

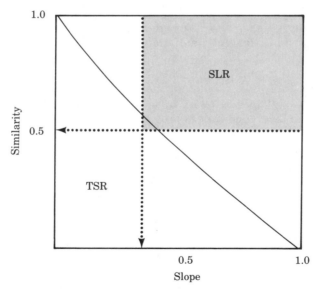

FIGURE 7. This diagram shows how the similarity in species composition (Jaccard's coefficient) between two equal-sized small reserves and the slope of the (log-log transformed) species–area relationship determine whether or not more species will be preserved in one large area or two smaller areas of the same total area. Above the solid line, a single large reserve (SLR) holds more species; below it two small reserves (TSR) hold more species. The dotted lines and shaded area define the expected parameter space when the SLR is small in absolute terms, and all of the reserves are geographically close (i.e. when conservation is operating on a local scale). (After Higgs and Usher, 1980.)

For each major habitat type within a given region, the result of applying these hierarchical strategies would be a small number of large national reserves, a network of medium-sized regional reserves, and a large number of small local reserves. With sufficient integration between organizational levels, primarily in regard to reserve placement (see below), a composite strategy such as this might be adequate to ensure the long-term persistence of those target species not already extirpated by fragmentation.

How close?

It will often be impractical to speak of clustering large national reserves such as national parks and forests, since they can be in widely separated regions of the country. On a local level, there may be great benefit to placing reserves close to each other. The large national preserves can serve as sources of colonists for the smaller local preserves, which themselves may serve reciprocally as stepping stones. These benefits will accrue only to the more vagile organisms such as birds (Lynch and Whigham, 1984), bats (Wilcox, 1980), and those species able to pass through the variety of habitats in the surrounding landscape (many temperate zone mammals). In terms of linking reserves, the value of corridors *per se* is debatable. They are unlikely to reduce the isolation of two distant reserves, and dispersal might occur anyhow if the reserves are close (Frankel and Soulé, 1981). More useful are land use practices which allow populations of many target species to exist at least marginally in the surrounding habitat. These populations can then diffuse into the reserves.

Reserve shape

Diamond (1975b) and Wilson and Willis (1975) have recommended that reserves be as nearly circular in shape as possible. The stated reason is to minimize dispersal distances within a reserve (but see Game, 1980). In the case of temperate zone forest reserves, we may add a second reason—to minimize the proportion of forest edge to forest interior. (By similar reasoning, clearings should not be allowed within the forest. If clearings must be created, they should be placed as close to the edge as possible and clustered together.)

Blouin and Connor (1985) have produced a detailed statistical analysis of data for oceanic islands which suggests that island shape is unimportant in determining the species composition of the islands studied. However, the analysis misses the point about application of the theory of island biogeography to the management of species in nature reserves. Essentially, circularity in the shape of forest fragments may be advocated purely to diminish the impact of edge effects. This is unlikely to be important in oceanic island habitats, where interactions between species across the border of the island are very infrequent. Indeed, at the risk of overgeneralizing, we suggest that the optimal shape for *any* habitat reserve is circular, so as to minimize contact between the protected interior and the surrounding habitat.

Management

Finally, we believe that over the long run virtually all temperate zone reserves will require active management to prevent or overcome the ecological imbalance created by fragmentation or human activity. Good reserve design will lessen but rarely eliminate the need for management (Gilbert, 1980). Such management may take several forms, including controlled treatment of the vegetation to preserve particular successional stages (open country for the large blue butterfly); the elimination of foreign species (wild boars in the Great Smoky Mountains National Park); or the culling of populations of "nuisance" animals (cowbirds in the breeding grounds of the Kirtland's warbler). Conservationists must realize that the battle is not over once the land has been saved. Indeed, it has just begun.

SUGGESTED READINGS

Burgess, R. L. and D. M. Sharpe (eds.). 1981. *Forest Island Dynamics in Man-Dominated Landscapes*. Springer Verlag, New York.

Harris, L. D. 1984. *The Fragmented Forest*. Univ. of Chicago Press, Chicago.

Verner, J., M. Morrison and C. J. Ralph (eds.). 1986. *Modeling Habitat Relationships of Terrestrial Vertebrates*. Univ. of Wisconsin Press, Madison.

Wilcox, B. A. 1980. Insular ecology and conservation. In *Conservation Biology: An Evolutionary-Ecological Perspective*, M. E. Soulé and B. A. Wilcox (eds.). Sinauer Associates, Sunderland, MA.

CHAPTER 12

EDGE AND OTHER EFFECTS

OF ISOLATION ON AMAZON

FOREST FRAGMENTS

*T.E. Lovejoy, R.O. Bierregaard, Jr., A.B.
Rylands, J.R. Malcolm, C.E. Quintela, L.H.
Harper, K.S. Brown, Jr., A.H. Powell, G.V.N.
Powell, H.O.R. Schubart, and M.B. Hays*

The Minimum Critical Size of Ecosystems project (Lewin, 1984; Lovejoy, 1980) of the World Wildlife Fund and Brazil's National Institute for Amazon Research (INPA/CNPq) is a long-term study (more than 20 years) of the effects of fragmentation on Amazonian forest as it is cleared to create pasture. Many groups of organisms and a variety of processes are being examined quantitatively both before and after isolation of forest plots. Hence this is a special opportunity to examine the process of change in forest remnants.

From one perspective this binational exercise is providing data on which to base design and management recommendations for national parks and reserves of particular relevance to the Amazon forest. From a second perspective the project is testing island biogeographic theory (MacArthur and Wilson, 1967). From a third perspective, that of community ecology, it will provide basic information as to whether there are critical species-specific interactions, and on the degree of structure (nonrandomness) of species composition in natural communities. The answer to these questions is obviously likely to vary with the particular ecosystem, taxon, or guild.

This is the third in a series of overall reports (Lovejoy et al., 1983, 1984) designed to highlight emerging results and insights. Of 25 anticipated forest research reserves, 22 are under active study, 10 have been isolated, and two have yet to be demarcated in as-yet continuous forest (Figure 1). Two reserves were isolated in 1980, five in 1983, and three in 1984 (Table 1).

The original focus of the project was almost exclusively on the area effects with almost no attention to edge-related changes. As the magnitude and complexity of the latter have become apparent, more attention has been accorded them. This paper is also the first to include results from reserves isolated in 1983 and 1984; these provide indications of the generality of trends observed from the two reserves isolated in 1980 (Lovejoy et al., 1983, 1984).

EDGE EFFECTS

The effects we are discussing here are not to be confused with the more traditional meaning of edge effect—namely, the increased number of species encountered where two major habitat types intergrade (Leopold, 1933). Rather the concern is with changes occurring in previously undisturbed forest by the abrupt creation (due to deforestation for pasture) of a very sharp edge at a reserve margin (Figure 2). It is likely, however, that effects such as described herein may stabilize eventually, producing a band of modified vegetation rather like a traditional ecotone. Only then will it be possible to determine the extent to which an Amazonian pasture/forest ecotone displays increased species number in the traditional sense of an edge effect.

Microclimatic changes

Following the isolation of a forest reserve, the understory interior is exposed to drastically different microclimatic conditions. Vegetation

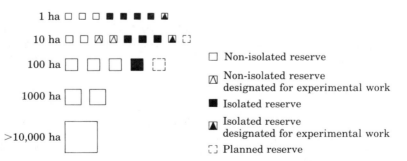

FIGURE 1. Reserves demarcated starting in 1980 for the Minimum Critical Size of Ecosystems study north of Manaus, in the Brazilian Amazon.

FIGURE 2. Margin of Reserve #1112 in 1983, just after clearing. Note the sharpness of the edge created by deforestation for pasture. (Photo by R. Bierregaard.)

within two meters of the edge is visibly affected within days. Increased insolation (particularly early in the morning and late in the afternoon) and hot, dry winds coming off the surrounding clearing alter the physical environment, with consequences for the flora and fauna within the reserve.

Preliminary data taken by V. Kapos (personal communication) demonstrated differences in relative humidity between reserve edges and 100 m inside ranging from 5 percent in the early morning to 20 percent at midday. Air temperature difference, in the shade, between the edge and 100 m into a 100-ha reserve was as great as 4.5°C. These data suggest such gradients from the edge to the center are greater

TABLE 1. Reserves demarcated for the minimum critical size of ecosystems study.

Reserve	Date established	Date isolated	Description
		1-HA RESERVES	
1104[a]	12/79	8/80	Completely flat.
1105	4/80	—	Plateau, NE corner slopes down in beginning of a stream valley.
2107	1/81	9/84	Gulley leaves reserve on the W. Unusual patch of understory bamboo.
2108	3/81	9/84	Gentle topographic relief sloping down to NW, includes dense stand of bamboo in SW corner.
1109	4/81	—	Little topographic relief.
1112	4/82	8/83	Some topographic relief and unusual vegetation, apparently of a single age class as if entire area disturbed recently. (Experimental)
1113	7/82	prob. 86	Hill crest runs E to W through N of the reserve. A steep drop to the S.
3114	2/83	7–8/83	Gentle slope E to W.
		10-HA RESERVES	
1202	9/79	7–8/80	No stream, plateau at W, border slopes steeply to E, back up to hill crest 100 m in from N border. One ha in SE corner burned by fazenda grass fire, now regenerating.
1204	6/80	—	No stream, topography varied. Flat in SW and most of W border, dry stream gulley slopes gently SW to NE. Unusual dense vegetation on N border where drainage is poor.
1205	9/80	—	Stream head drainage, curving from SW corner, E, N, then back W, leaving reserve at the NW corner. Slope to stream very steep from E border. Contains some particularly sandy soils.
2206	1/81	8–9/84	Double stream head drainage, main stream arises 75 m N of reserve with second gulley feeding in from E border. Very steep stream slope on W side. Stream leaves through S border.
1207	2/82	8/83	Gentle relief except for steep gulley to SE. (Experimental)
1208	2/82	—	Steep hill in SW corner sloping to N and E, smaller hill in NW corner sloping to E. (Experimental)
3209	8/82	8/83	Dry, gentle relief. One poorly drained area devoid of understory palms.

TABLE 1. (*Continued*)

Reserve	Date established	Date isolated	Description
1210	5/84	prob. 86	Flat with small slope, several treefall gaps. (Experimental)
100-HA RESERVES			
1301	1/80	—	Stream enters at SW corner, traverses S part of reserve, and leaves at SE corner, joined midway by smaller stream starting at N border. Topography varied, steep slopes to streams, high ridge cuts N to S through middle of reserve. Large poorly drained plateau in E half occasionally floods.
1302	7/80	—	Probably most diverse reserve. Three-way stream head drainage, very steep gullies, very high plateau in S central area, and a broad swampy area with many buriti palms (*Mauritia flexuosa*) where stream leaves at NW corner.
2303	12/80	—	Rolling bisected terrain, high hill in NW, drainage with gullies to SE. Very steep gulley leaves NE corner. Swampy area along S border. Soil sandy, canopy tree density high. Several ha in NW with poor drainage and a hummock effect.
3304	7–8/82	8/83	Undulating terrain, triple-headed stream system along E half. Unusual Moraceae stand in NW. Also some buriti stands in poorly drained areas.
1000-HA RESERVES			
1401	10/79	—	Irregularly shaped. Many streams, accentuated local relief, 7–10 streams drain into a large stream 2 m wide draining to NE. Soils in S less sandy than in north. Large swampy areas in NE.
3402	8/84	—	Irregularly shaped. Large plateaus, ample hydrographic basins, NE border a steep hill with high humidity. Some streams drain flat areas and don't have slopes on margins.
10,000-HA RESERVE			
1501	8/84	—	Still being delineated.

[a]The initial digit of a reserve's number indicates the fazenda (1 = Esteio; 2 = Dimona; 3 = Porto Alegre) on which the reserve is situated. The second digit indicates the size of the reserve (1 = 1 ha; 2 = 10 ha; 3 = 100 ha; 4 = 1000 ha; 5 = 10,000 ha).

in reserves large enough for a core to maintain normal humidity. For reserves of 10 ha and smaller, the whole reserve is affected. These altered conditions may explain the elevated tree mortality already recorded in reserves of 1 and 10 ha (Lovejoy et al., 1984; Rankin, unpublished data).

Over time the sharpness of this transition diminishes as trees fall at the edge of the reserve (Lovejoy et al., 1983) and a dense skirt of vines and other secondary vegetation provides a 10–25 m wide windbreak, and also shades the forest interior. Five years after isolation of the 1980 isolates, the reserve interiors were noticeably darker and more humid than they had been in the first years after isolation. Quantitative measurements of these changes will be initiated in early 1986.

Plants

Previous papers have noted a marked increase in tree mortality subsequent to isolation (Lovejoy et al., 1983, 1984). Data from 1983 and 1984 isolates confirm this pattern. Mortality rates of trees over 10 cm dbh in isolated 1- and 10-ha reserves are 2.6 percent over an initial two years as contrasted to 1.5 percent for a continuous forest sample. The elevated mortality is probably related to the microclimatic change through alteration of soil ecology and perhaps mycorrhizal relationships (see Jordan, Chapter 20).

Bierregaard (unpublished data) noted a dramatic increase in leaffall in the 1980 and 1983 isolates and quantified the effect in 1984, immediately after the burn of the surrounding felled forest. Such leaffall (now being monitored on a regular basis) might be a serious nutrient drain and a cause of elevated mortality (E. Tanner, personal communication).

With increased light at the forest margins there is a response in vegetative growth such as sprouting, which potentially could increase food for certain folivores.

Crowding and persistence in understory birds

One observation reported from the first two reserves isolated in 1980 (Lovejoy et al., 1983, 1984) has been replicated in subsequently isolated reserves: an influx or crowding effect. In understory birds this is measured as an increase in capture rates in mist nets when birds take refuge in reserves after the surrounding forest is felled. This experiment was repeated by the isolation of three additional reserves in 1983 (two of 1 ha each and one of 10 ha; Figure 3; Bierregaard and Lovejoy, in preparation). As was the case in the 1980 isolates, the 1-ha reserves showed higher capture rates than the 10-ha reserve.

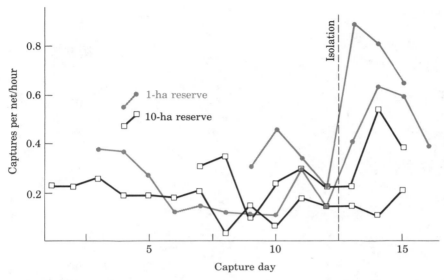

FIGURE 3. Capture rate of understory birds, showing increase coincident with isolation (dashed line) in three of four reserves. This is due to a crowding effect as refugee birds flee from their destroyed habitat into remnant forest. For an explanation of the absence of crowding effect, see Figure 4. (Data from R. Bierregaard.)

The extent of the crowding effect in an isolated reserve appears to be a function of the number of refugees and the area of habitat available to absorb them. For example, a second 10-ha isolate did not show any increase in capture rates after isolation (Figure 3). We believe this is because the total displaced understory bird population was small (Figure 4); the cleared band of forest was only 100 m on three sides. Furthermore, the adjacent forest area was large and nowhere more than 100 m from the cleared area. In addition, a 100-ha reserve established in 1983 did not show any influx effect. This is not surprising given the greater size of the reserve and the initial presence of the corridor of forest connecting it to a huge expanse of virgin forest (Figure 5).

Subsequent analysis of banding data in isolated reserves (Bierregaard and Lovejoy, in preparation) indicates additional effects of influx beyond the simple increase (and subsequent collapse) of capture rate. One of these effects is the persistence of birds banded in the isolated reserves *prior* to their isolation, which is low when compared to the persistence of similar size cohorts followed through time in nonisolated forest tracts (Figure 6).

FIGURE 4. This 10-ha reserve (#1207) did not show a crowding effect in understory birds. Only 100 m of forest was cleared on three sides, so the refugee population was small. There was also a large area of surrounding forest to absorb the birds. (Photo by R. Bierregaard.)

Persistence was calculated for the cohort of birds banded prior to isolation of the 1980 10-ha reserve and compared to that for birds banded immediately after isolation. New arrivals show greater persistence (Figure 7). This suggests residents are at a disadvantage compared to refugees. Perhaps residents invest so heavily in territorial defense against numerous invaders that they suffer higher mortality or abandon the reserve—a situation analogous to aggressive neglect (Ripley, 1959, 1961).

Our initial impression was that this depression of persistence was related to the crowding effect. For example, in the 10-ha reserve, which did not show a crowding effect (Figure 3, bottom), persistence is higher than in other 10-ha reserves (Figure 8). Therefore we anticipated persistence in the 100-ha isolate (where banding and mist net data demonstrate no crowding effect) would be relatively high. Paradoxically, however, all three net lines show low persistence levels (Figure 8). One of the three lines yielded a persistence level lower than in any 10-ha isolate. This seems to rule out the crowding theory as a universal explanation of low persistence.

FIGURE 5. A 100-ha reserve (#3304; center) connected by a corridor to continuous forest one year after being otherwise isolated. (Left) A 10-ha reserve (#3209) connected to a larger patch. (Top-center) An isolated 1-ha reserve (#3114). (Photo by R. Bierregaard.)

Edge effects in birds

The edge, by providing more resources, may initially attract certain species of birds. In one instance involving the 1980 1-ha isolate, netting data showed the "disappearance" of mixed-species flocks of birds about six months after isolation. Field observations, however, indicated the birds had not left the reserve, but had shifted foraging activity to the reserve edge. This was only transitory, but it may indicate the birds were able to persist longer because of increased availability of phytophagous insects capitalizing on sprouts and second growth which appeared in response to increased sunlight. Flocks in the larger 10-ha reserve that was isolated simultaneously continued to use forest interior rather than edge.

We had anticipated an invasion of the edge of isolated reserves by second-growth bird species, but both netting and visual data indicate this invasion did not take place. In over 12,000 net hours in seven isolated 1- and 10-ha reserves there were only 20 captures of only two species of second-growth birds. One was the silver-beaked tanager,

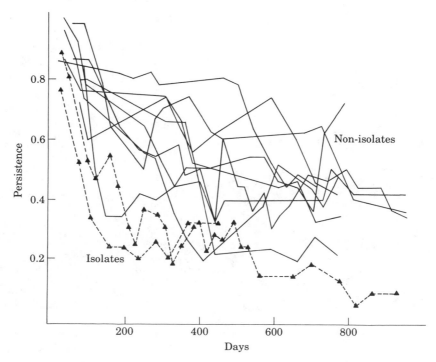

FIGURE 6. Persistence (percentage of recaptured birds from an initial cohort of about 100 individuals) in isolated versus nonisolated reserves. Persistence is higher in nonisolates. Data are 3-point running means and were collected in 10-ha reserves. (Data from R. Bierregaard.)

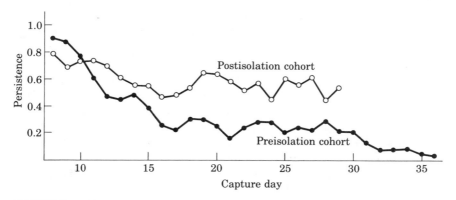

FIGURE 7. Persistence (percentage of recaptured birds) from an initial cohort of about 100 resident birds banded before isolation of a 10-ha isolate was lower than that of a similar size cohort banded after isolation (refugees). Data are 3-point running means. (Data from R. Bierregaard.)

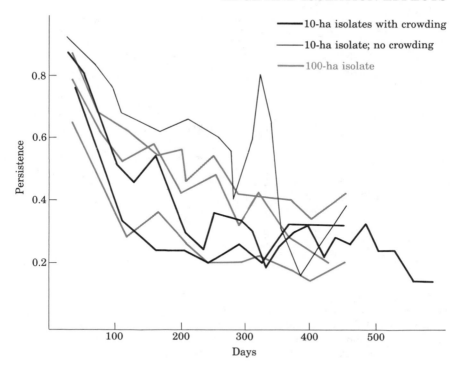

FIGURE 8. Persistence (percentage of recaptured birds) from an initial cohort of birds in an isolated 10-ha reserve (#1207) without crowding was close to levels for nonisolated reserves. Persistence for cohorts in an isolated 100-ha reserve (#3304) was as low as that in the isolates with crowding, although no crowding was detected. Data are 3-point running means. (Data from R. Bierregaard.)

Ramphocelus carbo, which was only caught where large treefalls had brought second-growth habitat into the reserve. On the other hand, understory birds often appear to avoid artificial edges.

Mist nets were deployed in one reserve beginning about one year following isolation, as well as in a nonisolated reserve deep within continuous forest. Nets were set in natural forest gaps (treefalls) and in forest without them, and at a third site where nets were 10 and 50 m from the forest edge (Figure 9). Twenty-four nets were used under each set of circumstances and run for comparable lengths of time, so that total net hours sampled under each set of conditions were similar.

Netting data (Figure 10) show that 10 m from a large manmade clearing 38 percent fewer birds were caught than 50 m into the forest,

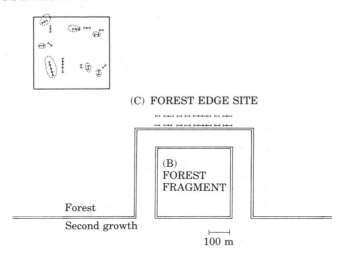

FIGURE 9. Net locations for studies of treefall gap (natural edge) birds and those of artificial edge. Nets were placed in treefall gaps (dotted lines) and in undisturbed forest in (A) an intact forest site (Reserve #1208, located 1 km deep into continuous forest); (B) a 10-ha forest fragment (Reserve #1207); and (C) a third site with nets set 10 and 50 m from the forest edge. (Data from C. Quintela.)

and 60 percent fewer than were caught 1 km into undisturbed forest. Mist nets deployed at natural edges (i.e. treefalls) within virgin forest did not yield depressed numbers of birds compared to surrounding virgin forest (no gaps). Species number was also affected: the nets at 10 m yielded 28 species, versus 47 species at 50 m and 50 species in deep forest. Even 50 m from the forest edge the commonest forest interior species are rare.

The 10 m and 50 m edge sites combined yielded roughly as many birds (in individuals and species) as the nets in the isolated reserves (gap and nongap combined). In this particular isolate, where no crowding effect was recorded, both this edge effect and the area effect must be contributing to the impoverishment of the bird community.

If 50 meters is used as the limit to which the edge effect extends into a forest, a 10-ha reserve will be 53.2 percent edge. This means that the decline in species number of understory birds documented for the 1980 10-ha isolate (Lovejoy et al., 1984) is the consequence of a mixture of edge effect, crowding effect, and area effect. We consequently no longer expect to learn a great deal about area effects from studies of 10-ha reserves and must turn our attention toward those of 100 ha. It is important to note, however, that while edge and crowding

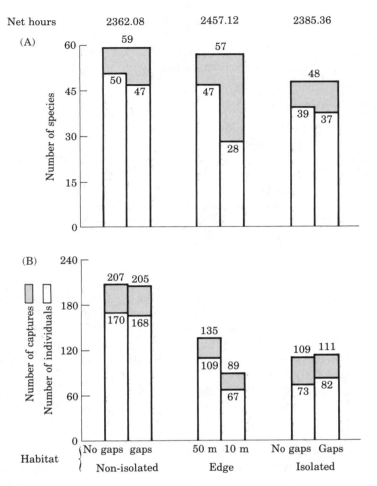

FIGURE 10. Mist net samples of birds netted in forest locations shown in Figure 9. (A) The shaded bar indicates total species shared between the two habitat types. (B) The shaded bar is the number of captures, and the open bar is the number of individuals. Results indicate a distinct depression of birds (species, individuals, and captures) in a zone up to 50 m from the artificial margin. (Data from C. Quintela.)

effects are in a sense confounding factors, they do in themselves have implications about reserve size in the sense that their effects can be mitigated by increasing reserve size.

J. Terborgh (personal communication) reports a similar "negative edge effect" from within forests of the Manu National Park in eastern Peru. Censuses from transitional zones between the seasonally flooded high forest and permanent terra firme forests consistently show fewer species than censuses taken on either side of the ecotone. Such a difference suggests that a high level of habitat specificity has been evolved by neotropical forest bird species.

Invasion by second-growth insect species

Immediately following the isolation of the 1980 reserves and the burning of the surrounding forest, Brown (in preparation; see also Lovejoy et al., 1983, 1984) noted a decrease in species number of Ithomiinae, Satyrinae, and other forest understory butterflies. However, within a year—in contrast to results from bird studies—butterfly species numbers increased to a level considerably higher than that recorded prior to isolation. The increase was due to the invasion of second-growth, light-loving species of butterflies. The number of species encountered in these reserves has continued to increase during the four years since isolation.

In the 1984 isolates, the decrease in species numbers after burning around the reserves was virtually identical to that recorded in the 1980 isolates. The situation was different, however, around the 1983 isolates, where there was no burning and the number of species encountered with standardized sample effort simply increased steadily after a slight initial depression. Without a burn around the reserve, the shade-loving species and the species generally encountered in undisturbed forest (about 100 species) remained almost constant in number, or increased.

Brown estimates these "edge" species of butterflies penetrate 200–300 m into the forest, competing for space and disturbing butterflies of the forest interior. This disturbance can affect access to nectar sources and oviposition sites, and can interfere with courtship behavior. This is a pattern distinct from the influx of refugee birds. The new butterfly arrivals, which are species common to openings and second growth, follow increased light levels into the reserves. In contrast, newly arrived understory birds in isolated reserves are primarily forest interior species fleeing forest destruction and, apparently, are seeking the reduced light levels available in the reserves.

In fact, the effects of increased light (including the consequent increased growth of new plant food species) explain almost all the major tendencies seen in the restructuring of butterfly communities. The list of species encountered in a day's field work can be effectively predicted by a combination of census effort, edge and second growth effects (which can double the number of species encountered), and time elapsed since cutting and burning (Figure 11). Burning greatly depresses species number in 1-ha reserves and slightly depresses species number in 10- and 100-ha reserves.

AREA EFFECTS

The following accounts of changes induced by isolation all relate to area to some degree, but not necessarily entirely. Because edge-related changes have been dominant in the smaller isolated reserves, area effects can only be studied in larger reserves. In large reserves rates of change are slower, so most area effects are yet to be documented.

Primates

Data on primates suggest a mixture of area and edge effects. Three species are evidently strongly affected by area (Table 2), and require reserves of more than 100 ha. Another three species are surviving in the 10- and 100-ha reserves. The extent of forest edge may have an important influence on the feeding ecology and survival of the latter in these small forest fragments.

Both the black spider monkey *Ateles paniscus* and the bearded saki *Chiropotes satanas* were absent from the 100-ha reserve (#3304) at the time of its isolation in June of 1983, although it contains likely habitat. Two *C. satanas* persisted for only a few months in the 1980 10-ha isolate (Ayres, unpublished data; Lovejoy et al., 1983). One group of 11–12 tufted capuchin monkeys *(Cebus apella)* remained in the 100-ha reserve for one year but disappeared at the beginning of the dry season in June 1984. The scarce and elusive black-and-white saki *Pithecia pithecia* still survives in the reserve, although one of the two original family groups has disappeared. The remaining group bred

TABLE 2. Primates in isolated reserves.

Primate	Home range	Number of groups present in reserve				
		1983[a] 100 ha	1980 10 ha	1983 10 ha	1983 10 ha	1983 10 ha
Ateles	> 200 ha	0[b]	0	0	0	0
Chiropotes	> 200 ha	0[b]	0	0	0	0
Cebus	> 100 ha	1–0[c]	0	0	0	0
Pithecia	10 ha	2–1	0	0	0	0
Saguinus	30 ha	4–3	1–0	1	0	1
Alouatta	Unknown	5	1–2	1	1	1

[a]Date is the year the reserve was isolated.
[b]Rylands (personal communication) speculates they were present prior to isolation but were excluded during deforestation.
[c]Initial number of groups at isolation, changing to second number as of May 1985.

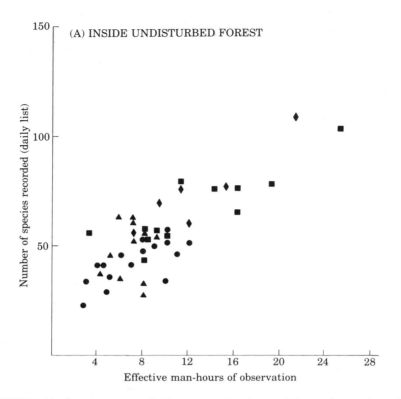

FIGURE 11. Species accumulation curve showing positive and negative edge effects in butterflies. (A) The relationship between species diversity and the number of man-hours of observation inside undisturbed "control" forests. (B) Diversity is depressed immediately after the felling and burning of the sur-

during the wet season of 1983–1984 and now numbers four individuals. Four groups of the golden-handed tamarin *Saguinus midas* were isolated within the large reserve. One of these groups left three months after isolation, and the three remaining groups occupy the majority of the reserve, each with home ranges of approximately 30 ha (Figure 12). Five groups of red howler monkeys *(Alouatta seniculus)* have been observed in the reserve since its isolation.

One or two groups of *A. seniculus* survive in each of the four isolated 10-ha reserves. A group of 10 howlers was isolated in the 10-ha reserve in 1980 (Lovejoy et al., 1983). The group numbered eight in December 1984 and two of the three adult females are breeding. A single male, probably originating from outside the reserve, established himself in a 3-ha section of the reserve in early 1984 and by May 1985 had achieved a mate and one offspring approximately 3 months old. Two of the four 10-ha reserves retain *S. midas* groups. A group of 5–

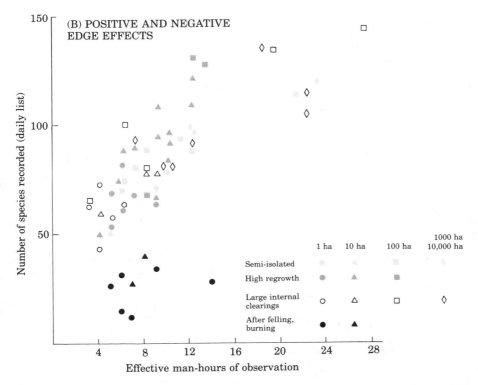

rounding forest. Butterfly diversity is enhanced by edge effects that increase the quantity of secondary growth (upper cluster of points). (Data from K. Brown.)

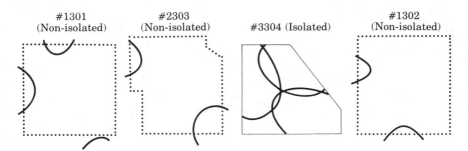

FIGURE 12. Home ranges (heavy curved lines) of golden-handed tamarin *(Saguinus midas)* troops in 100-ha reserves. Dotted margins indicate a non-isolated reserve, solid margins an isolate. In the isolated reserve three troops are contiguous, as compared to their wide separation in intact forest. (Data from A. Rylands.)

6 S. midas was isolated within the 10-ha reserve in 1980 but has since disappeared. No S. midas groups have been observed in the 1984 10-ha isolate, although it does contain a group of five P. pithecia. The other two 10-ha isolates contain groups of 4–5 individuals and 6 individuals, respectively. In the latter the reserve is isolated by only 100 m and the tamarins regularly leave and enter the reserve.

The survival of A. seniculus, P. pithecia, and S. midas in the 100-ha reserve is not surprising when one considers their diets and home range sizes. A. seniculus occupy very small forest patches in the Venezuelan llanos, with home range sizes of 3–7 ha, and they are the most persistent survivors in the 10-ha reserves (Sekulic, 1982). The persistence of S. midas and P. pithecia is more precarious in isolated forest fragments of this size. However, there are reasons to believe that the fragmentation of the forest and the creation of forest edge and second growth may be beneficial to S. midas and, perhaps to a lesser extent, to A. seniculus. In the following, the ecology of the three species which survive in these forest fragments are discussed, with emphasis on the possible benefits of forest margins for them.

Saguinus midas

Golden-handed tamarins are small (450–550 g) frugivore-insectivores. Group sizes range from 4–10 individuals occupying home ranges of 10–30 ha. In tall primary forest, tamarin densities typically are low (0.6 groups/km^2, or 3.9 individuals/km^2; Rylands and Keuroghlian, 1985) and groups may be separated by as much as 1 km (Figure 12). This is believed to result from a lack of their preferred habitat, which is a mix of tall primary forest with edge and second growth, such as is found along the margins of streams and swamps and in treefall areas (Thorington 1968; Mittermeier and van Roosmalen, 1981). This habitat preference has been noted for other tamarin species and marmosets (Saguinus fuscicollis and S. imperator, Terborgh, 1983; S. oedipus, Dawson, 1979; Moynihan, 1970; S. leucopus, Bernstein et al., 1976; Callithrix humeralifer, Rylands, 1981; Branch, 1983; and C. penicillata, Stevenson and Rylands, in press). Moynihan (1970) indicates that S. oedipus populations on Barro Colorado Island, Panama, have been declining since the 1930s because of the gradual diminution of second-growth vegetation.

Several explanations of this habitat preference are apparent, including predator protection, fruit types, fruit abundance and distribution, and insect abundance. The dense vegetation typical of edge and second growth are undoubtedly important in providing protection from predators (Dawson, 1979). This is reflected in the sleeping sites most commonly used by marmosets and tamarins—dense vegetation.

Of 32 sleeping sites recorded for a *C. humeralifer* group, 30 were in dense vegetation in tree crowns or tangles of leaves or lianas (Rylands, 1982 and in preparation); 23 were in disturbed primary forest which contained much dense vegetation and second growth, and only 9 were in tall primary forest with a closed canopy and sparse undergrowth typical of the continuous forest in the reserve areas.

The few studies of Amazonian marmosets and tamarins in the wild (*S. fuscicollis* and *S. imperator,* Terborgh, 1983, Chapter 15; *C. humeralifer,* Rylands, 1981, 1982, and in preparation) have shown that they depend on small fruits from abundant trees, which ripen over long periods. The small size of the fruits and their low yields exclude their use by larger primates, but for the marmosets and tamarins they are ideal (Terborgh, 1983). Highly clumped trees with small fruits and prolonged fruiting seasons are typical of second-growth forest following the early stages of succession. In tall primary forest, tree species tend to be rarer and more widely dispersed and have larger fruits produced in large crops over shorter and more irregular periods (Opler et al., 1980).

Enhanced insect abundance may benefit tamarins and marmosets. Secondary and pioneer vegetation support high densities of arthropods, particularly herbivorous insects. The pioneer plants of successional forests have leaves which tend to have less chemical defense, and the leaf lifespans are shorter. In addition, second growth shows a higher net productivity in comparison to mature forest (Cates and Orians, 1975; Opler, 1978).

A comparison of the *S. midas* populations in the three nonisolated 100-ha reserves with that of the Porto Alegre reserve isolated in June 1983 (Figure 1) supports the hypothetical advantage of edge habitat for this species. Unfortunately the size of the *S. midas* population in the reserve prior to the four groups noted at the time of isolation is not known, so the possibility remains that the floristic communities in the reserve have been more favorable to *S. midas* even before isolation.

Alouatta seniculus

Red howlers live in groups of 3–10 individuals comprising an adult male, several females and their offspring, and 1–2 subordinate or subadult males. They can occupy home ranges as small as 3–7 ha (Sekulic, 1982), although the range sizes in the reserve areas are 7–20 ha. *Alouatta* are the most folivorous of the New World primates and, although they prefer ripe fruits and young leaves, they also eat mature foliage in times of food shortage. They can evidently survive in the 100-ha reserve in densities similar to those in continuous forest.

They can survive in the 10-ha reserves because of their folivory. It is possible that they include more leaves in their diet when confined to these small areas.

The increase of second-growth vegetation resulting from the creation of forest margin and the increase in treefalls (Lovejoy et al., 1983) in the 10-ha reserves may benefit the remaining howler groups because of the foliage characteristics and higher net productivity typical of pioneer vegetation, as discussed above. In addition, despite the reduction in home range size, the groups may benefit from the loss of the other large frugivorous primates *(Ateles, Chiropotes,* and *Cebus)* which compete for large fruit.

Pithecia pithecia

Although *Pithecia* family groups are believed to occupy home ranges of 10 ha or less (Buchanan et al., 1981), the rarity of these animals means that their presence in a 10-ha reserve at the time of its isolation is a matter of chance. Colonization of isolated reserves is unlikely for this species because their mode of locomotion is largely saltatory, preferring small supports (Fleagle and Mittermeier, 1980), and they rarely if ever go to the ground. Too little is known of this species to indicate any definite benefits or disadvantages resulting from the creation of forest margins. However, groups are known to thrive in areas where secondary growth predominates (Oliveira and Lima, 1981) and it is possible that the fruit types, fruiting patterns, and the abundance and clumped distributions of pioneer trees may benefit *Pithecia* as they do *Saguinus.* The protective cover provided by the dense vegetation may also be an important factor.

In summary, a number of features of secondary vegetation, which will increase in the 10-ha and possibly the 100-ha isolated reserves, may be beneficial to the three species that survive there.

Other mammals

Initial results of small mammal trapping in 1982 indicated low densities in forest fragments and intact forest alike. The absence of some species from forest fragments (most notably the absence of the otherwise abundant terrestrial spiny rats, *Proechimys* spp.) was possibly attributable to isolation (Emmons, 1984; Lovejoy et al., 1984). Subsequent trapping in 1983–1984 by Malcolm (in preparation) yielded much higher capture rates for small mammals, both in isolated and nonisolated reserves. For example, the third most abundant mammal in Malcolm's sample was *Oryzomys macconnelli,* a species of rice rat never caught during the initial six months of sampling in 1982. This suggests the small mammal communities in isolated reserves and intact forest are quite dynamic, making it difficult to generalize at this stage about the effects of isolation on small mammals.

Malcolm's preliminary analysis of the small mammal species abundance patterns suggests that small mammal communities may take longer than a few months to respond to forest fragmentation. There is no more difference among samples from reserves of 10 ha and 100 ha isolated for 9 months than there is among samples from nonisolated reserves. On the other hand, a 1980 10-ha isolate (Reserve #1202) was significantly different from all other reserves, primarily due to a high capture rate of the mouse opossum, *Marmosa cinerea*.

At present, it seems that small mammal communities within small isolates exhibit high turnover rates. The abundance of some species may largely be determined by low rates of immigration from intact forest (possibly the case for the wooly opossum, *Caluromys philander*), whereas the abundance of others (such as *M. cinerea*) may be determined by their ability to utilize the early successional stages surrounding and gradually invading the small isolates.

Army ants and ant-following birds

The system that involves army ants *(Eciton burchelli)* and a number of species of birds that follow them (Table 3) is a typical feature of Amazon forests. The birds benefit from the flushing of insect prey from the leaf litter and adjacent habitat by the swarming ants. Although most families of these birds show mixed susceptibility to extinction, ant-following birds have been cited as extinction-prone (Willis, 1974, 1979).

Harper quantitatively established the significance of army ants to general insect availability. In both isolated and nonisolated reserves, sticky traps (plexiglass plates covered with Tanglefoot) set out in the forest caught a significantly greater quantity of insects by dry weight when army ant swarms were present (Figure 13).

The cyclical nature of army ant colonies means that the colony is a useful resource to the birds only during that part of the 35-day cycle when it is swarming. An army ant colony uses about 30 ha in continuous forest, so it was not surprising that the first 10-ha isolate failed to support the system for more than a couple of weeks; the birds left before the ants did (Lovejoy et al., 1983). It was intriguing that when the ants later returned briefly, so did some large ant-following birds.

To investigate the relationship between the quantity (area) and the quality (resources) of a forest fragment, as well as the tendency of ant-following birds to cross areas of deforestation, Harper introduced a number of birds at various times into isolated reserves, in

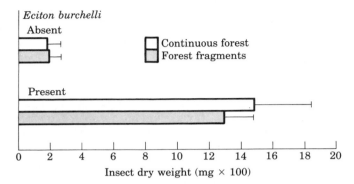

FIGURE 13. Availability of insects in the presence and absence of army ants *(Eciton burchelli)* as measured by sticky traps. (Data from L. Harper.)

both the presence and absence of army ants.[1] Of 36 individual obligate ant-followers released in the absence of army ants (Table 4), only one was recaptured in the 10-ha reserve. However, 18 were later recaptured in adjacent forest. These birds crossed 100–150 m of deforested land; one was caught 1 hour and 45 minutes after release, at almost precisely the point where it had been caught prior to release in the experimental reserve. In a second experiment, Harper recaptured 17 white-plumed antbirds, 4 rufous-throated antbirds, and 3 white-chinned woodcreepers in the continuous forest after they were released in an isolated reserve that did not harbor army ants. Each of these birds crossed over 300 m of deforestation which separated the isolated reserve from the forest.

In a different experiment Harper transplanted army ants into a 10-ha experimental reserve (Table 5). Colonies did not remain long, perhaps in part because they were disturbed by pheromone trails left by previous resident colonies. When Harper attempted to stabilize a transplanted colony by confining the queen in a cage, the colony ultimately abandoned the queen. When 34 birds were released in the reserves that had been temporarily enriched with army ants (Table 6), 7 were recaptured within the reserves and 16 in nearby forest. The difference in the persistence (as measured by recapture) of introduced ant-following birds in reserves with ants as opposed to those without ants is significant (Chi-square $p < 0.025$). However, this difference can be attributed to the number of recaptures of rufous-throated antbirds *(Gymnopithys rufigula),* which in itself is significant (Chi-square $p < 0.025$). The null hypothesis that recaptures of introduced individ-

---

[1] A small number of reserves (see Figure 1) have been especially designated for manipulative experimental work of this sort.

TABLE 3. Weights of ant-following birds.

Species	Weight (g)	Range (S.D.)	Sample size (N)
OBLIGATE			
Dendrocincla merula	52.5	4.2	384
Pithys albifrons	20.3	1.4	1593
Gymnopithys rufigula	29.1	2.1	867
OCCASIONAL			
Dendrocincla fuliginosa	40.2	3.4	144
Hylexetastes perrotii	114.5	5.6	66
Dendrocolaptes picumnus	78.4	6.5	14
Dendrocolaptes certhia	66.4	5.6	44
Xiphorhynchus pardalotus	37.8	2.8	525
Percnostola rufifrons	28.7	2.5	410
Hylophylax poecilonota	16.7	1.6	1460

TABLE 4. Birds released in 10-ha reserve without army ants.

Species	Number introduced	Recaptures	
		In reserve	In continuous forest
Pithys albifrons	19	0	8
Gymnopithys rufigula	12	1	7
Dendrocincla merula	5	0	3
Total	36	1	18

TABLE 5. Introductions of army ant (*Eciton burchelli*) colonies.

Date introduced	Estimated colony size	Colony phase	Birds introduced also?
Sept. 14, 1983	146,000	Nomadic	No
April 6, 1984	595,000	Nomadic	Yes
May 14, 1984	228,000	Statary	Yes
May 29, 1984	529,000	Nomadic	Yes

TABLE 6. Birds released in 10-ha reserve with army ants.

| | | Recaptures | |
Species	Number introduced	In reserve	In continuous forest
Pithys albifrons	18	1	13
Gymnopithys rufigula	12	6	1
Dendrocincla merula	4	0	2
Total	34	7	16

uals would be equal in both the presence and absence of army ants is rejected, and strengthens the case that the absence of birds in the absence of ants is a meaningful ecological link.

Can the antbird–army ant system persist in a 100-ha reserve? An isolated reserve of that size can theoretically support three army ant colonies, so the likelihood of one of them being in a swarming state and thus an adequate and reliable resource is fairly great. With the first 100-ha isolate this appeared to be the case, with a representative ant-following bird community persisting for a year after its partial isolation. After the first year, however, a 2-km long corridor connecting the reserve to extensive virgin forest was severed for 300 m at a point 1.5 km from the reserve. Subsequent to this complete isolation, ant-following birds declined and disappeared. Since home range sizes for the white-plumed antbird *(Pithys albifrons)* are in excess of 200 ha in the project reserves (Bierregaard, unpublished data; Harper, in preparation) and at Reserva Ducke, only 50 km south of the project reserves (Willis and Oniki, 1978), this was not a surprise. These results suggest that forest corridors linking isolated reserves to larger tracts of forest may serve to maintain obligate army ant followers in areas which otherwise might not support them, even though Harper recorded some species crossing large clearings to forage over ants in small isolated reserves.

Euglossine bees

Euglossine bees are important pollinators in tropical forests. In 1982 and 1983 during the dry season (June to September), Powell and Powell (in press) used three types of chemical bait to attract male euglossine bees at seven locations in both continuous and isolated

TABLE 7. Male euglossine bees collected at chemical baits.

Species	Field code	Bait[a]
Eulaema meriana	Eul.A	C, MS, V
E. bombiformis	Eul.A	MS, V
E. mocsaryi	Eul.B	MS, V
Euglossa chalybeate	Eug.A	C
E. stilbonota	Eug.B	C
E. crassipunctata	Eug.C	V
E. iopyrrha	Eug.D	MS
E. viridis[b]	Eug.E	MS
E. prasina	Eug.F	V
E. augaspis	Eug.F	V
E. piliventris[b]		MS
E. gaianii[b]		MS
Exaerete frontalis	Ex.A	C, MS
Eufriesea laniventris	Euf.A	MS
E. xantha	Euf.A	MS

[a] C = Cineole, MS = Methyl Salicylate, V = Vanillin.
[b] Single individuals.

forest. Fifteen species of euglossine bees were attracted to the scents during 39 sampling days (Table 7).[2]

Visitation rates among three sites in continuous forest were high and similar in 1982, but more variable among four sites in continuous forest in 1983, as deforestation was initiated in the vicinity. Rates declined following isolation of forest fragments of three size classes (Table 8). For several species of *Euglossa* the rates were positively correlated with reserve size (Figure 14). This pattern was not apparent in *Eulaema*, *Exaerete*, and *Eufriesea*.

Some species were favored by clearings. *Euglossa prasina* and *E. augaspis* were present only in clearings and *Eulaema mocsaryi* had higher rates in the clearing than in the forest. These species are unlikely to be useful as alternate pollinators.

Forest fragments too small to support forest euglossine bees may also experience declines in plant species. Male euglossine bees are the

[2] Two species of *Eulaema* and two of *Euglossa* are lumped as they are not easily distinguishable in the field.

TABLE 8. Visitation rates of euglossine bees.[a]

	1982	1983	
Site category	Pre-isolation	Pre-isolation	Post-isolation
100 ha		54.0 ± 0.0 ($N = 2$)	29.2 ± 12.4 ($N = 5$)
10 ha	57.0 ± 18.5 ($N = 4$)	18.5 ± 9.6 ($N = 4$)	13.6 ± 2.5 ($N = 5$)
10 ha	69.0 ± 14.1 ($N = 2$)	38.4 ± 15.3 ($N = 5$)	—
10 ha	—	24.0 ± 1.8 ($N = 4$)	—
1 ha	—	—	11.5 ± 4.4 ($N = 4$)
1 ha	56.3 ± 21.5 ($N = 3$)	—	7.0 ± 1.4 ($N = 4$)
Clearing	—	—	26.5 ± 4.9 ($N = 2$)

[a]Mean number of bees per hour ± standard deviation. Arrivals measured from 9 a.m. to noon.

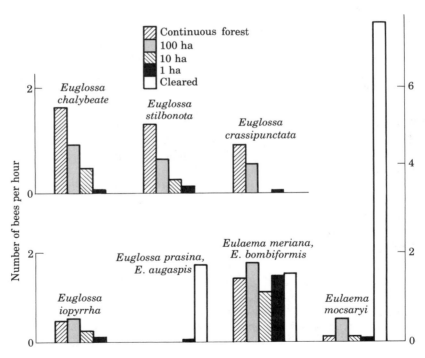

FIGURE 14. Visitation rates at chemical baits for male bees in species of the genera *Euglossa* and *Eulaema* in five kinds of habitat: continuous, undisturbed forest; isolated fragments of 100 ha, 10 ha, and 1 ha; and cleared (recently deforested 500-ha area). (Data from H. Powell and G. Powell.)

only pollinators for many orchids and are important in reproduction in 30 or more plant families (Zucchi et al., 1969; Williams and Dodson, 1972; Dressler, 1982). Since female euglossine bees pollinate a different set of plant species from those served by males, an additional number of plant species might be affected. These results suggest that a landscape dotted with woodlots might not maintain viable populations of forest euglossine bees.

SUMMARY

Many of the results of the studies cited above indicate a major set of effects triggered by newly created edges (Table 9). Prominent among these effects are abiotic changes (temperature, relative humidity, light, and exposure to wind). First-order biological responses include elevated tree mortality, increased leaffall, a flourish of new plant growth in response to increased light, and depression of bird populations in both number of individuals and number of species. Second-order biological responses will include insect populations responding to plant growth. In butterflies this leads to the disturbance of forest interior butterflies by the invasion of light-loving species characteristic of second-growth vegetation. A possible, but undocumented, third-order biological response is a higher density of insectivorous species.

TABLE 9. Classes of edge-related changes.

Class	Description of change
Abiotic	Temperature Relative humidity Penetration of light Exposure to wind
Biological: First order	Elevated tree mortality (standing dead trees) Treefalls on windward margin Leaffall Flourish of plant growth near margins Depressed bird populations near margins Crowding effects on refugee birds

Biological:
  Second order
	Increased insect populations (e.g., light-loving
	  butterflies)

Biological:
  Third order
	Disturbance of forest interior butterflies, but
	  increased population of light-loving species
	Enhanced survival of insectivorous species at
	  increased densities (e.g., tamarins)

To these must be added changes caused by mobile organisms fleeing destroyed habitat and a seed rain of animal-dispersed second-growth plant species.

The effect of an edge between habitats in temperate regions is often an increase in species richness relative to either habitat; yet the initial results from this study suggest the reverse. Species impoverishment is the rule, at least for neotropical forest birds. J. Terborgh (personal communication) has noticed the same depression of bird species number in natural ecotones at Manu in Amazonian Peru. This suggests that when a zone of modified vegetation has developed along the reserve margins, depressed bird populations may persist.

In contrast, for some groups of butterflies there is increased species richness at the forest edge after some vegetation regrowth. The increase, however, is primarily accounted for by the invasion of second-growth species, probably at the expense of those typical of the deep forest.

The implication for reserve design of these edge effects is that even larger areas will be needed. Reserves must include extensive buffer zones hundreds of meters in extent, where edge-related changes can take place without affecting the large, stable core areas.

Just as edge-related changes can occur in sequence, so too can first-order area-related changes lead to second-order biological responses and in a similar fashion to a cascade of effects. Second-order effects could include changes in dispersal and pollination biology caused by first-order effects such as the loss of a primary disperser or pollinator, or overexploitation of a food resource by a confined species. Initial impressions suggest canopy birds such as macaws are unaffected by fragmentation as compared to birds of the dark forest interior. This is similar to Karr's (1982a) analysis of avian extinction on Barro Colorado Island. Canopy birds might, in fact, be able to persist in a checkerboard landscape. Yet the first-order changes in the woody plant community may eventually diminish the capacity of the forest fragment to support some of these species.

Results with army ant-following birds suggest corridors can be of real value, and merit further study and consideration. The strength of edge effects indicates it is unlikely that corridors of less than a few hundred meters in width will long remain representative of primary forest. This may not be of great importance compared to the presence of vegetative cover *per se*. The significance of such altered cover, as opposed to primary forest, may vary with corridor length.

Data for primates and the army ant–ant-following bird system indicate natural home range size may have predictive value as to area requirements for species that are confined to forest fragments. The increased density of tamarin troops in the isolated 100-ha reserve suggests caution in applying such an approach.

Moreover, there may be two distinct phases of adjustment to forest fragmentation. During the first, the species adjust to the changing ecological conditions. During the second, the species will be subject to demographic stochasticity and genetic problems. (The latter subject, Minimum Viable Populations, is discussed in Chapter 2.) Consequently, a species that survives the basic environmental response to isolation may in the long run still not survive. For example, with regard to the army ant–ant-following bird system, chance alone could produce several days in a row when none of the three or four colonies would be swarming, and the ant-following birds might leave. The examples in this chapter primarily relate to changing abiotic conditions—the first phase. Even in relatively small reserves second- or third-order effects may take considerable time.

CHAPTER 13

THE ETERNAL EXTERNAL
THREAT

Daniel H. Janzen

Once a preserve is established and once its neighbors accept its presence as a geographic entity, there is a strong inclination to turn to other conservation projects and leave the preserve to the managers and users. At this point the chainsaws and hunters can be stopped, but how do you keep out the moths that breed in a cotton field many kilometers distant? Who will stop agriculturally-induced rainfall reductions from drying out a tropical park? Who is going to remove the European ruderal herbs from the riparian succession in a New York state park? Who is going to stop the fast-growing legume trees that are being introduced for firewood from spreading into a tropical preserve? Are *you* going to tell the peccaries in your park they cannot maintain their density ten times higher than normal while feeding in the secondary succession around your park?

I do not have solutions to the class of problems that I discuss here. My only hope is that those in direct contact with this class of problems will encounter the solutions, case by case. Every case will have unique solutions that will be evident to the persons who are privy to the natural histories of the particular organisms and habitats in their care. I can say with certainty that this area of conservation has only one general principle: the only insular preserves are those that were insular to start with—and even these are subject to long-distance movements (or the lack of them) from human-generated habitats.

Preserve managers turn to ecologists (among others) for methodologies and principles that will give them a greater chance of management success. My opinion is that ecologists have only one rule to offer:

get to know the organisms and habitats in your care so well that you will recognize this class of problems at inception or earlier. In fact, knowledge of other systems and of general rules can even be counterproductive if they lead to watchfulness only for problems that have already been identified. Every nature preserve is guaranteed to have one or more major problems of this type that are unique to it and its communities.

There are two groups of external nonhuman and unintentional human threats to a preserve. First, there are biotic problems, based on wild or free-ranging domestic organisms to which the preserve boundaries mean little or nothing, or are just another habitat discontinuity. Second, there are human-induced changes in the physical world outside of the preserve, changes that wash over the preserve as well. In both cases, the threat is as much (if not more) to the interactions within the preserve as to the simple persistence of a species. The best guarantee of the latter is the maintenance of the former.

THE BIOLOGICAL WORLD

Animals and plants move. While a preserve's boundaries may serve well enough to stop direct human transgressions, the boundaries *per se* will mean nothing to most organisms. There are many biological facets to this problem, and I list a few below. A number of these facets are managerially contradictory, yet occur simultaneously in nature. As I emphasized in the introduction, each preserve will have its own variants and combinations. Preserve managers will have to come to their own decisions about which to ignore and which to attempt to modify.

Normal dispersal

Organisms normally leave a site by eruptive migration—fledging juveniles, seed and pollen dispersal, etc. If the site is an arbitrary patch in a continuous ocean of habitat or mosaics of habitat, this outflow is eventually matched by inflow. However, this statement becomes less true the more the site is on the margin of a major population discontinuity for the organism concerned. Preserve boundaries are very often major population discontinuities (and are becoming more so). Even if there is no habitat discontinuity at the preserve boundary, human harvest activities often generate a population discontinuity at the boundary. This means that for most populations on margins, the preserve boundary is a unidirectional filter. It lets members out, but few return. An animal that moves out of the preserve and disappears into the outside world is just as dead as one that is shot by a poacher 2 km inside the preserve.

Such potentially severe losses occur without a human ever stepping foot in the preserve. How deeply these effects penetrate depends entirely on the behavior of the animal or the potential for pollen dispersal of a plant. For wide-ranging animals such as a herd of several hundred white-lipped peccary *(Tayassu peccari),* "the boundary" may be 10–30 km deep; for a howler monkey troop *(Alouatta palliata),* it is perhaps only a few meters from the forest–pasture habitat discontinuity. Imagine a large wind-pollinated beech tree *(Fagus)* that is growing on the margin between a 2-ha pristine forest preserve and an apple orchard. Half the tree's pollen is ecologically dead simply because it lands in the orchard. Furthermore, the beech tree's incoming pollen represents only half as many parents as does the pollen load that the tree would receive were it imbedded in a large pristine forest canopy.

There are two important components to this boundary effect. First, population density along boundaries may well be lower than in central areas simply because the rate of reproduction in this "habitat" is effectively reduced. This effect will be difficult to separate from other boundary effects to be discussed below. Second, this boundary effect depends on the habitat types surrounding the preserve. If the external habitats discourage animals from moving into them, they may have a less disruptive effect on the preserve than if they allow animals to go out but not to return. For plants, a preserve boundary is a sponge rather than a wall.

Migration

Anyone concerned with songbirds understands the omnipresence and importance of seasonal migrations in the biology of a preserve (e.g. Howe, 1984). It is likewise fully appreciated that if such migrants are to arrive, they have to come from somewhere. We all know the story: no matter how magnificent the waterfowl refuges along migratory flyways, if you eliminate the prairie pothole nesting grounds you eliminate the ducks. Conservation thus becomes a game of simultaneous and unerring maintenance of distant habitats connected by biological threads tens to thousands of kilometers in length. The interruption of either end of the cycle leads not only to the loss of the migrant population, but may also result in the loss of "cultural" knowledge of migratory endpoints for the species as a whole. A California monarch butterfly *(Danaus plexippus)* will have no idea of how to get to the traditional overwintering grounds of the eastern monarch in the Mexican highlands (e.g. Brower et al., 1985; Urquhart and Urquhart, 1976). This means that it is unlikely that the former, if introduced into the eastern United States, would become easily established.

We are only now beginning to dimly focus on the fact that migrations within the tropics are an integral part of the biology of many tropical organisms. This has been appreciated by biologists in Africa for some time; seasonal movements (and forced cessation of seasonal movements) of big game and birds have been a constant source of anguish for conservationists on that continent. Biologists are just beginning to discover the intratropical migration patterns of the Neotropics (e.g. Stiles, 1983; Leck, 1985; Janzen, 1984a, 1986), but I suspect that for many species the discovery is coming too late. A very great external threat from agriculture is that it will eliminate one of the end points in a migratory cycle. Here I offer a single example from Santa Rosa National Park in northwestern Costa Rica.

Large moths of the family Sphingidae are prominent visitors (and presumed pollinators) of flowers in the Park (Haber, 1984; Haber and Frankie, 1982). Their large caterpillars are conspicuous consumers of foliage and they are major dietary items for certain species of insectivorous birds and parasitic insects. There are at least 64 species of sphingids that breed in Santa Rosa during the first two months of the rainy season (Janzen, 1984a and in press), and it is clear that they interface with many points in the Park's ecosystem. There are at least 40 species that appear to have the following life history. At the beginning of the rainy season (mid-May) the adult moths arrive in the Park in large numbers. They visit flowers and oviposit on more than 100 species of plants (Janzen, 1984a and in press). Within three months, they have passed the first generation in the Park and emerged from their pupae. During this first generation, the carnivores that feed on them have built up from their annual low at the end of the dry season to their annual high at the end of this first generation. At this time many individuals of most species of sphingids leave the park. A few appear to stay there and have a second generation, but it appears that the majority fly to the rainforest side of Costa Rica, 15–50 km to the east.

By the end of the rainy season, all the adults of at least 40 sphingid species seem to have left the park, and they leave no residual dormant pupae or eggs. Also at this time, these adults of the migrant species are common on the rainforest side of Costa Rica. Because seemingly freshly emerged adults appear at lights for the following six months in the rainforest, I assume that one or more generations occur there. At the end of the Santa Rosa dry season, these adults fly back across the central mountain ranges to the dry forest on the Pacific side of Costa Rica for their (next) generation in their food-rich and relatively predator-poor habitats.

I suspect that if such a sphingid population had to maintain itself solely on the recruitment that is possible in the rainforest (where suitable food plants appear scarce and carnivores are always abundant), such recruitment might well not be adequate for persistence of the population. However, the population gets a large transfusion, so to speak, from the generation in Santa Rosa's dry forest. This increase boosts the population to a level whereby six months of decline in the rainforest does not push it to extinction levels. On the other hand, the carnivore level is very high indeed at the end of the first Santa Rosa generation, and these carnivores continue feeding throughout the dry season. In the face of such predation, I doubt that a sphingid population could persist on the low level of recruitment that is possible with a second or third generation in Santa Rosa's dry forest. Worse, the population would have to survive the dry season, a time when there is no opportunity for recruitment; this is due to the lack of leaves (most sphingid host plants are deciduous in Santa Rosa) and to the hot dry weather (which is prejudicial to caterpillars).

The conclusion for conservation is clear. Eliminate Costa Rica's rainforest, and the bulk of the sphingids disappears from Santa Rosa without ever going near the Park. The same happens if Costa Rica's dry forests are eliminated. The potential repercussions in pollination systems, energy chains, and other linkage networks are enormous. In fact, I suspect that many have already occurred. For example, the agricultural lands occupying Costa Rica's dry forest habitats still contain many small patches of badly disturbed secondary succession. These are currently major breeding sites for sphingids during the first rainy season generation. The same applies to Costa Rica's rainforest side. It would be silly to assume that Santa Rosa's 10,000 ha, plus the other tiny dry forest remnants, currently generate the number of sphingids that originally moved (during the second half of the rainy season) eastward across the mountains to occupy Costa Rica's rainforest sites. The same demographic statement applies to the moths returning to the dry forests at the beginning of the rainy season.

There is, however, a complication. At present, some sphingid species are probably more common as larvae than before, since numerous sphingid host plants are vines and shrubs of secondary succession. On the other hand, some sphingid species may be now more rare as adults because the nectar sources of many adult sphingids are restricted to the remnants of old secondary succession and pristine forest.

The situation should finally stabilize with a few sphingid species annually migrating between one tiny dry forest habitat island called Santa Rosa and several small rainforest habitat islands 30–100 km to the east. I cannot possibly guess which few species will be able to perform this feat, and what they will contribute to Santa Rosa's ecosystems. I can state, however, that Santa Rosa's 10,000 ha is so small that any increase in its size would be a step towards alleviating the problem. I can also stress, as has Myers (1985), that the surviving species are most likely to be the "weedy" species such as *Agrius cingulatus,* whose larvae feed on roadside morning glory vines (Convolvulaceae) and whose adults are found from Canada to Argentina and Chile. It is these species for which there is the lowest threat from habitat destruction.

Cross-boundary subsidy

A preserve is often centered on the most pristine habitats in the area. Such habitats often contain densities of generalist animals that are substantially lower than are those in mildly disturbed successional habitats (e.g. Johns, 1985). The causes are complex, but often related to the lower and more sporadic rates of production of harvestable plant materials in pristine than in successional habitats. In general, animals do not build up to a high density in food-poor habitats. The consequence is that if a pristine area (the preserve) is embedded in an ocean of secondary succession (in or out of the preserve), the animals maintained at high density in the secondary succession will occur, by virtue of short-term movements, at a higher than normal density in the pristine area. The "pristine" animal density is being subsidized by the secondary succession.

Since a high density of animals has its obvious attractions for some aspects of conservation, wherein lies the problem? Quite simply, the feeding, trampling, and seed dispersal of animals at a high density can destroy the pristine area as thoroughly as can a chainsaw (Janzen, 1983). Not only is there greater impact from feeding as conventionally understood, but mast seeding (and flowering) adds a further complication. Many species of primary forest trees have a type of reproductive behavior whereby they synchronously produce very large seed crops at supra-annual intervals (Janzen, 1971, 1974, 1978; Silvertown, 1980). When such trees are part of an ocean of pristine forest, the enormous pulse of seeds satiates the seed predators and recruitment occurs. Satiation occurs because the density of seed-eating animals is relatively low in the intervals between seed crops, owing to the relatively small amount of food present for them in pristine forest. However, most pristine preserves are small relative to the secondary successional forest that is rich in seed predators and in which they are imbedded. When their trees mast seed, the animals simply move in from neighboring secondary successional areas and consume the entire seed crop. The trampling effects of animals of secondary succession that are using the pristine habitat for shelter from the sun and weather require no elaboration.

Many animals eat fruits and defecate (or regurgitate) the seeds (Janzen, 1984b). This seed rain falls wherever the animals go. If a pristine preserve is embedded in secondary succession, the pristine habitat fragment is bombarded by a far higher density of these propagules than it would be if that site were simply part of a large piece of pristine vegetation (Janzen, 1983; Ranney et al., 1981; Estrada et al., 1984). Since microsuccession in treefalls and other kinds of natural disturbance sites is dependent as much on the numbers of seeds as on the kinds of seeds that arrive, this bombardment has a high potential for altering the course of succession and eventually the overall structure of the pristine vegetation.

The alteration is exacerbated not only by the seed dispersal activities of the animals, but also by the occurrence of the secondary successional plants around the pristine area at much higher densities than they were prior to human intervention. Rare plants and very patchily distributed plants in natural disturbance sites become the common ones in old field, pasture, and roadside regeneration (Marks, 1983; Janzen, 1984b). This means, for example, that a Santa Rosa treefall is not only getting traversed by more collared peccaries (*Tayassu tajacu*) per week, but those peccaries leave more seeds from secondary successional plant species per defecation than would peccaries living in a very large expanse of pristine vegetation.

Denizens of the crop habitat

Crop and pasture lands (abandoned or in production) generally contain large populations of many species of animals and plants. Many of these are exotics, and they are strongly dependent on the crop and its microhabitat. However, there are also many species that eruptively migrate (or colonize) into neighboring preserves from agricultural land (e.g. Marks, 1983). There are also indigenous species on agricultural land that still have noncrop populations in more pristine neighboring preserves.

Species input from agricultural habitats has two possibly severe impacts. First, as implied above, migrating or colonizing species may simply consume preserve organisms (and serve as food for them), thereby altering the preserve (see the discussion of nest predation in habitat edges by Wilcove et al., Chapter 11). This applies to parasites and diseases as well as to the traditional carnivores. Second, and much more insidious, these large widespread agricultural populations may homogenize the gene pools of the preserve's "wild" conspecifics. There may be a quick loss of whatever genetic adaptations there have been to local biotic and physical conditions (see Templeton, Chapter 6). Pesticide resistance, high reproductive rates, peculiar dormancy physiologies, and other traits that are strongly selected for in agricultural species should appear in the wild population where they may depress fitness. We know very little about the swamping of the gene pool of a small and locally adapted population by a large and widespread population that is subject to somewhat different selective pressures (but see Ledig, Chapter 5).

The crops themselves

There is a widespread lack of fear of the spread of crop organisms and horticultural varieties, both because they are "good" and because they are widely thought of as being so heavily dependent on humans for protection that they will not be a threat to relatively pristine areas. There are two conspicuous kinds of exceptions. First, when crop organisms have been introduced to islands (or to other naturally impoverished areas) there is little ecological resistance to their spread; this has been documented extensively (see Pimm, Chapter 14).

Second, there is the widely unappreciated problem that today's agriculturalists (in the broad sense) are busy looking for new organisms to cultivate as crops, organisms with properties that are not represented among traditional cultivars. Such neocultivars are but little different from truly wild phenotypes. The introduction of the Africanized honeybee into the Neotropics is perhaps the most widely popularized example. Introduction of *Eucalyptus* into the Neotropics is a threat in progress (somewhat less so, however, because its wind- and gravity-dispersed seeds are slow invaders, and because particular human-induced burning regimes are probably necessary to maintain a resident *Eucalyptus* breeding population).

A storm cloud on the horizon is the well-meaning but potentially catastrophic widespread introduction of fast-growing tropical firewood trees (Hughes and Styles, 1984). These are mostly ruderal legumes, and are the sorts of trees that will become prominent members of natural disturbance sites in relatively pristine vegetation. The introduced plants have a strong chance of influencing the outcome and speed of succession and competitively eliminating some of the habitat's members.

As humanity becomes ever more imaginative in building agroecosystem habitats that are constituted of a diverse array of organisms, the intercontinental and interhabitat movement of species will be ever more widespread. The steps are largely irreversible. Except for very large (and slowly reproducing) organisms, or those with pecularily susceptible individuals (e.g., the dodo, *Raphus cucullatus*), our chances of eliminating introduced and established organisms is nil except at very high cost. As a general statement, a large pristine habitat is probably the only effective barrier to such external threats, and even that will be penetrated by a variety of introduced organisms (Africanized honeybees, for example, maintain established breeding populations in Costa Rican pristine forests).

The threat from the preserve

As any African farmer next to a national park well knows, there are animals in a preserve that can be incompatible with contemporary agriculture. Indeed, if the Pleistocene hunters had not eliminated the New World herbivorous megafauna, the post-Pleistocene farmers certainly would have; glyptodonts, ground sloths, and gomphotheres (Janzen and Martin, 1982) would not have been miscible with corn fields. Complaints that protected animals damage crops range from Swedish wolves killing the occasional sheep (for which the owners are compensated by the government), to ducks feeding in North American grain fields, to vampire bats in Santa Rosa National Park feeding on cattle on neighboring ranches. As agricultural intensity and refinement increase around a preserve, the chance for such interaction creating ill will toward the preserve increases. Simultaneously, increasing educational sophistication of the neighboring landowners should result in greater tolerance to damage and the suggestion of intricate solutions. The balance between these two processes can only be determined on a case-by-case basis, but this balance does need to be considered if one is to understand the interaction of a preserve with its surroundings.

THE PHYSICAL WORLD

The threatening physical environment around a preserve ranges from that which is under the direct control of humans (e.g., pesticides and fire) to uncontrollable weather modification by humans.

Pesticides

The avoidance of pesticide and other agrochemical contamination of a preserve appears to be little more than a direct struggle between the humans that apply them and those concerned with the preserve. Spray plane overflights and nearby pesticide applications on windy days are generally a legal matter between the preserve and surrounding areas, as if the preserve were simply another farm. Pesticide contamination of wide-ranging vertebrates that sometimes feed outside of the preserve is a battle that hardly needs mention. However, there are three aspects of pesticide dynamics that deserve emphasis.

First, in developing countries the concern over pesticides is generally directed at human health hazards. As concern becomes more sophisticated, it is generally directed at vertebrates, and especially those found dead after severe sloppiness in pesticide applications. Insects and other arthropods often escape this concern. They are small (their carcasses are inconspicuous), generally feared and disliked by most people, and not a focus for preserve establishment. It has been my experience that very substantial pesticide damage to insect populations in a preserve can occur in areas marginal to agricultural land, with no notice taken by preserve managers who are quite eager to eliminate fires, poachers, and timber thieves. At least part of this oversight is due to a misconception about pesticides. Bulk containers identify not only the chemicals, but give warnings such as "mildly toxic," "extremely toxic," etc. Suspicious preserve managers who investigate pesticide operations near the boundaries are likely to have their fears diminished by reading that a pesticide being applied is only "mildly toxic." However, pesticides are designed to kill invertebrates; the warning labels refer to toxicity to humans and other warmblooded vertebrates, not to the target organisms.

Second, many preserves are established simply by taking in as much preserve-worthy habitat as is available. When more choice is possible, a time-honored criterion is the inclusion of entire drainage systems within the preserve. This is more than an ecological nicety and must be done even if the drainage basin contains croplands which have to be abandoned. Recent incidents with pesticide threats to Santa Rosa National Park in northwestern Costa Rica have underscored the need for a more thorough look at a drainage system.

Santa Rosa contains the upper drainage basin for several large dry river systems. However, those who established the Park boundaries did not notice that a tiny number of feeder streams and small slopes above those streams were left outside of the park. In no case does the area outside the Park make up more than 1–3 percent of the total catchment of the drainage systems. However, in recent years the cattle pastures that previously occupied the southern boundaries of the park have been converted to cotton and other croplands. The outcome is that each river system gets a major pulse of pesticides (all correctly applied on calm mornings by spray planes) and silt via rainwash surface runoff from the fields. One of the dry river systems (Quebrada Guapote) has been sterilized, and a second is very likely to be destroyed unless some kind of alternate land use can be assigned. A tiny error in boundary demarcation can have a very large effect on a preserve if extremely toxic chemicals are being pushed into the habitat.

Third, as any farmer knows, pesticide applications in fields do not eliminate pest populations but rather (it is hoped) depress them enough to allow a crop to develop. Associated with this, the populations of insects in fields are often resistant to pesticides. This often leads to the impression that pesticides are a nuisance to insects but not likely to exterminate them. What such observers generally do not know is that when the same pesticides are applied to wild vegetation, they are effective at eliminating wild insect populations. Perhaps the most dramatic examples of this come from small woodlots and low forest strips between cotton, sorghum, and other crops in Costa Rica and Nicaragua. Airplanes commonly spray this remnant vegetation at the same time that the crop is sprayed, and the small bits of wild vegetation become impressively free of insects. The foliage of the plants growing within the bits is so free of herbivory that it appears made in a florist's shop. However, the field itself is often quite rich in insects (which are presumably resistant to the pesticides). I suspect the cause of this pattern is that the wild insects are initially at a low density, as is their food. They are also not likely to be parts of large crop-adapted resistant populations. When hit with a pesticide, their numbers are severely depressed and the survivors have the added problem of locating food. When a population of a crop insect is hit with a pesticide, even if there are only a few survivors, those few are sitting on mountains of food.

Fire

This is not the place to enter into a discussion of whether inflammable preserves should be maintained so as to simulate precolonial fire regimes (or the lack of them). However, if the decision has been made to exclude fire, firing practices in neighboring grasslands (pasture or otherwise) are a severe threat. Fire moves with the wind and the wind knows no boundaries. Decades of a successful fire control program can be lost with a single error; even more threatening, the probability of catastrophe rises with each year since the last fire (at least until the vegetation has developed enough of a woody closed canopy to have a nonflammable understory). Perhaps worse, though inconspicuous, is that agrofiring regimes are generally different in intensity and seasonality than are natural ones (if there are such), and can as surely lead to gross alterations of habitats as can the elimination of fire from a naturally burned preserve.

In areas where pasture and woody secondary vegetation are frequently burned, the local attitude is often that fires are inevitable. Associated with this, unless special relationships exist with neighbors of the preserve, fires will usually be allowed to burn directly up to the preserve boundaries, thereby placing the onus of stopping them on the preserve. Fires, unlike cattle, are not viewed as belonging to anybody, and nobody retrieves them when they wander into your park.

It has been my experience that a fire moving across a large inflammable area will slightly penetrate or at least kill the margins of unburnable forest at the line of contact. The fire-damaged edge will then grow substantial amounts of herbaceous material in the following growing season, thereby extending the margin of the inflammable area. If the process continues annually, at least in the tropical dry forest areas of Costa Rica, Africa and Australia, grassland will gradually replace the forest (except for portions on exceptionally moist sites). Even these moist areas may be burned in dry years or in dry seasons that follow exceptionally heavy fuel accumulation in the grassland. In short, a forested preserve surrounded by pasture or young to middle-aged secondary succession in a tropical area dry enough to burn (and this includes almost all of the lowland tropics) is at exceptional fire risk because the large external inflammable area will carry a large fire to the preserve boundaries.

Even if the preserve boundaries are controlled by firebreaks that are wide and burned annually, more distant fires are also a threat. Wind carries glowing embers. When they land on forest litter or living trees, they are usually harmless. However, if they land on a dry rotting log or standing dead tree (or even on a dead branch of a living tree) within the forest, they can and do start that material on fire. It is this kind of fuel that accumulates in large amounts in seasonally dry forests, and such fuel is exceptionally abundant following a dry rainy season (due to slow decomposition rates). In normal years such burning material produces nothing more than a pile or line of ash, standing out strongly from the surrounding unburnable leaf litter. However, on occasion a small patch of forest understory burns. This creates a light-rich site that grows more herbaceous material. This site in turn has a higher chance of catching fire or sustaining a fire than does an intact forest. Likewise, if one of these small fires encounters a patch of secondary succession within the forest (e.g., an old roadside or home-site), the patch may contain ample fuel to accelerate the process of producing a sheet of highly inflammable vegetation within the forest.

Finally, there is an insidious trait of the small and seemingly trivial fires that enter a forest preserve. A light fire creeping along through the leaf litter appears to kill only seedlings and small woody plants, leaving the larger trees unscathed. However, in Costa Rican dry forests, such fires are often hot enough to kill small areas of the cambium near ground level, resulting in almost unnoticeable sloughing of bark and dead tissue at these points in subsequent years. If there is another light fire in the litter within the 5–20 year period that it may take a tree to grow over these small fire scars, the fire then has access to the internal dead wood of the tree trunk. A light litter fire, then, cuts down large living trees as surely as if done with a chainsaw. What would appear to be a trivial fire can convert a closed-canopy forest into a tangle of secondary succession that is ideal fuel for a serious fire in a subsequent dry season.

Climate

The most conspicuous external threat to a preserve is climate modification by humans, who are clearing forests and other pristine habitats. Speaking of the Amazon basin, Camara (1983) noted that "large scale destruction of it could change dramatically the amount of water in the drainage system of the basin and profoundly affect the ecosystems of all protected areas, indirectly. If such a situation arises, the aims of the creation of these areas may be frustrated by human action beyond their border." While it may as yet be undecided as to when regional climate changes (as expressed in standard weather records) occur following deforestation, there are clearly many local effects of deforestation that impinge directly on the climate of a preserve. Before discussing the details, there is a philosophical point that needs elaboration.

It may be argued that climate modification by clearing of forest is not necessarily bad for the people living on that land; whatever the climate and soils, humans will eventually grow (or select for) the crops that grow best there. In only a few cases will the habitat be so close to the margin of usability that climate modification will eliminate all agricultural possibilities. But reserves are not as flexible as agriculturalists. Let us say that the climate change induced by forest clearing is a 20 percent reduction in annual rainfall and a two-week shortening of a six-month rainy season. This kind of a change will have a dramatic effect on the habitat. There will be local extinctions, modifications of population densities, changes in microhabitat distributions, and changes in proportional representation of species.

Why should we worry about human-induced changes in climate? Hasn't every point in space and every species been subject to such changes in the past? Natural changes generally fall on continuous gradients of habitats. As species change geographically and demographically in response to climate changes, other species move into the site from previously wetter or drier areas. In short, arrays of organisms and their component parts get pushed around over the countryside as climates change. However, when the climate changes on a small habitat island in an agricultural ocean, the species simply get shoved into the ocean, and the island is so far from "land" (if there is any left) that there are no incoming species to constitute new arrays.

Species that inhabit seasonal climates (virtually all of the world's terrestrial organisms in virtually all habitats) are periodically subject to multiple-year runs of exceptional deleterious weather. While a local population may be decimated by this event, a number of years of deleterious weather are generally required to cause extinction of a species. This is due to on-site population residuals, including seed banks (e.g. Baskin and Baskin, 1978), robust adults, ratooning root stocks, refugia with exceptional peaks in resource abundance, and dormant pupae. However, when climate modification induced by humans mimics such a naturally-occurring weather event, the climate does not return to the original state after a few years. Population residuals can persist only so long.

In the same context, when a local population is extinguished in "nature," it is commonly reestablished by haphazard immigration or even by traditional migration movements. As preserves become ever more insular and islands are pushed ever further from other islands or from the "mainland" (if any exists), the opportunity for such reestablishment declines.

There is a second major way in which agricultural activities influence preserves via the physical climate. Most preserves have vegetation that is grossly different in overall structure from that of the agricultural land that is immediately adjacent. The difference is usually in the direction of more extreme fluctuations and absolute values in the agricultural land. They are hotter in the day, colder at night. It is drier and sunnier in the day. It is windier. This means that, quite irrespective of the overall climate modification by agricultural lands, lands adjacent to a preserve will create maximal edge effects within the preserve and thereby render small preserves even smaller than they seem to be. Ten preserves of 10 ha each may be all edge (see Lovejoy, Chapter 12 for a tropical rainforest example), while one preserve of 100 ha may contain a small core area that is relatively free of physical edge effects. The core would probably not persist in the face of biological interactions from the area that is rich in edge effects, however.

The oak and Santa Rosa

The demise of the population of the lowland tropical oak, *Quercus oleoides*, in Santa Rosa National Park is instructive on some of the above points. The park covers an area of 10,000 ha from the Pacific Ocean to about 350 m elevation along the gradual rise up to the steep sides of the recent 1650 m Volcan Cacao in the northwestern corner of Costa Rica (11° N latitude). Prior to Spanish occupation of the site in the late 1500s, a semi-evergreen forest composed primarily of *Quercus oleoides* occupied the northeastern and most upland portions of the park and continued up the slopes of the volcano to at least 500 m elevation. This region is the southernmost portion of the *Q. oleoides* population, a population ("superspecies") that extends through dry-forest Central America north to the Texas coastal border and through the eastern coastal United States (under the name of *Quercus virginiana*).

In Santa Rosa, *Q. oleoides* occupies original and only slightly degraded volcanic tuff substrates. These soils are so poor that agriculture has traditionally failed on them. Part of their low quality derives from the fact that when the rainy season stops or falters, they dry out very rapidly owing to their high porosity. When the northeastern part of Santa Rosa had a continuous moderately-to-highly evergreen oak forest canopy, the incessant winds that accompany the end of the rainy season and the first half of the six-month dry season blew across the top of this canopy and the shaded soil remained relatively moist for at least the first several months of the dry season. The oaks drop their acorns during the last two months of the rainy season. If the litter-soil surface is moist, the acorns germinate within weeks. An acorn's shoot tip grows straight down, however, and comes to rest several centimeters below the soil surface. The contents of the acorn are then rapidly transferred to a tuberous swelling on the underground stem. If the soil remains moist, an above-ground shoot is produced and the seedling grows. If the soil is very dry, however, the seedling remains dormant and produces an above-ground shoot only when the soil is again moistened. If the litter-soil surface is dry, the acorn does not germinate; after several months it is dead, its seed contents dry and rock-hard. In short, the acorn is self-burying during the moist end of the rainy season, the function of burial being escape from the upcoming dry season (and escape from the very active acorn harvest by a variety of mammalian seed predators). This portion of the story is inferred and observed from minute remnants of intact oak forest that still remain in the park.

At present, the northeastern portion of the Park is occupied by a mosaic of fragments (usually a fraction of a hectare in area) of the original oak forest, interspersed with grass pastureland and patches of woody succession ranging from one to hundreds of years of age. The pastures were cleared between the late 1500s and 1940s, and converted from native grasses to jaragua (*Hyparrhenia rufa,* an introduced East

African grass; Pohl, 1983) in the 1940s. The pastures are maintained by fire, though the park is currently engaged in a fire control program that will lead to their eventual demise through woody succession. The oaks are well known in the region, and there is a strong desire to see the Santa Rosa oak forest return to its original state. The prospects for this, however, are grim.

Consider a representative large adult Santa Rosa oak within one of these small patches of forest at the end of the rainy season. Its neighboring adult conspecifics are 5 to 50 m away; the intervening area is filled with vegetation that ranges from a 1–2 m stand of jaragua to closed-canopy forest understory. It obtains enough pollen to set a crop of tens of thousands of acorns at several year intervals. Such a crop is usually synchronized with a year of massive acorn production by many of its conspecifics. The acorns begin to fall in late October, and the bulk fall in November. In late October and early November, the rainy season wanes, with more and more breezy sunny days of blue skies. The sun has direct access to the soil in open areas and the winds are everywhere at ground level. Where there is forest, the edge is within 100 m of any oak, and the wind blows directly through the forest; even dense coniferous forests are viewed as needing at least 100 m of width before winds blowing into the interior from an edge are stopped (Harris, 1984). The forest leaf litter dries rapidly and the acorns lie on the dry soil without germinating, either in the forest or under the grass. The grass grows in small tight clumps, with bare soil and a few oak leaves between. Acorns are slippery and smooth; they end up on the soil and not beneath a moisture-retaining grass turf or mat. The herbaceous vegetation dries very rapidly, and when not excluded, the human-generated fires begin spreading by December. Because of the very high fuel load (since 1978 there has been essentially no livestock in the Park), even an early dry season fire is hot enough to kill any acorn, as well as any above-ground seedling or sucker shoot. The dry season continues for another five months, with wind, insolation and fire prominent throughout.

At present there are *no* oak seedlings in the disturbed parts of Santa Rosa. Prior to fire control but after the removal of cattle, the Santa Rosa oak population was steadily diminishing as root systems were gradually killed through the production of sucker shoots that were killed annually by fire, and large adults were gradually eliminated by fires burning through their bases at access points in old fire scars. The fires have been stopped, but there is no sign of seedlings; they are killed by the dry season starting 1–2 months early, from the viewpoint of the acorn exposed on the soil surface. As the vegetation gradually succeeds back to forest within the Park, will soil conditions at the end of the rainy season remain moist long enough to allow acorn germination? Perhaps in some sites, but the situation is complicated by the fact that the first several hundred years of succession in this dry forest is performed almost entirely by dry season-deciduous species of trees, and they do little to prolong the moistness of the rainy season.

Assuming that the Santa Rosa forest gradually returns to its semi-evergreen original status (there are slow-growing evergreen trees in the area that may aid the evergreen oaks in this context), will the overall climate of the region allow oak regeneration? All of the area around the park has been or is being cleared for agricultural purposes. The late rainy season winds and the early dry season winds sweep for 10–20 km across fully insolated dry pastures and fields before hitting the 1000-ha area that may someday again become an oak forest. It is highly doubtful that such a patch of forest, even if it has a closed canopy, can sustain an internal blanket of moist soil and air for the length of time necessary for oak seedling recruitment. Furthermore, the terrain is highly dissected, a situation resulting in numerous edges that are exposed to the drying wind. Santa Rosa's upcoming oak forest has as much chance to influence the overall climate of the Park as does an ice cube.

IN CLOSING

If nuclear winter threatens all of us and all those things we work to save, then the nucleotide summer is surely on the other side of the coin. Yes, genetic engineering can undoubtedly produce all kinds of fantastically useful organisms. But if you are worried about what the rabbit did in Australia, how the sea lamprey clobbered Lake Michigan, and how European diseases exterminated the original human occupants of the New World, "you ain't seen nothin' yet." The metazoans and microbes that humanity is gearing up to produce are without doubt the largest threat of all to nature as we know it. In Australia humans have managed to destroy easily 90 percent of the habitats and organisms in the past 40,000 years, first by hunting and burning, then following up with a small number of introductions of relatively ordinary animals and plants from other continents. When humans start releasing organisms that have been explicitly engineered to eat the world to the ground, turning it into hamburger, cotton, and honey, then you can kiss our "natural" preserves goodbye. You can legislate nuclear power plants out of existence. But there is no way to recall a genetically engineered product that is found to have seriously destructive effects on preserves (and it is highly doubtful that the commercial world would have the incentive to even attempt a recall).

And what do I have to offer of a concrete nature to a preserve manager? Within a tropical forested habitat, expect edge effects anywhere within 5 km of the preserve boundary. However, for wide-ranging vertebrates and the seeds they carry, the edge effects will be much greater. Know the details of the natural history of your organisms well enough that you can both anticipate their interactions and know at what distance their populations can be perturbed by a preserve edge. Be as concerned about the composition of the surrounding habitat as about the area you wish to conserve. Above all, be a field biologist who works with your preserve's neighbors to keep out what should be kept out.

SECTION IV

COMMUNITY PROCESSES

Some optimistic ecologists, especially those with backgrounds in physics or mathematics, feel they ought to be able to write down dynamic mathematical models of biological communities. But this enterprise has met with many difficulties and limitations, although some useful generalizations have emerged in the last 25 years. More important than the construction of such general equations, this effort has led to a much more comprehensive and realistic appreciation of ecological communities.

The authors of the chapters in this section help us to see the uses of theory as well as the dangers of overgeneralization. Pimm (Chapter 14) explains why resistance and resilience (stability) may or may not be dependent on the structural diversity of a community; but he does suggest where to look for the weak points in the biosphere, and which communities are most prone to invasion and collapse.

Terborgh (Chapter 15) helps to destroy the myth that lush tropical environments provide their animal inhabitants with a constant, dependable supply of resources. Indeed, he shows how uniformity can actually cause scarcity in both ecological and evolutionary senses. Terborgh also develops the theme of "critical links" in the sense of a handful of keystone species that sustain most of the mammalian and avian biomass during episodes of scarcity.

Dobson and May (Chapter 16) demonstrate the utility of mathematical theory when dealing with two or three species locked in life and death struggles of parasitism and immunity, and how such theories can yield counterintuitive results that are relevant to the control of disease by managers. They also illustrate the many strange and important linkages of disease to ecological and evolutionary processes.

How is this section related to the preceding ones? It is partly a matter of scale. Some of the topics developed in this section have been described at more macroscopic scales in earlier chapters. For example, Dobson and May (Chapter 16) point out that disease can cause a population to collapse to the point where the genetic drift that accompanies such bottlenecks can erode its level of heterozygosity, which in turn can compromise its immunological responsiveness, further endangering the population. This is an example of the F Vortex described in Chapter 2. Terborgh's (Chapter 15) discussion of the migratory behavior and area requirements of frugivore communities extends Janzen's (Chapter 13) and Lovejoy et al.'s (Chapter 12) treatments of fragmentation. Pimm (Chapter 14), in developing a macroscopic theory of stability based on connectivity and trophic level, provides a context for some of the earlier observations on biotic collapse following fragmentation.

None of the chapters attempts to address in detail the controversial issue of random versus deterministic contributions to community structure, but Johannes and Hatcher (Chapter 17) point out that the problem is partly one of scale: randomness decreases as size increases.

The accompanying illustration may assist in integrating the material in this section with those in earlier ones. This flow diagram summarizes the consequences of four categories of human impacts. *Additions* (including the introduction of competitors, predators, disease organisms, and vectors) can lead through various pathways to the decline or local extinction of indigenous species. These declines are exacerbated at some point by the stochasticities and extinction vortices described in Chapter 2, and by other perturbations (pollution, disturbance) described in Section V. Quite often the local collapse or extinction of a species is followed by changes (particularly increases in the density of other species) with ensuing ripple effects, some of which behave like further additions (e.g., an increase in the abundance of small predators; Chapter 11).

Density compensation, or the increase in population size of a species favored in early successional habitats, is a frequent consequence of *fragmentation*. Sometimes such habitats favor the vectors of pathogens and parasites, such as certain ticks, insects, and mollusks. This can be important because the immunosuppressive effects of a parasite can multiply its effects on fitness, and also because of the threshold nature of many disease processes, as described by Dobson and May in Chapter 16. As they suggest, the consequences may also include the spiralling collapse of host populations or the onset of outbreak–crash cycles in host and pathogen.

Whether local extinctions are the indirect effects of additions or fragmentation, or are the result of direct human impacts such as *deletions* and *harvesting*, the ripple effects can be both profound and unpredictable (Myers, Chapter 19), especially if the victims are participants in mutualistic interactions or are keystone resources (Terborgh, Chapter 15; Culver, Chapter 21). Both human harvesting and the less direct modes of population reduction (e.g., the effects of introduced predators and disease) can provoke episodes of genetic drift,

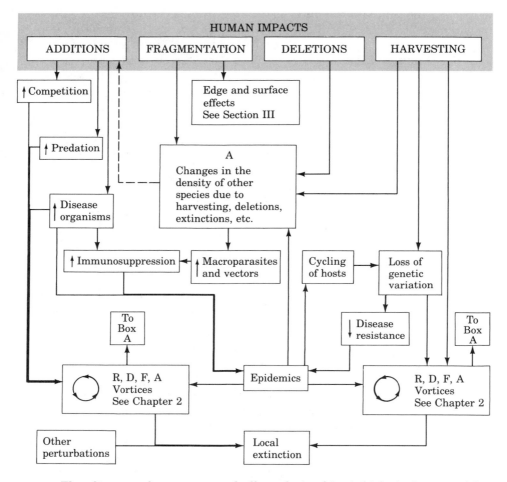

Flow diagram of some cause-and-effect relationships in biological communities that are consequences of certain human impacts. The dashed arrow from Box A to "Additions" indicates that an increase in the density of a species may behave ecologically like a species addition. The information in this diagram is based largely on Chapters 14, 15, and 16.

which in turn may reduce fitness by several mechanisms, including the reduction of immunological competence.

In the figure, the role of disease and epidemics is emphasized because of a growing conviction that it is the most underrated component of community dynamics relevant to management.

MANAGEMENT IMPLICATIONS

Here, as in previous sections, many of the management prescriptions are both self-evident and specific to the situation. I mean that the chapters can suggest the categories of phenomena to monitor, but that each site will have its own particular problems. The consequences of introductions, deletions, harvesting, and epidemics will vary from place to place and from time to time. This heterogeneity is just one more reason why there is no substitute for collaboration between managers, field workers, and other experts, including theoreticians and modelers.

EDUCATIONAL IMPLICATIONS

It seems justified here to point out some of the educational implications of this section. Just as an astronaut is more broadly trained than a commercial jet pilot, so will the manager of the future be an expert in more subjects than are contemporary managers. The future manager will belong to the cadre of conservation biologists, and will have mastered many techniques and disciplines that are absent from today's management curricula. Not only will these managers be trained in economics, anthropology, and human group dynamics, but they also will have learned the basics of population genetics, island biogeography, community ecology, and epidemiology.

This education cannot and should not replace the more traditional, resource-oriented curriculum. (Most astronauts still train initially to be pilots.) Conditions change, however, and we cannot continue employing educational models developed for temperate zone conditions that exist in the relatively benign economic, demographic, and biological conditions of the United States, the Soviet Union, Germany, France, and Great Britain. These mid-20th century, northern curricula evolved during a period of authoritarianism that was dominated by an anthropocentric, instrumentalist management philosophy, a premise of which is that nature is merely a resource for humans. In other words, the orientation was one of managing for commercial and recreational objectives (logging, grazing, hunting and fishing, and camping potential). In addition, there was little need to consider the resource needs, cultures, and sensitivities of local indigenous peoples.

This is a far cry from the situation in the world today where most of the action will be in the difficult and often chaotic social, economic, demographic, and biological conditions of the tropics. Even in the temperate zone, social and biological conditions are changing as a result of accelerating anthropogenic stresses and a growing awareness and involvement on the part of citizens.

Coming full circle, one of the most fundamental implications for managers that emerges from this section is that they will be handicapped to the degree that their training does not prepare them to deal with the social and biological challenges inherent in the control of exotic species, poaching and harvesting, and disease—to mention only a few of the relevant issues here or on the horizon. The big question is, who will take the lead in developing a management curriculum for the 21st century?

M.S.

CHAPTER 14

COMMUNITY STABILITY AND STRUCTURE

Stuart L. Pimm

Most of our planet is dominated neither by pristine ecological communities, nor by species on the brink of extinction. Rather, most ecological communities we observe are fragmented, harvested and polluted—stressed in a variety of ways—by humans and their technology. We must understand how we should protect communities against these stresses. One important approach has been the application of island biogeographic ideas. These ideas have been valuable in understanding the changes in species richness during the fragmentation of natural communities when, for example, we convert continuous forest to isolated forest tracts surrounded by farmland (Lovejoy et al., Chapter 12). But useful as these ideas are, they do not tell us the whole story: they say little about how species interact with each other.

This chapter addresses theories of community ecology that are more explicit about species interactions than the theory of island biogeography. The scheme of this chapter is illustrated in Figure 1. Communities may be changed by either the addition or removal of a species, and these additions and removals can, in turn, cause further changes. Such a scheme forces us to ask a number of questions: Just what should happen to communities when we stress them? What species losses should we expect on the basis of the community's structure? Are some communities more susceptible to species invasions than others? What kind of species will invade? Will the loss of one species cause other losses to cascade through the community and, if so, how far? Are certain species in key positions in the food web so that their loss will be particularly critical? How do successful inva-

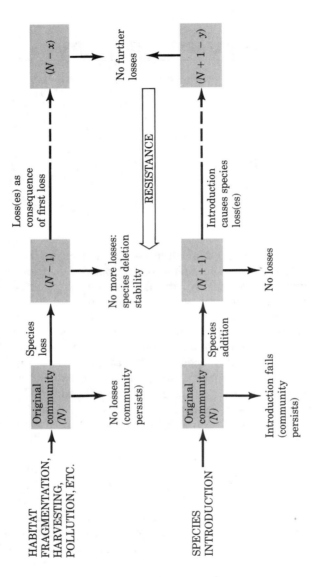

FIGURE 1. The consequences of factors which potentially remove (above) or add (below) species to communities. Questions discussed in this chapter include whether species will be lost and the consequences of those losses; and whether a species can invade a community and the consequences of a successful invasion.

sions effect changes in the existing community? And, if we lose substantial numbers of species locally, will we recover the original community composition when we restore the original conditions?

EARLY IDEAS ABOUT COMMUNITY STABILITY

The origins of these questions are clear in the debate about community complexity and stability. Ideas on complexity and stability have been influential and yet have seemingly flip-flopped every decade. Some synthesis of the apparent contradictions is needed if we are to proceed in an orderly fashion. Elton (1958), MacArthur (1957), Margalef (1968), and Hutchinson (1959) had a simple answer: more complex communities were likely to be more stable. Thus our goal, in conservation, should be to maintain complexity. The consequences of simplification should be to cause instability, species losses and further simplification. Clearly, complex communities should be more robust than simple ones. These authors argued so persuasively that complexity enhanced stability that this idea became a central feature of ecology texts. Watt (1973) states as one of his core principles of ecology that ". . . the accumulation of biological diversity . . . promotes population stability."

Elton (1958) proposed six lines of evidence for more complex systems being more stable. The first was a theoretical argument: simple models are characterized by oscillations; Elton expected more complex models to fluctuate less. Another theoretical argument, that of MacArthur (1957), was that the more pathways for energy to reach a consumer, the less severe would be the consequences of the failure of any one pathway. As we shall see, this argument is correct but incomplete: energy may flow up food chains, but effects (prey on predator, predator on prey) go both ways.

Another of Elton's lines of evidence involved the increased chance of pest outbreaks in agricultural systems; yet another argument was the apparent absence of pest outbreaks in tropical (but not temperate) forests. Elton also noted the prevalence of population cycles in Arctic ecosystems (which are species-poor) and, finally, the ease with which species could invade small, remote (and hence species-poor) oceanic islands.

These early arguments were heterogeneous and incomplete (Goodman, 1975). Yet this heterogeneity only points up the fact that there are many questions we need to answer. Further, these early studies used "stability" to mean at least four different things: what I define to be resilience, persistence, resistance, and variability (Table 1).

TABLE 1. Definitions of stability.

Stable (Units are nondimensional; either the system is stable or else it is not)	A system is deemed stable if and only if the variables all return to the initial equilibrium following their being perturbed from it.
Resilience (Units are those of time)	How fast the variables return to equilibrium following a perturbation.
Persistence (Units are those of time)	The time a variable lasts before it is changed to a new value.
Resistance (Units are nondimensional)	The degree to which a system is changed by a perturbation is the reciprocal of resistance.
Variability (Coefficient of variation is dimensionless)	The variance of population density (or similar measures like standard deviation or coefficient of variation).

MATHEMATICAL STUDIES OF COMMUNITY STABILITY

By the early 1970s, other researchers—Gardner and Ashby (1970), for example—were arguing that more complex systems were *less* stable than simple ones. This was a trend followed by most of the subsequent studies which defined stability in its strict mathematical sense (Table 1; May, 1979; Pimm, 1984a). These studies imply a different kind of comparison to those involving resilience, persistence, resistance, and variability (Pimm, 1982, 1984a). If a system is not stable, we will probably not observe it, except in transition to a new, stable equilibrium. This is because special dynamics are required to permit species to persist with population densities which do not return to an equilibrium, but instead cycle indefinitely. Cycles cannot be of too great an amplitude or species densities will fall to levels from which they cannot recover. Simply, we should expect to observe stable systems, only rarely observing unstable ones that contain species which are quickly going extinct. When we investigate stability and the features that affect it, we are implicitly comparing existing systems with hypothetical or extinct systems and not making comparisons among existing systems. Yet we can compare existing systems in terms of their resilience, persistence, resistance, or variability.

It is certainly interesting to contrast which community features are likely to enhance stability, and hence occur in nature, with those structures that are likely to cause instability, and so not occur. Such an approach has been used in the study of food webs—the diagrams depicting which species in a community interact. A large catalogue of food web features has now been assembled (Pimm, 1982), and these features are in remarkable agreement with the features that enhance stability in models. But these comparisons have a relevance that seems less than immediate to the questions asked in the introduction and to which I shall now turn.

SPECIES LOSSES IN STRESSED COMMUNITIES

If we kill enough organisms it is hardly surprising that we will eventually cause species extinctions. We do not even have to kill all the individuals, merely reduce the numbers to low levels, where the various problems besetting small populations make recovery unlikely (Gilpin and Soulé, Chapter 2). A major practical problem facing conservation biologists is to halt continuing population declines due to increased pollution, habitat loss and habitat fragmentation. But there are two issues, associated with progressively increasing death rates, that are less obvious. First, which species are going to be lost first? The answer must have something to do with which species are stressed most, but we might also expect species in different food web positions (high or low in the food chain, generalized or specialized, for example) to be differentially vulnerable. Diamond (1984a,b) has discussed this issue at some length.

Second, some species losses may not have been readily predicted by extrapolating previous population trends: they may occur rapidly, following long periods of superficially "healthy" population levels. This section considers questions involving the ability of populations to recover from perturbations ("resilience" in Table 1) and how stress, in diminishing this ability, may cause unanticipated species losses.

For range lands where we harvest domestic cattle or sheep, and for fisheries where we exploit wild populations, we increase death rates of species above normal levels yet have reasonably good intentions to preserve the communities in something like an original state. Harvesters often talk of achieving maximum sustainable yield (MSY); at whatever population level this occurs (about half of normal in many models), it is a level that may frequently be encountered naturally by a population.

How successful have harvesters been in maintaining their populations? The answer would seem to be: not at all. The last century has provided an extraordinary list of fishery collapses (Gulland, 1983) and range destructions (Noy-Meir, 1975). Some of these collapses may have been due to overharvesting because of the inability to regulate harvesting; to factors—like pollution—outside the control of those managing the population (Regier and Hartman, 1973); or to economic considerations which place little value on future resources (Ehrlich and Ehrlich, 1981). But the sheer number of examples makes one wonder whether there are other explanations as well.

One possible explanation is that there are breakpoints (Jones and Walters, 1976; May, 1977; Peterman et al., 1979). For quite simple population models, densities may return to a high equilibrium provided the population density remains above a certain threshold. But densities will move to a low equilibrium—or even extinction—if the density falls below the threshold. Such behaviors pose a special problem for conservation biologists because, without very detailed knowledge of the population, it may be very difficult to predict when species are going to "fall off the edge" and be lost.

Another explanation involves the problems inherent when we try to simultaneously harvest more than one species in our system. Treating each species separately and pushing each to its MSY may be expected to lead to the system's collapse (Beddington and May, 1980).

But the explanation to be considered here in most detail involves how increasing death rates diminish a species' ability to recover from natural perturbations. Moreover, because species interact, a harvested species' inability to recover may affect the other species in the system. One argument is that as we increase a species' death rate by pollution or harvesting, we may not only alter the resilience of the species but that of many of the community's other populations as well. With changes in resilience there will be changes in variability, and with increased variability a greater chance that a species will be lost. If this argument is reasonable, it offers the promise that we can detect a population in danger of crashing before it actually does so.

Harvesting and resilience

The first component of this argument is the effect of harvesting on resilience. For a wide variety of single-species models, May et al. (1978) have shown that resilience decreases with increasing death rates, and does so dramatically when the system is pushed beyond its maximum sustainable yield. This result is consistent with a number of theoretical studies (Doubleday, 1976; Sissenwine, 1977; Beddington, 1979) as well as with the claims of various empiricists (SPOF, 1983; Rapport et al., 1985).

Unfortunately, for two or more species, the theoretical result is easily reversed. For two-species systems, increasing a harvest can *increase* resilience to a point sometimes well beyond the MSY (Pimm and Hyman, 1986). At very high levels of harvesting, however, resilience will begin to decrease again. With other parameter combinations, increasing the harvest can also cause a continual decrease in resilience. Which alternative we should expect depends on the details of the model. Suppose the predator–prey interaction is "tightly coupled"—meaning that the predator is reducing the prey well below the prey's equilibrium without the predator. Following a natural perturbation, predator and prey densities will oscillate as they return to equilibrium. A harvester of the predator in this tightly coupled system will initially cause the system to oscillate less and return to equilibrium faster. Considerations of the parameters that lead to this circumstance suggest this is not an unlikely event. For more complex multispecies models, increased resilience with increasing harvest also seems a probable outcome. In contrast, simultaneously increasing the death rates of more than one species makes it more likely that resilience will decrease.

Resilience and variability

The second stage in the argument involves the effect of resilience on a population's variability and hence its chances of falling below some threshold from which it will not recover. We face the dilemma of two seemingly reasonable arguments that give diametrically opposite results.

The first argument is that species whose densities return quickly to their equilibrium level following a perturbation will vary less than those that return to equilibrium slowly (May, 1973); a "stiff" population (one that returns slowly) will still be at relatively low levels when the next perturbation comes along and so is likely to be driven to low densities. The second argument is that the carrying capacity is itself variable. Resilient populations will closely track this fluctuating carrying capacity and so be more variable than stiff populations which average out these fluctuations (Turelli, 1978).

These two alternatives thus lead to opposite predictions about the relationship between resilience and variability: the first that decreased resilience increases variability, the second that it decreases variability. Deciding which alternative is the better description of a natural population will not be easy. Some laboratory experiments on protozoa and bacteria (Luckinbill and Fenton, 1978) and data on noctuid moths (Spitzer et al., 1984) support the second alternative; data on bird populations in English woodlands support the first (Pimm, 1984b).

Which alternative is the better description probably has a lot to do with the relative time course of the perturbations and population recovery. In the first alternative, a resilient population is less variable than a stiff one because it recovers faster from an occasional, short duration drop in resources. This is clear in the woodland bird data, where the perturbation was an unusually cold winter. In the second alternative, a stiff population will be less variable because it does not track a quickly varying resource level. When a resource level changes very slowly, both resilient and stiff populations will track those changes.

Resonance

Finally, we should consider the possibility that harvesting makes population densities more likely to resonate. In some simple food web models, population densities tend to oscillate as they return to equilibrium. Moreover, the period of this oscillation can increase with increasing harvesting. Thus there is the possibility that increasing harvesting forces a match between the system's intrinsic period of oscillation and those of the natural perturbation. This will cause the system to resonate and density oscillations of relatively large amplitude will result. The frequency of natural perturbations in nature has received some attention recently from Steele (1985). In terrestrial systems, the distribution of these frequencies is uniform—no frequency predominates so the spectrum lacks "color." Marine systems, in contrast, are "red"—the amplitude of natural perturbations increases with the period of the oscillation. Thus, when harvesting causes increases in the period of population oscillation towards the long periods of natural perturbations, this may produce erratic population changes.

Species losses: Summary

There is much uncertainty about the changes in a community's ability to recover from stress as we increase a species' death rate. First, models are equivocal about the direction of the change in resilience with increasing death rates. Under some circumstances, resilience can increase. Second, the chance that populations will oscillate, as well as the amplitudes and frequencies of the oscillations, may either decrease or increase with harvesting. Third, increasing resilience can either increase or decrease a population's variability depending on the relative time courses of the perturbation and the population's recovery. Simply, the theory that increasing harvesting leads to lessened resilience and greater variability seems sensitive to a variety of parameters and assumptions whose respective values and validity are unlikely to be known in nature. The response shown by any species is going to depend on its pattern of interactions with other species. As yet, there do not appear to be any obvious rules that allow us to pinpoint which species are going to respond the most at any given level of stress.

The theory also needs to be reconciled with the observed population patterns. Although increased population variability seems widely accepted as a correlate of population stress (SPOF, 1983; Rapport et al., 1985; Cairns, Chapter 23), a comparative study of the data available for the Great Lakes, for example, does not show a predominance of the pattern that combines decreasing yields with increasing variance in yields just prior to collapse (Pimm and Hyman, 1986). Superficially, the data show a wide range of behaviors prior to collapse.

What is certain is that even when there are powerful economic reasons to maintain the species in a community, we do not understand enough about the effects of increased death rates on population dynamics to enjoy a reasonable measure of success.

SPECIES ADDITIONS

The devastating effect of introducing pigs, goats, cats, and rats to oceanic islands has been observed so often that the inescapable recommendation must be to prevent future introductions to what few places these species have not yet reached. But the list of plant and animal introductions is enormous; for example, about 5 percent of all bird species have been introduced elsewhere (Long, 1981). Because introductions continue, we need to ask two questions:

1. Which introductions will succeed? This is a question of persistence (Table 1)—how long will a community last before an introduction is successful and so changes the community's composition?
2. Which successful introductions will merely add to an existing community, and which will cause extinctions? This is a question of resistance (Table 1)—how much will a community be changed by an introduction? (This problem is considered later in the chapter.)

The success of an introduction will depend on a variety of genetic, physiological and autoecological properties. This chapter is concerned only with community-level questions: Can we predict whether a species will succeed from a knowledge of community structure and the species' relation to that structure? A question to be asked later is, Are certain communities more vulnerable than others to species losses following a successful invasion?

Persistence: What makes a community hard to invade?

Among the first studies to consider persistence were those stimulated by Hutchinson's (1959) suggestion that species' morphologies seemed to differ by constant ratios. There might be a "limiting similarity:" species too similar morphologically and, by extension, ecologically, were predicted to compete too strongly and so hasten species extinctions. In contrast, communities with widely spaced species should be invaded quite easily. Simply, competition should lead to morphologies (and, by extension, ecologies) that are "patterned"—by which is meant (by analogy to spatial distributions) that they are overdispersed rather than random or aggregated.

That Hutchinson's paper has been the motivation behind much of the animal community ecology in the last 25 years is apparent from the papers in three recent volumes devoted to community ecology (Cody and Diamond, 1975; Case and Diamond, 1985; Strong et al., 1984a). Early results seemed encouraging. Theoretical studies (MacArthur and Levins, 1967; May and MacArthur, 1972) suggested a tight bound on limiting similarity, and empiricists found widespread evidence for community patterns in overdispersed ecologies and morphologies.

Later theoretical studies, however, have shown just how complex the patterns of morphological differences could be, even in communities where competition is important (Rummel and Roughgarden, 1983). Most of the theories suggest that a community will be difficult to invade if the potential invader faces too many competitors; but how many is too many is going to depend on a wide variety of factors.

But what of the data that spawned these studies? As Hutchinson suggested, if communities are highly patterned, is it due to their rejecting those species that are too ecologically similar to fit in? If so, the existence and clarity of the patterns in species' morphologies may be a good predictor of the community's persistence. Deciding whether species morphologies are overdispersed, however, requires the statistical prediction of the "ghosts"—those species which have attempted to invade a community and failed because they were too similar to existing species. Not surprisingly, no one has a universally accepted recipe for ghost prediction, and the literature on what we can deduce about communities from their current composition involves a heated debate (Lewin, 1983a,b).

Certainly, there is good experimental evidence that some species compete (reviewed by Schoener, 1983), just as there is equally good evidence that many species do not (Strong et al., 1984b). But experimental demonstrations of competition between a few species do not tell us if competition is intensive and extensive enough to structure entire communities by applying a strict filter on community membership.

Simply, studies of existing community patterns are unsure ways of understanding what factors affect community persistence. To predict whether a species can invade a community, and if not, why not, we must use an experimental approach. Interestingly, the literature on man-made species introductions is rich in examples and implications for community theory and from it I have selected some case histories.

Island vertebrates

Introductions of vertebrates to islands allow us to look at two effects. First, across a large range of islands, we can examine the effects of native species on the chance that an introduction will succeed and, if so, how abundant it will become. Second, in lowland Hawaii there are virtually no native vertebrates: there is only one native land mammal (a bat), no amphibians or reptiles, and the lowland birds have been removed by humans from the time of the Polynesians onwards. There are, however, an unusually large number of introduced species. So, within this island group, we can examine the effects of introduced species on the success of other introductions.

The effects of native species on introductions. Of direct concern to conservation biologists is the penetration of native habitats by introduced species. How does the species richness of a community affect its chances of being changed by a successful invasion? Diamond and Case (1985) found that the success rate of bird introductions on tropical and subtropical islands declined steeply with the species richness of the extant native avifauna *(S)*. Success varied from 90 percent on Ascension *(S = 0)*, to 60–80 percent for Lord Howe, Bermuda, Rodriguez, and Seychelles *(S = 3–19)*, to 0 percent for Borneo *(S = ca 420)* and New Guinea *(S = 513)*. For the Hawaiian islands, with 6–10 native species and about a 58 percent success rate, the result is comparable. But I wonder if competition from the native species on the introduced species is the sole cause for these results. In the Hawaiian archipelago, for example, few introduced species ever meet a native species; the 6–10 native species are restricted to upland forests and most introductions are to the lowlands. Certainly, there are more introduced and fewer native species on the island of Oahu than on Kauai (Moulton and Pimm, 1985; Berger, 1981) but this must reflect, in part, the greater number of people on Oahu. More people lead to greater destruction of the native habitat and to more bird introductions.

What about comparisons between different vertebrate classes both within and between island groups? Between island groups, the number of native insular species decreases in the sequence: birds, lizards, mammals, freshwater fish. Diamond and Case quote as-yet unpublished results that invasion success appears to increase in the same sequence. Moulton and Pimm (1986) found that for Hawaiian vertebrates, 29 of 50 (48 percent) introduced passerine birds succeeded, yet 17 of 20 (85 percent) mammals and 12 of 15 (80 percent) reptiles succeeded. However, in an admittedly much smaller and less well documented data set on amphibians, 4 of 8 (50 percent) introduced species succeeded.

Another test of the ability of natural communities to resist introductions would be to look at the population densities of the introductions in native habitats as a function of the number of native species. Rather than tables of data, all I can offer is some rather subjective impressions. In the species-rich continent of North America, the European starling and house sparrow, although widespread, are not common away from human settlement. Similar examples include the skylark around Vancouver, spotted dove in Los Angeles, and red-vented bulbul in Miami. Where introduced, all have restricted ranges, yet all three have very large native ranges. A similar situation is found on the relatively species-rich island of Viti Levu in Fiji. It has 48 species of land birds and introduced species are strictly confined to man-made habitats (Diamond and Case, 1985).

The species-poor Hawaiian Islands are very different. Certainly, most of the introduced species are confined to the lowlands. But in the upland rainforest (above 1000 m) of the larger islands, the Japanese white-eye is often the commonest species and the red-billed leiothrix can also be common (personal observations). In upland grassland habitats, the introduced skylark is often the only passerine bird species to be found in the islands.

For mammals, Diamond and Case (1985) report that in New Guinea, where there are about 200 native mammals, *Rattus norvegicus* is very rare and *R. rattus* is a commensal of man. In contrast, New Zealand has no flightless land mammals, and has 33 successfully introduced species of mammal, none of which is restricted to man-made habitats. Some mammals have reached densities in native forest which lead to serious damage to the vegetation (Gibb and Flux, 1973). Similarly, in Hawaii, even remote remnants of the native rain forests contain rats, a mongoose and substantial numbers of pigs. In open areas, even in the national parks, goats achieve extraordinary densities, despite intensive hunting efforts to remove them.

The effects of introductions on introductions: Lowland Hawaii. Lowland Hawaii (below 1000 m) has been so extensively abused by man that finding a native land vertebrate requires considerable patience. The lowland fauna has been replaced with 29 species of passerine birds, two doves, a variety of game birds (whose fate is difficult to ascertain), 17 mammals, 12 reptiles, and 4 amphibians. Using historical sources, Moulton and Pimm (1983, 1985, 1986) compiled lists of which bird species were introduced when to which islands, and if and when they became extinct. Records go back to 1860 and about half the species failed. In contrast to studies of natural communities, where we must predict species failures using debatable statistical procedures, the Hawaiian record gives these failures directly.

Moulton and Pimm found that (1) a species is more likely to fail, the more species there are already present in the community, and (2) as the morphological similarity between an introduction and an already present congener increases, the chance of failure increases. Both effects are predicted by competition theory and the latter is an experimental demonstration of limiting similarity.

We must conclude that interactions with other species (including but not restricted to competition) play a major role in determining how many species can invade a community and their abundance after invasion. In the introduced Hawaiian birds, we can even detect an effect of morphological limiting similarity. Other things being equal, species-poor systems would seem to be less persistent. But we have only looked at vertebrate introductions; what of other groups? For insect introductions we shall find that the conclusions are very different.

Insect introductions

Simberloff (1986) has reviewed insect introductions to the United States using data compiled by Sailer (1978, 1983). Although there are no records of how many species failed, Sailer (1978) found patterns that resemble those for birds. The 1554 successful introductions to North America represent about 1.7 percent of the continent's insect fauna; the 1476 species introduced to Hawaii are 29 percent of the islands' total. The remote South Atlantic island of Tristan da Cuhna has suffered a fate similar to Hawaii's: it has only 84 native insects, and 32 introduced ones—38 percent of the total (Holdgate, 1960). Simberloff urges caution in attributing these patterns to interactions with competitors or predators, suggesting instead that the patterns reflect different degrees of habitat destruction and introductions of alien plants.

In one data set, the fate of all the species introductions has been followed. It involves the species introduced to control homopteran, coleopteran, and lepidopteran pests (Hall and Ehler, 1979; Ehler and Hall, 1982). Again the results seem generally consistent with the hypothesis that competition affects success rate. Thus, given 5–9 introduced species already present, introductions of two or more species are much less likely to be successful than single introductions in the absence of any other introduced species. But Simberloff (1986) points to three features of biological control procedures that would also explain why few introduced species will appear to succeed better than many introductions, irrespective of any competitive effects the introductions may encounter. First, we can recognize two extreme strategies in species introductions: introduce only one species after very careful evaluation of which species will work best; or, use a "shotgun" approach of introducing many species in the hope that one will succeed. Second, if a species is successful there will likely be no more introduction attempts. Third, successful species will be among the more probable choices for attempted control in other areas.

In sum, even though the patterns of insect introductions might appear superficially like those for vertebrates, predicting the success of insect introductions from a knowledge of community structure seems a much less certain task.

Species additions: Summary

There are two broad strategies in examining the persistence of communities in the presence of continued attempted species introductions. The first involves looking at natural communities we assume are subject to constant attempted invasions by species present in neighboring communities. If existing communities are strongly patterned (for example, if species morphologies are highly overdispersed) then it is probably because they are excluding many species invasions (by inference, of intermediate morphologies). Detecting these patterns is statistically painful and their interpretation is difficult.

A second approach is to examine how well communities persist when subjected to species introduced by man. For some taxa both the successes and failures have been recorded, thus circumventing the problems with the indirect methods used for natural communities. Some generalizations appear possible for vertebrates. Bird introductions, for example, are less successful when there are more native or introduced species already present, and when the introduced species include morphologically similar species. Moreover, successful introductions seem to achieve less high densities when more species are present. The patterns of insect introductions, although superficially similar, may be more readily interpreted in ways that do not reflect differences in community persistence.

THE CONSEQUENCES OF SPECIES LOSSES

Theoretical studies

If we remove a species from a community and all the other species persist (probably at different densities), then we call that system "species deletion stable" (Pimm 1979, 1980). This definition was developed to investigate MacArthur's (1957) argument that more complex systems were more stable. He supposed that the less a community changed in response to a species at some abnormal abundance, the more "stable" it would be. ("Stable" is being used in the sense of "resistance;" Table 1). If we consider the disturbance in abundance to be a reduction (and, moreover, a reduction to zero), and the variable of most interest to be community composition, then we arrive at species deletion stability.

Removing a species from a community either causes further species losses or it does not. But by examining a large number of models with the same structure but different parameters, we can come up with a probability that a species removal from a given position in a given web structure will lead to species deletion stability. We can then average these probabilities over all the species in a web, or some more interesting subset of species.

Within a web, species deletion stability varies widely: some species are in critical positions, where their removal will certainly cause further losses; other species are in positions where their removal will have little effect. Despite this variability, there are some very broad patterns. The most important ones depend on the trophic level from which species are removed. If we remove plant species, then increasing web connectance[1] *increases* species deletion stability. This matches the aphorism not to "put all your eggs in one basket." It supports Mac-Arthur's argument that the more pathways via which energy can flow to the top from the bottom of a web, the less severe will be the consequences of failure of any one pathway. Connectance measures the number of pathways. However, for the removal of top predators, species deletion stability *decreases strongly* with connectance. Why should this be so?

Consider a three trophic level example with just three species, one plant, one herbivore, and one carnivore. Removing the carnivore will certainly lead to an increase in the density of the herbivore and a decrease in that of the plant species. But it is unlikely that the herbivore will exterminate the plant—its only food source. Next, suppose we have a polyphagous herbivore feeding on many plant species. Now when we remove the carnivore, the herbivore can flourish even if it eats all but one of the plant species to extinction. Alternatively, the predator's removal may release a competitively dominant species which can then eliminate other species.

---

[1]Connectance measures community complexity and is the number of actual species interactions in a food web divided by the number of possible interactions. Interactions may be between predator and prey, competitors, mutualists, etc. Of course, many species pairs do not interact at all.

For the models, we usually find that when we remove a species, another will be lost and its removal will cause other losses. How far will the effects cascade? The extent to which they do is an inverse measure of the system's resistance (Table 1). The modelling results do not tell us directly, but they give a suggestion. If we remove top predators, systems that are species deletion stable have low connectance and few species; these are likely to be the end points of the cascade. Thus, when we remove top predators, species-poor, low connectance systems are probably not going to change as much as species-rich, highly connected systems. If we remove plant species, we might expect the low connectance systems to change the most. Both results, however, represent averages of values that will vary greatly depending on the position in the food web of the species we remove.

Field results on species removals

It is clear that species removals usually have a dramatic effect on the composition of the species at lower trophic levels; very few natural systems appear to be unchanged when we remove top predators. Examples include a variety of marine and freshwater studies reviewed by Pimm (1980); McNaughton's (1985) fencing experiments, which excluded large mammalian grazers from portions of the Serengeti plain of Tanzania; and Brown's (1985) removals of various combinations of granivores from a desert community in the southwestern United States, which caused visually spectacular changes in the vegetation.

But can the extent of these losses be related to richness and complexity? One experiment would be to remove species from species-poor and species-rich systems and to look at the effects on lower trophic levels. In doing this, McNaughton (1977) found that, as theory predicts, species composition changed more when a grazing herbivore (the African buffalo) was excluded from a species-rich grassland than from a species-poor grassland.

Another experiment would be to remove plant species from species-poor and species-rich systems and to measure the effects on higher trophic levels. I know of no such direct experiment, but a suggestion of the differential sensitivity to species removals on a large scale comes from a comparison of the effects of plant species removal by goats in Hawaii with the effects of chestnut blight in the eastern United States.

The blight is able to survive in tree species other than the chestnut, and it has reduced the chestnut from being once the locally most abundant tree species (Krebs, 1984, p. 44), with a range over most of the eastern United States, to the verge of extinction. To my knowledge, no extinctions in the top trophic levels (carnivorous and insectivorous birds, mammals, and reptiles) of the forests have been attributed to this dramatic removal (but see Terborgh, Chapter 15).

The effects on higher trophic levels of removing plant species from the species-poor Hawaiian flora were probably much more dramatic. Goats and other mammalian herbivores have endangered a wide variety of Hawaii's plant species. Among these are species that provided nectar for various species of Hawaiian honeycreepers (Drepanididae). There have been many extinctions of Hawaiian birds, including species with a variety of food and habitat requirements. But the effect of herbivores in removing such rich nectar sources as species of *Clermontia* and *Hibiscadelphus* must be considered a plausible explanation for extinctions and rarity of the more behaviorally dominant nectarivores (Pimm and Pimm, 1982).

Species losses: Summary

The resistance of model communities to species removals depends on their species richness and complexity and also from which trophic level the species are removed. For removals of predators, species-rich, complex systems are likely to be affected the most; these systems are likely to simplify considerably. But for removals of plant species, the reverse is likely to be true: we may remove a plant species from a complex system and notice relatively less effect than if the system is a simple one. Experimental studies show that few natural communities are species-deletion stable; most removals cause further species losses. The data also show wide variation in the consequences of species removals. Communities are differentially resistant and, for small scale experiments, the patterns of resistance match the models. But drawing conclusions of the effects of plant removals from a comparison of removing plants from Hawaii and the eastern United States would be premature. While the results are consistent with the theory, even in the models there is considerable variability in resistance within a trophic level, depending on which species is removed. While in general species-rich systems may be more resistant to plant species removals, we might expect that the removal of species of *Ficus* and *Combretum* from the spectacularly species-rich Amazonian forest (Terborgh, Chapter 15) would have a major effect on species composition.

THE CONSEQUENCES OF SUCCESSFUL SPECIES INVASION

Given a successful introduction, there could be no, few, or many species losses at the same trophic level as the introduction and at the levels above or below it. What do our models say about how community properties (species richness, connectance, etc.) affect these losses, and how do the data relate to these models? To my knowledge, there are very few attempts to answer these questions. The only synthesis of empirical results I can find is that of Simberloff (1981) who reviewed 10 papers covering over 850 plant and animal introductions. Although the data differ widely in quality, there are three patterns which apply to all the studies.

1. Less than 10 percent of the introductions caused species extinctions (71 extinctions). Greenway (1967) reported the highest percentage (30 percent; this involved 55 of the 71 extinctions) and he was principally concerned with avian extinction. Introductions tend to add species to a community rather than to cause losses.
2. Of Greenway's 55 extinctions, over 90 percent were on islands.
3. Both overall and within Greenway's study, predation was the principal cause of extinction (50 of 71 and 42 of 55, respectively). Habitat change accounted for 11 of the 71 extinctions; competition from the introduced species was a distant third, with three extinctions.

Are island communities more likely to loose species following a successful species introduction? We might expect that animal communities lacking the experience of predators and plant communities lacking the experience of grazing herbivores will be damaged more by introductions of predators and herbivores than mainland communities, which already have these species. But is this readily observable in the data? For many reasons already discussed, islands have proportionately more introduced species, so the high percentage of island extinctions caused by introductions may be caused by the greater number of introductions. And with a greater number of introductions, we are more likely to encounter an introduction that has a devastating impact on a community—as cats, rats, pigs and goats certainly have on islands. Showing an increased vulnerability of islands to introductions—and separating increased vulnerability from other possible explanations—may not be particularly easy.

Successful species invasion: Summary

On average, natural communities appear highly resistant to successful species introductions; most introductions add species to, rather than replace, the existing community. But, although communities appear to be resistant "on average," that average hides an enormous variance. Some introductions—goats on islands, for example—clearly have a devastating impact on the community. So while recognizing that communities are generally resistant, we might ask under what circumstances are they so spectacularly vulnerable.

FRAGMENTED MAINLAND COMMUNITIES

In developing my objectives, I have often relied on comparisons of island systems with mainlands. Many of the world's endangered vertebrate species are on islands. It is worth considering the full range of hypotheses for why islands are so vulnerable. These hypotheses include the greater potential for habitat destruction on islands; differences in community persistence (species may more readily be introduced to islands); and resistance to species removals (island animal communities may be less resistant to plant species removals).

None of these results directly considers the fate of mainland communities that are relatively species-poor because of species losses following habitat fragmentation. Treating fragmented habitats as islands is certainly not a new idea in conservation biology, but I suspect we can push the "habitat fragments as islands" model beyond that embodied in simple island biogeography. Indeed, in viewing habitat fragmentation, I think we *must* go beyond the seductively simple predictions of the species *(S)*–area *(A)* relationship $S = cA^z$, $z = 0.25$. Such simple predictions suggest that if we reduce the area of our habitat by a factor of 20, we should reduce our species list to $(\frac{1}{20})^{0.25}$, or about 50 percent of its former level. The area of forests in the eastern United States has been reduced by about this amount (Greenway, 1967), and the total bird species losses from all causes is barely 2 percent. The Hawaiian forests have not been reduced so much. Areas higher than 600 m are relatively undisturbed by agriculture, and 24 percent of Kauai, 41 percent of Maui, and 68 percent of Hawaii are above 600 m *(Atlas of Hawaii,* 1983). Yet over two-thirds of the native bird species have been lost since the early 1800s—even more if we count the effects of Polynesians (Olson and James, 1982). Simply, oceanic island communities are not just small pieces of mainland communities; their structures are different and so are their persistences and resistances. The fragmentation of habitats may well create communities that are not only island-like in their species richness, but also in their structures. Isolated, species-poor habitat fragments may surprise us in their lack of resistance to further species losses— particularly of long-lived plant species which may take some time to be lost. Finally, we might expect complex communities to be changed much more dramatically by predator removals than relatively simple communities. Thus, our experience in observing the loss of large predators from species-poor temperate communities may be of little use in predicting the consequences of similar losses in species-rich tropical communities.

SPECIES REASSEMBLY

Can we reassemble the pieces of communities that were once species-rich? Species losses from fragmented habitats will make their communities unlikely to persist in the face of species invasions. But we should not take any comfort from the fact that increased susceptibility to invasions could lead them to recover lost species, because the invasions may be of different species.

From field studies, alternative community states have been suggested as an explanation for highly patchy species distributions under relatively constant environmental conditions. Thus, Sutherland (1974) has argued for the existence of multiple community compositions, each persistent in the face of attempted invasions from species not belonging to the particular community. Theoretical studies show how easy this result is to obtain. I have already noted the literature on multiple stable states in the discussion of fishery and range collapses. Such results comes from highly nonlinear models of a small number of species—typically one or two. The existence of alternative communities among larger numbers of competing species has been discussed by Gilpin and Case (1976) who used much simpler models. Similarly, Drake (1985) modelled the successive assembly of species from a finite species pool of 125 species, again using simple models. After a time, the communities became totally persistent with a richness of about 15 species; none of the other 110 species could invade. In different simulations with randomized assembly sequences, different sets of 15 species formed the persistent communities. About 50 percent of the species were shared between any pair of persistent assemblages. Drake also asked if the sets of 15 species could be assembled in any order on their own. Generally they could not; some of the species that were subsequently lost were essential for the final community's composition. In the second part of his study, Drake demonstrated these effects experimentally using replicated laboratory microcosms with three trophic levels.

THE SUBJECT OF COMMUNITY ECOLOGY

Much of community ecology concentrates on communities relatively undisturbed by man. And, indeed, there is much that we can learn from near-pristine communities. Yet in concentrating on natural communities we may be missing an important opportunity. Community differences in resilience, persistence, and resistance are basic issues in ecology; it is obvious that we need to perform experiments to test our ideas about them. What appears less obvious is that many of those experiments have already been done for us in our reckless destruction of natural communities. In contrast, these theories are often hard to test by looking at natural communities, as the vigorous debate over community patterns demonstrates. Of at least equal importance is our need to learn about community destruction from the destruction that has already taken place. We need to pinpoint which communities are most vulnerable and channel our efforts accordingly. Independent of any theoretical explanations, the losses of managed rangelands and fisheries need to be understood if we are to prevent more losses from increasingly managed ecosystems of a wide variety. Similarly, the high rates of species loss on islands, the success rates of alien species on islands, and their occasional destructive effects are all likely consequences of low species richness. And low species richness is the likely end point for many mainland communities. Simply, we need to know more about how communities are destroyed so we can anticipate future destructions. While comparative studies of community destructions are few, there is no shortage of material on which to base those studies.

CHAPTER 15

KEYSTONE PLANT
RESOURCES
IN THE TROPICAL FOREST

John Terborgh

The time-honored stereotype of the tropical forest as a Garden of Eden offering a benign and constant climate and year-around abundance of fruit and other plant resources has been shattered by recent research. The myth of benign abundance has been exploded by the realization that a constant supply of resources is by no means a direct concomitant of a lack of marked seasonal variation in temperature and rainfall. Even the most notoriously uniform of tropical environments, such as those of equatorial Southeast Asia, are far from uniform when one examines the production of fruit by the rainforest (Raemaekers et al., 1980; Leighton and Leighton, 1983) or the nesting cycles of birds (Fogden, 1972). Animals can experience alternating periods of feast and famine even though the climate may appear monotonously constant.

The reasons for this seeming paradox are just now becoming clear. Tropical trees are under selection to flower synchronously to assure cross-pollination, and to time the ripening of their fruits so as to maximize dispersal while avoiding as much as possible the depredations of seed predators (Janzen, 1967). The requirement for synchronous flowering means that interbreeding populations of trees must flower in response to some universally recognized signal in the environment. The obvious signals that serve this purpose in seasonal environments—such as the transition from winter to spring, or changes in day length—are absent, and alternations between wet and dry seasons are more subtle or less reliable in the ever-wet equatorial tropics. The signals that do occur in this environment—such as sharp drops in nighttime temperatures due to a succession of clear, less humid nights, or an abrupt dry spell in the midst of a normally rainy period—are signals that come unpredictably and at irregular intervals (Janzen, 1974; Ng, 1981). Long periods, lasting up to a year or more, may pass without any effective signals. These spells may lead to a prolonged dearth of resources for the animal community, followed by a burst of abundance when a major fruiting episode finally does occur (Leighton and Leighton, 1983).

A new generalization seems to be emerging about tropical forests, one that is at odds with the conventional wisdom: the more uniform the climate, the more unpredictable and irregular will be the production of resources by vegetation. Tropical forests, from the point of view of their animal inhabitants, are not what they seem to the outside observer. Scarcity and abundance is the key consideration to understanding their ecology. This chapter develops the theme of how scarcity mediates the ecological interactions of vertebrate consumers in a tropical forest.

THE ANIMAL COMMUNITY OF A TROPICAL FOREST

Tropical grassland ecosystems, as exemplified by African savannas, support large numbers of grazing herbivores. Although less productive photosynthetically than tropical forests, the African grasslands can support animal biomasses as great as 12 tons per km^2 (Bourliere and Hadley, 1970). For reasons that are not yet fully understood, the foliage of tropical forests is not as edible as the grasses and forbs of savannas, and, being located at the tops of trees, neither is it so accessible. Forest ecosystems, in spite of their high productivity, thus do not generally support large numbers of herbivores. Instead, frugivores are the dominant group of animals in most tropical forests (Emmons et al., 1983; Terborgh, 1983). Because frugivores subsist on plant reproductive parts rather than foliage, their collective biomasses are correspondingly less than those of grazing species. Their diversity, however, can be exceedingly great. Birds, bats, primates, and other mammal groups contribute to this diversity, which in some forests may surpass 100 species—far more diversity than that of vertebrate herbivore communities.

Due to the extreme difficulty of observing (much less counting) frugivores in tropical forests, very few data exist on the composition and biomass of frugivore communities. In fact, there are only two localities for which reasonably complete and accurate figures exist. One of these is Barro Colorado Island, Panama (Eisenberg and Thorington, 1973; Glanz, 1982). The other is Cocha Cashu Biological Station in the Manu National Park of southeastern Peru. Located at the base of the Andes in the upper Amazon basin, Cocha Cashu lies in a region of exceptionally high species diversity. Unlike Barro Colorado, the fauna of Cocha Cashu has not suffered extinctions, and a full complement of predators assures that animal populations are regulated by entirely natural processes.

Over a period of several years members of the Cocha Cashu research group have undertaken a series of censuses of the bird and mammal communities. The results indicate that frugivores dominate the animal community, contributing about three-quarters of the total bird and mammal biomass (Table 1). Among the mammals, primates comprise the predominant group, contributing more than one-third of the total biomass. Rodents and peccaries follow in importance. Flying frugivores (bats and birds), while undoubtedly crucial in their roles as seed dispersers, are overshadowed in a quantitative sense, as they represent only about 5 and 10 percent, respectively, of the total frugivore biomass. This gives us at least a sketchy picture of how the higher vertebrate community is structured at Cocha Cashu and points out the overriding prevalence of frugivory as a way of life in the Amazonian forest.

PRODUCTIVITY AND PHENOLOGY OF FRUIT RESOURCES

Now that we have established the importance of frugivory in the tropical forest ecosystem, we should like to know how much fruit the forest produces, how fruit production varies seasonally, and how the production relates to the energy needs of the frugivore community.

So far it has not been possible to obtain precise measurements of the amount of fruit produced by any tropical forest. This is because most of the fruit is produced high in the canopy where it is inaccessible to terrestrial primates like ourselves. Workers have meticulously estimated the fruit crops of individual trees, but never of a whole forest (Howe, 1982; Howe and vande Kerckhove, 1979, 1981).

The best estimates of fruit production have been made with what are called fruit traps: sheets or buckets placed on the forest floor to catch fruit falling from the canopy. The problem with this method is that it results in a systematic underestimation of actual production because fruit eaten by arboreal frugivores does not reach the traps. As we shall see, this consumed amount can at times represent a sizeable fraction of the probable crop.

Nevertheless, fruit trap data provide us with the only available estimates of fruit production by tropical forests, and these estimates

TABLE 1. Biomass of mammals and avian herbivores at Cocha Cashu, Peru by diet class (in kg/km^2).[a]

Taxon	Plant reproductive parts		Foliage	Other	Total
	Fruit pulp	Seeds			
Mammals[b]					
Marsupials	65				65
Bats	75				75
Primates	650				650
Rodents	140	140	50		330
Procyonids	120				120
Deer and tapir			230		230
Others[c]			+	80	80
Total mammals	1050	140	280	80	1780
Frugivorous birds [d]					
Tinamous		30			30
Guans and curassows	23	22			45
Trumpeters	12				12
Pigeons	7	3			10
Parrots		15			15
Toucans	20				20
Others	22	6			28
Total frugivorous birds	84	76			160

[a]Dietary categorization is based on known or presumed major food type; e.g., foliage if more than 50 percent foliage, etc.
[b]Mammal data modified from Terborgh, 1983.
[c]Other mammals include sloths, armadillos, anteaters, mustelids, and felids, many of which are too scarce to census accurately. The plus (+) indicates the presence of sloths, which are rare at Cocha Cashu.
[d]Only frugivorous birds are included because data on the rest of the avifauna are not yet available. However, frugivores, broadly construed to include both pulp and seed eaters, certainly contribute more than 50 percent of the avian biomass at Cocha Cashu.

exist for, essentially, only two localities: Barro Colorado Island, Panama (Smythe, 1970; Foster, 1982a) and Cocha Cashu, Peru (Terborgh, 1983). Surprisingly, annual fruit production in these two localities, as measured in fruit traps, is almost identical (2200 kg/ha per year in Panama and 2000 kg/ha per year in Peru). Actual production is of course higher than this amount by the annual consumption of arboreal frugivores, a quantity that can be crudely derived from estimates of frugivore biomass (see below).

It is rather meaningless to talk about annual levels of fruit production except in the context of its seasonal variability, because periods of scarcity will have far greater impact on the frugivore community than periods of superabundance. Fruit traps provide a reasonably faithful record of seasonal variability since the underestimation of production is fairly constant throughout the year and will not result in major distortions of the temporal position of peaks and troughs in the record.

The overall patterns of variability in fruitfall at Barro Colorado and Cocha Cashu are similar, though there are differences in detail, both between the two localities and between years at each locality. A major peak in fruitfall invariably occurs in the first month or two of the rainy season (Figure 1). A second major peak occurs later in the rainy season, though its timing can vary. In both localities the period of minimum fruitfall occurs in the transitional period between the wet and dry seasons (November–January in Panama and May–July in Peru). During this period the amount of fruit falling from the canopy may be only one-twentieth or one-fiftieth of that falling at the time of maximum production.

Further insight into the significance of the seasonal rhythm in fruitfall can be gained by estimating the approximate intake of the frugivore community. Although this requires a rather adventuresome extrapolation, the outcome is nevertheless sufficiently robust to be of some heuristic value. The daily food intake of wild frugivores can be estimated from measurements made on Barro Colorado Island on both free-ranging and captive howler monkeys. Howlers eat the equivalent of 20 percent of their weight per day in fruit (Leigh and Smythe, 1978; Nagy and Milton, 1979). If this rough value is extended to the entire frugivore community at Cocha Cashu using the relationship $M \propto W^{0.75}$ where M is metabolic rate and W is weight (Peters, 1983), one emerges with a figure for community-wide consumption in the range of 4–8 kg/ha per day. Referring again to Figure 1, we note that this value falls at an intermediate level in the annual production cycle. For 8–9 months production is in excess of the estimated daily consumption, and for the remaining 3–4 months it is less. Even if one adds an amount equivalent to the probable consumption by arboreal

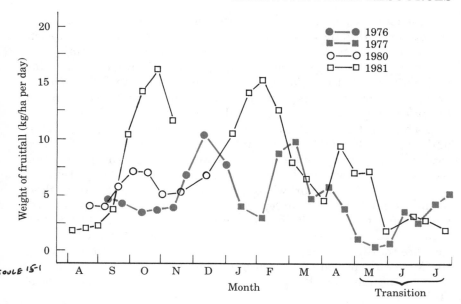

FIGURE 1. Weight of fruitfall in mature forest at Cocha Cashu, Peru in 1976–1977 and in 1980–1981. Measurements are based on biweekly collections from 100 fruit traps. The period from May to July represents the wet to dry season transition, when fruitfall is minimal. (See Terborgh, 1983 for further details.)

frugivores to the fruitfall measurements, it would not alter the conclusion that consumption is precariously close to production for 2–3 months in the wet–dry transition period. Obviously, consumption cannot exceed production, so we are left wondering how the frugivore community weathers the period of scarcity.

DIET SHIFTS IN PRIMATES IN RELATION TO FRUIT ABUNDANCE

The bird and mammal communities at Cocha Cashu have been under study for 12 years. Over this period investigators have accumulated information on the feeding habits of many species, including most of the principal mammalian and avian frugivores. Although it is not practical to review all this information here, it can be stated in general that the majority of frugivores in the community either migrate or modify their diets during the period of food scarcity.

As examples of this behavior, we shall examine the cases of five primate species I have studied myself (Terborgh, 1983). These species encompass a tenfold range in body size, and, as shall later be apparent, the dietary shifts they show during the time of scarcity are representative of a wide cross-section of the overall frugivore community.

The five species consist of two capuchins, the brown *(Cebus apella)* and white-fronted *(C. albifrons);* the squirrel monkey *(Saimiri sciureus);* and two members of the marmoset family, the emperor tamarin *(Saguinus imperator),* and the saddle-backed tamarin *(S. fuscicollis).* Capuchins weigh 3–5 kg as adults, squirrel monkeys 800 g to 1.2 kg, and tamarins 400–500 g. Primarily as a result of these size differences, capuchins, squirrel monkeys, and tamarins diverge sharply in their feeding habits when faced with a shortfall of sweet fruit pulp (Table 2).

At times of the year when soft fruit is abundant, all of these primates eat little else (except for arthropods and other small prey which do not rival fruit in the quantities consumed). This pattern holds for the entire nine-month period in which the fruit trap data indicate that fruit production is well in excess of consumption. But in the May–July period of minimum production, the proportion of fruit pulp in the diets of all five species drops, drastically in some cases (Table 2).

The capuchins turn to breaking open palm nuts, which they can crack by virtue of sturdy teeth and heavy jaw musculature (Kinzey, 1974). Fruit pulp consumed during this period is almost entirely in

TABLE 2. Diets of five primate species (by percentage of feeding time) during the wet and dry seasons at Cocha Cashu.

Species	Fruit pulp		Palm nuts		Nectar		Other	
	Wet	Dry	Wet	Dry	Wet	Dry	Wet	Dry
Brown capuchin *(Cebus apella)*	99	66		25		1	1	8
White-fronted capuchin *(Cebus albifrons)*	99	53		42		3	1	3
Squirrel monkey *(Saimiri sciureus)*	100	91				9		
Emperor tamarin *(Saguinus imperator)*	97	41			1	52	2	7
Saddle-backed tamarin *(Saguinus fuscicollis)*	96	16				75	4	9

the form of figs. These are available only irregularly, but when found are harvested avidly by capuchins at the rate of 30–40 per minute.

Tamarins, being no bigger than squirrels, are incapable of breaking palm nuts and are unable to compete with larger monkeys for feeding positions in fig trees. Instead they resort to lapping nectar from the specially modified flowers of certain plants (Janson et al., 1981). In the month of July they appear to have no fruit available to them at all, because nectar accounts for over 90 percent of their feeding time. One cannot assume, however, that nectar is an adequate substitute for fruit, since the animals lose weight during this period (J. Terborgh, unpublished data).

The only one of the five species that maintains a high proportion of fruit pulp in its dry season diet is the squirrel monkey. The figure of 91 percent cited in Table 2 is misleading, however, because it does not take into account a drop in total feeding time. Figs provide nearly all of the fruit pulp consumed by squirrel monkeys during the dry season, but fig crops appear with notorious irregularity (Janzen, 1979), a ld the monkeys are consequently obliged to endure periods of a week or more when they are unable to find fruit of any kind. At such times the animals travel little and search unceasingly for insects, probably enduring negative energy budgets until the next fig crop ripens (Terborgh, 1983).

Palm nuts, figs, and nectar as keystone plant resources

Having now reviewed the means by which five primate species survive the period of minimum fruit production, the question arises as to whether the use of palm nuts, figs, and nectar is peculiar to these primates or is more widespread in the frugivore community at Cocha Cashu. If additional resources are available to other frugivores, then there is nothing particularly significant about our observations on the five primates. But if other consumer species are relying on the same resources, then the plants that produce these resources can be recognized as playing a critical role in the ecosystem.

Dry season feeding observations at Cocha Cashu on marsupials, procyonids, the remaining primates, peccaries, and many birds strongly imply that from May through July the forest produces little else in the way of plant reproductive parts other than palm nuts, figs, and nectar (Janson et al., 1981; Kiltie and Terborgh, 1983; Terborgh, 1983). A quick look at these plant resources will help to bring the situation into perspective (Table 3).

1. *Palm nuts.* Because they are well protected, palm nuts escape the depredations of all but a specialized set of seed predators who

TABLE 3. Keystone plant resources in the flora of Cocha Cashu.

Plant resource	Period available	Eaten by
Palm nuts		
Astrocaryum sp.	April–June	Capuchins; peccaries; squirrels; agoutis; other rodents; macaws.
Iriartia ventricosa	May–July	Capuchins; titi monkey; spider monkey (exocarp only); peccaries (seed).
Scheelea sp.	All year	Capuchins; squirrel monkey (mesocarp); squirrels (seed).
Figs		
Ficus erythrosticta	Irregularly	Nearly all monkeys; marsupials;
Ficus killipii	throughout	procyonids; guans; trumpeters;
Ficus perforata	season	toucans; many passerines.
Nectar sources		
Combretum assimile	July	7 spp. primates; marsupials; procyonids; >20 spp. birds.
Erythrina ulei	July–August	Spider monkey; squirrel monkey; capuchins; many parrots and other birds.
Quararibea cordata	Aug.–Sept.	Capuchins; squirrel monkey; tamarins; marsupials; procyonids; birds.
Miscellaneous non-fig fruits		
Allophylus scrobiculatus	May–June	Spider monkey; capuchins; trumpeters.
Calatola sp.	May–June	Titi monkey; trumpeters.
Celtis iguanea	March–August	Titi monkey; tamarins; parrots.

(Modified from Terborgh, 1983.)

possess appropriate adaptations—either a large size and powerful jaws for crushing (Kiltie, 1982) or a capacity for gnawing. The list of consumers is thus short, consisting of peccaries, capuchins, agoutis, squirrels, and macaws. Peccaries and capuchins, however, are both common and large so that in the aggregate the species that consume palm nuts make up a biomass of nearly 500 kg/km^2, thereby accounting for approximately 30 percent of the total frugivore biomass at Cocha Cashu.

2. *Figs*. As a dry season staple for birds and mammals, figs play a preeminent role in the ecosystem. Though ripe crops appear at widely scattered and unpredictable points in time and space, the total production is large, as is the biomass of consumers supported by it. All the larger primates use figs heavily, as do procyonids, marsupials, guans, trumpeters, toucans, and many other birds. Figs sustain a high diversity of species ranging in size from spider monkeys (10 kg) to tanagers (20 g) and constituting up to 60 percent of the frugivore biomass at Cocha Cashu. Subtract figs from the ecosystem and one could expect to see it collapse.

3. *Nectar*. Nectar is important, not so much for the biomass of the animals supported by it (less than 10 percent of the total), but for the diversity of species dependent on it. These include marsupials, procyonids, and eight species of monkeys, as well as parrots, icterids, honeycreepers, and many other birds. Interestingly, hummingbirds, bees, wasps, butterflies, and other obligate nectarivores are not attracted to the few species (e.g. *Combretum*) that provide nectar to the larger vertebrates. The flowers of these plants show unusual features that appear to be specializations for pollination by opportunistic—or, perhaps better put, seasonal—nectarivores (Janson et al., 1981).

4. *Miscellaneous non-fig fruits*. Our observations have revealed the presence of three additional species of trees (Table 3) that fruit in the dry season. The fruits are eaten by only a small number of consumers, and none of them seems particularly important in the overall economy of the forest. In their absence, the community would probably survive intact. These three non-fig fruits have been identified as keystone resources to emphasize the completeness of the list. There may be, in addition, a few rare species that have not been noticed as yet, but if so, it would matter little because their contributions to the overall energy budget of the community would be insignificant.

Keystone plant resources are thus characterized as playing prominent roles in sustaining frugivores through periods of general food scarcity. An additional key feature of these resources is reliability: the plant species listed in Table 3 show little year-to-year variation, either in the amount of resource produced or in the seasonal timing of its availability. (Figs might be considered an exception because of their irregular fruiting habit, but in the aggregate over a large population, they constitute a reliable resource.) Finally, it is remarkable that less than 1 percent of the plant diversity at Cocha Cashu (a mere 12 species in a flora of approximately 2000) sustains nearly the entire frugivore community for three months of the year. This fact has far-reaching implications for the future management of tropical forests, as we shall see.

KEYSTONE PLANT RESOURCES IN THE NEOTROPICS

We have examined the issue of seasonal scarcity from the points of view of both the consumer organisms and the plants that provide resources for them. Palm nuts, figs, and nectar have been identified as key links between the plant and animal communities, but our perspective has been limited to a single locality. It remains now to ask whether the results from Cocha Cashu are representative of a broader cross-section of the tropics. Unfortunately, the literature offers little help on this point. Fragments of information can nevertheless be assembled into a reasonably coherent picture for the Neotropics.

Palm nuts are repeatedly mentioned as a critical resource in the literature on Barro Colorado, particularly for rodents such as agoutis (Smythe, 1970; Smythe et al., 1982) and squirrels (Glanz et al., 1982). It is particularly relevant that the nuts of palms and a few other species can be scatterhoarded by rodents for later consumption during periods of scarcity (Smythe, 1978). The scatterhoarded nuts are then available to other terrestrial mammals, such as peccaries (Kiltie, 1981). Capuchins have been reported to crack palm nuts in two Colombian localities, though without reference to season (Struhsaker and Leyland, 1977; Izawa, 1979). In view of the nearly universal occurrence throughout the Neotropics of the genera of palms listed in Table 3, and of the species that consume them, there seems a good likelihood that palm nuts may serve as keystone resources over much of the region.

As for figs, there can be little doubt of their crucial importance as a lean season food supply for birds and mammals throughout the Neotropics. Figs occur, often commonly, in nearly every type of forest vegetation in the region (Gentry, 1986). The flora of Barro Colorado includes 17 species (Croat, 1978), and over 30 species have been collected at Cocha Cashu. Their nonsynchronous fruiting habit assures that crops will ripen at all times of year, and numerous reports affirm the appeal of figs to a wide range of birds and mammals (Leck, 1972; Morrison, 1978; Milton, 1980; Foster, 1982b).

The information available on nectar as a dry season food resource is less compelling, although a number of authors stress its importance to birds in various Central American localities (see for example Janzen, 1967; Wolf, 1970; Leck, 1972; Toledo, 1977; Feinsinger, 1980; Schemske, 1980). Corroborative data from South America are relatively scarce. In an isolated observation, capuchins have been reported to feed on nectar in Brazil (Prance, 1980), but at Cocha Cashu nectar is used extensively by many other diurnal and nocturnal mammals. Comparable observations have not been reported from other localities, presumably because feeding observations on mammals are so difficult to obtain. It is thus entirely possible that nectar does provide a universal keystone resource in the Neotropics, but the evidence needed to support such a contention is incomplete.

KEYSTONE RESOURCES IN THE OLD WORLD TROPICS

What about elsewhere in the tropics? Have keystone resources been identified in the forests of Africa and Asia? With one exception, to which I shall return shortly, the answer is, not yet. What has been thoroughly established as universal feature of tropical forests is the tendency for fruit production to fluctuate, both seasonally and from year to year. Phenological studies documenting these patterns are now available for all three sections of the humid tropics: the Neotropics (Smythe, 1970; Frankie et al., 1974; Croat, 1975; Foster, 1982a; Terborgh, 1983); Africa (Gautier-Hion et al., in press); and Southeast Asia (McClure, 1966; Fogden, 1972; Medway, 1972; Raemaekers et al., 1980; Leighton and Leighton, 1983).

For Africa and Asia there is little evidence to suggest what particular resources might be important in sustaining animal communities through periods of scarcity. One can suspect that figs play such a role, however, both on the *a priori* grounds of their aseasonal fruiting behavior (Janzen, 1979), and because they are mentioned by numerous authors as a major food source for many birds and mammals (Africa: Waser, 1977; Struhsaker, 1978; Gautier-Hion, 1980; Waser and Case, 1981. Asia: McClure, 1966; Terborgh and Diamond, 1970; Fogden, 1972; Raemaekers et al., 1980; Leighton and Leighton, 1983).

Fortunately, there is one site in Southeast Asia for which there is explicit information on keystone resources at the community level. This is the Kutai reserve of West Kalimantan, Indonesia, the focus of a two-year investigation of fruiting phenology and fruit use by Leighton and Leighton (1983). There is little seasonality of rainfall at this site, and very high plant diversity. Fruiting was recorded in over 600 tree species and 170 lianas and stranglers. As in other rainforest localities, fruiting was highly episodic and periods of high abundance were separated by long intervals of relative scarcity. During these periods both avian and mammalian frugivores subsisted on a limited subset of plant species, just as at Cocha Cashu. Within this subset, figs were the most important, from the standpoints both of absolute abundance and evenness of supply. A set of Annonaceous climbers provided a somewhat more fluctuating supply of fruits to birds and primates, and a number of species of Myristaceae and Meliaceae were of critical importance to large avian frugivores, especially hornbills. In this Bornean forest, squirrels and primates placed greater dependence on foliage when fruit was scarce, and pigs migrated out of the area to unknown destinations (Leighton and Leighton, 1983). In outline the situation is similar to the one at Cocha Cashu, but the details are different. Nectar and nuts do not seem to be available and, perhaps in consequence, frugivores are more prone to eating foliage, and even bark in the case of the orangutan (Rodman, 1977).

MIGRATORY FRUGIVORES: A CHALLENGE TO CONSERVATION

In Central America (Wheelright, 1983), South America (Terborgh, 1983), Malaysia (Medway and Wells, 1976) and Borneo (Leighton and Leighton, 1983) obligate frugivores have been found to migrate in response to periodic food shortages. We know that in Costa Rica these migrations are between different elevational zones in the mountains (Wheelright, 1983), but in the other regions, details of the migratory routes are totally unknown. Obviously, in an essentially flat landscape such as that of Amazonia, elevational migration is out of the question. At present we cannot even say whether these migrations are local— that is, between different forest types on a relatively small scale of a few kilometers—or long distance, over hundreds or even thousands of kilometers.

Migratory frugivores are extremely worrisome from a conservation biologist's viewpoint because it is unlikely that reserves will be large enough to contain their wanderings. This concern is reinforced by evidence that tropical frugivores are especially prone to extinction in isolated tracts (Willis, 1979; Terborgh and Winter, 1980).

Even nonmigratory frugivores require larger home ranges than herbivores or omnivores (Clutton-Brock and Harvey, 1977), a fact that makes their conservation especially challenging. At Cocha Cashu we found that several primates shifted their ranging seasonally to take advantage of temporally staggered fruiting peaks in a series of distinct vegetation types (Terborgh, 1983). If habitat diversity on a large scale is necessary to maintain migratory frugivores, it also seems necessary on a more moderate scale to sustain nonmigratory frugivores. Gilbert (1980) made the same point about habitat diversity on an even smaller scale in discussing the conservation of the food web that supports *Heliconius* butterflies.

IMPLICATIONS FOR MANAGEMENT

The small percentage of their national territories that most countries seem to be willing to devote to parks and reserves is simply inadequate to preserve tropical diversity over the long run (but see Diamond, Chapter 24). Larger areas are needed, and those areas can be provided by managed forests, presuming that management policies take into account the goal of perpetuating high species diversity. This can be done, in principle, without major economic sacrifice by selectively increasing the densities of high-value timber species. The means of doing this is an important topic, about which much needs to be learned (see Jordan, Chapter 20), but it is a digression from our immediate concern with keystone plant resources.

From the evidence reviewed above it is becoming clear that keystone plant resources play a vital role in sustaining the animal communities of many, if not all, tropical forests. If, as some authors believe (Leighton and Leighton, 1983; Terborgh, 1983), the abundance of these resources establishes the carrying capacity of the environment for most of its animal biomass, the implications for management are quite positive. First, it might be possible to enhance the animal carrying capacity of some forests by interventions that increase the abundance of keystone plant species. Second, because keystone species, by definition, constitute only a few percent at most of the diversity of any forest, retaining them under management will not be costly in terms of forfeited timber production. Third, since fruit is superabundant for much of the year, the densities of species that bear during the season of plenty can be reduced without consequence for the animal community. And last, the same can be said of the numerous species in most tropical forests that produce neither valuable timber nor fruits that are eaten by birds and mammals. A great deal of the space in most tropical forests could potentially be given over to timber production without reducing the carrying capacity for vertebrate animals.

The word "potentially" here should be emphasized, because at present we do not know how to manipulate the species composition of tropical forests on a commercial scale, nor do we know what the long-term consequences of such manipulations might be. In sum, there are grounds for hope that one day tropical forests can be managed, but we have a long way to go in learning how to do it.

SUMMARY

Tropical forests around the world have been shown to experience marked fluctuations in their production of fruit resources. The availability of fruit may greatly exceed the consumptive capacity of the frugivore community during periods of abundance, but these periods alternate with times of scarcity when many frugivores are obliged to feed on substitute resources of presumably inferior quality.

Regardless of their quality, the resources that sustain frugivores during times of general scarcity are of great ecological significance, because they appear to set the carrying capacity of the community. In one neotropical forest the resources that fill this role were identified as palm nuts, figs, and nectar. Although these "keystone plant resources" were provided by species that represented less than 1 percent of the local plant diversity, for a period of three months annually their fruit, flowers, and seeds sustained most of a frugivore community consisting of roughly 1600 kg/km^2 of mammals and birds.

A review of the literature suggests that such keystone plant species may be a widespread feature of tropical forest ecosystems. If so, the implications for forest management are rather profound, for it may prove possible to dedicate much of the space of tropical forests to economic activities without seriously altering their carrying capacities for large vertebrates.

SUGGESTED READINGS

Foster, R.B. 1982. Famine on Barro Colorado Island. In *The Ecology of a Tropical Forest: Seasonal Rhythms and Long-Term Changes,* E.G. Leigh, Jr., A.S. Rand and D.M. Windsor (eds.). Smithsonian Institution Press, Washington, DC, pp. 151–172.

Leighton, M. and D.R. Leighton. 1983. Vertebrate responses to fruiting seasonality within a Bornean rain forest. In *Tropical Rain Forest: Ecology and Management,* S.L. Sutton, T.C. Whitmore and A.C. Chadwick (eds.). Blackwell Scientific, Oxford, pp. 181–196.

Terborgh, J. 1983. *Five New World Primates: A Study in Comparative Ecology.* Princeton University Press, Princeton, NJ.

CHAPTER 16

DISEASE AND CONSERVATION

Andrew P. Dobson and Robert M. May

It is not frivolous to begin a chapter entitled "Disease and Conservation" by observing that our concern is with the problems that pathogens can pose for the conservation of "higher" species in zoos, arboretums or in the wild, and not for the conservation of pathogens themselves. No conservation group mourned the passing of the last wild smallpox virus, and no organization speaks for the conservation of gut nematode species. This juxtaposition is an extreme example of a more general aspect of the way the conservation ethic is applied differently to different taxonomic groups, in a way that to our knowledge has received little systematic treatment. For most people, any member of our own multitudinous species deserves more concern than the last wild members of other great ape species; large mammals excite more compassion than small ones; mammals outrank other vertebrates (except some birds); and vertebrates outrank invertebrates. Harvard's dormitory rules say it well: students may not keep "animal, bird or reptile" pets. Intensity of public opinion tends to rise along a continuum that matches the deliberately oversimplified characterization of "r-selection" to "K-selection" in life history, in a way that invites the attention of a philosopher or ethicist. We resist the temptation to pursue these ideas.

Throughout this chapter we define the term "parasite" to embrace a range of pathogenic organisms from viruses, bacteria, protozoans, and fungi through to the more conventionally defined helminth and arthropod parasites. We distinguish these groups by differences in their population dynamics rather than on taxonomic grounds. The term "microparasite" will be used in those cases where the host population may be reasonably regarded to be subdivided into a few distinct classes (susceptible; infected but latent; infectious; recovered and

345

immune). The majority of the viral, bacterial and protozoan infections are of this kind. In contrast, for "macroparasites" the pathogenic effects upon the host, the egg output per parasite, and most other dynamic considerations depend on the number of parasites harbored by the host individual in question. Macroparasites are also, almost invariably, distributed in an aggregated or clumped fashion, with a small proportion of the host population harboring most of the parasite population. While host–*microparasite* associations may be modeled by studying the flows among the relatively small number of different classes of hosts, the modelling of host–*macroparasite* systems requires a full description of the distribution of parasites among hosts. (For a more detailed discussion of these ideas see Anderson and May, 1979, and May and Anderson, 1979.)

DIRECT LIFE CYCLE MICROPARASITES

Direct life cycle microparasites are communicated from one host to the next by contact, via droplets, or—in the dynamically most complicated case—by free-living infective stages. These diseases cause many of the most dramatic impacts on populations. The classic example of a disease of this type is rinderpest or cattle plague. This myxovirus is closely related to human measles, and is the most lethal and potentially dangerous infectious disease of wild artiodactyls (Plowright, 1982). Although endemic in Asia and parts of Europe, rinderpest did not occur in much of Africa until its introduction with the cattle of European settlers. The first, and most devastating, pandemic to be observed started in Somalia in 1889 after the introduction by the Italian army of herds of Zebu cattle. The infection rapidly spread through the wild game and domestic cattle populations of Africa to reach the Cape almost 10 years later, having killed over 5 million cattle and an untold number of wild game animals. No real figures are available, but Plowright quotes estimates of mortality of greater than 90 percent for African buffalo, *Syncerus caffer*. Although other species were less catastrophically affected in the first wave of the epidemic, later recurrences produced strains of the virus which proved equally devastating to other species. When the epidemic eventually died out in the Serengeti, the annual survival probability of wildebeest *(Connochaetes taurinus)* doubled from 0.25 to 0.50 or higher.

Examples of such large scale epidemics are not restricted to pathogens of vertebrates. The recent mass mortality of the long-spined sea urchin *Diadema antillarum* in the Caribbean appears to have similar dynamics (Lessios et al., 1984a,b), as does the fungus *Endothia parasitica*, the species responsible for effectively removing the chestnut tree from deciduous forests of the eastern United States. In all cases these diseases have temporarily or permanently removed one species from an ecosystem. They may therefore have a considerable indirect effect on the dynamics of species which interact with the directly affected species (e.g. predators, prey, and competitors).

Thresholds for parasite transmission

The establishment and maintenance of most parasites can be related to a threshold population number of hosts in which the parasite can just maintain itself. In rinderpest, for example, the virus tended to persist only in areas with large aggregations of wild ungulates; it died out in smaller game communities (Plowright, 1982). For this and the other simple direct life cycle microparasites considered in this section, the threshold density of hosts, N_T, may be given in the simplest case (Anderson and May, 1979) by:

$$N_T = (\alpha + b + v)/\beta \tag{1}$$

Here α is the disease-induced host death rate, b is the per capita death rate from all other causes, v is the recovery rate, and β is a parameter measuring the intensity of transmission. If the host population N exceeds the threshold density ($N > N_T$), the parasite has a basic reproductive value greater than unity, and can maintain itself. Conversely, if $N < N_T$, the basic reproductive rate of the parasite is below unity and the host population is too small for the parasite to become established within it. If the microparasitic infection is highly virulent (large α), or if recovery is rapid (large v), or transmission low (small β), then N_T will tend to be large. Thus direct life-cycle microparasites typically infect organisms that are found in dense aggregations (colonial birds, herding ungulates, many fish and marine organisms). Table 1 qualitatively lists the characteristics of several common directly transmitted microparasites of wild and domestic animals.

A variety of refinements can alter Equation 1 in particular situations (Anderson and May, 1979). One of the most important with regard to conservation studies is the presence of an incubation, or latent, period when the animals are infected but not yet infectious. If this period is of duration σ, then the minimum host population required to sustain the pathogen is:

$$N_T = (\alpha + b + v)(b + \sigma)/\beta\sigma \tag{2}$$

Long incubation periods will tend to increase the threshold population size required to sustain the disease. Conversely, if there is "vertical transmission" such that a fraction f of all offspring of infected females are born infected, then threshold densities will have the lower value:

$$N_T = (\alpha + b + v - f\alpha)/\beta \tag{3}$$

Here a is the per capita birth rate. Indeed, if $fa > \alpha + b + v$, such a vertically transmitted infection can maintain itself in an arbitrarily small population.

Such considerations of threshold densities for the transmission and maintenance of infections can become important in zoo collections of animals, where population densities are often at an artificially high level and the introduction of one animal with an undetected latent infection can rapidly lead to the eradication of an entire breeding population. The density required to maintain a disease will often be further reduced in such collections by the presence of a variety of species, each of which shows a different response to the pathogen. This problem is further compounded in primates, because the presence of human visitors to the zoo artificially raises the effective pool of hosts and enormously increases the number of potential infectives. Thus measles (rubeola) and herpesvirus are often transmitted to primates, both when the animals are already present in zoos and (often) when they have just been captured and are awaiting transport to the zoos (Cicmanec, 1978; Hime et al., 1975). This leads to increases in the mortality rates of nonresistant animals freshly captured in the wild and also seriously compromises the credibility of many schemes to reintroduce such species into the wild (Jones, 1982).

Many kinds of environmental disturbances (pollutants, runoff, thermal effluent, pesticide application, and other agricultural practices) tend to disrupt natural patterns in the relative abundance of species, producing new patterns that resemble those in the early stages of succession in that a relatively few species are highly abundant. That is, man-made perturbations often cause "outbreaks," with one or two species rising to unusually high levels of abundance (for a review see May, 1981, Chapter 9). It is often observed that invasions by pathogens or parasites follow such disturbances. While such events may often be "correlation without causation" (humans made the disturbance, and simultaneously brought the invaders), it is probable that the unusually high abundance of hosts creates the "above-threshold" density necessary to establish and maintain microparasites with direct life-cycles. Likewise, the unusually high densities of some species that may arise when they crowd into unmanaged wildlife refuges can create a situation that helps a pathogen or parasite to establish itself.

Diffusive spread of microparasites

Once a microparasite has been introduced into a host population, its spread can depend on many factors. Figure 1 illustrates the geographic spread of three diseases. In all cases the pathogen moved as a wave of infection which rapidly destroyed a large proportion of the host population as it spread. Kallen et al. (1985) have recently developed some diffusion models to illustrate how the spread of such epidemics might be modeled. Although they apply their analysis to rabies in Europe, only minor modifications are needed to adapt the model to the data illustrated for rinderpest and sarcoptic mange. Figure 2 illustrates the simulated spread of a wave of rabies through a fox population. It should be noted that the initial wave of infection is followed by a number of smaller waves which continue to appear for some considerable time after the first wave has passed through. These patterns have been qualitatively observed in studies of both rabies and rinderpest (MacDonald, 1980; Plowright, 1982), and are predicted by other detailed models of the dynamics of rabies (Anderson et al., 1981).

The model of Kallen et al. (1985) indicates that the velocity of spread (c) of the infection depends on the rate of mortality (u) caused by the pathogen; on the basic reproductive rate (R_o) of the pathogen within the host population; and on a diffusion coefficient (D), defined below. The explicit relation between the velocity of spread and these three parameters is:

$$c = 2[Du(R_o - 1)]^{1/2} \qquad (4)$$

Kallen et al. estimate the parameters for the fox–rabies case. Here we will attempt to derive them for the case of rinderpest among ungulates, and specifically for the African buffalo (Sinclair, 1977). The parameter u can be easily determined from the data in Table 1, and R_o can be crudely estimated from the proportion of animals killed in the first wave of the epidemic. The diffusion coefficient, D, is harder to estimate: it has dimensions (length^2/time), and is essentially an estimate of the area covered by a wandering infected animal within some appropriate time interval. However, the velocity of spread of the disease may be estimated from the historical records of the first major pandemic, and it should thus be possible to obtain estimates of D by substitution into Equation 4.

The data in Table 1 suggest that an animal lives for a total of 14 days following infection, which gives an estimate for u of around 26 yr^{-1}. Simon (1962) quotes figures that suggest that at least 90 percent of the buffalo were killed in the first wave of infection. If we assume the actual value lies between 0.90 and 0.95, we get estimates of R_o in the range 2.6 to 3.2 (Kallen et al., 1985)[1]. The map in Figure 1 and

---

[1] This crude estimate follows from the Kermack–McKendrick (1927) relation, which can be written as $1 - I = \exp(-R_o I)$, where I is the fraction infected and (in this case) killed. Thus we have $R_o = [-\ln(1 - I)]/I$—whence the estimates given in the main text. If a significant number of infected animals recover, I is larger than estimated by the fraction killed, and R_o is concomitantly bigger. In short, our estimates are crude.

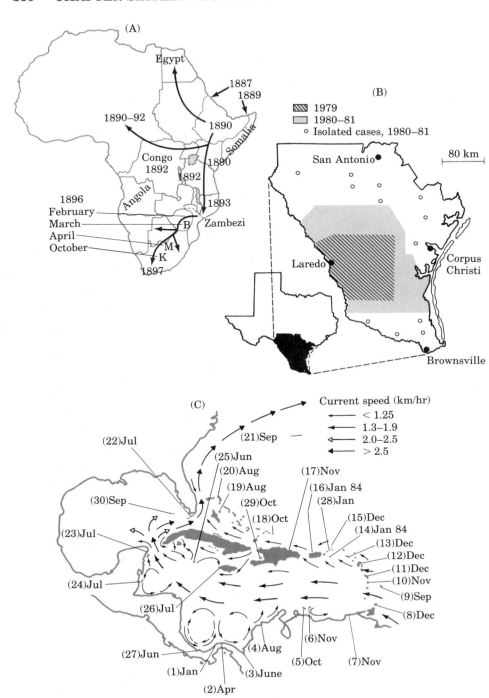

(A)

Egypt

1887
1889

1890–92

1890

Somalia

Congo
1892

1890

1892

Angola

1893

1896
February
March
April
October

Zambezi

B

M

K

1897

(B)

1979
1980–81
∘ Isolated cases, 1980–81

San Antonio

80 km

Laredo

Corpus
Christi

Brownsville

(C)

Current speed (km/hr)
< 1.25
1.3–1.9
2.0–2.5
> 2.5

(22)Jul

(21)Sep

(25)Jun
(20)Aug
(19)Aug
(30)Sep
(29)Oct
(18)Oct

(17)Nov
(16)Jan 84
(28)Jan

(15)Dec
(14)Jan 84
(13)Dec
(12)Dec
(11)Dec
(10)Nov
(9)Sep
(8)Dec

(23)Jul

(24)Jul

(26)Jul

(27)Jun

(1)Jan

(2)Apr

(3)June

(4)Aug

(5)Oct

(6)Nov

(7)Nov

◀ **FIGURE 1.** Geographic spread of three epidemics. (A) Rinderpest in Africa, 1889–1897. (From Mack, 1970.) (B) Sarcoptic mange in south Texas coyotes. (From Pence et al., 1983.) (C) Spread of *Diadema* mass mortality through the Caribbean and the western Atlantic. (From Lessios et al., 1984a.)

the data of Mack (1970) suggest that it took the epidemic 10 years to spread from Somalia to the Cape. If we assume that this distance is about 5000 km, then $c = 500$ km per year. Substitution into Equation 4 then gives an estimate for D of around 900 to 700 km^2 per year, or just over 2 km^2 per day. Although this estimate seems high, infected herds tend to fragment, move more quickly, and cover greater distances than usual in search of water with which to slake the thirst induced by the disease. Furthermore, our estimate of R_o may well be low, and the two parameters (R_o and D) essentially complement each other.

The information can be used to calculate how wide a "firebreak" (within which susceptible animals are either killed or vaccinated)

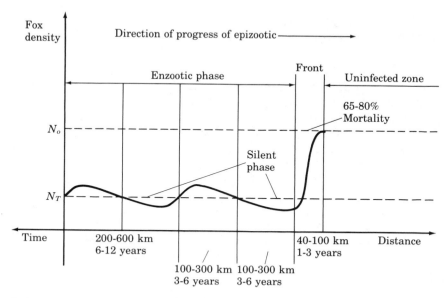

FIGURE 2. Simulated spread of a rabies epidemic. Fluctuations in fox density as a function of the passage of a rabies epizootic. N_o is the population density in the absence of rabies and N_T is the threshold population density below which the infection dies out. (From MacDonald, 1980, after data from the Centre National d'Etudes sur la Rage, 1977.)

TABLE 1. Population characteristics of some common directly transmitted microparasites.

Characteristic	Rabies	Rinderpest	Foot and mouth disease
Incubation period (days)	20–30	3–15	?
Duration of infectiousness (days)	4–10	4–6	7–10
Infectiousness	High	High	Very high
Duration of immunity (days)	?	Lifelong	?
Lifespan of infective stage	Short	Short	Long
Case mortality rate (pathogenicity)	High	High (but very species dependent)	Low

(Data are from the papers in Davis et al., 1970.)

must be in order to halt the spread of infection. Kallen et al. (1985) show the critical width, l, is given by:

$$l \simeq L[D/(uR_o)]^{1/2} \tag{4}$$

Here L is the dimensionless width of the barrier; its value depends on the exact value of R_o and on assumptions about how large a fraction of susceptible animals should be removed from the "firebreak." In an illustrative example where host density within the "firebreak" is reduced to 20 percent or less of its original abundance, and for $R_o \simeq 2.6$ to 3.2 with 90 to 95 percent of host animals killed, then L is around 16 to 13. Upon substitution these give an estimate for the width of the firebreak of around 50 to 70 km. This estimate closely matches the width of the break that prevented the epidemic entering Rhodesia in 1937–1941 (Scott, 1970). However, we note that an attempt to prevent a new spread of rinderpest across, say, the Serengeti, would require the clearing of an area larger that many of the more famous African game parks.

The estimates given above are crude, but they indicate the possible uses of this kind of analysis. The elimination of rinderpest from southern Kenya and Tanzania, mainly by the vaccination of cattle, has removed one of the main sources of density-dependent regulation from the wildebeest and buffalo populations (Talbot and Talbot, 1963; Sinclair, 1977; Sinclair and Norton-Griffiths, 1979). This has allowed the game populations, and the natural predators which they support, to recover to a relatively healthy level (Figure 3a). Now, however, there is no natural immunity to the disease among most of the wild ungulates (Figure 3b), and the occasional outbreaks of the rinderpest among domestic cattle in the northeastern countries of Africa present a continual threat that another epidemic could start. Indeed, evidence is presently accumulating that suggests that a strain of rinderpest is at present infecting both domestic cattle and wild game in many parts of East Africa (Plowright, 1985; Rossiter et al., 1983). A "firebreak" may be the only way of stopping such an epidemic reaching the game parks of Kenya and Tanzania (Plowright, 1982). Obviously, the above models would require some refinement before thay could be applied. This would require more information on the spatial distribution of the different species involved, as well as more data on the susceptibility of the different species to infection. The seasonal movements of both game species and domestic cattle would also be important in determining the spatial distribution of spread, as would the presence of natural breaks such as lakes, rivers, and mountain ranges.

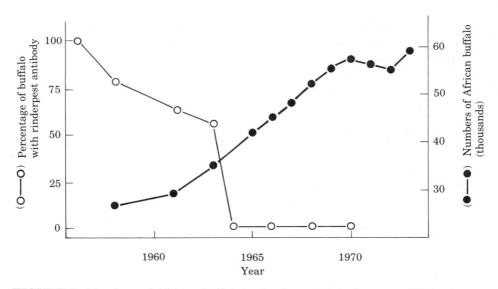

FIGURE 3. Numbers of African buffalo in the Serengeti in the years 1955–1975 (○), and proportions of buffalo with rinderpest antibody for the same period (●). (Data from Sinclair, 1977.)

DIRECT LIFE CYCLE MACROPARASITES

Direct life cycle macroparasites are usually passed between hosts via free-living infective stages. These stages may often be fairly long-lived, as may the adult stages which live in the host. The expression for the host threshold density is usually messier than for microparasites (May and Anderson, 1978, 1979). A rough approximation of N_T for a typical macroparasite can be written:

$$N_T \simeq (\alpha + b + u)/\beta \qquad (5)$$

Here α is the per capita host mortality rate induced by each individual parasite (assuming the overall host death rate rises roughly linearly with worm burden), b is the per capita death rate of hosts from all other sources, u is the death rate of adult worms, and β measures the transmission rate. The transmission parameter β can be a lot higher than for microparasites, particularly where the free-living stages are relatively long-lived, as they are for many nematodes, or where they actively seek out the host, as occurs with many ectoparasitic arthropods. Conversely, α will typically be much smaller than the disease-induced death rate α of virulent microparasites. Finally, u is typically much smaller than the recovery rate v of Equation 1; many helminths live for years in their hosts, while recovery times for most microparasitic infections are measured in days. In combination, these three factors result in N_T characterisitically being significantly lower for macroparasites than for microparasites (Anderson and May, 1979). We therefore expect to see macroparasites in much lower density host populations than we do microparasites. Several types of macroparasites may present problems for conservationists.

Direct life cycle nematodes are often found in low density herbivore populations. These parasites have high fecundity rates and relatively long-lived, free-living larval stages. These attributes of the parasites presumably evolved as adaptations to facilitate transmission between members of different small migratory groups of hosts. Domestic ungulates rapidly build up large burdens of these species as their restricted movements and frequent use of the same pasture allow transmission to proceed very efficiently (Michel, 1969, 1976). When game animals are restricted in their movements by placing them in zoos or small game parks, it also seems likely that large burdens may rapidly build up. Similarly, the numbers of species of macroparasites associated with a herd of animals increases dramatically with group size (Figure 4). The best control for infections of this sort would be to continually rotate the sections of pasture used by the animals, as is done in many agricultural systems.

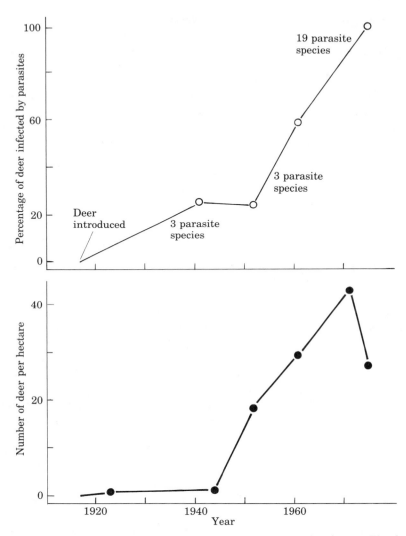

FIGURE 4. The number of roe deer in the Uzysmansky forest. The lower figure depicts the proportion of the deer infected with parasites and the number of parasite species recorded from the population. (Data from figures given in Romashov, 1977.)

Botflies and other species of flesh- or blood-feeding insects are examples of macroparasites that are particularly well adapted to parasitizing small groups of hosts. In an infestation of the lions in the Ngorongoro crater by *Stomoxys calcitrans*, many of the animals became emaciated and covered with bloody patches before either dying or migrating out of the crater (Fosbrooke, 1963; Schaller, 1972). This reduced the population of lions from 70 to 15 in the period between April and June of 1962. Although the population has now returned to its previous level, the reduction in effective population size may still affect the genetic structure of the population (Pusey and Packer, 1985). Botfly infections are also often sustained by small groups of primates, where they again cause significant mortality (Milton, 1982).

When host population size is sufficiently large we may also see epidemics of macroparasites that spread through the hosts with devastating effects similar to those observed for microparasites. This most frequently occurs with the parasites of fish. One of the most dramatic example occurred in the late 1930s when the transplantation of the gray sturgeon *Acipenser stellatus* from the Caspian Sea into the Aral Sea led to the introduction of the monogenean *Nitzschia sturionis*. This caused mass mortalities in the population of the local endemic species of the sturgeon *A. nudiventis*, which had no natural resistance to the parasite (Dogiel, 1966). Other examples of invasions of this sort are given in Dobson and May (1985).

When the free-living stages of macroparasites are long-lived and the parasite is detrimental to the host's fecundity as well as its survival, cycling patterns of host and parasite abundance may occur. Hudson et al. (1985) suggest that the nematode *Trichostrongylus tenuis* may be responsible for the cycling patterns of abundance observed in red grouse in the north of England and Scotland (Figure 5). These 4 to 5 year cycles have been observed in over 75 percent of the moors for which data are available. On the other moors, climatalogical data suggest that conditions are too dry to allow the free-living stages of

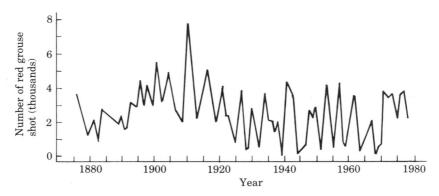

FIGURE 5. The numbers of red grouse shot on a moor in the north of England. (From Potts et al., 1984.)

the parasites to live for long. May and Anderson (1978) show that self-sustaining limit cycles in host and parasite abundance occur when:

$$1/k < \text{(constant)}/\gamma \qquad (6)$$

Here $1/\gamma$ is the life expectancy of the free-living stage of the parasite, which is much lower on dry moors than on damp moors. The length of the cycles scales inversely with host fecundity; this may explain why we observe an increase in cycle length as we move further north from England into Scotland. Similar mechanisms may be involved in driving the 10-year cycles of abundance observed in the snowshoe hare in Canada and North America. Here both a trichostrongyle nematode and several other macroparasite species may be important in manipulating host fecundity and abundance (Boughton, 1932; Erickson, 1944).

The ability of pathogens to produce long-term patterns of cyclic abundance in their hosts is often overlooked by biologists concerned with conservation or management. A well-managed grouse population can be a valuable asset both to the British land owner, to whom it supplies an acceptable financial return, and for the welfare of the specialized moorland birds of prey and waders which are able to utilize the habitat managed for grouse. When fluctuations in grouse density become too large, the financial return to the moor's owner becomes erratic, and often the only viable alternative is afforestation. This is often detrimental to the increasingly rare upland bird species that utilize grouse moors in Britain, and may also have long-term repurcussions on the area's environment, landscape value, and the social fabric of human upland communities (Hudson and Watson, 1985). Although culling is unlikely to be a successful method of controlling the disease (the grouse populations examined are all harvested at a range of different intensities with no apparent effect on the period of the cycle), other means of control, such as the introduction of anti-helminthics with artificial food supplies, may allow the host and parasite population densities to be regulated at some more stable equilibrium density.

The tendency to cause regular fluctuations in population density is not restricted to direct life cycle macroparasites. Many virus parasites of moths and other insects have relatively long-lived free living stages, which may be responsible for the long-term patterns of cyclic abundance shown in these arthropods (Anderson and May, 1981). Although the insect species themselves may not be of great concern to conservationists, the impact they have on their habitat may be of prime importance in the years of their peak abundance.

INDIRECT LIFE CYCLE MICROPARASITES

Indirect life cycle parasites must use one or more species of intermediate hosts, or vectors, to complete their transmission cycle from one definitive host to the next. Diseases of wildlife of this type that are important to conservationists include trypanosomiasis, leishmaniasis, malaria, yellow fever, and myxomatosis. The population dynamics of these pathogens typically are much more complex than those of direct life cycle parasites. Essentially, the inclusion of a second species of host into the life cycle tends to reduce the threshold population size of definitive hosts required to sustain the life cycle. Expressions for the threshold magnitude of the primary host population needed to maintain an indirect life cycle microparasite are similar to those given in Equations 1 and 2, except that now the transmission parameter β involves the population dynamics of the intermediate host. The resulting expressions can become quite complicated (Dietz, 1975; Anderson and May, 1978, 1979; for a simple overview, see May, 1984).

The life history of the vector, or intermediate host, is particularly important in determining the relationship between the threshold population density of the primary host, N_T, and the density of the intermediate host, V. Most indirect life cycle microparasites are transmitted by arthropods (usually insects, but occasionally ticks), where the vector acquires or transmits an infection in the course of taking a blood meal from the definitive host. As most vectors have some roughly constant biting rate, it becomes important to consider the average number of vectors per definitive host, rather than simply the abundance of vectors. This leads to a different expression for the threshold population density:

$$N < N_T \simeq \text{(constant)}V \tag{7}$$

Here, to attain the threshold for persistence, the host population density needs to be *below* some threshold, which is roughly proportional to intermediate host density. Although initially counterintuitive, this result is sensible enough: if an insect vector makes a fixed number of bites per day, there may not be enough vectors to ensure that infected hosts are bitten sufficiently often to spread the disease when the host population density is too large in relation to the number of vectors.

Trypanosomes in African cattle and game

Trypanosomes can be highly pathogenic to humans and to domestic livestock in many parts of Africa. The complexity of the life cycles of these parasites usually means that where they occur naturally they have been associated with both their mammalian hosts and vectors (the tsetse fly, *Glossina* spp.) for long periods of time. They are only mildly pathogenic to the wide range of game animals that act as natural definitive hosts (Molyneux, 1982; Molyneux and Ashford, 1983). Concomitant infections with other disease organisms are probably a major cause of pathogenicity, because such infections suppress the immune response and may render the host incapable of keeping a previously cryptic trypanosome infection in check. Conversely, trypanosomes themselves are immunosuppressive and death can result from exacerbation of previously occult infections. Drought or other detrimental environmental changes may also lead to immunosuppression resulting in disease and increased mortality rates.

The data in Figure 6 illustrate the range of species infected in the Serengeti. They also outline the dynamic complexities that arise when a pathogen infects a wide range of species. Thus we see that although predators have much higher proportions of their populations infected by trypanosomes, their low population densities reduce them to only a small proportion of the total pool of potential infectives. This picture will be further complicated by the differential preferences of the vectors for each of the species, and the different ability of the disease to develop in each host, as well as varied susceptibilities of different trypanosomes to different species of *Glossina*.

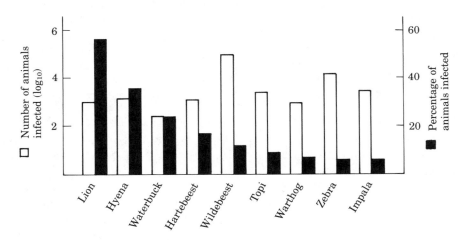

FIGURE 6. (A) Numbers of individuals of different species infected with trypanosomiasis (sleeping sickness) in the Serengeti region of Tanzania. (B) Proportion of individuals of each species infected with trypanosomiasis. (Data from Bertram, 1973.)

Avian malaria in Hawaii

When a vector is able to establish in an area where there is no innate resistance to a disease it carries, the effects can be devastating. A notable example of this kind is presented by Warner (1968), who contends that the extinction of roughly half the endemic land birds of the Hawaiian Islands since their discovery by Europeans in 1778 is due mainly to introduced diseases. In particular, he argues that avian malaria, birdpox, and other unidentified diseases "swept through the lowland bird populations, causing a widespread disappearance of birds from even those forests where vegetation had not been disturbed by man," following the introduction of the mosquito vector *Culex pipiens fatigans* in 1826. This tropical subspecies of the night mosquito is restricted to elevations below 600 m; despite the existence of apparently habitable regions at lower elevations, the extant *Drepaniidae* species are found only above 600 m elevation. Warner observes that the temperate zone subspecies of the night mosquito, *C. pipiens pipiens*, is not restricted to lowland elevations, and warns that its accidental introduction to the Hawaiian Islands could well allow malaria and birdpox to extinguish the remaining drepaniids.

INDIRECT LIFE CYCLE MACROPARASITES

Parasites of this type are very well adapted to utilizing low density host populations. This is mainly due to decreases in the mortality rate of the parasite during transmission from one definitive host to the next, and increases in the efficiency of transmission.

If the passage of infection from definitive to intermediate host, and from intermediate back to definitive host, depends on the frequency of encounters between such hosts, then establishment requires:

$$N > N_T \simeq \text{(constant)}/M \qquad (8)$$

Essentially, this suggests that a high density of intermediate hosts reduces the density of primary hosts required for successful establishment of the parasite.

Many indirect life cycles include a predator–prey link, where the definitive host actively preys upon intermediate hosts in order to complete the transmission cycle. In life cycles of this sort, the parasite may be much longer lived in the intermediate host than in the definitive host. Models of these life cycles suggest that the threshold number of definitive hosts may drop below unity. This will adapt the parasite for persistance in habitats that are only occasionally visited by a migratory definitive host. Models of life cycles of this type suggest that the dynamics of the intermediate host are of prime importance in determining the occurrence of sudden outbreaks of the disease. Dobson and Keymer (1985) illustrate that when intermediate host growth rates are high enough, even low densities of definitive hosts can rapidly amplify the parasite density to produce an epidemic. Such outbreaks are characteristic of many acanthocephalan infections (Itamies et al., 1980; Jennings et al., 1961), where they can cause severe mortalities. An epidemic of this sort in a collection of wildfowl could be disastrous, as these species are usually very nonspecific in their choice of definitive host. Obviously, this type of epidemic can be prevented by careful monitoring of the populations of intermediate hosts in the enclosures in which the birds are kept.

INDIRECT EFFECTS OF DISEASE AND DISEASE CONTROL

Pesticides and DDT

Despite the ban on their use in most of the developed countries of the world, DDT and other organo-chlorine insecticides are still heavily used in many parts of the developing world today. Ironically, many of these compounds are supplied under the guise of "foreign aid" from countries where their use is now illegal on conservation grounds (Chapin and Wasserstrom, 1981; Jones, 1982). The identical patterns of reduced fecundity and eggshell thinning that were observed throughout the 50s and 60s in the bald eagle (Grier, 1982) and the European sparrowhawk (Cooke, 1979) are now being observed in African fish eagles and other tropical species (Tannock et al., 1983).

Birds are generally more susceptible to pesticides than mammals, due mainly to their relatively smaller liver sizes and higher body temperatures, as well as to the complications that arise from the process of egg-laying (Walker, 1983). However, detailed studies of areas of high pesticide use indicate that mammals are also being affected, as are many fish species (Koeman et al., 1971; Mattheisson, 1981). The last is a particularly undesirable side effect in areas where fish provide a high proportion of the protein needs of the local human population.

The rate of evolution of resistance to pesticides depends directly on the generation time of the organisms infected (May and Dobson, 1985). Thus, although we see large-scale patterns of resistance in many of the insect species that the pesticides are specifically targeted against (Dover and Croft, 1984), it is unlikely that the vertebrate species incidently affected will evolve resistance until long periods of time have passed (Frankel and Soulé, 1981). Although the benefits of pesticide usage are usually only short-term, the costs are twofold. First, the costs of control increase as resistance appears in the organisms the insectisides are targeted against; and second, mortality rates of the vertebrate predators at the higher levels of the food chain increase. The different aims of the groups of people manufacturing and utilizing these compounds lead to problems that are social and political rather than purely biological (May, 1985). Their resolution is a problem that conservationists should be increasingly concerned about.

Control of the tsetse fly

Similar ethical problems arise when we consider the possible means available to control the tsetse fly, *Glossina*, in Africa. These species, vectors of trypanosomes, still infest around 40 percent of tropical Africa, where they cause disease problems in both humans and their domestic cattle (Ford, 1971; Molyneux and Ashford, 1983; Rogers and Randolph, 1985). Although the widespread destruction of the game species known to harbor trypanosomes has now been largely curtailed, one method widely used today is the ruthless clearing of vegetation by bulldozers dragging pairs of heavy chains. Such operations are undertaken in open *Acacia* savanna, in woodland, and in secondary forest (Molyneux, 1982). Their impact on the game that utilizes this habitat can only be mildly less deleterious than the original direct control. Ormerod (1976) has even proposed that tsetse control by habitat destruction and human resettlement is responsible, at least in part, for the large-scale climatic changes associated with the Sahelian drought. However, his analysis applies only to an area of West Africa where tsetse control is not carried out on a scale that seems likely to realistically generate longer term changes in climate. More detailed economic analyses of trypanosomiasis and tsetse control that take into consideration the game animals as a tourist attraction suggest that the combined use of insecticides, chemotherapy, limited vegetation clearing, and human population resettlement may provide more acceptable solutions to these problems (Habtemariam et al., 1982, 1984).

Trypanosomiasis may be thought to be somewhat of an aid to conservation in many parts of Africa, as cattle and humans may be excluded from areas of high endemicity, leaving the land free for the resident species of resistant game. However, attempts to control the disease by removing the tsetse flies often leads to devastating effects on both the local wildlife and its habitat. Resolution of these partially conflicting interests will again require integrated management which considers not only biological but also political, economic and social factors.

Disease may destroy habitat

In cases where an epidemic almost entirely wipes out its host, the impact of the disease may be felt indirectly by species which were unaffected by the disease. A classic example would be the indirect effect of the *myxomatosis* epidemic in Britain on the large Blue butterfly, *Maculina arion*. Here the decimation of the rabbit population produced a reduction in the grazing pressure on heathlands. Although this was not directly detrimental to the butterfly, the ant species that tend the butterflies' caterpillars were unable to nest in the longer, ungrazed grass (Ratcliffe, 1979). Ultimately this led to the extinction of the butterfly in Britain. Similarly in Britain, Dutch elm disease had a considerable effect on the species of birds that utilized elm trees as preferred nest sites (Osbourne, 1982).

Disease and the subdivision of nature reserves

The subdivision of natural populations often acts as a powerful mechanism to prevent the spread of pathogens. In rock hyrax, *Procavia johnstoni*, for example, occasional outbreaks of sarcoptic mange due to the immigration of infected individuals may lead to the complete eradication of the group occupying any individual rocky outcrop, or kopje (Hoeck, 1982). However, the total population within any area is relatively safe from eradication because immigration rates between groups on different outcrops is low.

As species become more endangered, their last remaining populations are likely to be concentrated into a few reserves and zoos rather than to be subdivided as they are in their pristine state. The artificially high densities of hosts that occur when animals are constrained in their movements by reserve boundaries will increase the rates of transmission of all types of direct life cycle pathogens. As each one of these populations could be eradicated by the sudden outbreak of a disease, the subdivision of the remaining population into a number of different reserves will ensure that only a fraction of the remaining individuals will be affected by any epidemic (Frankel and Soulé, 1981).

In areas such as East Africa, where the protected species are not only a valuable tourist asset but also a potentially enormous source of nutrition, it may be sensible to establish buffer zones around the reserves. In these regions the game might be harvested at a rate allowing the intrinsic density-dependent regulatory mechanisms to maintain the population at a level below the natural carrying capacity and above the "maximum sustainable yield." The animals within the core zone would not be harvested at all in order to allow tourists and scientists to study them in their natural habitats. The width of the buffer, or "multiple use," zone might then be determined by the considerations outlined above for the "firebreak" needed to stop the spread of rinderpest.

Although movements of individual animals between reserves may be necessary for genetic management of populations where densities are low and inbreeding coefficents high, it is essential to check for the presence of pathogens in potential mates. The introduction of an infectious strain of a disease from a resistant subpopulation into a susceptible one could obviously have disastrous consequences. This argument will also apply to schemes to reintroduce animals from zoo stocks into the wild. When this was attempted with orangutans *(Pongo pygmaeus)* in Indonesia, a last-minute check revealed that the animals that were about to be reintroduced had contracted human tuberculosis. Mixing between the wild, nonresistant populations and the reintroduced carriers could have had disastrous consequences for the wild population (Jones, 1982).

COEVOLUTION: GENETICS AND DISEASE

Epidemics will often reduce the population size of populations to sufficiently low levels to produce severe genetic bottlenecks. O'Brien et al. (1983) have suggested that bottlenecks may have been responsible for the extremely low levels of genetic heterozygosity observed in cheetah populations. A more recent paper by the same group of workers suggests that increased disease susceptibility may be one of the prime costs associated with reduced genetic heterzygosity. O'Brien et al. (1985) describe how an isolated incidence of feline infectious peritonitis (FIP) in a captive cheetah colony rapidly spread through the population, ultimately killing 18 animals. The authors attribute the rapid spread of this disease to the lack of genetic variation in the cheetah population, and suggest that the FIP virus had managed to circumvent a specific MHC (major histocompatibility complex) haplotype present in the entire cheetah colony. These viruses also contribute to the high mortality rates of cheetah cubs in the wild (Murray, 1967).

SUMMARY

Pathogens and parasites are one of the most important though frequently unconsidered aspects of conservation biology. In the past epidemics due to diseases such as rinderpest and myxomatosis have devastated the populations of one or more species on a continental scale. Today, as populations become concentrated into both zoos and nature reserves the threat of epidemic outbreaks continually increases. This is primarily due to artificially high population densities increasing the transmission rates of pathogens and thus lowering the threshold population number of host individuals required to support an outbreak of the disease. We have attempted to illustrate how recent theoretical developments in the population biology of infectious diseases allow us to elucidate the biological factors which are important in determining the magnitude of these thresholds for different types of parasites and pathogens. Consideration of these factors by conservation biologists should reduce the potential of disease to eradicate populations of endangered species in the wild and in captivity.

Attempts to control the variety of diseases that affect humans and their domestic animals also create a wide range of second-order problems associated with the destruction of habitat or the misuse of pesticides. The resolution of these more complex problems requires the development of integrated management techniques based on knowledge gathered by conservation biologists, epidemiologists, and economists.

SUGGESTED READINGS

Anderson, R.M. (ed.). 1982. *The Population Dynamics of Infectious Diseases.* Chapman and Hall, London.

Anderson, R.M. and R.M. May (eds.). 1982. *Population Biology of Infectious Diseases.* Dahlem Workshop No. 25. Springer-Verlag, Berlin.

Cooper, J.I. and F.O. MacCallum. 1984. *Viruses and the Environment.* Chapman and Hall, London.

Edwards, M.A. and U. McDonnell (eds.). 1982. *Animal Disease in Relation to Animal Conservation.* Symposia of the Zoological Society of London, No. 50. Academic Press, London.

SECTION V

SENSITIVE HABITATS: THREATS AND MANAGEMENT

Where does one draw the line when listing the "sensitive" or "threatened" habitats on this planet? Given the infinite degrees of threat and the finite length of any book, the answer can only be "arbitrarily." Therefore, some readers will be disappointed by the absence of their favorite threatened ecosystem. My only excuses are that I tried to choose systems that have been ignored in other compilations, or that are especially critical from the viewpoint of species diversity. Additional systems not explicitly discussed in this section are referenced in other chapters: Mediterranean habitats in Cody (Chapter 7) and Gentry (Chapter 8); temperate forests in Ledig (Chapter 5) and Wilcove, McLellan, and Dobson (Chapter 11); islands in Gentry (Chapter 8), Pimm (Chapter 14), and Diamond (Chapter 24).

TOO MANY PEOPLE AND TOO LITTLE PLANNING

Whereas some of the earlier chapters have considered the destructiveness of mankind, this was not their main focus. At this point, we step, perhaps gingerly, into territories that are near the frontier between biology and the social sciences—the effects of too many humans on communities and ecosystems.

If the population bomb has lost some of its political bang, it isn't because Paul Ehrlich and others were wrong. Most of the authors of this section point out that there are already too many people on the planet. This is especially clear in Chapters 19 by Myers and Chapter 22 by Le Houérou and Gillet. Given the expectation of a doubling or more of human numbers in many parts of the tropics (Myers, Chapter 19), there is little room for optimism in the short run.

Where do all these people settle? Unfortunately for nonhuman life, people and their by-products obey the law of gravity, settling into lowlands and coastal zones because this is where the water supply is most dependable, and where rich alluvial soils are deposited. The problem is that these are also the regions where biological diversity reaches its zenith.

Human by-products (sewage, trash, toxic chemicals, heated discharges, and topsoil) also follow the pull of gravity, and deliver the most powerful environmental blows to lowland wetlands, rivers, lakes, and coastal marine habitats. Of all the major ecological systems, freshwater habitats are probably the most perturbed. The dilution of by-products in rivers, lakes, and coastal waters is no longer a viable solution because of the tremendous increase in effluents with the growth of populations and technology.

Another set of environmental problems is caused by society's perceived need for energy and water. In Chapter 18 Sioli shows that the future looks grim for tropical aquatic life if humans continue to pursue short-term economic gains by diking and damming every river in sight.

The situation is not quite as critical for coastal marine habitats. But as Johannes and Hatcher (Chapter 17) explain, tropical marine organisms exist much closer to their intrinsic physiological limits for oxygen and temperature than temperate species do. For this reason and others, mangrove, seagrass, and coral reef communities are more fragile than might be apparent.

Ideally, we might learn from our past errors in the management of water and land, though the opportunity for short-term profits often clouds cognition and conscience. Failure to learn from past mistakes is nowhere more evident than in flood control, irrigation, and hydroelectric projects. Sioli (Chapter 18) reviews this and gives an example of "technological desertization" in the region of the Upper Rhine. There, as a result of straightening and diking of the river, erosion of the riverbed was so great that the water table in the surrounding land was lowered 9 m. Most trees cannot reach this far down, and the landscape was transformed into a steppe. The mining of underground aquifers for agricultural water has caused similar deforestation in the arid southwest of North America.

Desertization should not be viewed in isolation from other forms of habitat destruction. It is often the penultimate stage of a process that begins much earlier, with population growth and deforestation. Myers (Chapter 19) and Jordan (Chapter 20) describe the first stages: deforestation → savannazation. Le Houérou and Gillet (Chapter 22) extend the causal chain: savannazation → desertization → starvation. This chain of events, though by no means universal, is common in many temperate and tropical regions.

CONSERVATION, DEVELOPMENT, AND GREED

There is now a growing awareness of the disastrous social and environmental results of many huge development projects, and more attention is being paid to environmental impacts and to the mitigation of some of these impacts. This is good. But typical mitigation measures, including small nature reserves and more "integrated," less obtrusive forms of development, may not be sufficient to prevent mass extinctions. If there is any message that comes through loud and clear from this book and the weight of research in the last 20 years, it is that the only feasible way to maintain the vast majority of species is with very large, very secure nature reserves. Vest-pocket reserves and integrated agro-ecosystems cannot ultimately preserve more than a tiny fraction (probably much less than 10 percent) of the biota.

Of course, integrated land-use planning and appropriate agricultural technologies are essential for sustained soil fertility, fishery productivity, and the maintenance of other ecological services. In addition, farmers would be better off if they minimized the use of chemicals and poisons and maintained some of their traditional practices. Finally, the common sense management of erosion and pollution in river basins and coastal zones is critical (Chapters 17 and 18). But candor requires that we point out that even the most harmonious forms of farming, when practiced by too many people, still will destroy the vast majority of the flora and fauna in a region.

Traditional practices and values are often a rich mine of conservation policy (Johannes and Hatcher, Chapter 17). Indeed, their consideration and adoption in many cases is simply pragmatic. On the other hand, some cultural practices become lethal when they are combined with certain medical, religious, and technological features of the dominant consumerist culture. Le Houérou and Gillet (Chapter 22) point out, for example, that pastoral practices that work when the population density is low become disastrous above a certain threshold of grazing intensity.

Why do humans continue to pillage the biosphere when this behavior can be shown to be socially and economically counterproductive in the long run? I think it is because short-term, selfish interest (fitness) usually has priority over behaviors that benefit future generations and the biosphere as a whole. More bluntly, the exploitation of natural resources is motivated by simple greed. The brave politician who opts for long-term economic growth and sustainable exploitation of natural resources is in danger of losing the support of the wealthy and powerful, not to mention the military.

MUTUAL RESPECT

What can we do, then, to slow, stop, and reverse these trends? Complacency is obviously lethal, but so is impatience. The greedy desire for quick results only creates hostility in others and frustration in ourselves. As scientists, we must continue to study the basic processes. That is what we do best. Conservationists still depend on science for the foundation of hard facts and models that provide their best arguments. As Diamond (Chapter 24) illustrates, governmental action on a comprehensive system of national parks in Indonesia only came after the conservationists could marshal the scientific evidence for the necessity of many large parks.

As conservationists, we must not repeat past errors, including the disciplinary and institutional alienation of "applied" and "pure" sections (a false dichotomy) of our field. Managers must be as conversant with genetics, ecology, and biogeography as the biologists are with the social and economic rules that must influence management policies. Managers and agency personnel must avoid "putting down" science. The tolerance that permits scientists to "do their thing" is what has produced many of the most interesting and important contributions to this and other books.

Nevertheless, managers are justified in insisting that research in reserves not be destructive (see Culver, Chapter 21) at the least, and preferably be aimed at resolving management conundrums. Human nature being what it is, "town–gown" type problems will never vanish; but to indulge in convenient stereotypes can only slow us down.

M.S.

CHAPTER 17

SHALLOW TROPICAL MARINE ENVIRONMENTS

R. E. Johannes and B. G. Hatcher

Rather than attempting a general review of conservation in shallow tropical marine communities, this chapter focuses mainly on those aspects for which temperate zone experience offers comparatively little guidance. Although tropical communities are subject to many of the same stresses as those at higher latitudes, the relative importance of these stresses differs. Excessive nearshore sedimentation due to bad land management, for example, is a bigger problem in the tropics, whereas industrial pollution is more widespread in the temperate zone.

Humans use tropical marine resources in a number of distinctive ways that dictate different management strategies. Much shallow water fishing in the tropics, for example, involves low technology methods. In addition, patterns of customary resource use are very different from those of industrial fisheries. For managers of tropical fisheries this creates some unique problems which will be outlined below.

Physically, tropical marine environments differ from their temperate zone counterparts. As a consequence, the biological responses to stresses are also different (reviewed in Johannes and Betzer, 1975). For example, the release of heated wastewater is more stressful in the tropics because tropical marine organisms live at environmental temperatures closer to their upper thermal limits than do temperate ones. Tropical organisms also live closer to their lower oxygen limits. Not only are their metabolic rates higher, but also dissolved oxygen concentrations are lower; seawater saturated with air contains 35 percent less oxygen at 30°C than at 8°C. Thus any environmental perturbation

371

which lowers the oxygen concentration is likely to exert a greater effect on tropical biota.

In addition, the unique structures of some tropical marine communities result in different responses to a given stress. For example, the aerial root systems of mangrove trees render them sensitive to certain stresses which have comparatively little impact on salt marsh grasses, the closest ecological counterpart to mangroves at higher latitudes.

Incident ultraviolet radiation increases with decreasing latitude. Recently it has been demonstrated by Jokiel (1980) and others that ultraviolet radiation (which penetrates seawater more readily than has generally been recognized) is lethal to many shallow marine organisms at low and intermediate latitudes. The harmful effects of reductions in the atmospheric ozone layer would thus apparently not be limited to the terrestrial environment.

The notion that tropical marine communities exist in physically benign environments has lost favor. It is now well established that storms, flooding, or thermal shock can be major structuring forces in coral reef communities, especially in shallow water (Dollar, 1982). Likewise, large temperature or salinity excursions, storm-generated winds and waves, or excessive sedimentation are often sufficient to slow down, set back, or reverse succession in mangrove communities (Lugo, 1980). Cyclical mortality and expansion of mangroves in response to natural periodic disturbances appears to be a common occurrence, especially along arid coastlines. In short, "baseline" environmental conditions vary significantly from year to year, complicating the problems of measuring the impact of human activity on these communities.

CORAL REEF COMMUNITIES

The best-studied tropical marine community type is undoubtedly the coral reef. It has attracted an explosion of research in the past 15 years. Out of this work has grown, among other things, an awareness of the importance of scale. A number of controversies concerning reef community structure and function seem largely resolvable when it is recognized that processes that are important on one temporal or spatial scale may be unimportant on another. For example, stability may be high in terms of trophic complexity at the scale of the community, but low in terms of species composition at the habitat scale (King and Pimm, 1983). Current emphasis on heuristic models which relate processes of different spatial and temporal scales is likely to provide useful approaches to understanding the bewildering complexity of reef systems (Reichelt et al., 1983).

Within the past few years it has also become apparent that events occurring in the pelagic realm—dispersal and recruitment of the pelagic larvae of reef fishes, corals and other reef fauna—are important in determining reef community structure (Sale, 1980; Rogers et al., 1984). This has generated increased interest in the effects of behavior and of wind and current patterns on larval distribution.

It is clear that currents often sweep larvae of reef animals well beyond the point of no return. For example, reef animals are often reported in temperate waters off both the east and west coasts of the United States and off the south coast of Australia, where it is unlikely they are able to reproduce. However, there is also evidence that the larvae of some species at some locations recruit to reef areas near where they were released as eggs or larvae. Whether recruits to a reef community are the progeny of local populations or come from distant ones obviously has major implications for reef conservation and much current research is focusing on this question.

So far the application of ecological theory to specific coral reef management or conservation problems has proven rather unhelpful. It is now generally accepted that different models are needed for accurate ecological prediction than for ecological description (Bradbury and Reichelt, 1981). Although our descriptive ecological models are improving, our predictive models are not generally reliable much beyond a point we can reach using simple observations and common sense. From the standpoint of conservation our most useful information still derives mainly from simple descriptions of the nature, distribution, and generalized impact of human activities on coral reef communities. These have been reviewed extensively (for example, see Johannes, 1975; Pastorok and Bilyard, 1985) and include information such as the following.

1. Increased sedimentation—due to dredging, filling, sand and coral mining, and, most importantly, bad land management in nearby watersheds—almost certainly degrades more coral reef communities than all other human activities combined. Topsoil may be precious to a tree but it can be fatal to a coral. The successful conservation of coral reef communities thus starts with coastal land use management.
2. The effect of liquid hydrocarbons on reef communities is important because of the overlapping distributions of coral reefs and oil deposits. The main threat to reefs from oil pollution appears to be from chronic, low-level leakage near production and loading facilities rather than from massive spills (Loya and Rinkevich, 1980). As in temperate waters, the use of dispersants often increases environmental damage.

3. The ecological effects of sewage depend greatly on the nature of the discharged material. Particulate organic loading generally has a worse effect that dissolved nutrient loading, and can bring about drastic alteration of both water column and benthic communities. However water column communities respond very rapidly to reduction in sewage input, while benthic communities revert to their prepollution status much more slowly and may never regain their original condition (Smith et al., 1981).

4. Evidence is accumulating which suggests that benthic community structure can assume at least two stable forms on reefs, one coral-dominated, the other macroalgae-dominated (Hatcher, 1984). The fact that reef communities may not necessarily revert to their initial state with cessation of stress complicates management and conservation strategies.

5. The fact that the majority of stresses and pollutants depend upon water motion for their delivery to the reef explains some of the variability observed in their effects. For example, if sediments are merely advected over the benthic community by brisk currents, the damage will be much less than if they are deposited by slower currents. Alterations in local currents as a result of human activities such as harbor construction or channel dredging may have far-reaching consequences.

6. Finfishes and many other organisms are harvested from coral reef communities. The impact is often more difficult to predict than in temperate marine communities because of the higher species diversity and trophic complexity and the typically smaller sizes of both populations and individual harvesting sites (Grigg, 1979). While reefs are highly productive at the community level, local populations can be easily depleted. Beyond the obvious direct effect of local extinction of target species, the broader impact of coral, shell and aquarium fish collection cannot readily be generalized, but must often be assessed on a case-by-case basis.

The threat to the conservation of coral reefs posed by any one of the impacts outlined above cannot be considered in isolation because they rarely occur in isolation. For example, coral mining is accompanied by sediment resuspension and sometimes also by altered water circulation. Synergistic effects are inevitable, making single-factor laboratory studies of limited value to managers.

MANGROVE COMMUNITIES

Mangrove trees and shrubs are characterized by adaptations to loose, wet soils, saline habitats and periodic tidal submergence. They fringe about 25 percent of the tropical coastline. Once mangroves were widely viewed as being fit only for "reclamation;" throughout the tropics vast areas of mangrove forests have been removed in order to accommodate farming, aquaculture, salt mining, housing, port and airport facilities and industrial sites. Today there is growing recognition that mangrove communities serve a number of valuable functions. They sustain important populations of fish, invertebrates, and birds, and are valuable sources of timber and fuel. In some nearshore tropical waters 60–80 percent of marine species of recreational or commercial value use mangrove habitat during some phase of their life cycle.

The leaves of mangroves contain large amounts of tough, comparatively undigestible cellulose, lignin, and wax; very little of the living leaf is eaten by terrestrial herbivores. The leaf becomes available to fish and larger invertebrates as food after it drops, rots, and is transformed into protein-enriched and more digestible microbes and meiofauna (Heald and Odum, 1970; Odum and Heald, 1972).

Several authors have found direct relationships between the area of mangrove forests and the catch of shrimp in adjacent waters. But, "one vital question cannot yet be answered: how much mangrove can be removed in any particular area without adversely affecting adjacent coastal fisheries? . . . the widespread establishment of aquaculture ponds [in reclaimed mangrove areas in Southeast Asia] may be a case of robbing Peter to pay Paul" (Ong, 1982).

In peninsular Malaysia the timber value alone of mangroves is greater than that of all landward forests. According to Ong (1982), sustained-yield timber harvesting as practiced in Malaysian mangrove forests today has comparatively little impact on adjacent fisheries. In apparent support of this Odum and Heald (1975) found that leaffall, which is a measure of detrital food production, appears to be almost as great in stands of young, low mangroves as in mature mangrove stands. But in many tropical countries clearcutting of mangroves is still practiced rather than sustained-yield forestry.

Mangrove communities appear to have the potential to serve as unusually efficient natural sewage treatment plants due to their ability to absorb and transform large inputs of organic material. But more research is needed on this subject.

Mangrove trees and bushes possess extensive aerial root systems. The survival of the below-ground roots in anaerobic mangrove soils is due to the delivery of atmospheric oxygen via pores in the aerial roots. But this same structure, which permits survival in such soils, is the tree's Achilles' heel. Anything that prevents gas transport through the pores kills the plant. Although mangroves commonly occur in sediment-laden water, they cannot tolerate the heavy loads of fine, flocculent materials in industrial effluent (such as bauxite, sugar cane, and pulp processing wastes), nor of oil, which coats the roots. Nor can they tolerate excessive sediment deposition, which buries the aerial roots. Reduction in gas transport caused by lack of periodic tidal exposure to air is one of the reasons mangroves die when exposed to long-term flooding as a consequence of impounding or diking.

The single biggest threat to mangrove forests is their extirpation by humans. Mangrove forests have been reduced in extent by more than 90 percent in India, Bangladesh, and Singapore and by more than 75 percent in the Philippines and Puerto Rico. The remedy is simple—in principle: leave them alone, or harvest the trees on a sustained yield basis. But this requires educating government planners and the public to their value. International agencies such as IUCN and UNESCO have performed valuable services in the regard during the past decade, but much more needs to be done at the local level.

SEAGRASS COMMUNITIES

Seagrass beds are widespread and extensive in the tropics; they are typically located in shallow waters between mangrove communities along the coast and coral reefs further offshore (Figure 1). Like mangroves, they provide shelter and food for small fishes and the juveniles of commercial shrimps and other invertebrates. They also trap and stabilize sediments. Their distribution appears to be less constrained by water movement than that of mangroves. They can create and stabilize near-vertical sediment walls, greatly reduce the height of storm surges, and in some instances seem hardly affected by hurricanes which severely damage nearby mangroves and coral reefs.

Dredging and filling appear to have caused the destruction of more tropical seagrass habitat than any other human activity. Dredging removes habitat; filling buries it. Both activities also result in increased turbidity and reduced photosynthesis in adjacent waters. Once lost, seagrass beds are extremely difficult and expensive to reestablish by either natural or artificial processes, leading to the hypothesis that seagrass destabilization as a result of human activity is "anti-catalytic" (Larkum and West, 1983).

CORAL–MANGROVE–SEAGRASS COMMUNITY INTERACTIONS

Important interactions occur between coral reef, mangrove and tropical seagrass communities (Ogden and Gladfeller, 1983). Barrier reefs function as self-repairing breakwaters, creating the low energy con-

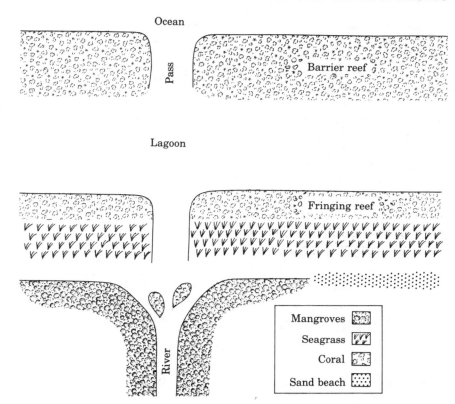

FIGURE 1. Diagram of typical spatial relationships of coastal mangrove, seagrass, and coral reef communities.

ditions which favor mangrove development along thousands of miles of coastline. Large quantities of calcareous sediments produced by calcifying reef organisms create the sand and sediment in which mangroves and seagrasses grow. In turn, both mangrove and seagrass communities reduce the offshore transport of terrigenous sediments, thus reducing their impact on adjacent reef communities. Mangrove and seagrass detritus may serve as energy and nutrient sources in nearby reef communities, although the quantitative significance of this link has yet to be assessed.

All three community types exhibit considerable faunal overlap, especially of fishes. The presence of seagrass beds enhances nearby reef fish biomass because of the foraging grounds these communities provide to complement the shelter provided by the reefs. Fishes migrate between all three communities to feed, seek shelter, or spawn.

Juvenile reef fishes and lobsters use seagrass beds and mangroves as nursery areas.

Obviously, anything that adversely affects one of these communities may ultimately affect the others. Coral reef communities, no matter how rigorously managed, will decline if adjacent mangroves are cleared or seagrass beds dredged and the resulting sedimentation envelopes the reefs. Food fishes and lobsters feeding in seagrass beds will suffer if adjacent coral reefs, where many of these species shelter, are degraded. Mangrove communities will be threatened where reefs which protect them from high wave energies are removed to create channels or anchorages.

NEARSHORE TROPICAL FISHERIES

During the past few years there has been a rapidly expanding awareness of the importance of "traditional," "artisanal," or "small scale" fisheries in nearshore tropical waters. Although the catch-per-unit-effort of these artisanal fishermen is very low, their numbers are very great—over 8 million throughout the tropics—and their total harvest amounts to almost one-half of the world's food fish catch. (Only about one-eighth as much fossil fuel energy is expended on capturing this portion of the catch as on the remainder caught by industrial fishermen.)

Artisanal fisheries possess a number of features which dictate a rather different approach to research and to conservation of stocks than that taken in typical industrial fisheries. Artisanal fisheries usually involve, per unit of catch, far more fishermen, boats, methods used, habitats fished, species caught, landing sites, and distribution channels than industrial fisheries. The cost of monitoring such complex fisheries in order to obtain the information necessary for conventional management would often exceed the economic benefits by a great margin. What is needed, therefore, are research shortcuts—"quick and dirty" methods which do not require the kinds and amounts of data we cannot realistically expect to obtain. One simplifying approach has been to treat groups of species as single management units, and various theoretical multispecies fisheries models have been proposed (see for example Pauly and Murphy, 1982). It will be some years before the practical utility of such models can be widely tested.

In order to manage tropical small-scale fisheries it is necessary to study not only the biological resources, but also the methods and relevant customs and values of the people who use them. Prescriptions for management that work in western cultures may fall flat elsewhere.

Tens of millions of dollars in tropical fisheries aid have been wasted through lack of appreciation of this fact. In a wide-ranging analysis of tropical small-scale fisheries, Emmerson (1980) concluded that anthropologists and biologists working together offer the best hope for improving management. But although workers in tropical agriculture have for some years recognized the important contributions that social scientists have to make to sound resource management, the message seems to be taking longer to get through to tropical marine conservationists and fisheries managers.

Some tropical fishermen recognized the need for conservation of fish stocks centuries before westerners. Such awareness was widespread in Micronesia and Polynesia, for example, where seafood was the main source of animal protein and accessible marine resouces were limited largely to narrow fringes of reef and lagoon. Under such conditions overharvesting was easier and the consequences more obvious than in continental areas with their wide continental shelves and significant terrestrial sources of animal protein. Thus for centuries islanders employed size restrictions, closed seasons, closed areas, taboos on exploiting spawning aggregations, limited entry—in fact every basic marine conservation measure used in the west (and employed there only since about 1900) (Johannes, 1978a). Similar customs have been found in parts of Africa, South America and elsewhere. But such practices are weakening in the face of westernization, political upheavals, and well-meaning but uncomprehending expatriate fisheries managers.

Research is urgently needed to document such practices. Modern management regimes in which traditional management customs are recognized and incorporated (where practical) are likely to gain greater local support and thus be easier to enforce. Traditional authority and indigenous environmental regulations often carry considerably more weight in the hundreds of isolated fishing villages in the tropics than do remote government edicts.

Such traditional practices, however, have lapsed (or may never have existed) in some tropical regions. Here other approaches to conservation must be sought. Whatever they are, it is becoming increasingly clear that they must be culturally as well as environmentally sensitive.

Tropical fishermen can sometimes provide much important information on such things as local fish behavior, the timing and location of spawning, and migration pathways of various species. For example, the number of species throughout the tropics as known by biologists to form lunar spawning aggregations more than doubled a few years ago as a consequence of information provided by such fishermen (Johannes, 1978b). These aggregations provide biologists with excellent opportunities to monitor stocks, because they occur at predictable times and locations. They can also provide useful foci for management.

MARINE PARKS AND RESERVES

Marine conservation areas are being designated with increasing frequency in the tropics. Unfortunately a significant number are conservation areas in name only, and little effort is devoted to management. Criteria for selection and delineation of such areas are not well developed (Bakus, 1982/83).

Island biography theory has formalized the recognition of the importance of setting aside natural reserves of sufficient size and number so that a balance between local extinction and immigration of species is maintained. The crucial question remains, what size? Two estimates of critical minimum core areas for coral reef reserves have been made. Goeden (1979) recommends 3470 ha based on studies of Great Barrier Reef fish populations, whereas Salm (1984) suggests 300 ha, based on studies of Chagos Archipelago corals. It is not known to what degree this order-of-magnitude difference is due to either ecological differences between the two areas, differences in distribution of reef fish and reef coral populations, or methodological differences between the two studies. More research is needed to determine the utility of island biogeography theory in the delineation of marine protected areas.

Whereas most shallow water tropical marine animals and plants are sessile or have limited home ranges, most also have floating or planktonic juvenile or reproductive stages. In general, therefore, dispersal of populations from protected tropical marine communities can be assumed to be wider than in terrestrial communities. The larvae of corals typically drift with the currents for from one day to several weeks before settling to the bottom. Coral reef fish larvae take up residence in a reef community only after spending two weeks to several months in the plankton. Thus overexploited tropical marine communities may recover relatively quickly when protected, owing to rapid colonization via the plankton. They may in turn export juvenile colonists to areas ranging up to hundreds of kilometers away.

The significant movement of organisms and materials between seagrass, coral reef, and mangrove communities where they occur together, means that these communities are often not defined simply by their physical boundaries. The protection of one community type will be less effective if the integrity of neighboring communities with which it interacts, and of adjacent watersheds, is not also secured.

The great dispersive abilities of tropical marine organisms and the interconnectedness of adjacent communities thus complicate decisions concerning reserve sizes, spacing, and buffer zones. Criteria for reserve size and buffer zone width derived for coastal fringing reefs could be quite inappropriate for oceanic atolls. Human use patterns will also influence what constitutes a viable reserve.

The Bali Action Plan, arising from the 1982 World National Parks Congress, contains a recommendation to "investigate and utilize the traditional wisdom of communities affected by conservation measures, including implementation of joint management arrangements between protected area authorities and societies which have traditionally managed resources." This marks a substantial, if belated, shift in the perception by tropical resource managers of traditional users of protected areas. Seen less now as intruders in their own lands and waters, they are becoming accepted as integral parts of the ecosystem. Traditional fishermen cannot be expected to cooperate with marine reserve planners in the absence of some consequent benefits. These could include lease payments [fishing rights in tropical coastal waters are often traditionally owned by local fishermen (Johannes, 1978b)], employment as rangers or guides, protection from outside encroachment, income from tourism, or political advocacy at the national level. A sympathetic resource manager can convert local resistance to his objectives into active support, and in the process gain a constituency with unparalleled knowledge of local areas and their natural resources.

SUMMARY

For the biologist the precise effects of human impact on tropical marine communities are difficult to predict with satisfactory rigor. But it would be misleading to characterize these effects as typically subtle. The general environmental consequences of coastal clearcutting, dredging, mangrove clearing, coral mining, and dynamite fishing (widespread in the tropics) are often extreme and obvious even to the casual observer.

Freqently the physical solutions are also obvious. For example, an examination of the recent UNESCO Coral Reef Management Handbook (Kenchington and Hudson, 1984) reveals that many of the physical solutions to reef community degradation are of a common sense nature, often requiring little additional research to implement.

Much of the degradation of tropical marine communities can be traced to the rapid increase in the populations of developing countries. Here human activities have proved more destructive than those of similar-sized populations in developed countries. According to Gomez (1982/83), this is due to the loss or lack of a conservation ethic. Gomez states, "Apparently deep-seated in the social fabric is the inability to reconcile drive for economic gain with the obvious wisdom of long term planning and conservation."

This observation should perhaps be qualified by acknowledging that conservation issues cannot be expected to rate high concern among people whose basic economic and nutritional needs are not being met. Nevertheless we believe that expanded, culturally specific educational programs to motivate people to conserve these resources (in developed as well as developing countries) are critically needed today, even (in some cases) where the money to support them must be taken away from research programs. Human nature and the scientific establishment being what they are, however, the emphasis will probably remain firmly on understanding nature better rather than on conserving it.

SUGGESTED READINGS

Hamilton, L.S. and S.C. Snedaker. 1984. *Handbook for Mangrove Area Management*. International Union for the Conservation of Nature and Natural Resources, Gland, Switzerland.

Kenchington, R.A. and B.E.T. Hudson (eds.). 1984. *Coral Reef Management Handbook*. UNESCO, Jakarta.

Wood, E.J.F. and R.E. Johannes (eds.). 1975. *Tropical Marine Pollution*. Elsevier, Amsterdam.

CHAPTER 18

TROPICAL CONTINENTAL
AQUATIC HABITATS

Harald Sioli

It is a fundamental of ecology that life in all its manifestations does not occur in isolation but thrives only in an environment with abiotic and biotic components. Organisms and environment together form what we now call ecosystems in which in principle every participant directly or indirectly, actively or passively, acts upon and is acted upon by all the others. Changes inflicted upon the abiotic or biotic environment of an organism are reflected to one degree or another in the organism's life. In turn, changes in populations alter the environment.

This complex network of mutual interactions and interdependencies includes frequent feedback which is often overlooked and unperceived. An unbiased look into nature shows a constant interplay of the genotypes and phenotypes of organisms and the environment resulting in a higher unit: a geobiocenosis or landscape (Sukačev, 1964), or an ecosystem.

Human cultural diversity can be understood only as a result of the interplay and feedback between two such partners: the mentalities of the concerned human populations, and their geographical environments. Thus, we speak of "forest cultures," "steppe cultures," and "mountain cultures," including their specific spiritual worlds (Sioli, 1973, 1984a.) Such fundamental concepts are the starting point when considering problems in conservation biology.

HUMAN INTERFERENCE IN NATURE

Human interference produces impacts of varied magnitude, from slight touches, easily absorbed and compensated for by what we may call the "buffering capacities" of the respective ecosystems, up to the complete destruction and irreparable breakdown of these systems. That means that qualitatively and quantitatively different human interferences result in different extinction rates of organisms and, eventually, of whole taxa. And extinction of these species is not only forever (Prance and Elias, 1977) but the whole sequence of evolution is being changed or amputated by the alteration or annihilation of the original environments.

In the present decade we have the dubious privilege of watching the most destructive human interferences in all regions of the biosphere—the greatest impact on our globe ever accomplished by man. These impacts affect the atmosphere, the oceans, and the terrestrial and aquatic habitats of the continents. All these subdivisions of the biosphere must be looked upon and understood as being linked to each other by a dense network of mutual interactions and feedback. It is only in recent times that we have begun to comprehend parts of this global network.

THE CONTINENTAL AQUATIC HABITATS

The continental freshwater habitats of the tropics—lakes, rivers, and wetlands—are extremely vulnerable to human activity. They differ, among other ways, in their shapes, their internal dynamics, their active and passive relations to the geomorphology of their surroundings, their roles in the context of the landscape to which they belong, and in their biotas. Classical limnology characterized lakes as "microcosms" (Forbes, 1887) and stressed their self-contained independence (Thienemann, 1953). Rivers, however, are passively and actively enmeshed in their surroundings (Sioli, 1975, 1982).

Seasonality in tropical waters is caused not so much by thermal seasons, which are practically absent in the equatorial belt, as by alternating wet and and dry seasons. The corresponding oscillations of the water level are most profound in rivers and adjacent floodplains and lagoons (Junk, 1980, 1984). Another peculiarity of the warm tropics is the fact that most biological processes, including anthropogenic ones, occur more rapidly than in cooler climates.

Material metabolisms of aquatic ecosystems

Ecosystems can be grouped into three main types with regard to their exchange and metabolism of matter. These types, of course, represent points on a continuum. In *closed systems* matter is not exported. The matter is contained mostly in the living biomass and is continuously recycled through the system's organisms. There are almost no losses.

Such a system maintains itself virtually independently of its environment, be it the surroundings or the subsoil, and can exist infinitely as long as the outlet remains locked. A good terrestrial example is the tropical rainforest on extremely nutrient-poor soil, as is the case on the terra firme of central Amazonia.

An opposite type of system is *continuous throughflow* of matter. Its prototype is the river. The drainage basin constantly delivers fresh dissolved and particulate matter into the river, the products of weathering and of biotic processes in the terrestrial surroundings. In the river this matter is partly processed by chemical, physical, and biotic acitivities, transported downstream by the current, and finally released into the ultimate recipient, the sea. The role of the running waters is therefore not merely to drain excess water; rivers also eliminate the end and side products of the basins' abiotic and biotic processes. In this way the running waters are analagous to a kidney system. Throughflow ecosystems are theoretically of unlimited existence as long as the inputs and the outlets stay open and the gradient persists.

Between these two opposite system types there is another one, open at the input side but closed at the outlet. The prototype of such *intermediate system* is the lake. Dissolved and particulate matter accumulates within the system, eventually changing its conditions. Such a system is represented by inland lakes. In the extreme form, lakes have no outlet, or at least reduced throughflow, so that all particulate matter, the introduced inorganic as well as the detritus of the lake's own bioproductivity, accumulates on the bottom, making the lake increasingly shallow and finally turning it into a swamp and firm land. Where there is no outlet at all, the salt content of the lake water increases and eventually, especially in dry, warm climates, the salt precipitates along the shore or on the bottom. The lifespan of this type of ecosystem is necessarily limited.

Wetlands is a collective name for terrains permanently or temporarily waterlogged or covered by shallow water. They include peat bogs, swamps and marshes, alluvial and nonalluvial floodzones, deltas of rivers, "sudd" formations as in the Nile (Rzóska, 1974), and many other types (Gore, 1983). They may be of different origins.

Wetlands vary with regard to their processing of matter (their material metabolism). Those associated with rivers are influenced by floods which may periodically introduce nutrient-rich suspended particles deposited on alluvial floodplains. The water of some "blackwater" rivers, common in central Amazonia, contain only dissolved substances and deposit no sediment on their floodplains.

Sedimentation rates do not necessarily dictate the life span of wetlands. Alluvial floodplains of sediment-rich rivers may be constantly worked over by the river's lateral erosion, and all matter brought from the headwaters may eventually reach the sea. Indeed, the biota of the Central Amazonian "igapós"—the seasonally flooded forests along stretches of the shores of clear and blackwater rivers— shows some peculiar adaptations that indicate a very long existence of the igapó forests (Irion and Adis, 1979; Adis, 1984). Also, floating sudd formations may exist as long as the river offers adequate conditions for their life. Wetlands with no outflow, however, often represent the final stages in the development of former lakes. Their lifespan is thus finite.

Some tropical wetland habitats are highly fertile. In South America, examples of such fertile wetlands include, among others, the "várzea" (floodplains) of the Amazon and its whitewater affluents (Junk, 1984), and the "esteros" around the junctions of Rio Paraguay, Rio Pilcomayo, and Rio Bermejo in northern Argentina and southern Paraguay. Floodplains of great turbid (whitewater) rivers in Amazonia are the most important sources of primary production for the basin (Sioli, 1968). Cattle introduced in such wetlands evidently do little harm. Many of the fertile wetlands accomodate an extremely rich and numerous fauna of fishes, aquatic reptiles, and waterbirds; terrestrial mammals, although common, are not so abundant.

Other wetlands, however, are not flooded by rivers of nutrient-rich sediments and may be of low fertility. Such a situation is found in the Iberá region of northern Argentina, which has its own small catchment basin separate from Rio Paraná. Some eastern parts of the Pantanal de Mato Grosso, Brazil, are also relatively infertile. Peat bogs, with only rainwater to supply them, are extremely infertile. Thus, wetlands are anything but uniform.

HUMAN INTERFERENCE IN AQUATIC HABITATS

Historically, there is little to suggest that human exploitation in aquatic habitats and ecosystems caused perceptible damage. Preindustrial fisherman may have played a role similar to that of other fish-eating animals such as cormorants and herons. Subsistence fishing was most successful in nutrient-rich shallow and more or less stagnant waters, such as lagoons in the floodplains of certain rivers. In some places landscapes were altered and new fisheries were created by the construction of paddy fields, fertilized ponds, ridged fields (Parsons, 1966), and man-made lakes.

This situation has changed with the increase of the human population, with the advent of motorized energy-intensive technology (including cold storage), and with the intensification of large-scale commercial operations.

Current human interference in tropical aquatic habitats does not differ in principle from those in the temperate zones. But while the surfaces of temperate waters freeze in winter, killing or setting back most emergent or floating vegetation, tropical waters frequently support perennial mats of "macropleuston," or floating plants such as *Papyrus* in Africa (Carter, undated), or of several grasses, and of *Eichhornia, Pistia,* and *Salvinia* in South America and elsewhere. These mats may even grow to vast coherent and solid areas such as the sudd of the Nile, or to floating islands called "matupá" in Amazonian várzea lakes (Junk and Howard-Williams, 1984).

Tropical lakes

Human disturbance of tropical lakes is not yet as evident and serious as in tropical rivers. Where it occurs, the first assault is usually pollution. The sources of this pollution include sewage from urban sources and vacationers, and runoff from fertilized fields. These lead to accelerated eutrophication and, later, to hypertrophication. The consequences are algal blooms, oxygen depletion, production of hydrogen sulfide in the hypolimnion, and subsequent mortality of the fauna.

There is an extensive literature on these calamities, most of which is devoted to lakes in the densely populated and industrialized countries of the temperate belt of the northern hemisphere. There are many fewer lakes in the tropics because of the absence of Pleistocene glaciers. Indeed, in many tropical countries, very few natural lakes exist. The largest and deepest of these are the Rift Valley lakes in East Africa; so far they have escaped eutrophication.

On the other hand, tropical lakes have not escaped the effects of industrial and agricultural pollution. One of the smaller Rift Valley lakes, Nakuru, has no outlet and was threatened by toxic wastes from a copper industry. Most of the volcanic crater lakes of Indonesia—the first objects of limnological investigations in the tropics (Thienemann, 1931–1957)—are now threatened with encroaching numbers of people, expanding agriculture, and deforestation of the surrounding mountain slopes. The ecology of several tropical lakes has also been upset by overfishing and by the introduction of exotic species of fish (Zaret, 1984; FAO/UNEP, 1981).

Tropical wetlands

More threatened than lakes are wetlands, which are often considered useless impediments to development projects (Gore, 1983). It is still not widely understood that many wetlands play crucially important roles in regulating the water regime of landscapes and of the rivers connected with them, and in degrading organic wastes. It is rarely appreciated that many such wetlands have the highest productivity on earth and are extremely rich in bird and other animal life (Benthem et al., 1971).

The greatest threat to their biological diversity consists in draining wetlands for industrial or housing purposes and for agriculture. In addition, wetlands associated with river floodplains and deltas are being diked to protect low-lying lands from floods and tides, or to submit them to controlled irrigation for extensive monocultures.

In the case of the Pantanal de Mato Grosso in Brazil, a unique and vast wetlands region of about 140,000 km^2 (Wilhelmy, 1957, 1958) has been linked by construction of roads to the expanding industrial megalopolis of São Paulo. These roads offer easy access for the exploitation of the wetlands' rich fish resources, which are trucked to the consumer cities. Commercial and sports hunters are also endangering its most spectacular birds and mammals.

The Convention on Wetlands of International Importance Especially as Waterfowl Habitat (RAMSAR), which has been signed by 35 countries since 1971, designates suitable wetlands for inclusion in a list of areas of international importance—nothing more concrete. Still, this convention is a sign of the growing awareness of the alarming situation of wetlands in general (Braakhekke and Drijver, 1984). In 1985 the World Wildlife Fund and the International Union for Conservation of Nature and Natural Resources together launched "The Wetlands Campaign" as a part of their efforts to save important ecosystems, species and habitats.

Tropical rivers

In many human cultures and since ancient times, creeks and rivers were not only populated with nymphs and spirits, but were also used as dumps for domestic wastes, to wash clothes, and to discard any unwanted object (including dead bodies). Such procedures were not harmful as long as population densities were low, because they did not exceed the rivers' capacities as "kidneys of the landscape."

With exploding cities and growing industrialization, not only creeks but also great rivers became increasingly polluted. After the Second World War bathing in most rivers in Germany was prohibited because of health risks. In the 1950s and 1960s urban and industrial pollution of rivers and lakes grew, to become a major calamity in most of the highly developed regions of the northern hemisphere, and the overcrowded cities and industrialized zones in certain parts of the tropics soon followed.

The explosive growth of tropical cities had predictable results. Rio Tieté at São Paulo, Brazil became a stinking sewer in a canalized bed. Rivers in the sugarcane regions of Brazil are polluted with the heavy liquid residue ("vinhoto") of numerous alcohol distilleries.

The straightening and diking of riverbeds to control flooding has been practiced for almost two centuries in the temperate zone. The Upper Rhine, formerly a meandering river, has deepened 9 m in the course of 150 years and has drawn the groundwater level with it, transforming the landscape of the Upper Rhine plain into a steppe. The fruit trees couldn't reach the groundwater table and died.

Dikes have been built in the deltas of tropical rivers and along some inland floodplains to prevent periodical or occasional flooding of the fertile neighboring alluvial land. Besides the biological effect, diking hinders the natural annual fertilization of the former floodland by the sediments brought and deposited with the floods. The dikes indeed prevent floodwater from spreading over wide areas, but they also force them to stay in the riverbed itself, augmenting the current velocity and causing higher floods in undiked lands.

The construction of dams and reservoirs is occurring in the name of progress and industrialization. Practically all major development projects, at least in Amazonia, involve the export of raw materials. But raw materials and their export require a lot of energy for transportation and refining—especially the processing of bauxite to metallic aluminium.

In such cases rivers are considered to be sources of plentiful and cheap energy. In mountainous regions a little water can produce a lot of energy. Manmade reservoirs in mountains are generally small compared to the extensive ones in relatively flat countries, such as are found in much of the African and Brazilian tropics.

Enormous dams and immense reservoirs have been built in the last 30 years. The first reservoir built in Africa for energy purposes was the famous Kariba Dam on the middle Zambezi River, completed in 1958. Balon and Coche (1974) describe the more recent major projects:

In 1964 both the Akosombo Dam (Volta River, Ghana) and the Aswan High Dam (Nile River, Egypt) were put into operation. Four years later the Kainji scheme (Niger River, Nigeria) was completed while work was being started in Ivory Coast on the Kossou Dam (Bandama River, closed in February 1971). Today dam construction rapidly progresses as for example downstream from Kariba (Cabora Bassa, Mozambique), in the Orange River Valley (Republic of South Africa), and on the Congo River (Inga, Zaire). It is considered that in West Africa the basic and most economic source of electric energy for the future lies in its waterways, and plans have been made for building dams on the Niger, Benue, Volta, Senegal, Bandama, Cavally, Konkouere, Sassandra and Comoe Rivers.

All these reservoirs are in relatively flat country, and their sizes are therefore mostly enormous: Volta Lake, 8849 km^2; Lake Nasser, 6276 km^2; Lake Kariba, 5344 km^2; Lake Kossou, 1600 km^2; and Lake Kainji, 1280 km^2.

This series of big African reservoirs is being matched by a no less ambitious one in South America, mainly in Brazil. The first dams were built in the Brazilian states of São Paulo and Minas Gerais, and more are continually being added (CESP, 1979). The tributaries of the east side of the whole upper Rio Paraná system are now interrupted many times, and most of these affluents are heavily polluted.

The first reservoir situated entirely in a tropical rainforest was Brokopondo Lake in Surinam, 1500 km^2, filled in 1971. It caused serious ecological problems (Leentvaar, 1973). In the Amazonian rain forest of Brazil, the first dams to be built were small ones: Curuá-una, south of Santarém State of Pará, about 100 km^2 and operational in 1977; Paredão, in the Territorio do Amapá, also about 100 km^2 and operational in 1978. Both are of only regional importance for energy generation. Curuá-una developed problems similar to those caused by Brokopondo—over 20 percent of the lake surface was covered by aquatic weeds, anoxia occurred at depths greater than 2 m, high levels of hydrogen sulfide accumulated in deeper water, and several fish species disappeared (Junk et al., 1981).

In the Paraná River, between southern Brazil and Paraguay, a mighty dam has been built in recent years: Itaipú, which started to produce electricity in 1983. The reservoir covers 1460 km^2 and has a potential output of 12,600 megawatts—the largest in the world. It drowned the famous Sete Quedas waterfalls. Its power serves the most industrialized and populated part of Brazil, including the complex of São Paulo.

Now it is the turn of the great Amazon system. Fortunately, the biggest such project ever conceived—"A dam across the Amazon" that would have created a true freshwater sea of 250,000–350,000 km^2 (Panero, 1969)—was rejected by the Brazilian government soon after it was proposed by the Hudson Institute. Nevertheless, Brazil plans to dam up most great rivers of the enormous Amazon system (Figure 1). The first reservoir of this vast series is that of Tucuruí on Rio Tocantins, which covers an area of about 2500 km^2 and is foreseen to produce 8000 megawatts. In November 1984 the sluice gates opened for the first time—and with the water there came many tons of dead fish, killed by lack of oxygen and by hydrogen sulfide that had developed in the deep water of the new lake from decomposition of the

FIGURE 1. The known hydroelectric projects in the Amazon region.

drowned forest (which had not been removed before the reservoir was filled).

Most of the energy to be produced by Tucuruí will be used for exporting the mineral resources of eastern Amazonia: 60 percent of the final output is for aluminium production from the nearby bauxite deposits. Most of the rest goes to the Grande Carajás mining project and to the transport of its ore on an 800 km-long electrified railway to the ore harbor of Itaquí. Only the small remaining fraction is destined for the city of Belém.

Many other resevoirs are planned or already under construction along the Amazon's tributaries. They include Balbina on Rio Uatumã (2100 km² or much more); Samuel on Rio Jamarí (642 km²); Porteira on Rio Trombetas (1400 km²); Babaquara and Cararão on Rio Xingú

(6100 and 1200 km²) (Junk, 1983). These projects will ultimately disrupt the coherence and continuity of the Amazonian river system.

THE EFFECTS OF DAMS

"So the more we dam the rivers, the sooner we are damned . . ."
(K. E. Boulding in Balon and Coche, 1974, p. 679)

Rivers are intimately associated with their catchment basins, and they reflect the physical and biological properties of these basins. They shape their beds and their adjacent lands by bottom, lateral, and retrograde erosion, and by sedimentation. In many cases rivers build up their floodplains and deltas, whose fertility and life processes are maintained and regulated by flooding and other effects of the oscillations of their water levels.

Dams interrupt the flux of the river and transform sections of rivercourses into lake-like units of more or less stagnant water. Dams choke off most of the dynamic attributes of rivers, quenching the flowthrough of sediment and other matter. Beyond the dam, where the current is restored, there is a lack of sediment load, the temperature and oxygen regimes are changed, and lake plankton are added. Within the reservoir, the sedimentation of particulate matter limits its lifespan to periods of a few decades to some centuries.

In the new manmade lake the rotting of the drowned forest consumes dissolved oxygen. This, together with the temperature stratification of the lake, produces an anoxic hypolimnion. Frequently hydrogen sulfide is generated, killing all life, corroding the metal of the turbines, and attacking the concrete of the barrage (Leentvaar, 1984; Ambrokwa-Ampadu, 1984).

The aquatic biota of the former river suffers the most. The lacustrine conditions of the reservoir are not tolerated by many of the fluviatile species. Those species, particularly fishes, that depend on the uninterrupted river for breeding migrations and the completion of their life cycles perish. In tropical South America, where there are no true great natural lakes (Junk, 1983), there are few indigenous species of fish to replace the riverine species. The introduction of exotic species is always a risk, as previous disasters demonstrate (e.g., Lake Titicaca and Gatún Lake; Zaret, 1984; FAO/UNEP, 1981).

Damming up rivers also affects the nearby terrestrial habitats. In most cases the fertile land is lost by drowning of the alluvial floodplain, where there is one. The people who may live along the river or on the site of the reservoir are driven away. Many times indigenous peoples, with their wealth of traditional biological knowledge, are displaced. Rarely does modern civilization on its advance in the name of "progress" respect the rights of other groups of human beings to cling to their traditional lifestyles. The human toll is often misery and death (Aspelin and Dos Santos, 1981).

There are economic benefits, of course, but these tend to fall into the pockets of the rich. The costs are apt to fall upon the shoulders of the poor. Cost–benefit analyses will nearly always justify the building of a solid concrete fact (K.E. Boulding in Balon and Coche, 1974).

The human destruction of the aquatic habitats in the tropics is part of the general aspiration to exploit these regions, to make them "useful," and to profit from them. That greed is coupled with the honest belief in a moral, missionary obligation to bring our lifestyle to all other peoples and to make them "finally happy." Only by a fundamental change in this attitude can humanity avoid "the sooner we are damned."

SUGGESTED READINGS

Rzóska, J.. 1978. "On the nature of rivers, with case stories of Nile, Zaire and Amazon." Dr. W. Junk, The Hague. 67 pp.

Lowe-McConnell, R.H. (ed.). 1966. *Man-made Lakes.* Symposium 15. Institute of Biology, London.

Obeng, L. (ed.). 1969. *Man-made Lakes.* The Accra Symposium. Ghana University Press, Accra. 398 pp.

Ackermann, W.C., G.F. White and E.B. Worthington (eds.). 1973. *Man-made Lakes. Their Problems and Environmental Effects.* American Geophysical Union, Washington DC.

Sioli, H. 1984. Present "development" of Amazonia in the light of the ecological aspect of life, and alternative concepts. In *The Amazon: Limnology and Landscape Ecology of a Mighty Tropical River and Its Basin.* H. Sioli, ed. Dr. W. Junk, The Hague. pp. 737–747.

CHAPTER 19

TROPICAL DEFORESTATION
AND A MEGA-EXTINCTION
SPASM

Norman Myers

Tropical forests contain far greater biotic diversity than any other biome. They probably harbor around half of all species, even though they cover only 7 percent of the planet's land surface. It is all the more regrettable, then, that tropical forests are being degraded and destroyed faster than any other biome. Already sizeable sectors are gone. If present trends of exploitation, or rather overexploitation, persist—and these trends are likely to accelerate—there may be little left except isolated remnants by the middle of the next century (at the latest). The ultimate consequence will surely be that millions of species become extinct. In terms of the number of species involved, and the compressed time scale of the phenomenon, this could be as great a biological debacle as any since the first emergence of life on Earth.

This chapter starts out by looking at the present extent of tropical moist forests. It considers the ecological character of these forests and the scope of their biotic diversity. It goes on to examine their species richness in both the quantitative and the qualitative sense; and it reviews the distribution of species, with particular reference to exceptional concentrations of species and centers of high endemism. The chapter then assesses the status of tropical moist forests: how much has been lost already, how fast they are being converted right now, and what we can anticipate for the foreseeable future. Then we move on to look at the issue of extinctions: how many have probably occurred to date, how many we can expect during the next few decades, and what may be some repercussions for the future course of evolution. Finally the chapter looks at our conservation response: what we are doing, what needs to be done, and what overall prospects face tropical forests.

CHARACTER AND EXTENT OF TROPICAL FORESTS

I shall define "tropical moist forests" as evergreen or partly evergreen forests, in areas receiving not less than 100 mm of precipitation in any month for two out of three years, with mean annual temperatures of 24°C or more and essentially frost-free; some trees may be deciduous. Such forests usually occur at altitudes below 1300 m, though they often occur up to 1800 m in Amazonia, and generally only up to 750 m in Southeast Asia. Mature forests contain several more or less distinctive strata (Myers, 1980; Raven, 1980). This definition includes seasonal or monsoonal forests, but tropical dry forests are left out of the picture. The definition also reflects a highly generalized approach, and does not take into account many significant subclassifications of the biome.

Tropical moist forests are the most complex and diverse biome on Earth. Since environmental conditions are relatively constant (although not necessarily uniform), a high level of dynamic stability can persist within a narrow amplitude of environmental fluctuations. We should not suppose, however, that stability is interrelated with complexity in a simple and positive fashion. Indeed, some theoretical and empirical research suggests that "Communities with a rich array of species and complex web of interactions (the tropical rainforest being the paradigm) are likely to be more fragile than relatively simple and robust temperate ecosystems" (May, 1975; see also Pimm, Chapter 14). Overall we can conclude, so far as we understand the situation, that "The humid tropical environment permits the systems to persist in spite of their fragility, because perturbations are relatively small and restricted to small areas. . . . Tropical forests are in a constant turmoil of phasic development . . . and they are stable only within a relatively small domain of parameter space" (UNESCO, 1978).

What is the present extent of tropical forests? Curiously, we do not have a precise figure. Remote sensing technology is still not advanced enough to differentiate between various vegetation types, especially along forest fringes with their degraded formations (bush, scrub, etc.). A recent FAO/UNEP report (1982) postulated an expanse around 10 million km^2 in 1980, but on-ground investigations reveal that this is on the high side. In India, for instance, Landsat surveys suggest around 23 percent of territory is forested in one way or another, yet detailed on-ground checks have revealed less than 10 percent. Similar discrepancies have been noted in Burma, Kenya, Cameroon, Guatemala, Peru and a number of other countries. So a more realistic estimate of tropically forested area is probably some 9.5 million km^2 or less, in accord with the figure produced by the National Academy of Sciences survey (Myers, 1980). Since we have been losing tropical forests at a rate between 75,000 and 92,000 km^2 per year to outright destruction, and a further 100,000 km^2 each year to gross disruption (see below), the 1985 figure for largely undisturbed tropical forests (i.e. primary or near-primary forests) is more like 8.5 million km^2.

BIOTIC RICHNESS

There is far greater abundance and variety of species in tropical forests than in any other ecological zone of extensive scope. Extrapolation of figures for well-known groups of organisms in closely inventoried areas suggests that the forests could well contain half of all species on Earth (Hemming, 1985; Gentry, 1984; Myers, 1984; National Research Council, 1981; Raven, 1980; Sutton et al., 1984).

We are much better informed about plants than animals. Systematic taxonomic surveys reveal that tropical moist forests contain at least 90,000 of Earth's 250,000 identified species of higher plants—and another 30,000 could well await discovery. This total of 90,000 is to be compared with only 50,000 in the entire northern temperate zone, with its vast territories in North America and Eurasia. The single richest large bloc of tropical forests, Amazonia, is believed to support at least 30,000 plants, as compared with only 10,000 in all of temperate South America, and the Choco area of Colombia and Ecuador probably contains at least another 8000 plant species. Colombia as a whole, with the majority of its plants in its Pacific coast and Amazonian forests, possesses around 25,000 plant species in its 1.1 million km^2, which means that the country, only a little larger than Texas and New Mexico combined, contains at least 5000 more species than the eight times larger United States. Ecuador probably harbors as many as 20,000 plant species, the bulk of them in its tropical forests; this total is to be compared with Europe's 13,000 plants in an area 31 times greater. Costa Rica possesses 8000 plant species in its 52,000 km^2, in contrast to Great Britain with 1443 species in 244,000 km^2.

We know much less about animals. The proportion of total tropical species that we have identified is a mere one-sixth at most. Nonetheless we can come up with some informed estimates of animal species' totals. If we assume a global total of five million species of plants and animals, we can realistically assume that the tropical forests of Latin America support one million, those of Southern and Southeast Asia three quarters of a million (most of them in Southeast Asia), and those of Africa one third of a million. The great bulk are insects.

Again, it is instructive to look at some comparative figures. The Sunda Shelf sector of Southeast Asia, roughly the western half, supports 297 mammal species and 732 bird species, whereas Europe west of the Soviet Union—an area almost four times as large—has only 134 and 398 respectively. The La Selva Forest Reserve of Costa Rica, only 7.3 km^2 in size, supports 320 tree species, 394 birds, 104 mammals, 76 reptiles, 46 amphibians, 42 fish and 143 butterflies—a tally that is, broadly speaking, half again as large as that of California with its 410,000 km^2.

Yet even these figures for animal totals could turn out to be gross underestimates. According to Erwin (1983), drawing on field research in Panama and Peru, there conceivably could be as many as 30 million insect species alone in tropical forests. Virtually all these species occur in the forest canopy, which represents the last great unexplored frontier of systematics and ecology.

Remarkably enough, and fortunately for conservation biologists, forest species, whether plant or animal, are far from evenly distributed throughout the biome. As we have seen, the wettest portions of tropical Latin America contain the greatest numbers, while the Zaire basin, with its expanse of roughly one million km^2, is relatively depauperate. In tropical Latin America, moreover, there appear to be a number of localities—at least 26 (Figure 1)—that feature unusual concentrations of species, many of them coinciding with the so-called Pleistocene refugia (Prance, 1982a). There are similar ultrarich localities in the Atlantic coastal forest of eastern Brazil; three apparent refuges in central Africa; the montane forests of East Africa; the eastern strip of forest in Madagascar; the forests of Sri Lanka, peninsular Malaysia, and northwestern Borneo; certain of the islands of Indonesia and the Philippines; and New Caledonia in the South Pacific. Certain of these localities, furthermore, feature exceptional levels of endemism. In Sri Lanka, 800 endemic plant species occur in only 2500 km^2 of primary forest, most of them in the uplands; in Madagascar, 80 percent of 10,000 forest plants are endemic; in the Choco forest of Colombia, probably one quarter of 8000-plus plants are endemic; and in New Caledonia, 80 percent of 3500 plants are endemic.

Let us bear in mind, furthermore, that each species represents a unique manifestation of life's diversity, with its own "genetic fingerprint." This means that, to cite Eisner (1983), "Extinction does not simply mean the loss of one volume from the library of nature. It means a loss of a loose-leaf book whose individual pages, were the species to survive, would remain available in perpetuity for selective transfer and improvement of other species."

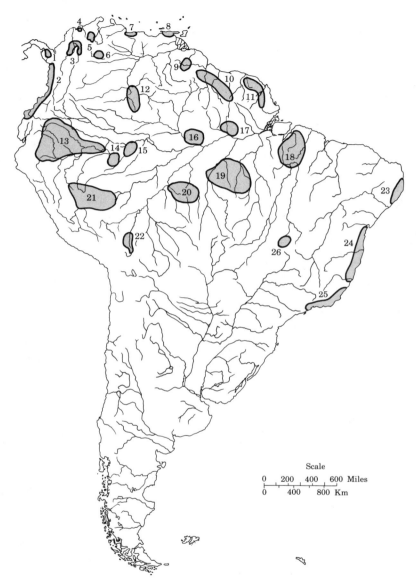

FIGURE 1. Twenty-six regions with exceptionally high levels of endemism in tropical South America. (From Prance, 1982a.)

STATUS OF TROPICAL FORESTS

Past depletion

As we have noted, there are probably no more than about 8.5 million km^2 of tropical forests remaining in primary or little disturbed form. How great was the expanse before it began to be significantly depleted by commercial logging, planned agriculture, slash-and-burn farming, and other severely disruptive human activities? Timber exploitation was making pronounced inroads into forests of the Indian subcontinent by the middle of the last century; rice cultivation eliminated much of the lowland forests of Burma, Thailand and Indochina by the start of the present century; and slash-and-burn farming had eliminated much forest in Madagascar, West Africa, the Indian subcontinent, and Java by the middle of this century.

How much forest has actually been accounted for is very difficult to determine. According to bioclimatic data, the extent of the humid tropics which could theoretically have supported moist forest totals as much as 16 million km^2—though it is likely that some sectors would have featured savannas rather than forests for edaphic reasons alone (as we see today in parts of eastern Amazonia, such as the area north of Santarem). In all events, we can probably give credence to the estimate that at least one-third of former tropical moist forests have already disappeared.

Present depletion rate

How fast are the forests being depleted today? This contentious and consequential issue has received much attention in recent years. Two major reports have recently appeared (FAO/UNEP, 1982; Myers, 1980, updated in Myers, 1984; for some comparative analyses, see Houghton et al., 1985; Melillo, 1985; and Molofsky et al., in press), which present systematized findings, much of them based on remote sensing surveys, which, whatever their shortcomings, certainly represent a major advance on previous inventorying methods, insofar as they present methodically and objectively derived data. The conclusion is that at least 76,000 and possibly as much as 92,000 km^2 of forest are being completely and permanently cleared each year and another 100,000 km^2 or so are being grossly disrupted. This means that around 1 percent of the biome is being destroyed each year, and rather more than 1 percent is being impoverished.

What are the main depletive processes in question? Roughly 25,000 km^2 are being destroyed or degraded through nonsustainable fuelwood gathering. Another 20,000 km^2 are eliminated through cattle

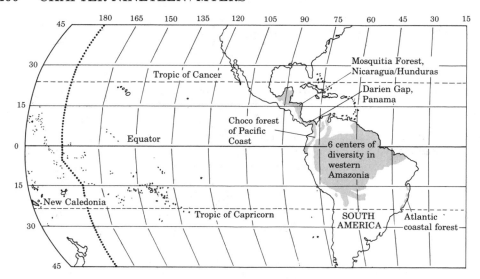

FIGURE 2. Tropical moist forests of the world, areas that feature exceptional concentrations of species and/or exceptional levels of endemism. These environments are greatly threatened. By the end of this century there will be virtually no primary forest left in the lowland territories of Asia, except for parts of Papua New Guinea; the same for East and West Africa and Madagascar; and the same again for Central America, the Choco Forest, the Atlantic coast forest of South America, and much of eastern and southern Amazonia. All that are likely to be left are two large blocs, one in equatorial Africa and the other in western Amazonia.

ranching. Another 45,000 km² are disrupted through commercial logging. The rest is due to the activities of small-scale cultivator communities, who total as many as 250 million persons. All of this means that by the end of the century there will be virtually no primary forest left in most tropical nations, at least in lowland territories. A detailed review is presented in Figure 2.

To give an idea of the rate at which deforestation can proceed, let us look at the case of Rondonia in southern Brazil, where agricultural settlement, in the form of landless migrants from other parts of Brazil, began to gather pace in the mid 1970s (Fearnside, 1985; Tucker et al., 1984; Wilson, 1985). In the total State, measuring 243,000 km², 1200 km² had been cleared by 1975, more than 10,000 km² by 1982, and as much as 16,000 km² by early 1985. During the second half of the 1970s, the population was growing at an average rate of 15.8 percent per year, from 111,000 to almost 500,000. Deforestation increased at an average annual rate of 37 percent. By mid 1985, the population surpassed one million.

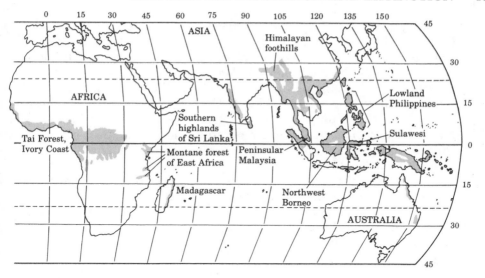

Depletion patterns in the future

As for the future, the rate of ranchland expansion in Latin America will surely slow down as governments, development banks, and other support agencies find that the production costs (let alone other costs such as environmental costs) are too great to allow a continuation of what amounts to "shifting ranching" (Hecht, 1981). So we can hope for a switch to enhanced management of existing ranches, which, through a modicum of upgraded supervision, could generate a sustainable increase in beef output. By contrast, the fuelwood problem is likely to persist, if not expand, due to the basic needs of vast throngs of impoverished peasantry. As for timber production, there is a move toward plantation forestry rather than natural-forest logging.

With respect to the most destructive agent of all, the small-scale farmer, it is difficult to see how his impact is going to be stemmed, let alone halted, without massive efforts undertaken forthwith to enhance conventional agriculture in areas already occupied by farmers. The aim must be to transform his cultivator lifestyle from extensive to intensive, from migratory to stabilized, and from wasteful to permanently productive. Otherwise, the remaining forests will be increasingly overrun by farmer populations that in many countries are increasing at rates of 4 percent and even 5 percent per year, due not only to natural increase but to in-migration (the phenomenon of the "shifted cultivator").[1]

[1] For details of the situation throughout the humid tropics, see Esman, 1979; FAO,

Many countries of the tropical moist forest biome already possess large populations in relation to available cultivable land, and these are the countries where the problem of forestland farmers is most pronounced, *viz.* the Philippines, Indonesia, Vietnam, Thailand, Bangladesh, India, Madagascar, most countries of West Africa, Colombia, Peru, Ecuador, and all countries of Central America. Unless these large-population countries produce alternatives to forestland farming for landless people, the numbers of forestland farmers will continue to grow at ever-increasing rates. Whereas population growth in the countries concerned is projected to produce an increase of almost 50 percent during the last two decades of this century, the number of forestland farmers could double.

As for the long-term future, we should note that none of the countries in question is projected to reach zero population growth, even with stepped-up family planning programs, until late in the next century, some of them not until early in the twenty-second century. (For some country-by-country projections of population growth, see Table 1.) One relatively hopeful projection is that Zaire, Gabon, and Congo, with a tropical forest area of 1.2 million km^2, are projected to grow from a 1980 total of only 30.5 million to an eventual total of "only" 168 million.

All this means that even if commercial logging, cattle ranching, and fuelwood gathering were to be contained forthwith, forest depletion will proceed progressively unless the issue of the small-scale cultivator can be resolved. Much can be done to assist this cause through agroforestry, intensive tree–crop cultivation, "garden farming" systems, and a Gene Revolution (Myers, 1985), but to accomplish this will require an international development effort of a nature and scale to match that of the Green Revolution. Since the farming communities in question rank among the poorest of the poor, they tend to be bypassed by conventional development processes.

SPECIES EXTINCTIONS

Susceptibility to extinction

As forest habitats disappear, so do species. This much is obvious. Not so apparent is what happens when the depletive process stops short of outright elimination of the forest, and a good number of trees remain standing. In these circumstances, the life-support systems of certain

1980; and International Labour Organization, 1977. For the specific situation in Southeast Asia, see Jones and Richter, 1981; in tropical Latin America, Barbira-Scazzocchio, 1980; and in Africa, Persson, 1977.

species can disappear just as surely as if the forest were entirely destroyed.

The latter problem is due to the ecological complexity—the intricacy of food webs and the like—of tropical forests. If, for example, a plant that serves as a significant food source for a crucial pollinator or seed disperser (a bat, a bird, an insect) is severely reduced by virtue of its value to, say, the commercial logger, the process can set in train a process of "linked extinctions" (Pimm, Chapter 14; Gilbert, 1980; Roughgarden, 1979; Start and Marshall, 1976).

A notable illustration is the durian–bat relationship in forests of the environs of Kuala Lumpur. The durian fruit is considered a delicacy, and is a commercial crop worth $35 million a year in Southeast Asia. The tree is pollinated by a single species of bat, *Eonycteris spelaea*, which appears to find its main source of nectar food in durian trees. The bat spends a good part of its day roosting in caves; large numbers of them occupy caves about 40 km from coastal mangrove swamps in which grows a particular flower, also favored by the bats. The swamps are being reclaimed for building land, which reduces a significant food source of the bats. Moreover, the caves have been long exploited for their limestone. In the wake of this double assault, bat populations have declined. Fortunately the blasting of the caves has recently been halted. But as long as the swamplands continue to be eliminated, the bat populations steadily fade away.

We can speculate that many other such "obligate" relationships exist between forest plants and bats, whether nectar-consuming or fruit-eating bats. Bats pollinate hundreds of genera of tropical trees and shrubs; and frugivorous bats are often as numerous as all birds, mammals, and other creatures that consume fruits combined. In many instances, a plant is pollinated or its seeds are dispersed by just a single species of bat. These bats, together with birds and insects, supply critical pollinator services that form part of larger food webs. In this vital sense, we can view the animals as "mobile link" species, and their plant hosts, who supply food to extensive associations of mobile links, as "keystone mutualists," to use the graphic terms of Gilbert (1980; see also Terborgh, Chapter 15).

If a keystone mutualist is eliminated as a result of human disturbance of forest ecosystems, the extinction of several other species will follow inevitably. Still more to the point, these additional losses may, in certain circumstances, trigger a cascade of linked extinctions. Eventually a series of forest food webs can become unravelled, with shatter effects throughout their ecosystems.

Moreover, a good number of forest species feature only localized distributions. A species that is confined to a relatively small area is all the more susceptible to summary extinction if its tract of forest is

TABLE 1. Population projections for some tropical forest countries.

Country	Population in 1980 (in millions)	Rural population (%)	Population in 2000 (in millions)	Year of net reproduction rate	Year of stationary population	Total of stationary population (in millions)
			LATIN AMERICA			
Brazil	118.7	39	176.5	2010–15	2075	281
Colombia	26.7	40	39.4	2005–10	2065	60.2
Costa Rica	2.2	59	3.3	2000–05	2065	4.8
Ecuador	8	58	13.6	2020–25	2085	28.4
Guatemala	7.2	64	12	2020–25	2085	24.2
Guyana	0.8	60	1.2	2000–05	2065	1.8
Honduras	3.7	69	6.7	2025–30	2090	16
Mexico	69.8	36	115	2010–15	2075	202.6
Nicaragua	2.6	51	4.7	2025–30	2090	10.8
Panama	1.8	50	2.8	2005–10	2070	4.3
Peru	17.4	38	27.3	2015–20	2080	48.8
Surinam	0.4	34	0.6	2005–10	2070	0.9
Venezuela	14.9	25	23.8	2005–10	2075	38.6

SOUTHERN/SOUTHEAST ASIA

Bangladesh	88.5	91	141	2030–35	2125	338.2
Burma	34.8	78	54	2025–30	2095	89.7
India	673.2	78	994.1	2015–20	2115	1,621.5
Indonesia	146.6	82	216	2015–20	2110	388.4
Malaysia	13.9	73	20.8	2000–10	2070	30
Papua New Guinea	3.1	87	4.4	2030–35	2125	9.4
Philippines	49.1	68	76.9	2010–15	2075	125
Thailand	46.9	87	68	2000–05	2070	102.5
Vietnam	54.2	78	87.9	2010–15	2075	153.4

AFRICA

Cameroon	8.5	71	14.2	2035–40	2110	40.8
Congo	1.5	60	2.7	2035–40	2100	10.5
Gaon	0.7	68	0.9	2035–40	2130	2
Ivory Coast	8.2	68	14.8	2035–40	2110	47.4
Madagascar	8.7	84	16.1	2035–40	2110	51.1
Zaire	28.3	70	51	2035–40	2110	156.1

(From Vu and Elwan, 1982.)

subject to gross disruption or destruction. We have already noted the high levels of endemism among plants (Gentry, Chapter 8). Much the same applies to many of their associated animal species. For example tropical birds, despite their exceptional dispersal capacities, feature high levels of endemism: some 450 species of land birds in South America, mostly in the tropical forest zone (about one quarter of all birds on the continent), have ranges of less than 50,000 km^2. This contrasts with only 8 bird species, or 2 percent of the total, which have similarly restricted ranges in North America (Terborgh and Winter, 1983).

Furthermore, many species of tropical forests exist at very low densities (Gentry, Chapter 8; Hubbell and Foster, Chapter 10; Gilbert, 1980; Janzen, 1975). This leaves them extremely vulnerable to sudden elimination. If a park is established in a tropical forest without expanse enough to account for long-term viability of its small-scale "ecologically fragile" species, the park might be unable to sustain a complete forest community, even though it totals hundreds of square kilometers. In tropical forests, then, we need to consider that a park should usually cover at least 1000 km^2, sometimes several times as much, to do its job.

Three illustrative areas

Let us now consider three geographic areas where deforestation is unusually extensive, and where large numbers of extinctions have surely occurred in the recent past—and where we must anticipate that there will be sizeable extinction spasms within the foreseeable future. These areas are: the Pacific coast forest of coastal Ecuador; the Atlantic coast strip of Brazil; and Madagascar. In each of these areas, so far as we can judge, there were at least 10,000 plant species before broadscale deforestation intervened, with endemism levels between 50 and 80 percent. We can reasonably assume that each area must have supported a complement of animal species that would bring their totals to some 200,000 species each, with similar levels of endemism overall (Raven, 1985).

In each of these areas, all too little primary forest now remains. In coastal Ecuador, primary forest was virtually undisturbed before the area was first penetrated by roads (around 1960), whereupon banana and oil palm plantations started to replace the forest rapidly. Within 10 years, the wet lowland forest had virtually disappeared except for a few small patches, one of which has been converted into the Rio Palenque Science Reserve. There, in a mere 1.7 km^2 of original forest, at least 1025 plant species exist, probably the highest recorded plant diversity in the world; one quarter of them are endemic to coastal

Ecuador (see Gentry, Chapter 8, for details; also see Dodson and Gentry, 1978; Gentry, 1982b). As many as 100 new plant species have been discovered in the Reserve; some of them exist in the form of a single individual plant, which means that extinction cannot be far away. How many species must have become extinct as a result of the 1960s deforestation is unknown, but must number many thousands at the very least.

Remnant scraps now constitute less than 10 percent of the original Atlantic coastal forest of Brazil, which formerly amounted to about one million km^2, extending from Rio Grande do Norte to Rio Grande do Sul in a strip ranging from just a few to 160 km wide (Mori et al., 1983). The forest has been heavily and repeatedly logged for centuries, and extensive sectors have completely given way to plantations of sugarcane, coffee and cocoa. One of the largest remaining sectors, in southern Bahia, is a mere 7000 km^2 or so, mostly disturbed. Well over half the tree species are endemic to the area, while another one-eighth are endemic to the area plus adjacent territories in Brazil and neighboring countries. Again, we can only surmise how many species have been lost, but the total must surely be many thousands.

In Madagascar, the eastern moist forest that may once have covered as much as 62,000 km^2 (and possibly far more) has been reduced to only 26,000 km^2 at most, at least half of which has been impoverished through shifting cultivation (Jolly et al., 1984; LeRoy, 1978; Rauh, 1979). Currently the rate of loss is about 3000 km^2 per year. The forests contain—or rather, they once contained—at least 10,000 flowering plants, of which 80 percent are or were endemic. Again, we have only a vague notion of how many species have already disappeared. But if there were originally 200,000 species of plants and animals together, more than half of them endemics, the loss must already total in the thousands. Of course, as in the cases of coastal Ecuador and coastal Brazil, these estimates for extinctions are rudimentary in the extreme, and they are advanced with the sole purpose of indicating the scale of species loss in the recent past.

Some further causes and consequences of extinction

In certain circumstances, deforestation may entrain climatic consequences that modify, if not impoverish, remaining forest. In Amazonia, for instance, more than half the basin's moisture remains within the ecosystem (Salati and Vose, 1984). This means that as the forest expanse is reduced, much more of the rain is lost to the system as runoff during the rainy season. Accordingly, much less remains to be transpired into the atmosphere. Therefore the climate in the remaining portion may well become drier, and the process could make for a steadily desiccating ecosystem—with all that implies for the survival prospects of remaining biotas. Further climatic dislocations for forest ecosystems could be engendered by the albedo effect (Henderson-Sellars, 1982; Pinker et al., 1980; Sagan et al., 1979). Additional meteorological disruptions are likely to follow in the wake of carbon dioxide buildup in the global atmosphere due in part to tropical deforestation (Woodwell et al., 1983). Both albedo and carbon dioxide induced changes may well prove more pronounced in areas outside the equatorial rain belt.

CONSERVATION STRATEGIES

By far the best way to safeguard biotic diversity is by the establishment of parks and other protected areas. The last decade or so has seen a regular outburst of protected areas in certain tropical forest countries. In 1972 there were only 120,000 km^2 of parks and reserves in tropical forests, whereas today there are more than 400,000 km^2. Of course, this still falls far short of the 10 percent figure that is advanced by some observers as the minimum necessary to safeguard the majority of tropical forest biotas (provided the conservation areas are established in select localities, covering distinct ecosystem types and centers of endemism).

In the Amazonia region, Venezuela, Peru, Ecuador, Bolivia, and especially Brazil have thus far designated about 110,000 km^2, an area about as large as Mississippi or Austria. Still more encouraging, Brazil entertains ambitious hopes of expanding its network year by year, in accord with a methodical plan that should take account of all eight distinctive plant zones in Brazil's sector of Amazonia, each with its characteristic faunal communities (Prance, 1982a; Wetterberg et al., 1976). All these localities contain exceptional concentrations of species and large numbers of endemics, making them worthy targets for priority conservation. Brazil has already established several large parks in Amazonia, one of which measures 10,000 km^2. At the same time, some of Brazil's parks are already being eroded because of inadequate legal protection and enforcement.

Let us also note some encouraging news from Indonesia, which accounts for roughly one-tenth of the world's tropical forest biome. Indonesia's forest parks and reserves now account for 112,000 km^2 out of a primary forest expanse covering no more than 750,000 km^2. Although not all the protected areas are parks (many are reserves, and some are conservation units), they amount to 6 percent of national territory, to be compared with a world average of only about 2 percent. Also in Southeast Asia, several new protected areas have been established in Malaysia and Thailand; we can say as much also for Zaire, Gabon, Congo, and Cameroon in Africa, and the record of Costa Rica is surely outstanding by any measure. Unfortunately, some of these protected areas suffer from extreme mismanagement.

Will future development continue along the current path of misuse and overuse of forest resources—in effect a "mining" operation that effectively eliminates the forests? Or will it be sustainable development that takes account of all the forest's many outputs, both material goods and environmental services, and that reaches beyond the few conventional products (such as timber or beef) whose production results in the severe degradation, if not the destruction, of forest ecosystems (Jordan, Chapter 20)?

Fortunately the World Resources Institute, in association with the World Bank, has put together a comprehensive and systematized Action Plan to confront the challenge of deforestation (World Resources Institute, 1985). The Plan proposes broad scale initiatives on each of five fronts: timber plantations; fuelwood plantations; watershed rehabilitation; conservation and protected areas; and research, education, and training. Supporting all of these initiatives will be a new push for enhanced agriculture, both within forest lands and far outside, as a forceful way to tackle the problems presented by the small-scale cultivator. The total bill for the Action Plan is an estimated $1.3 billion per year until the end of the century. Large as this sum sounds, it is being perceived as a sound investment by leaders in many developing and developed countries alike. And even if, during the start-up phase at least, it does not generate more than one-third of the funding proposed, this will still amount to a doubling of current expenditures on tropical forestry. Especially promising is the emphasis on protection forestry as opposed to production forestry. In my opinion, this is the most hopeful piece of news to have emerged about the tropical forest scene in the past several decades.

SUMMARY

Tropical moist forests are the Earth's richest biome—by far. They are also being depleted faster than any other biome—by far. In these forests, amounting to a mere one-fourteenth of Earth's land surface, there are surely occurring many more extinctions than in all other habitats put together. If we do not do a far better job of conservation in the next few decades, there could be an extinction spasm in these forests to match and even exceed the greatest extinction events of the prehistoric past. Fortunately, there are now signs that governments and international agencies are prepared to come to grips with the problem. If they do indeed prove ready to confront the challenge, they will need all the guidance from conservation biology that we can supply.

CHAPTER 20

LOCAL EFFECTS OF TROPICAL DEFORESTATION

Carl F. Jordan

In temperate zones, following deforestation and establishment of crops, pasture, or tree plantations, the annual yield of usable biomass is often higher than that of the undisturbed forest. Even if total net primary productivity is considered instead of only productivity useful to man, annual crop plants often have higher yields than native forests. Relatively high productivity on deforested temperate soils can continue for many years. Eventually, of course, without fertilization productivity will decline.

In contrast, in the humid tropics deforestation followed by cultivation results in only one or two years of high crop yields. The rate at which deforested cultivated soils in the humid tropics lose their fertility depends on the type of cultivation and the type of soil. Even in the better tropical soils, total net primary productivity and crop yields drop sharply after two or three years, to the point where cultivation is no longer worthwhile. Soils in the humid tropics usually are relatively infertile, and even fertile soils rapidly lose their nutrients following clearing.

Infertility of tropical soils results from heavy rains and biological processes which go on year-round, resulting in high annual rates of decomposition, leaching of soluble nutrients, and fixation of phosphorus (details of these processes have been given by Jordan, 1984, 1985). Rapid loss and fixation of nutrients could also mean that once disturbance ends, recovery of forests may be slow, or may not occur at all.

Not all disturbances of tropical forests are equally serious. Some disturbances affect potential for production and recovery very little, while others may have a devastating effect. The seriousness of the disturbance is not necessarily related to the involvement of man. Many of the natural disturbances that occur in tropical forests are more serious than those caused by some types of agriculture.

This chapter examines the various types of disturbances which frequently occur in tropical forests, discusses how each type of disturbance affects the potential productivity of a site (including the ability of the forest to recover), and describes the factors important in these processes.

CLASSIFICATION OF DISTURBANCES

The impact of a disturbance on the productive potential of a site and on the ability of the site to recover depends on the intensity, size, and duration of the disturbance.

Intensity

The intensity of a disturbance can be light to severe. A light disturbance is one which does not seriously disrupt the structure and functioning of the naturally occurring forest. A treefall gap is a disturbance of light intensity. In a moderate disturbance such as conversion of forest to cacao plantation, the structure of the forest is altered, but the soil is not seriously degraded. The intensity of disturbance is severe when the forest is destroyed and the soil is degraded, for example when forest removal is followed by agricultural practices that cause severe erosion.

Size

In small disturbances such as treefall gaps, the primary forest can quickly reestablish. Seeds from trees surrounding the clearing can fall directly into the opening. Mycorrhizal symbiosis—an interaction between roots which supply carbohydrates to fungi and fungi which increase the nutrient supply to the roots—reestablishes readily in the clearing through contact with roots of trees surrounding the gap. In contrast, seeds and mycorrhizal spores in intermediate and large disturbances must be transported in by vectors. A convenient criterion for distinguishing moderate from large scale disturbances is based on the movement of seed vectors such as birds and small mammals across the disturbance. In intermediate size disturbances such as most shifting cultivation sites, the vectors usually move freely into and across the openings. In large disturbances, however, the distance from undisturbed forest through the clearing is sufficient to inhibit frequent movement of the animals which carry seeds and mycorrhizal spores. Revegetation may be by windborn seeds such as those of many grasses and herbs.

Duration

A short disturbance occurs and is over in a matter of hours or days. Examples are hurricanes and typhoons. A disturbance of intermediate duration is exemplified by shifting cultivation; when the disturbance ceases, there is little which inhibits natural regeneration. A long-term disturbance is one in which the disruptive effects continue long beyond the cessation of the original disturbance, as occurs in overgrazed pasture. In such pastures, trampling by livestock compacts the soil and prevents the infiltration of water. Consequently, reestablishment of vegetation is inhibited, erosion begins, and degradation continues long after the cattle have been removed.

EXAMPLES OF DISTURBANCES AND THEIR EFFECTS

Light intensity, small size, short duration: Treefalls

Treefall is a natural phenomenon in all forests. As trees age, their growth rates slow, and they become more susceptible to disease or insect attack. Eventually the trees die, and if they fall over, they may pull down several nearby trees when the canopies are tied together by lianas. These treefalls result in an opening in the forest which is usually called a "gap" (Hartshorn, 1978; Whitmore, 1978). Although the falling trees may uproot and pull up some mineral soil, the size of the disruption is too small to affect the fertility of the soil and the productivity of the forest. In fact, formation of the gap may result in local increases in net primary productivity. The old tree may have been senile and slow growing. Saplings already growing in the gap may show high rates of growth due to increases in available light, nutrients, and water.

Light intensity, intermediate size, short duration: Windstorms

A windstorm or hurricane sometimes results in multiple treefalls in a forest. For example, in 1932, a severe hurricane affected large areas of the Luquillo Forest in eastern Puerto Rico. Many large trees were blown over. However, the productive capacity of the site and its ability to regenerate forest were not affected. A series of surveys over the years following the hurricane showed a rapid increase in basal area and volume of wood in the blow-down area (Crow, 1980).

On August 29, 1979, Hurricane "David" hit the island of Dominica in the Caribbean. The damage was the most extensive ever reported for a hurricane. Forty-two percent of the standing timber was severely damaged. Nevertheless, regrowth of a large number of species was rapid (Lugo et al., 1983). Although hurricanes appear destructive because large trees are knocked over, their impact on site productivity and forest regeneration is usually slight (unless they cause landslides). As with tree gaps, there may be an increase in growth rate in saplings remaining in the affected area. Also as in tree gaps, the fertility of the soil is not affected, nor is there destruction of the pool of seeds and mycorrhizae in the soil.

Light intensity, intermediate size, long duration: Shelterwood forestry

During the first half of the twentieth century, management systems for the naturally occurring tropical forests of Africa and Southeast Asia were developed (Baur, 1964). In Africa, the systems were often referred to as the "tropical shelterwood system," and in Southeast Asia as the "Malayan uniform system" (Fox, 1976). These systems were designed to promote the establishment, survival, and growth of seedlings and saplings of desirable species by poisoning undesirable trees and removing vines and weeds. After several years, when surveys showed that the reproduction of desirable species was well established, the canopy trees of the timber species were harvested. The tract was then monitored, and weedings and thinnings took place as needed. The system was abandoned in the 1960s, partly because it did not make sufficiently intensive use of the land to compete with other forms of land use such as cacao or agricultural crops (Lowe, 1977).

Apparently no nutrient cycling studies were made in forests under the shelterwood or uniform systems of management, where size of cut might have affected nutrient loss. A recent study in a Costa Rican rain forest (Parker, 1985) illustrates the importance of size of cut on nutrient loss. In cuts resembling natural gaps, nutrient leaching due to water percolating through the soil did not increase above rates in the undisturbed control forest. In a cut of 500 m^2, however, nutrient loss per m^2 was significantly greater than in the gaps. Losses per m^2 in the 500 m^2 cut were almost indistinguishable from loss rates in a 2500 m^2 cut, suggesting a threshhold effect between gap-sized disturbances and larger clearings. A threshhold effect could be related to the extent to which roots of trees surrounding the cuts are functioning beneath the cut area and taking up nutrients released by the cuts.

Despite nutrient losses in the larger cut areas, nutrient loss apparently was not great enough to inhibit recovery of the forest. Only 18 months after the cut, leaf area index, a measure of the total photosynthetic surface, was close to half the value of the mature forest (Parker, 1985).

Variable intensity: Global pollution

In the early 1960s, before the atmospheric test ban treaty, potentially dangerous amounts of radioactive fallout occurred throughout the world, including the tropics (Whicker and Schultz, 1982). In the 1960s, when DDT was used heavily in the United States and other developed countries, traces of this insecticide were found in remote corners of the globe (Woodwell, 1967). The effect of such global pollution depends on the amount of accumulation. There is no evidence that radioactive fallout affected the structure and function of tropical forests. Insecticides also have not reached critical levels on a global scale. However, there is the potential for higher intensity damage. For example, DDT is still used to control insect pests in cotton fields in the Pacific lowlands of Guatemala. Just to the east, in the altiplano of Guatemala, the severity of the pine bark beetle damage is worse than can be recalled by any of the natives. Local naturalists suspect that one cause may be a reproductive failure and decrease of birds that are predators on the beetles. DDT, which is known to have such effects, is wafted up the mountain slopes from the cotton fields to the west.

Moderate intensity, small size, short duration: Experimental disturbances

There have been a number of experimental disturbances of moderate intensity, small size, and short duration to evaluate various types of stress on a forest, and the recovery of the forest from that stress. Three such experiments are reviewed here.

Radiation experiment. Experimental irradiation of a rainforest was undertaken in 1965 to determine the effects of a predetermined and controlled amount of radiation on a tropical ecosystem (Odum, 1970). The experiment was sponsored by the U.S. Atomic Energy Commission as part of a program to predict the consequences of accidental or purposeful release of radiation into the environment. In the experiment, a 10,000 curie source of radioactive cesium was used to irradiate a lower montane rain forest in Puerto Rico for three months (Odum and Drewry, 1970). To the layman, classification of radiation as a disturbance of moderate intensity might seem an underestimate, but this is because people are accustomed to thinking of tests of atomic weapons which produce huge craters. Cesium 137 emits gamma radiation, which resembles visible light in that shielding the source after the experiment "shuts off" the radiation and there is no residual radiation in the environment. Irradiation was carried out from January through April 1965. The forest was killed for a radius of about 15 m from the radiation source.

Immediately after the radiation ceased, nutrient concentrations in the soil increased due to the pulse of nutrients from leaffall as the forest died. By September 1965, leaching caused nutrient levels to decrease below levels of the undisturbed forest (Edmisten, 1970). However, the nutrient loss apparently did not affect primary productivity or recovery. By 1982, only 17 years later, above ground tree biomass was already 60 tons/ha (Silander, 1984). If this rate of biomass accumulation were to continue, it would take about 65 years for the forest to recover to the 228 tons/ha of the undisturbed forest.

Manual clearing experiment. Harcombe (1977a,b) manually removed all the vegetation from a series of secondary forest plots in Costa Rica, and then allowed some to revegetate naturally while others were kept bare by continual weeding. Bare plots lost more nutrients than revegetating plots, but even in the latter there was some initial nutrient loss. In comparison, undisturbed forests usually have little or no net nutrient losses from surface soils. Despite nutrient losses from the experimental plots, net biomass accumulation in the vegetated plots was very high—1551 g per m^2 the first year. Harcombe concluded that any nutrient loss due to manual clearing did not affect productivity. Rapid vegetative growth in the Costa Rican study may have resulted from a high nutrient stock in the soils. The experiment was carried out in volcanically derived soil in a mid-elevation valley in Costa Rica, one of the most fertile areas in the tropics.

"Cut, burn, and abandon" experiment. Uhl and Jordan (1984) carried out a study of a 0.5 ha plot of Amazon rain forest that was cut, burned, and then immediately abandoned and allowed to recover. They found that although there was some leaching of nutrients, the quantity lost was small compared to the amount in the slash remaining after the burn. The first year after the burn, only 55 g per m^2 of aboveground biomass accumulated on the site, but this poor growth probably reflected high seedling mortality due to exceptionally dry weather immediately after the burn rather than any lack of nutrients or soil degradation. During the second year, over 900 g per m^2 accumulated,

while living trees in the control forest accumulated biomass at about 600 g per m^2. In the third through fifth years, biomass accumulation in the experimental plot continued to be higher than that in the control forest.

Experimental disturbances: Summary. None of the above three experiments appeared to have any effect on the productive capacity of the site, nor the ability of the forest to recover. However, it is important to point out that in the radiation experiment, and in the cut, burn and abandon experiment, a large quantity of slash was left on the soil which, as it decomposed, supplied nutrients to the regenerating forest. Presence of decomposing slash as a nutrient source plays an important role in maintaining productivity. Presence of slash which can serve as a nutrient source is a basic difference between disturbances of intermediate size and short duration already discussed, and the intermediate to long duration disturbances discussed next.

Moderate intensity, intermediate size, intermediate duration: Shifting cultivation

Shifting cultivation is a common disturbance in tropical forests. It is a disturbance of moderate intensity, since the original vegetation is killed, but the soil is not degraded to the point that the site cannot return to forest. In areas where population pressure is not high, shifting cultivation is usually an intermediate scale disturbance, since the peasant farmers ordinarily do not clear and plant more than a few hectares at a time. Shifting cultivation sites are usually cropped for two to three years. Following abandonment of cultivation, "fallow" vegetation usually colonizes the sites.

Shifting cultivation is carried out on many different types of soils in the tropics. In the lowland evergreen rainforest area of the Amazon Basin, shifting cultivation frequently occurs on the highly weathered and leached Oxisols, or Latasols as they previously have been called. In the mountainous areas of Central America, and on volcanically derived islands of Southeast Asia, shifting cultivation can occur on much richer Andosols. Relatively fertile soils also may occur in valleys at intermediate elevations on tropical mountainsides, and on alluvial flood plains.

Despite the large variation in fertility of tropical soils on which shifting cultivation is practiced, intensive cultivation and harvest rarely last more than three years (Nye and Greenland, 1960; Watters, 1971; Norman, 1979). Jordan (1985) has discussed the most probable reasons for the decline in crop productivity on the various soil types. On the Oxisols and Ultisols, which are highly leached and usually occur at lower altitudes, decline in productivity of shifting cultivation is usually due to the decline in availablility of soluble phosphorus in the soil. Iron and aluminum are present in high concentrations in these soils because of the removal of more readily leached elements by the intensive leaching which occurs in the wet tropics. Iron and aluminum react with and bind phosphate at the naturally occurring low values of soil pH. The high concentrations of soluble aluminum at low pH also can cause toxicity problems in crop plants. These problems of phosphorus insolubility and aluminum toxicity affect crop plants much more than native tree species, since most crop plants have been selected for high growth rates under optimal soil conditions. Native species are much better able to tolerate the nutrient scarcity and aluminum toxicity of acidic and nutrient-poor tropical soils.

If the pH of the soils can be raised, the aluminum becomes insoluble and precipitates, phosphorus is released, and both the phosphorus and aluminum problems for crop species are relieved. The pulse of ash following burning during shifting cultivation raises soil pH, and permits high productivity of annual food crops for a few years. After the ash has leached away, crop productivity declines, and cultivation is abandoned. Native fallow or secondary successional vegetation invades the area. The native vegetation is better able than crops to exploit the stocks of relatively insoluble nutrients which are in the soil. This is because successional species have larger root biomass, slower root uptake kinetics, longer life, greater storage capability, and more effective mycorrhizae than crop plants (Chapin, 1980).

The native successional vegetation gradually enriches the organic matter stock of the soil by taking up the relatively unavailable nutrients from the soil, incorporating the nutrients in plant biomass, and then shedding leaves, fine roots, and other litter. Nutrients in this litter are released in soluble form as soil organic matter decomposes; thus the soil organic matter acts as a fertilizer.

Even though the total content of phosphorus and other nutrients is much higher in volcanically derived Andosols, and in many alluvial soils of the tropics, crop productivity in these soils still only persists for about two or three years. The problem is basically the same as on the Oxisols. Iron and aluminum content is high, and soluble phosphate is rapidly bound (Sanchez, 1976).

Although cultivation is possible for only a few years in both the nutrient rich Andosols and the nutrient-poor Oxisols, there is nevertheless a large difference between the two soil types in the amount of agricultural exploitation which each can sustain. Due to the higher nutrient stocks in Andosols, the buildup of nutrients in the fallow vegetation is more rapid. Consequently, on Andosols, the fallow period necessary to restore the stock of available nutrients may only be a few years. In contrast, the necessary fallow period on nutrient-poor Oxisols of the Amazon Basin may be several decades or more (Jordan, 1985).

Restoration of nutrients during fallow periods also may be relatively rapid in intermediate level valleys (800 m to 2000 m) in mountain ranges of the tropics. Such valleys often have optimal conditions for plant productivity. Leaching of nutrients is not as intense as in the wetter lowland forests, and alluvium from higher slopes can enrich the soil. Many agricultural experiment stations in the tropics are located in such valleys, and results disseminated from such stations give the unfortunate impression that the productive potential of the entire tropics is high. In reality, such fertile valleys comprise only a very small proportion of the tropics.

At altitudes above mid-elevation, soil and climatic conditions change and phosphorus fixation no longer seems to be the limiting factor for crop production. The cooler temperatures, in combination with sometimes wetter and sometimes drier conditions, result in slow rates of organic matter decomposition. Because of slow decomposition, nitrogen is mineralized slowly, and the availability of nitrogen in inorganic form may be the factor which limits primary productivity (Jordan, 1985).

In addition to slash-and-burn agriculture, there are other types of disturbances in the tropics which can be considered of moderate intensity, small size, and intermediate duration. Most of these disturbances are management systems for the production of food and fiber crops. Weaver (1979) has recognized eight major categories of management.

1. Shifting cultivation. In shifting cultivation, an area is cut and burned and crops desirable for subsistence or exchange are cultivated for a few years. When productivity declines, the plots are abandoned and soil fertility is restored by the vegetation which naturally invades the site. This system is practical only where population density is low.

2. The corridor system. In this system, developed in densely populated regions of Africa, strips 10–50 m wide are cut through the forest and planted with crop species. Because the distance to the forest edge was never great, litter from the forest could contribute to the fertility of the soil and the trees would hold the soil, and prevent large-scale erosion. When the area was abandoned, seeds and mycorrhizae could quickly and easily move into the previously cultivated area. As pressure for continuous use of land for cash crops increased, the system was abandoned.

3. Taungya. This system was initiated on public lands in Burma in the 1860s. It is a cooperative system between local farmers who

plant food crops simultaneously with the timber species desired by the government's forestry operation. The farmer can get several crops before the tree canopy closes and annual cropping must be abandoned. Annual or perennial crops growing between tree seedlings hold nutrients and prevent erosion until the trees are large enough to stabilize the soil (Figure 1).

4. Tree intercropping. In this system, the economically desirable crop is often an understory tree species, such as coffee or cacao, planted

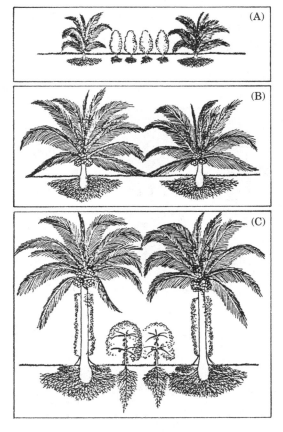

FIGURE 1. Schematic presentation of a Taungya type of cultivation using coconut palms as the dominant tree species. (A) Early phase, up to about 8 years; there is sufficient space between trees for planting of annual or short-term perennial crops. (B) Middle phase, about 8–25 years; there is greater ground cover by the palms, and no possibility for herb or shrub species. (C) Later phase, after about 25 years; as a canopy of palms is raised, there is opportunity for growth of understory species such as cacao beneath it. (Adapted from Nair, 1979.)

beneath larger trees which sustain soil fertility and provide shade. The same plot of land is used indefinitely.

5. Simulation of natural succession. In this system, species which establish naturally following intensive cultivation are replaced by analogous species with greater economic value. A sequence might consist of: (a) annuals and plants whose roots and stems are harvested; (b) bananas and plantains; (c) palms; and (d) cacao, rubber, or commercial timber species. Another sequence designed to include forage for swine production is shown in Figure 2.

6. Self-sufficient farms. These farms completely recycle nutrients within the management unit. Many of the units contain livestock, fruit and forest trees, home gardens, pasture, ponds stocked with fish, and fields of multiple crops.

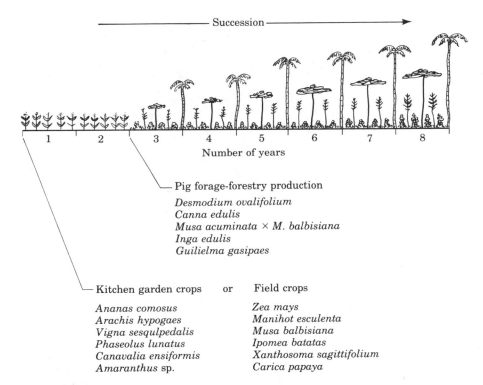

FIGURE 2. Representation of a managed successional mimic system that includes garden crops, field crops, pig forage, and forest species. (Adapted from Bishop, 1979.)

7. Scattered or row trees. Trees planted in rows between cultivated fields or along property boundaries help prevent erosion by wind and water. In some regions, nitrogen-fixing trees are planted in and around pastures to maintain the nitrogen supply for the grasses.
8. Forest blocs. Upper slopes and hilltops that are left undisturbed are beneficial for downslope cropped fields. They help prevent erosion, and leaf litter from the upslope forest improves the organic matter content of the cropland. In addition, when the cropland is abandoned, seeds from the upslope forest readily invade and forests can reestablish quickly.

These eight major categories of tropical land management subsume virtually hundreds of systems which differ in the crops planted, rotation time, cultivation methods, and many other details. Management systems in which trees are grown simultaneously with herbaceous crops are termed "agroforestry" or "agrisilviculture." Even within the general category of agroforestry, there are a wide variety of methods.

Disturbances of moderate intensity, variable size, and long duration: Savannazation

Savannazation is a long-term process in which the nature of the ecosystem is transformed from forest to grassland or savanna. Examples of both incipient and long-term savannazation are given here.

Incipient savannazation is occurring in South Vietnam and elsewhere. Prior to 1960, most of South Vietnam was covered with tropical moist forests. Between 1961 and 1971, during the Vietnam War, over 72×10^6 l of herbicides were sprayed from the air onto the forests of what was then South Vietnam (Westing, 1984). Damage to the forests depended on the number of times an area was sprayed. Several treatments usually were required to completely kill a forest. In areas where little or no disturbance followed the herbicide treatment, natural regeneration has occurred (Ashton, 1984). In many areas, however, peasant farmers moved into herbicide sprayed areas because leaf litter fall supplied a pulse of nutrients to the soil, and the defoliation allowed sunlight to penetrate the canopy. It was much easier to initiate cultivation in such areas than in areas where the trees had to be cut. Crops were planted directly beneath the decomposing trunks. When crop productivity declined, the fields were burned to encourage the growth of grasses for cattle (Ashton, 1984). Repeated burning gradually killed the remaining trees, and converted the area to grassland or savanna (Figure 3). Once dense grasses became established, recov-

FIGURE 3. Photograph of upland rice field taken in 1983, about 100 km north of Ho Chi Minh City (Saigon), Vietnam. The rice in the foreground is growing poorly, and the field will soon be abandoned. In the background are the remaining bare trunks of the original forest. On the hills are thickets of bamboo which will invade the field as soon as cultivation ceases. (Photograph by C. Jordan.)

ery of the forest was inhibited. One reason is the lack of a seed source. The second is that the grasses are so dense that they outcompete any tree seedlings which happen to become established. As a result, many areas of southern Vietnam are now densely covered with bamboo or relatively worthless grasses such as *Imperata* sp.

Long-term savannazation is occurring in much of Africa. Periodic fires in the Serengeti plains of Tanzania and Kenya are important in maintaining the grassland and the ungulates which graze there (Norton-Griffiths, 1979). In the absence of fire and grazing, the grassland and savanna would gradually be replaced by woodland. The burns kill tree seedlings which if allowed to grow would eventually shade out the grasses utilized by the grazers. Burning may also influence soil nutrients, and results in grasses which are more palatable. Burning, grazing, and subsequent savannazation in the Serengeti is an example of a case in which disturbance to tropical forests is viewed favorably by conservationists. Much higher densities of grazing ungulates can be sustained in this periodically disturbed habitat.

Disturbances of moderate intensity, large size, and long duration: The Jari project

One of the largest human-caused perturbations of tropical forest began in 1967 when billionaire industrialist Daniel Ludwig bought 1.6 million ha of virgin Brazilian rainforest to establish a pulpwood plantation (Fearnside and Rankin, 1982). This plantation, known as the Jari project, has been one of the most controversial forestry projects ever undertaken (Kinkead, 1981; *Time,* 1976, 1979, 1982). Early criticism of the project came from ecologists who warned that the infertile soils of the eastern Amazon Basin would limit productivity (Goodland and Irwin, 1975; Fearnside and Rankin, 1980, 1982).

Large-scale plantings of a broadleaf softwood, *Gmelina arborea,* commenced in 1969 on sites that were cleared by bulldozers. The slash was moved into long rows and burned. The *Gmelina* is reported to have grown well along the rows, but to have done poorly or failed in other areas (Greaves, 1979). Forest clearing by heavy machinery was abandoned for three reasons (Posey, 1980): it disturbed the already skimpy topsoil, it compacted the topsoil, and it was expensive. Heavy equipment was replaced by laborers with axes and chainsaws.

The best growth of *Gmelina* was reported to be about 14 tons/ha per year (Woessner, 1982). On plots studied intensively by Russell (1983), biomass accumulation was between 8 and 12 tons/ha per year. However, the average productivity in a large sample of plots throughout Jari was only 3 to 6 tons/ha per year (Schmidt, 1981). This is considerably less than *Gmelina* production rates in other regions of the world and is also below the average production rates of other pulpwood species (Wadsworth, 1983). *Gmelina* production at Jari was 40 percent below that originally projected (Kinkead, 1981), and as a result, trials with other species began. Many of the new sites were planted with species of pine and *Eucalyptus* (Posey, 1980). The best growth of pine was 12 to 16 tons/ha per year (Russell, 1983), but, as in the case of *Gmelina,* average production throughout the Jari plantation was low—less than 6 tons/ha per year after seven years (Schmidt, 1981). *Eucalyptus* showed better growth in some places, but there was considerable variation within the sites, and patches of very reduced growth were noticeable (Hartshorn, 1981).

Because of nutrients lost during the first rotation at Jari, it appears that wood production during the second rotation would be much lower if the project had been continued. In 1981, the project was sold at a loss of hundreds of millions of dollars (Fearnside and Rankin, 1982).

Disturbances of moderate intensity, large size, and long duration: Mayan agriculture

The history of the lowland forest areas of Guatemala and Mexico bears on the question of the long-term ability of long-disturbed ecosystems to recover. Most of the area in Guatemala which supported the ancient Mayan civilization is now covered by high forest, and much of the Mayan area in Mexico has been covered until recently. Less than 1000 years ago, these areas were intensively cultivated (Hammond, 1982; Wiley, 1982; Turner and Harrison, 1981; Matheny and Gurr, 1979). Although a decrease in soil fertility was probably only one of several factors that contributed to the decline of the Mayan civilization, the relevant point is that the existence of high forest today demonstrates its ability to regenerate even after large-scale, long-term disturbance.

High intensity disturbances: Soil removal

In disturbances of high intensity, not only is the structure of the ecosystem destroyed, but the soil is also severely affected. In the tropics, natural disturbances of high intensity may be more frequent than commonly realized. For example, in 1976, two earthquakes struck near the southeastern coast of Panama. Landslides associated with the tremors denuded about 545 km^2 of steep terrain originally covered by tropical rainforest (Garwood et al., 1979).

The time required for such areas to recover to the point where they have the structure of a forest depends on the physical and chemical nature of the exposed substrate. If there is any soil fine enough to hold water and nutrients, normal secondary succession can occur, and vegetation can reestablish quickly. This is probably what occurred following the landslide in Panama. On the other hand, if the substrate is rock or impervious clay, primary succession may have to occur before forest can reestablish. In primary succession, lichens, algae, and mosses colonize the bare surface. Respiration of these lower plants produces carbonic acid which hydrolyzes rocks and minerals, and provides nutrients for the roots of higher plants. The organic material of the lower plants themselves also provides substrate and nutrients for establishment of herbs, shrubs, and eventually trees.

High intensity disturbances: Pastures

The conversion of Amazon forests to pasture has been ranked the least desirable of all possible uses (Goodland, 1980) because of the land degradation that accompanies overgrazing. When too many cattle graze a tract of land, the vegetation cover is reduced, root biomass is decreased, and the soil becomes more susceptible to erosion. Trampling and compaction of the soil by too many cattle prevent the establishment and growth of grass. Water cannot easily infiltrate the soil, and when heavy rains occur runoff over the bare soil surface creates gullies, which often increase rapidly in size with each succeeding rainstorm (Denevan, 1981).

In dry areas, overgrazing can lead to desertization because cattle, goats, and sheep destroy the vegetation that holds the sandy soil in place (Le Houerou, Chapter 22). Approximately 65 million ha of previously productive land in the southern portion of the Sahara are estimated to have become desert during the last 50 years. In the Sudan, the desert is reported to have advanced 100 km in the 17 years prior to 1975 (Novikoff, 1983).

The degree to which conversion to pasture causes irreversible damage depends on the intensity of grazing. Buschbacher (1984) carried out a series of studies in the area of Paragominas, Brazil, in the eastern Amazonian region. In some areas following forest cutting and burning, pasture is established, and is weeded and burned every one to two years. Grazing pressure is intermediate (about one animal per ha). Abandonment occurs 6–8 years after pasture formation. Recovery of tree biomass is slow in these areas compared to areas which were cut and burned but not grazed. Nevertheless, it appeared that eventually a forest would reappear in these grazed areas.

At other sites in the Paragominas area, pasture management involves much more severe disturbance. After several cutting and burning treatments to control weeds, the areas are mechanically cleared. This entails removing all woody biomass including logs and standing trunks from the site. Subsequently, the sites are mechanically mowed and burned each year. Grazing pressure is heavy (greater than two animals per ha). Abandonment occurs after 10 or more years of use, but fires from actively grazed areas may escape into these areas and continue to suppress woody vegetation. In 1984, when the sites were studied, there was almost no establishment of trees (Buschbacher et al., 1984). The communities appeared to lack potential for growth and establishment of high forest, and the site may have become permanently altered into a heath-like community, resembling the heath communities of northern Europe. It is too soon, however, to conclude that highly degraded tropical pastures are permanently altered, and will never return to a community having the structure of a forest.

SUMMARY

Almost all disturbances to tropical forest ecosystems affect net primary productivity, and the ability of the site to recover to a forest community. The resilience of tropical rain forest structure however, is greater than is perhaps generally believed. Although productivity of intensely cultivated crop species may decline rapidly in converted rainforest ecosystems, the potential productivity of naturally occurring tree species, and the ability of the site to recover to a forest community, similar in structure to that which existed before cutting, is rarely completely compromised. Only when there has been extreme stress, as in overgrazed pastures, with subsequent mechanical clearing and periodic burning, is the question of structural recovery in doubt.

The eventual growth of trees, however, is not the same thing as reestablishing a tropical rain forest. The tropical rain forest is a highly diverse community, consisting of a large number of plant species and animal species which have coevolved with the plants. Reestablishment of the high species diversity and complex food webs of the predisturbance tropical forest may take much longer, or may never occur at all if disturbances have caused important extinctions.

SUGGESTED READING

Ecosystem studies of the type upon which this chapter is based are still relatively rare for the tropics. Much of the published work is still in dissertations, foreign publications, and other sources which are not readily available. However, a 1985 textbook pulls together over 600 recent references on ecosystem studies in the tropics and organizes them in a format suitable for advanced undergraduate courses and as a guide for graduate research.

Jordan, C.F. 1985. *Nutrient Cycling in Tropical Forest Ecosystems: Principles and Their Application in Management and Conservation.* Wiley, Chichester.

CHAPTER 21

CAVE FAUNAS

David C. Culver

Caves and cave faunas pose important and interesting problems for conservation biologists. It is generally not realized how ubiquitous caves are. Likewise, the variety of species that depend on caves during some critical time in their life cycle, such as the hibernation of bats, or that complete their life cycle only in caves, is impressive and usually underestimated.

Caves occur throughout the world, and major cave areas occur on all continents except Antarctica (Figure 1). Many islands contain caves as well; prominent examples are the extensive lava tube caves of the Hawaiian Islands and the marine-inundated caves of Bermuda and the Bahama Banks. The majority of caves are formed by the solution of limestone and similar rock such as dolomite, gypsum, and marble, but caves are also common in lava where they are formed when crusts develop over flowing lava. Caves are also found in ice and snow. Caves range in size from ones that one average sized human nearly fills up to the Flint Ridge–Mammoth Cave system in Kentucky with over 500 km of passages. Caves are also common and widely distributed. In the United States, caves are known in every state, and are very common in some states. For example, there are over 2300 caves in the state of Virginia.

Most caves contain a biologically interesting fauna. Twenty years ago, the prevailing view of cave faunas was that most cave-limited species were isolated in caves during the climatic changes occurring during the Pleistocene, such as warming trends that forced leaf-litter dwelling arthropods into caves. Accordingly, cave-limited species were thought to be rare in caves that were either covered by Pleistocene ice or located in the lowland tropics where, the influences of the Pleistocene were thought to be minimal. Work over the last two decades has changed this view. First, it is clear that cave-limited species

FIGURE 1. Major limestone cave areas of the world (shaded). (From Sweeting, 1973.)

occur in caves in glaciated areas. The most spectacular example of this are the cave-limited isopod and amphipod species found in Castleguard Cave in the mountains of Alberta (Holsinger et al., 1983). This cave, which is developed in limestone, ends in an active glacier. Second, cave-limited species are very abundant in tropical caves. The work of Howarth and his colleagues on the fauna of Hawaiian lava tubes has brought to the attention of the scientific community an incredible wealth of cave-limited species, including bugs, crickets, and crane flies (Howarth, 1973). Finally, due to the work of cave divers in the Bahamas and elsewhere, an ancient crustacean fauna has been found that clearly predates the Pleistocene by millions of years (Yager, 1981).

The biological importance of caves and the unique conservation problems they pose can only be understood by considering the interrelated problems of the nature of the fauna and the nature of the habitat. These will be the topics of the next two sections.

SPECIAL BIOLOGICAL FEATURES OF CAVES

A rather elaborate, specialized terminology is used to describe and categorize the fauna of caves. This terminology will be used sparingly in this chapter, but acquaintance with the rudiments of the terminology of cave biologists is necessary to pursue the matter further. The most widely used terms that characterize cave-dwelling species are "troglobite," "troglophile," and "trogloxene." A troglobitic species is one that is found only in caves; a troglophilic species is one that is found both in caves and in noncave habitats. A trogloxene is a species in which the individuals spend part of their life cycle in caves. The dependence on caves is obligate for troglobites and can also be so for trogloxenes. For example, the endangered gray bat, *Myotis grisescens,* is a trogloxene that has an obligate dependence on caves for both hibernation and maternity roosts (Tuttle, 1979; Brady et al., 1982).

Much more interesting than this rather obscure terminology is the fact that caves and cave faunas can serve as model systems for the study of a variety of geological and biological questions, including minerology, adaptation, speciation, regressive evolution, and species interactions (Poulson and White, 1969; Culver, 1982). As Poulson and White point out, caves are, in a sense, natural laboratories for the study of these and other problems.

The isolation of species in caves may have been due to a variety of causes, including climatic changes such as occurred in the interglacial periods of the Pleistocene, or changes in sea level (which may be especially important on oceanic islands). But whatever the causes of isolation, a series of evolutionary changes occur to isolated cave populations that make them of special interest to ecologists and evolutionary biologists, and that pose special problems for conservation biologists.

Once a population is isolated in a cave or a set of closely adjoining caves, population sizes are likely to be small. Many cave species are described from only one or two specimens, and extensive searching often reveals only a handful of individuals. A striking example of this phenomenon is provided by the cave pseudoscorpions of Appalachian caves: most of these species are known from less than ten individuals. The few mark–recapture studies that have been done on troglobites (reviewed by Culver, 1982) have resulted in population estimates in the hundreds and thousands. The largest recorded population of troglobites is 9090 *Orconectes inermis inermis* crayfish in Pless Cave, Indiana. Genetic variation may be less than normal for populations of such sizes because the population probably went through a bottleneck when originally isolated. In addition, migration between caves, especially for terrestrial species, seems rare, as judged by low levels of genetic similarity between populations of the same species (Laing et al., 1976) and by the large number of species endemic (limited) to a single cave. Single-cave endemism is especially common in regions of highly dissected limestones, such as the Valley and Ridge province in the eastern United States. The best studied case of speciation in cave animals is probably that of the carabid beetle genus *Pseudanophthalmus* in the eastern United States (Barr and Holsinger, 1985). In Virginia and the adjoining area of northeast Tennessee there are at least 47 species of *Pseudanophthalmus*, and 27 of these are known from single caves (Holsinger and Culver, 1985).

During the course of evolution in the cave environment, troglobites have experienced reduced capacity to withstand environmental fluctuations. Ahearn and Howarth (1982) document the diminished ability of Hawaiian lava tube troglobites to withstand desiccation, and Poulson (1964) summarizes data on the increased sensitivity of aquatic troglobites to temperature fluctuations. This loss of regulatory ability is an interesting question in its own right, but it also puts cave faunas at special risk from environmental perturbations.

Perhaps the main interest of cave fauna to the evolutionary biologist lies in the strongly convergent morphologies of cave-adapted organisms. This convergence is of two kinds. First, there is the widespread reduction and loss of eyes and pigments, commonly termed regressive evolution. Second, there is the widespread lengthening and increase in complexity of nonoptic sensory structures, such as antennae in arthropods and lateral line systems in fishes.

Caves are especially useful evolutionary laboratories because the selective pressures on cave organisms are easier to enumerate than for most habitats. These selective pressures include darkness with the accompanying problems of food gathering and mate location; generally low food supplies with the accompanying problem of starvation; and for terrestrial species, 100 percent relative humidity with the accompanying problem of water balance (Howarth, 1980; Culver, 1982).

Cave faunas are also useful ecological laboratories, especially in the study of species interactions (competition, predation, and mutualism). Cave communities are usually quite simple, with the number of potentially interacting species rarely exceeding four or five. Robert May (1976) has pointed out that theoretical ecologists, in common with the Australian Arunta tribe, count "one, two, many" in terms of models of species interactions. Since caves often have only three or four interacting species (Culver, 1982), cave communities provide an incentive to count "one, two, three, four, many." In a sense they occupy a place intermediate between a rigidly controlled laboratory environment with one or two species present, and the "world of light" with dozens of species present.

Because of the patchy nature of the cave environment, because of limited potential for dispersal or survival in surface habitats, and because of the importance of climatic changes in the Pleistocene for some cave faunas, biogeographic questions have often occupied center stage in cave biology. The best documentation of the effect of warming periods in the Pleistocene is Stewart Peck's study (1980) of terrestrial invertebrates in caves in the Grand Canyon. Using data from radiocarbon dating of cave rat middens and the taxonomic relationships and amount of morphological modification, Peck tentatively was able to assign the isolation of different cave rat species to different interglacial periods.

The general picture of terrestrial faunas in temperate areas is one in which most species have very limited distributions, often occurring in only a few caves. Those troglobitic species with widespread distributions, such as the cave collembolan *Pseudosinella hirsuta*, probably represent a cluster of species or incipient species (Christiansen and Culver, 1968), or are able to disperse through noncave subsurface habitats (Juberthie, 1983).

By contrast, aquatic troglobites tend to have broader ranges and occur in more caves than terrestrial troglobites. One example of this contrast is shown in Table 1, which summarizes data for terrestrial and aquatic invertebrates from Virginia caves. Levels of endemism are high for both groups, but it is especially striking that 42 percent of the terrestrial troglobites are known only from a single cave, whereas only 21 percent of the aquatic troglobites are known from a single cave. These data and others (Culver, 1982) suggest that not only are rates of movement of aquatic species greater, but indeed that there are other noncave subsurface habitats that are important for the aquatic cave species. This will be discussed in the next section.

Terrestrial and aquatic troglobites do share one most important characteristic: a nearly complete dependence on food carried in from

TABLE 1. Contrasts in endemism between aquatic and terrestrial troglobitic (cave-limited) invertebrates.[a]

		Number of troglobites	
		Aquatic	Terrestrial
Total number of species		43	97
Study area endemics:	N	30	79
	%	70	81
Basin endemics:	N	26	71
	%	60	73
Single-cave endemics:	N	9	41
	%	21	42

(Data from Holsinger and Culver, 1985)
[a] The caves studied were in Virginia and part of northeast Tennessee. Three scales of endemism are considered: (1) the entire study area, which comprises part of the drainage of seven major rivers—the Shenandoah, James, Roanoke, New, Clinch, Holston, and the Powell; (2) the drainage basin of each of the rivers; and (3) each of the approximately 450 caves that were sampled.

outside the cave. While there are chemosynthetic autotrophic bacteria in some caves, they are a minor source of energy. Food enters a cave in three main ways. Particulate organic matter, such as leaves and twigs, is carried in directly by streams. (This can be important for the terrestrial as well as the aquatic community because a layer of plant detritus is often left by receding floodwater.) A second source is dissolved organic matter, bacteria, and protozoa in water percolating into the cave through minute cracks and openings in the soluble rock. Finally, there are the feces (and eggs in the case of cave crickets) of animals that regularly enter and leave the cave. In some caves food may enter in more exotic ways, such as exudates from tree roots in Hawaiian lava tube caves (Howarth, 1973). With the exception of bat guano, which often harbors a fauna distinct from the rest of the cave fauna (Poulson, 1972), available food is generally scarce. Obviously, the quality and quantity of food in the cave can be adversely affected by changes on the surface, such as deforestation and water pollution.

Caves are no less critical for many species of bats than they are for troglobites. Unlike troglobites, bats routinely leave caves in search of food. Bats use caves as hibernacula, as maternity roosts, and as resting places during the day. Caves also serve as hiding places from the bats' predators. Of the 40 species of bats in the 48 contiguous states, 21 can be found with some regularity in caves during part of their life cycle (Tipton, in press). The species with the greatest dependence on caves is the endangered gray bat, *Myotis grisescens*. Its biology and status are described later in the chapter.

THE SUBSURFACE HABITAT

The International Speleological Union defines a cave as a natural cavity that is large enough for human access. While this definition is not entirely satisfactory, especially from a philosophical point of view, the number of caves so defined more or less corresponds to the number of habitable patches for terrestrial troglobites, and greatly exceeds the number of patches habitable by bats. But the actual number of caves, even in very well studied areas is difficult to predict. This is because there are many caves with no entrances. Curl (1966) points out that the formation of caves by solutional processes is independent of the formation of entrances, which are chance events whose probability of occurrence is related to the length of the cave. In a careful analysis of a series of cave regions, Curl concludes that the majority of caves in most areas have no entrances. The situation for aquatic subsurface habitats is more complex. Caves are not the only important subsurface habitat, and the aquatic cave habitat can be subdivided as well.

The word "karst," derived from the Slovene word for bare, stony ground, has come to mean a region where solutional features predominate, and which has underground drainage. Karst regions typically have caves and other solutional features such as sinkholes and blind valleys. The phrase "pseudokarst" is sometimes used to describe areas with features such as caves and sinkholes that are not the result of solutional process. Prominent examples of pseudokarst are some lava flows. Figure 2 is a schematic view of the major subdivisions of karst habitats and nonkarst subsurface habitats. In areas of high faunal diversity, each of the six habitats depicted in Figure 2 can harbor a more or less distinct assemblage of species, but most species can at

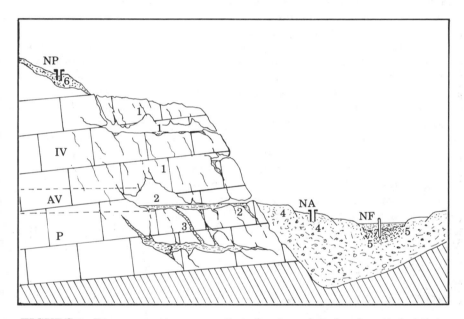

FIGURE 2. Diagrammatic cross-section of major subsurface karstic habitats. In caves formed by solution, there are three main zones: (1) inactive vadose (IV), a dry upper zone; (2) active vadose (AV), a periodically flooded zone; and (3) phreatic (P), a continuously flooded zone. Noncave subsurface habitats, which are numbered from 4 to 6 in the figure, are collectively called interstitial habitats in order to differentiate them from the karstic habitats. The major interstitial habitats are (4) *nappes phreatiques* (NA), the gravels along the sides of streams and rivers; (5) *nappes fluviales* (NF), the underflow of streams and rivers; and (6) *nappes perches* (NP), subsurface habitats above the water table. The parallel vertical lines for these three habitats represent the fact that they must be sampled indirectly, using various pumping devices (Bou and Rouch, 1967). (From Henry and Culver, 1982. Other more elaborate classifications of subsurface habitats are reviewed by Jurberthie, 1983.)

least survive in most of the six subsurface habitats (Culver, 1982). Of the noncave subsurface habitats, both the *nappes phreatique* and the *nappes fluviales* are more or less continuous, at least along stream courses. On the other hand, the *nappes perches* is extremely patchy and island-like. European biologists have been especially active in sampling these habitats (see Bou and Rouch, 1967) but they are largely unsampled in other areas. Consequently, some subterranean species known only from caves may in fact be primarily noncave species because the noncave subsurface habitats have not been sampled. Of the karstic habitats, both the active and inactive vadose zones are patchy and island-like, and indeed correspond to the accessible portion of caves. In contrast the phreatic zone, which is rarely accessible except to divers, is much less patchy.

Just as it is common to categorize surface streams into drainage basins, the same can be done with cave streams. In well-developed karst areas, there exist highly integrated underground drainage basins that often involve many caves, whose water exits from a single spring (Bögli, 1980). These underground basins may correspond to prominent surface features, for example, a limestone valley surrounded by impermeable rock such as shale and sandstone. This is the case with the karst ecosystem at Baget extensively studied by Rouch and his colleagues (summarized in Rouch, 1977). In other places, the underground drainage basin may not correspond at all to surface topography. Such is the case in Greenbrier County, West Virginia. Several karst drainage basins, ranging in size from 5 km^2 to 120 km^2 occur in the same river valley, and two of them partly lie one on top of the other (Jones, 1973). Whatever the local situation, the aquatic karst habitat is integrated into drainage basins and individual caves are not isolated from each other.

THREATS TO CAVE FAUNAS

Among the obvious threats to cave faunas are the following:

1. Caves are used for legal or illegal trash and garbage dumping (Iliffe et al., 1984, give a case study). Although relatively few animals can survive in a garbage heap, cave faunas are especially sensitive to these changes. The increased resource levels that result from dumping allow some species that are not usually found in caves to thrive and to outcompete the normal cave inhabitants. In many rural areas in the United States and elsewhere, a great deal of dumping occurs in caves and pits.
2. Inadequate sewage treatment in karst areas rapidly leads to pollution of the ground water and the fauna of karstic and other

subsurface habitats suffer as a result. One example (the pollution of a Virginia cave from septic tank overflow that was subsequently corrected) is discussed by Holsinger (1966). During the period when Banner's Corner Cave was polluted with septic tank overflow, the diversity of the aquatic fauna declined. In particular, the amphipod *Stygobromus mackini* disappeared, while the isopod *Caecidotea recurvata* increased at least tenfold in abundance. Since surface streams are rare in karst areas, pollutants seem to disappear. Of course, this disappearance is only an illusion.

Interstitial faunas are also very sensitive to pollution. Danielopol (1981), in a study of ostracods in wells in Greece, found that the specialized interstitial species were replaced by surface species in unprotected wells with large amounts of organic matter on the bottom and in wells from which large amounts of water were pumped. In fact, the ratio between surface species and interstitial (and karstic) species can used as an index of pollution.

3. Cave entrances are vulnerable to closure as land use patterns change. Entrances are bulldozed shut in housing developments, roads, and other construction activities. This is not merely a question of cutting off access to human visitation; it also profoundly affects the cave fauna. This is certainly the case for bats. An entrance closure will affect air circulation patterns and alter temperature patterns. Bats are incredibly sensitive to even the smallest microenvironmental changes, and the closure of an entrance, even if there are alternate entrances, often causes bats to abandon a cave (Tuttle and Stevenson, 1978). Terrestrial species are also greatly affected because major sources of food for troglobites come in through cave entrances.

4. Deforestation or changes in land use in general in a karst drainage system can have major consequences for water flow patterns. In general, there will be less buffering, and flooding will increase in frequency.

5. The use of pesticides in both agricultural areas and in some forested areas has a negative effect on bat populations. Most temperate zone bats that use caves are insectivores, and pesticide use results in population declines in some of these species (Brady et al., 1982).

6. The most irreversible threat is from mining and quarrying, which can completely destroy caves.

7. A very widespread problem, and one that has no easy answer, is human visitation to caves. The effect of human visitation on cave faunas ranges from obvious cases where youth groups amuse themselves by clubbing bats to death in caves, to what appear to be totally innocuous visits to a cave by a conservation-minded caver. This problem is discussed at length in the following section.

HUMAN VISITATION AND CAVE FAUNA

A very sobering view of human visitation is provided in Table 2, which documents the devastating impact of human visitation on the endangered gray bat, *Myotis grisescens*. As little as one visit per month to a cave over a period of five years has resulted in a 50 percent decline in the population of summer bat colonies in a series of caves in Kentucky and Tennessee. While some of these visits resulted in deliberate vandalism (Tuttle, 1979), even apparently innocuous visits can cause serious problems for bats, especially in hibernating colonies.

Myotis grisescens roosts in caves the year round. About 95 percent of the entire population hibernates in nine caves, all of which are deep, vertical caves with unusually cold winter temperatures (6°C to 11°C) for caves in the southeastern United States. These and other habitat requirements make most caves unsuitable hibernating sites (Tuttle and Stevenson, 1978). Upon arrival at a hibernating site, adults copulate and females immediately begin hibernation. Stored fat reserves must last 6–7 months, and during this time any disturbance can be catastrophic because of the energy reserves used by the bats when they are aroused (Stebbings, 1969; Tipton, 1982). A single visit by a human to a hibernating gray bat cave, with lights being directed at the bats, may cause the bats to abandon the cave, without any alternate habitat available (Tuttle and Stevenson, 1978). At the end of hibernation, the now pregnant females congregate in a maternity cave, which is usually the warmest one available. During the summer the gray bats occupy caves near water and forage for insects (Brady et al., 1982). Other bat species are less dependent on caves, but many of the 21 cave-associated species require caves for hibernation. The status of cave bats is indeed one of crisis.

Another case of unintentional harm being done to cave faunas by human visitation is in marine caves. Lucayan Cavern in Grand Bahama Island is a marine cave with a freshwater lens on top of the salt water. Such caves harbor a unique fauna, including a class of crustacean (Remipedia) found in no other habitat in the world. A distinct freshwater layer and a distinct saltwater layer is apparently a requirement for the survival of many of the species in this habitat. In Lucayan Cavern, one entrance, which opens into a large underground lake, was used by local dive shops for commerical tourist dives. This resulted in both a mixing of the water layers of the cave and a dramatic decrease in visibility. The subsequent decline in both the crustacean fauna and the bat population resulted in deeding of the entrance to the Bahamas National Trust, the establishment of a national park, and a temporary ban on all but research diving in the cave. This has

TABLE 2. Effect of human disturbance on summer colonies of the endangered gray bat, *Myotis grisescens*.

Disturbance category[a]	Number of gray bats		Percentage decrease
	1968–1970	1976	
< 1 visit per month			
1	10,200	10,200	0.0
2	6,100	6,200	−1.6
3	3,900	4,000	−2.6
4	12,200	12,200	0.0
5	19,000	18,700	1.6
6	10,000	8,000	20.0
7	12,200	9,700	20.5
Mean	10,514	9,857	5.4
1 visit per month			
8	10,200	6,500	36.3
9	15,600	9,200	41.0
10	46,200	18,900	59.1
Mean	24,000	11,533	45.5
2–4 visits per month			
11	174,700	127,500	27.0
12	32,500	18,500	43.1
13	12,200	6,100	50.0
14	19,100	8,700	54.4
15	26,200	9,100	65.3
16	31,100	9,000	71.1
17	9,400	0	100.0
Mean	43,600	22,700	58.7
> 4 visits per month			
18	28,600	4,100	85.7
19	111,400	5,100	95.4
20	13,600	1,900	86.0
Mean	51,200	3,700	89.0

(Date from Tuttle, 1979)

[a] Data gathered from 20 caves in Alabama and Tennessee. The colonies are divided into four groups based on the average number of visits to the cave per month in the summer. An analysis of variance of the data based on the percentage decrease from 1970 to 1976 indicated significant difference between the groups ($F = 22.7$, $df = 3.16$, $P < 0.01$), and the differences among the four groups account for 80 percent of the overall variance.

resulted in an increase in the crustacean fauna and the bat fauna, and apparently no irreversible damage had been done.

Other cave faunas are probably less adversely affected by human visitation, but two other problems are worth noting. Illumination of cave passages, a prerequisite for any commercialization, almost completely eliminates the cave fauna in that area, due to competition from surface species that can grow in the cave, behavioral avoidance of light, and perhaps the directly harmful effects of light on the nearly transparent cave animals. Recreational cavers (and cave scientists for that matter) often use acetylene produced by the reaction between water and calcium carbide as a light source. The by-product, calcium hydroxide, is very toxic. The backpacker's adage—pack it in, pack it out—also holds in caves.

The various IUCN Red Data Books provide a useful worldwide perspective on the dangers posed to cave and karst faunas, and this information is summarized in Table 3 (see also Figure 3). Many non-cave subsurface species not listed in Table 3 are threatened or endangered as well, including various species of desert pupfish in the genus *Cyprinidon*. While the species listed in Table 3 are certainly only a tiny fraction of the threatened cave species, it does provide a useful overview of the problems. Two kinds of threats are especially common. Groundwater pollution and overuse are the major short-term and long-term threats to endangered, rare, and vulnerable cave species. Second, an appalling number of species are threatened by overcollecting. The well-worn phrase "no collecting except for scientific purposes" is no longer sufficient. Every museum need not have a specimen of every species of cave fish; every cave ecologist need not examine the gut contents of cave salamanders. If these practices continue, the scientific community may be directly responsible for some extinctions.

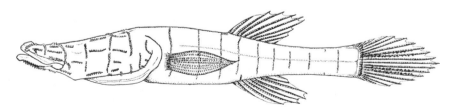

FIGURE 3. *Speoplatyrhinus poulsoni,* a cavefish found in Alabama. The species is threatened by groundwater pollution from agricultural runoff. (Drawing by J.E. Cooper.)

THE CONSERVATION BIOLOGY OF BATS

The problems associated with the conservation, preservation, and protection of bats and their habitats are distinct enough from other components of the cave fauna that they warrant a separate discussion. Education must play a major role in the conservation of bats and bat caves. As the pamphlets of Bat Conservation International point out, bats have received bad press. It is probably fair to say that even among conservationists, at least until very recently, bats and the caves they inhabit were not high on the priority list of species and habitats to be protected. In a paper that deserves wide distribution, Tuttle and Kern (1981) debunk the myths surrounding the negative effects of bats on human public health, especially the erroneous claims that bats are asymptomatic carriers of rabies.

The gray bat, *Myotis grisescens,* is probably the most endangered of all the American bats, and certainly the one most dependent on caves. The recovery plan for this species (Tuttle, 1979; Brady et al., 1982) can serve as a general model for other threatened species. The immediate objective is to reduce human disturbance in bat caves. Tuttle suggests the following three types of caves be given the highest priority to receive immediate protection:

1. Primary hibernating caves (those occupied now or in the past by more than 50,000 bats).
2. Primary maternity caves (those occupied now or in the past by more than 50,000 bats).
3. Primary bachelor caves (those used now or in the past by more than 50,000 male and nonreproductive female bats).

The total number of caves involved is relatively small. For example, there are only nine primary hibernating caves. The standard method of reducing human disturbance has been to install gates. Unfortunately, some gates have done more harm than good, either by altering the cave microclimate so that it is unsuitable for bats, or by subjecting the bats to high predation levels because of the difficulty bats have in getting through some gates (Tuttle, 1977). However, suitable designs for a cave gate that does not adversely affect bats are now available (Tuttle, 1977, Brady et al., 1982). In other caves where the threat of human disturbance is less severe, the entrance is posted with a sign asking people to keep out during the critical period, for example during winter for a hibernating site.

In some instances, gating or closing caves for hibernating or maternity colonies of bats has proved to be quite controversial within the

TABLE 3. Endangered, rare, and vulnerable cave species.

Species	Common name	Known localities	Status[a]	Threats
PHYLUM: ARTHROPODA CLASS: ARACHNIDA				
Troglohyphantes gracilis T. similis T. spinipes	Kocevje subterranean spiders	Caves near Kocevje, Slovenia, Yugoslavia	R	Industrial development
Adelocosa anops	No-eyed, big-eyed wolf spider	Koloa Cave and one other lava tube; Kauai, Hawaii	E	Groundwater pollution; withdrawal due to tourism, development
Banksula melones	Melones Cave harvestman	16 caves in Tuolumne County and 2 caves in Calaveras County, Calif.	V	Flooding from New Melones dam
PHYLUM: ARTHROPODA CLASS: CRUSTACEA				
Antrolana lira	Madison Cave isopod	Madison's Cave & Steger's Fissure, Augusta County, Va.	V	Cave visitors; industrial groundwater pollution
Mexistenasellus wilkensi M. parzefalli	Mexican stenasellids	Cueva del Huisache San-Luis Potosi, Mexico	R	Cave visitors; groundwater pollution
Thermosphaeroma thermophilum	Socorro isopod	Sedillo Spring, Socorro County, N.M.	E	Groundwater pollution; overuse
PHYLUM: ARTHROPODA CLASS: INSECTA				
Grylloblatta chirurgica	Mount St. Helens grylloblattid	Lava tubes on south slope of Mt. St. Helens, Wash.	V	Cave visitors
PHYLUM: CHORDATA CLASS: OSTEICHTHYES				
Speoplatyrhinus poulsoni	Alabama cavefish	Cave in Madison County, Ala.	V	Groundwater pollution from agriculture runoff

TABLE 3. (*continued*)

Species	Common name	Known localities	Status[a]	Threats
Amblyopsis rosae	Ozark cavefish	Caves in SW Mo. and NW Ark.	R	Overcollecting
Prietella phreatophila	Mexican blind catfish	Well in Muzquiz, Coahuila, Mexico	E	Groundwater pollution; overcollecting
Satan eurystomus Trogloglanis pattersoni	Texas blind catfish	Artesian wells, Bexar County, Tex.	R	Groundwater pollution
PHYLUM: CHORDATA CLASS: AMPHIBIA				
Typhlomolge rathbuni	Texas blind salamander	Caves and deep well, Hayes County, Tex.	E	Groundwater pollution; draining of wells
Proteus anguinus	Olm	Caves in Yugoslavia	V	Groundwater pollution; overcollecting
PHYLUM: CHORDATA CLASS: MAMMALIA				
Macroderma gigas	Ghost bat	Caves and mines in northern Australia	V	Human disturbance; quarrying
Myotis sodalis	Social bat	Caves in eastern U.S.A.	V	Human disturbance
Myotis grisescens	Gray bat	Caves in southeastern U.S.A.	E	Human disturbance
Plecotus townsendii ingens	Ozark big-eared bat	Caves in Ark., Okla., Mo.	E	Human disturbance
Plecotus townsendii virgineanus	Virginia big-eared bat	Caves in Ky., Va., W. Va.	E	Human disturbance

(Data from IUCN Red Data Books: Miller, 1969; Honegger, 1979; Thornback and Jenkins, 1982; Wells et al., 1983)

[a] IUCN definitions are:

Endangered (E): Taxa in danger of extinction, whose survival is unlikely if the causal factors continue to operate.

Vulnerable (V): Taxa likely to move into the endangered category in the near future if the causal factors continue to operate.

Rare (R): Taxa with small world populations that are not at present endangered or vulnerable but are at risk.

caving community. Trout Cave is one of several caves in West Virginia that were purchased by the National Speleological Society through a fund raising drive directed at its general membership.[1] Trout Cave is a very popular recreational cave in an area where several other caves have been closed either by land owners or to protect hibernating bat colonies. Trout Cave itself harbors a small hibernating colony of the endangered social bat, *Myotis sodalis,* and there is considerable evidence that it harbored a considerably larger colony in the past. There was a lengthy and acrimonious controversy within the National Speleological Society concerning what the Society should do. The compromise between sport cavers and conservationists was that Trout Cave would be closed during the winter months for one year, after which both environmental and population data will be assessed to determine its importance for *Myotis sodalis.*

The controversy points up two important problems that need resolution. First, biologists must establish which caves are essential and which are not for the viability of local bat populations. Not every cave that has any endangered bats in it should be closed. In the case of the large hibernacula of *Myotis grisescens,* there is no doubt from any reasonable point of view that access should be prohibited. Such caves are critical to the survival of the species, represent a tiny fraction of known caves, and are in general a rather unpleasant environment for humans to be in. On the other hand, the endangered subspecies of big-eared bat, *Plecotus towsendii virginiana,* occurs in much smaller colonies; there are numerous reports of single individuals being reported in caves (Bagley, 1984). It is unreasonable to close all of these caves.

Second, it is most important that all groups of people interested in caves agree on a conservation policy. A split between recreational cavers and cave conservationists would be a disaster, and it is a disaster that seems to loom on the horizon.

SUMMARY

Education must play an important role in the conservation of caves and cave faunas in general. In previous sections, I have tried to summarize some of the reasons ecologists and evolutionary biologists find caves so interesting. Caves have other values. Some caves contain important anthropological sites, including the famous cave drawings in France (Jackson, 1982). Finally, caves offer the best opportunity

---

[1] The National Speleological Society is the largest organization of cavers in the United States. Its membership of over 5000 is largely recreational cavers, but includes a considerable number of biologists and geologists as well.

for an adventurer to actually walk (or crawl as the case may be) where no one has gone before.

The comprehensive protection of karst areas is essential if cave faunas are to be saved. Aquatic faunas are especially vulnerable to groundwater pollution, but ultimately groundwater pollution will affect the terrestrial fauna (including bats) as well. Unfortunately, there has been little effort to protect karst drainage basins, and nearly all of the efforts of conservation organizations and other concerned agencies (including the Nature Conservancy, the National Speleological Society, and the U.S. Fish and Wildlife Service Office of Endangered Species) have been directed toward the protection of cave entrances by purchase and/or gating. This is only a first step and unless it is accompanied by protection of the above-ground habitat, it may all come to naught. Until the karst area is protected, there can be no insurance that even a gated entrance will, in the long run, do any good.

SUGGESTED READINGS

Bögli, A. 1980. *Karst Hydrology and Physical Speleology.* (J. Schmid, trans.). Springer-Verlag, Berlin. The best technical introduction to cave hydrology and geology.

Culver, D.C. 1982. *Cave Life: Evolution and Ecology.* Harvard Univ. Press, Cambridge, MA. A synthesis of a variety of problems concerning the evolutionary ecology of cave faunas.

Jackson, D.D. 1982. *Underground Worlds.* Time–Life Books, Alexandria, VA. A beautifully illustrated introduction to caves and cave faunas on a nontechnical level.

McClurg, D. 1980. *Exploring Caves. A Guide to the Underground Wilderness.* Stackpole Books, Harrisburg, PA. A thorough discussion of the specialized equipment and techniques used in cave exploration.

Tuttle, M.D. 1979. Status, causes of decline, and management of endangered gray bats. *J. Wildlife Manag. 43:* 1–17. A superb introduction both to the problems of cave bats and their conservation.

Tuttle, M.D., and S.J. Kern. 1981. Bats and public health. *Milwaukee Public Mus. Contrib. Biol. Geol.,* No. 48. An indispensable guide to bats and public health.

CHAPTER 22

CONSERVATION VERSUS

DESERTIZATION IN AFRICAN

ARID LANDS

Henri Nöel Le Houérou and Hubert Gillet

Arid lands, broadly speaking, are distributed in Africa on each side of the 30° parallels in both hemispheres, as well as in the region of the "Horn." Some 23 out of 40 countries in continental Africa contain arid lands. Some of these countries are totally arid, such as Mauritania and Somalia (Figure 1).

Arid lands are defined here as those that are unsuitable for permanent and sustained rainfed farming (Meigs, 1953). Sustained and dependable rainfed cropping is usually defined as a 75–80 percent probability of harvesting the most drought-tolerant traditional food crop, for example, barley in Mediterranean Africa and pearl millet in the tropics. Crop failure for these two crops usually occurs when the total year's precipitation falls below 250–300 mm and 350–400 mm, respectively. The mean annual precipitation levels that insure against crop failure for the two crops in four out of five years are 350–400 mm and 500–550 mm, respectively (Hargreaves, 1977).

Arid lands in Africa are roughly delineated by the 400 mm annual isohyet (precipitation contour) to the north of the Sahara Desert and 600 mm to the south of it. These two limits also correspond to clear-cut zones of natural vegetation. The 400 mm annual isohyet corresponds almost exactly with the northernmost extent of steppe vegetation in northern Africa (Le Houérou, 1969); the 600 mm isohyet is

Constraints on space required the deletion of 60 references from the original text.

444

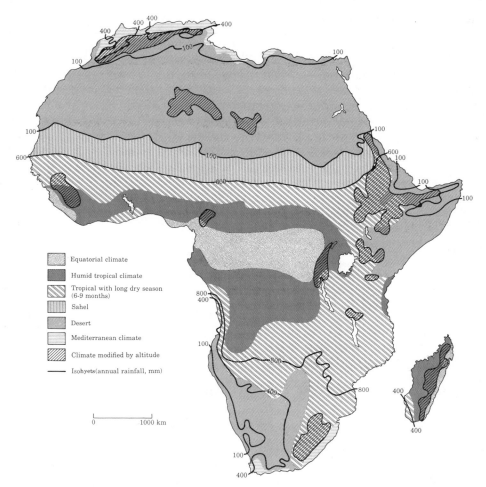

FIGURE 1. Major climatic regions of Africa. (Only selected isohyets are indicated.)

approximately the border between the Sahelian and the Sudanian ecological zones north of the Equator (Le Houérou, 1976c; Boudet, 1975, 1977). Such agreement between crop ecology data and the distribution of natural ecosystems is worth mentioning. The distribution of arid lands in Africa is shown in Table 1.

This does not imply, of course, that rainfed cropping is impossible at lower rainfalls. There is an ever-growing acreage of "gamble cropping" and a continuous encroachment of cultivation over rangelands. In many areas occasional cropping occurs in regions where the probability of a decent harvest is only 20 or 25 percent or less. A "decent

TABLE 1. Distribution of arid lands in continental Africa.[a]

Ecoclimatic zones (km²)	Rainfall belts (in mm)			Total 0–600	Percentage of continent
	Semi-arid 400–600	Arid 100–400	Hyper-arid 0–100		
Tropical[b]	1,284,028	2,022,883	3,240,347	6,547,258	21.60
Equatorial[c]	540,598	459,446	423,280	1,423,324	4.69
Mediterranean[d]	247,260	497,620	5,104,050	5,848,930	19.30
TOTAL	2,071,886	2,979,949	8,767,677	13,819,512	45.59
Percentage of continent	6.83	9.83	28.93	45.59	—

[a]For detailed information on ecoclimatic zonation by country, see Le Houérou and Popov, 1981.
[b]Tropical = summer rains, monomodal regime, one dry season.
[c]Equatorial = summer rains, bimodal regime, two dry seasons.
[d]Mediterranean = winter rains, bimodal/monomodal regime, one dry season, one cool season.

harvest" is when the market value of the crop is superior to the cost of cultivation and harvesting (this cost may be very low for family farming and seeds).

THE INITIATION OF DESERTIZATION

Causes

Desertization has been defined as the expansion of desert-like conditions and landscapes to areas where they should not occur climatically (Le Houérou, 1968) or where they did not occur in historical times. Desert landscapes and conditions include extensive pebble-strewn bare pavements, bare sand sheets, sand veils, barkhans fields, sand massifs and sand seas, bare-rock structural plateaus (Hammadas), very sparse perennial plant cover (less than 5 percent), and a very low primary productivity (less than 400 kg of dry matter/ha per year) and rain use efficiency usually below 1.0 kg dry matter per ha/year for each mm of rain (Le Houérou and Hoste, 1977; Le Houérou, 1984; Le Houérou et al., 1984).

Desert encroachment in Africa is not a result of a natural worsening of climate; detailed statistical analyses of rainfall and temperature records from many arid-zone weather stations having 100–150 years of data (the first weather stations were set up in Africa in the 1830s) do not allow any firm conclusions regarding long-term trends nor for predictable cycles. There are indeed clusters of years with rainfall above and below long-term averages, but these clusters occur randomly and are unpredictable. The same conclusions have been drawn for arid zones in other continents as well. The study of river discharge and lake levels confirms the conclusions drawn from weather records. Furthermore, investigations in many disciplines (such as geology, paleontology, palinology, prehistory, archaeology, and history) suggest that the climate remained stable in Africa for the past 3000 years, albeit with fluctuations of various lengths (Daveau, 1969; Toupet, 1976; Maley, 1981; Nicholson, 1978, 1981). Nevertheless, the length of the current drought is unprecedented since weather recording began, which is fueling speculation among climatologists that desertization itself is having an effect on the climate.

Desertization is essentially the result of human impact upon vulnerable ecosystems. This impact is worsened by temporary climatic crises, especially droughts that occur periodically several times per century. The impact may be so great that the ensuing environmental deterioration becomes irreversible (Le Houérou, 1968; Floret and Le Floch, 1973). These impacts result from the following activities.

1. *Cultivation of unsuitable terrain and soils that are excessively arid, too sandy, too stony, too shallow, too saline or too steep.* In African arid zones, agricultural crops are expanding over rangelands at approximately the same rate as demographic growth (2.5–3.0 percent per year; Le Houérou, 1973; Floret and Le Floch, 1973; Le Houérou, 1979). Indisputable evidence for this is provided by aerial photos and satellite imagery (Schwaar, 1965; Le Houérou, 1969; Long et al., 1978; Gaston, 1981; De Wispelaere and Toutain, 1976, 1981; Barral et al., 1983; Haywood, 1981). Because of the exponential population growth, lands less and less suitable are cultivated, initiating a self-catalyzing process. The result is lower and lower yields, which in turn necessitate larger and larger acreages under cultivation. The least arid rangelands are cleared first; therefore the pressure of livestock on the remaining mediocre pastures increases exponentially (Le Houérou, 1979). As population grows the traditional period of fallow is reduced or suppressed, fertility and therefore yields decrease, and more land is needed to produce a given amount of grain. The obvious way to break this vicious cycle is chemical fertilization, but this solution is impractical for technical and economic reasons.

2. *Cultivation practices that exacerbate wind erosion.* These include the use of disc ploughs in sandy soils, clear fallowing throughout the dry season, and the use of short-cycle annual crops without wind breaks.

3. *Excessive removal of woody plants for fuel.* Fuel removal causes soil denudation, wind and water erosion, and barren land. Except for cow dung, wood is often the only fuel available. The minimal fuel requirement is 1.5 kg per person per day, and since the rural population in the African arid zone is about 30 million persons, the consumption is about 16.5 million metric tons per year, equivalent to the woody biomass produced over some 200,000 km^2 of arid steppe or savanna each year. This is a potent factor of desertization. In addition, many urban dwellers rely on this type of fuel. Destructive woodcutting extends in concentric circles of 50–100 km radius around large towns such as Bamako, Niamey, N'djamena, Ougadougou, Khartoum, Nouakchott, and Mogasdishu.

4. *Faulty irrigation practices.* Large areas are sterilized by the use of excess water, or water that is too saline, especially on soils that are too clay-laden or have poor drainage, particularly north of the Sahara and in the Sahara itself. The fact that the situation is much worse in the Near and Middle East provides little solace.

5. *Inappropriate water and livestock management policies.* The digging of deep boreholes discharging several liters of water per second, without the enforcement of any range management or land use policy has too often resulted in large concentrations of livestock (20,000 to 40,000 head) during the dry season. This destroys the range in the vicinity of the well over a radius of 20 km (125,000 ha) in one or two seasons (the stock rate being 10–15 times the carrying capacity; Bernus, 1971).

6. *Careless mineral resource surveys.* The use of heavy vehicles for mineral exploration for ores or oil may destroy vegetation and disturb top soils. Locally, this may be a significant cause of desertization.

7. *Unregulated tourism development* may also be a local source of ecological trouble.

8. *Overstocking, overgrazing and poor range management.* These practices frequently result in the reduction of perennial cover, biomass, and productivity to less than one-tenth of their initial values. They are extremely harmful and warrant discussion at greater length in the next section (see also Rodin et al., 1970; Le Houérou, 1972; Le Houérou and Hoste, 1977).

Obviously, these various activities have different impacts. Generally speaking, crop expansion, overstocking, and firewood collection are the most harmful, and are responsible for 80 percent or more of the havoc.

Overstocking and overgrazing

The upper limits of browsing and grazing pressure are fairly well established. Detailed ecosystemic studies on the sahelian vegetation of Senegal for over a decade would suggest the amount of phytomass (forage and wood) levied by herbivores and man should not exceed 20–30 percent of the maximum standing crop in order to ensure sustained long-term productivity (Bille, 1977, 1978). This conclusion might be somewhat pessimistic on a long-term basis especially if conditions are less arid. The same rule of 20–30 percent would seem to apply in the Mediterranean steppes of northern Africa (Floret, 1981; Floret et al., 1983; Le Houérou, 1980a,b).

It should be kept in mind that the arid zone harbors about 55 percent of Africa's 550 million head of livestock, and that livestock numbers have increased by 75 percent from 1950 to 1985, in spite of the severe droughts that occurred in most African arid zones in the early 1970s and 1980s (Le Houérou, 1977c, 1985b). This is exponential growth of about 0.7 percent per year compared to the 1.0–1.5 percent demographic growth rate of African pastoralists (Bernus, 1974, 1981; Le Houérou, 1973, 1985b).

On a country-by-country basis it is quite clear that desertization grows apace with crop expansion and stock numbers. Two extreme cases are Botswana and Sudan. Botswana, in southern Africa, with a total surface of 0.5 million km^2, has an average population density of only one inhabitant per km^2, and a livestock density of less than 7 head per km^2; it fares rather well. Although the country is entirely arid, with rainfall between 200 and 600 mm, rangelands are in fair to good condition, croplands cover less than 20 percent, wildlife is still plentiful, and the livestock industry is prosperous; the country is still in ecological balance.

At the other extreme, the Sudan has undergone frightening desertization creep (Lamprey, 1975; Le Houérou, 1977a), with a 500 percent expansion of cropping in the Kordofan and Darfur provinces from 1960 to 1972, while the yields decreased by 75 percent. At the same time, livestock numbers increased by 189 percent between 1950 and 1973. As a result, the southern limit of the desert expanded 80–100 km to the south between 1958 and 1975 (Lamprey, 1975); this is an increase of some 100,000 km^2, which averages over half a million ha per year—an annual rate equivalent to one percent of the country's arid zone.

As frightening as it is, the situation in the Sudan is not unique. The same rate of desertization was found on a large test zone of 50,000 km^2 in southern Tunisia that was intensely studied for 10 years by Floret and Le Floch (1973). A rate of 0.6 percent was found in the steppes of Inner Mongolia by Zhu Zhenda and Liu Shu (1983). Although less accurately documented, the situation is not much different in Mauritania, Mali, Niger, Chad, Somalia, and northern Kenya, not to speak of the northern fringe of the Sahara.

The Kalahari Desert in Botswana bears many similarities to the Sahel in terms of climate, soils, pristine vegetation, and wildlife conditions (Monod, 1957; Aubreville, 1949). It therefore seems fair to say that the present situation in Botswana is probably a reflection of what the Sahel was like about 100 years ago, and the Sahara 3000–6000 years ago.

Subsequent effects and processes

Once the perennial plant cover has been destroyed by the human activities described above, the soil surface is left bare during most of the dry season and throughout the beginning of the rainy season. The bare soil is then subject to the following processes:

1. *Wind erosion on sandy soils.* Soil removal of 0.5–1.5 cm (150–450 metric tons/ha annually) has been reported from sandy steppes or savannas cleared for cultivation, on both sides of the Sahara (Le Houérou, 1962; Floret and Le Floch, 1973; Khatteli, 1983; Aubert and Maignien, 1949).
2. *The removal of the upper soft layers,* either by wind or by runoff erosion on shallow soils. This leaves various kinds of hardpans (calcareous, gypseous, iron) or encrustations, (calcareous, ferrugineous or gypseous) or hard geologic subtrata. All of these are very unfavorable to plant growth because of high runoff rates and low water storage capacities. Such hardpans often remain barren and desert-like, even under moderate climatic conditions.
3. *Development of desert pavements.* As the soft material (sand, silt, clay) is blown away, the coarser detritic particles (pebbles, stones, gravel, and cobbles) that are too heavy to be transported by wind remain on the ground and thus accumulate on the soil surface, giving way to desert pavements called "Reg," "Serir," "Tanezrouft," or "Tenere" in the Sahara. These pavements constitute very xeric environments. Plant life is always very limited on such surface material. This type of stony desert is called "Hammada" when the substratum is a hard rock structural surface of flagstones (limestone, sandstone, basalt, etc.). These are less xeric than desert pavements because of the presence of crevices, cracks, chasms, and other hollows where water and soft material accumulate and plants find suitable conditions.

4. *Dust and sand storms.* These take place more and more frequently as soil denudation progresses. Desert dust fallout may occur very far from the originating areas; Saharan dust falls in the Caribbean in the winter and in northeast Brazil in the summer (Morales, 1979; Lundholm, 1977). Red-mud rains from northern Africa and the Sahara occur over western Europe several times each year, as far north as Scandinavia. Saharan dust is held responsible for about half the troposphere's 200 million tons of mineral aerosols and may thus have an impact on atmospheric circulation and climates (Morales, 1977). (But Pleistocene dust winds are also responsible for the deposition of loessial soils that skirt many desertic regions in Africa, Asia, and America and sometimes constitute the bread baskets of a number of countries.)

5. *Sand deposition.* This is certainly the best known and most spectacular process of desertization, although not more than 20–30 percent of true deserts have their surface under sand seas. Sand deposits take various forms, including sand sheets, barkhans, and star dunes. Desertization is often accompanied by the reactivation of fixed sand dunes; this is the case for most of the Sahel. The Sahel was a very large sand sea some 15,000 to 30,000 years ago, established under climatic conditions much drier than the present ones (Ogolian dune system). These dunes became fixed during the later pluvial periods and supported a mixed grass–shrub savanna until the present time.

6. *Deposition of clay dunes and formation of "lunettes."* These result from accumulation of "pseudo-sands," made of little peds (aggregates) of highly saline flocculated clay, blown from the surface of dried out salt lakes (Sebkha, Playa, Salar).

7. *Water erosion, sheet erosion, rill and gully erosion.* These are serious problems in hilly country on soft geological substrata such as clay, marls, gypsum, anhydrite, chalk, and shale. Such effects may transform the landscape into "badlands" once the natural vegetation has been removed by man and his animals.

8. *Sealing of the bare surface* of medium- to fine-textured surface soil material by raindrop splash. This makes the soil surface almost impervious. As the rate of runoff increases, water intake is less, and the soils become very dry. Perennial vegetation, particularly shrubs and trees, die off in a very few seasons. This process is responsible for the development of "vegetation arcs" or "tiger bush" (Boudet, 1972), which is a potent factor of desertization in tropical savannas. Continuing deterioration leads to the destruction of the arcs themselves, and their replacement by bare ground or desert pavements. Vegetation arcs, which used to cover huge areas in the southern half of the Sahel and in the "Horn," have

considerably regressed for the past two decades, giving way to barren land (Peyre de Fabrègues, 1985; Peyre de Fabrègues and De Wispelaere, 1984).

9. *Bush fires.* Wildfires are no problem in the Mediterranean steppes because there is not enough fuel; perennial plants are too scattered to carry fire. The opposite situation prevails in the tropics, where 15–25 percent of the sahelian "scrub" is burned every year as evidenced from satellite imagery, while some 80 percent of the Sudanian savannas are the prey of wildfires (G. Wickens, personal communication). These different levels of burning result from differences in fuel load, vegetation types, clearing, cropping, hunting, grazing, and other traditional practices. In the Sahel and arid East Africa, herders lop the crowns of tall shrubs and trees to make browse available to livestock, particularly in time of drought. The lopped branches usually remain attached to the stems by the bark, forming a kind of half-closed umbrella. During the next rainy season annual grasses and forbs will grow up under the often spiny dried out branches. When a wildfire occurs the tree or shrub will be "cooked" (Piot et al., 1980). This process is a self-accelerating one; as desertization progresses, less grass is available and more trees and shrubs have to be cut (and "cooked").

10. *Fencing.* In the arid African tropics livestock are always herded and kept at night in enclosures ("bomas") made of thorn cut from *Acacia* and other spiny wood species (Lamprey, 1983) to protect them from predators and thieves. These enclosures are also used around houses, compounds, vegetable gardens, and staple food fields. As nomads move at least twice a year, and as they never re-use the same bomas (because of internal and external parasitism), and as the life span of these thorn fences hardly exceeds two years (due to termites), the local woody vegetation is gradually destroyed. Particularly affected are spiny trees such as *Acacia tortilis,* which are the best adapted to these harsh environments and which, in addition, produce large quantities of nutritive pods (Lamprey, 1983). Fewer pods means more lopping of shrubs, and less grass available in the following season. This is obviously an important aspect of the spiral of desertization.

11. *Changes in soil properties.* In addition to the changes occurring in the soil surface, the removal of perennial plants during the process of desertization has many feedback effects on morphology, structure, chemistry, and biology of soil components due to the drastic reduction in the production of organic matter. These effects include a sharp decrease in organic matter (from say 2–3 percent to 0.1–0.5 percent in the upper soil layers); changes in the distribution of organic matter in the soil profile; changes in the nature

of soil organic matter, in particular in the ratio between humic and fulvic compounds; reduction of structure stability; reduced porosity and therefore reduced permeability, water intake, storage capacity, and oxygenation because of altered structure; and, finally, a decrease in macro and microfauna and flora, and therefore in biological activity, hence in nutrient cycling, hence in fertility, hence in primary productivity, hence in organic matter production—another desertization spiral (for details see Le Houérou, 1969).

THE EXTENT OF DESERTIZATION

Detailed long-term studies on both sides of the Sahara and in Eastern Africa using remote sensing techniques (aerial photography, satellite imagery, and low altitude reconnaissance flights) have shown that desert encroachment over the arid zone involves 0.5–1.0 percent of the latter annually (Schwaar, 1965; Floret and Le Floch, 1973; Gaston, 1981; Le Houérou, 1962; 1968; Watson and Tippett, 1981; De Wispelaere, 1980; Barral et al., 1983; Barry et al.,1983).

It has been estimated that about 9 million km^2 of the world's arid lands have been turned into manmade deserts over the past half century; one-third of those are in Africa (Dregne, 1983). The phenomenon affects directly and immediately some 80 million persons in the world and 30 million in Africa. If the present human population growth rate remains substantially unchanged for the next 15 years, desertization will affect 4 million km^2 and 45 million people by the year 2000.

AN EXAMPLE: THE CASE OF THE SAHEL

The Sahel is the ecoclimatic region that borders the Sahara to the south in the 6000 km long, 500 km wide strip crossing the continent from the Atlantic Ocean to the Red Sea between the 100 mm and 600 mm isohyets of mean annual rainfall. The Sahel is undergoing a very long drought which started in 1969 and persists at this time, after a short and weak remission in the late 1970s. The drought actually worsened in 1983 and 1984; in 1983, for instance, precipitation in the Ferlo region of Senegal was only one-third of the long-term average (Tucker et al., 1985). In the Sahel of Niger, rain averaged only 42 percent of the 1951–1980 mean in 1984 (Peyre de Fabrègues and De Wispelaere, 1984). The mean rainfall in the Ferlo for the period 1969–1980 was only 60 percent of the long-term average since the the beginning of records (De Wispelaere, 1980) and probably not more than 50 percent for the period 1969–1984. In Niger the 1969–

1984 average was only 70 percent of the 1941–1970 value (Peyre de Fabrègues, 1984). Long dry periods extended over the whole Sahel several times in the past—in 1911–1916 and 1940–1946 (Bernus and Savonnet, 1973; Nicholson, 1978, 1981)—but none of those droughts reached the duration and intensity the area is presently experiencing. (For a detailed and concise account of the climatic history of the Sahel from the Pleistocene up to the present time, see Nicholson, 1978, 1981.) The 1984–1985 season was perhaps the worst in history. The discharge of the Niger River at Niamey in January 1985 was only 285 m^3 per second instead of the average 2000 m^3 per second for that month. In May 1985 the Niger River at Niamey went dry for the first time in living memory. The Chari River at N'djamena also went dry in May 1985, an event that has happened only once before, in 1914.

The number of cattle in Niger increased from 0.9 million in 1938 to 4.2 million in 1968 and was still 3.2 million in 1982. The number of cattle in the Sahel as a whole expanded fourfold between the Second World War and 1968 (FAO Production Yearbooks; Le Houérou, 1973, 1977c).

The impact of humans on the Sahelian vegetation may be described as follows. Woody species recede at a fast pace as a consequence of wood cutting, wildfires, overbrowsing, and sealing of silty soil surface, as mentioned previously. Various studies using remote sensing and field investigations in several Sahelian countries have shown that the woody cover has receded by over 1 percent annually between 1950 and 1975 (Gaston, 1981; Haywood, 1981; De Wispelaere and Toutain, 1976; Barral et al., 1983; De Wispelaere, 1980). In Chad, the border between the Sahara and the Sahel moved 50 km southward between 1954 and 1975 (Gaston, 1981). The process has apparently speeded up lately in Niger (Peyre de Fabrègues, 1984).

In pristine Sahelian conditions the canopy cover of woody vegetation may vary from 5 to 30 percent, depending on local factors such as mean annual rainfall, topography, and soil. Assuming an average of 15 percent, there are 500 to 1000 trees and shrubs per ha for the geographical zone (Poupon, 1980; Le Houérou, 1980a; Hiernaux, 1980). In depleted areas this may be reduced to 1–2 percent, with 100 individual trees and shrubs per ha (Piot et al., 1980), and zero in fully desertized areas. Generally speaking regression of woody vegetation is particularly acute on the upper parts of the topography: hardpan ferrugineous soils, slopes, surface-sealed silty soils, gravel plains, and pediments. But because of increased runoff, a temporary increase of woody vegetation may occur in the depressions around ponds and along water courses (Haywood, 1981). At later stages, denudation occurs even in these depressions because of high and permanent stock pressure. At the same time there is a turnover in species composition, when the desertization process is slow enough. Xeric spiny species such as *Acacia* spp. and *Balanites* tend to replace more mesic, soft, broadleaved species such as the Combretaceae. The regression of trees and shrubs has a substantial effect on the herbaceous layer; as described in the following section, the layer's productivity is usually much higher, and its growing season is much longer, under the shade than in the open.

The regression of the grass layer may be described as follows. First, perennial forage grasses such as *Aristida longiflora* and *Andropogon gayanus* are eliminated. Second, the mesic annual grasses of good fodder value such as *Brachiaria* sp., *Setaria* spp., *Panicum* sp., *Urochloa* spp., *Pennisetum* spp., *Digitaria* spp., and *Sorghum* spp. are replaced by more xeromorphic, less palatable species of mediocre feed value, such as *Aristida*, *Stipagrostis*, *Eragrostis*, *Cenchrus*, and *Chloris*. Third, more or less unpalatable forbs such as *Cassia* spp., *Borreria* spp., *Limeum* spp., *Gisekia* spp., *Mollugo* spp., *Tribulus* sp., and *Blepharis* spp. undergo expansion, along with one pantropical ruderal, a shrub-like asclepiad *(Calotropis procera)*, and a New World tobacco-shrub pest, *Nicotiana glauca*. Finally, denudation occurs in large tracts around villages, and around ponds, boreholes, rivers, and irrigated perimeters, which, on silty or gravelly soils, become barren and sterile.

THE PROSPECTS FOR RURAL DEVELOPMENT AND RECOVERY FROM DESERTIZATION

Experiments and small-scale development efforts throughout the African arid zones have shown beyond doubt that desertized land can be largely rehabilitated, at least in principle. Many successful environmental protection schemes—tree planting projects, for example—have been carried out. There are, however, a number of constraints. One of these is fundamental: no successful project can be implemented unless the desertization pressure exerted by humans is lessened or removed. Unfortunately, removing pressure from one area means increasing it somewhere else.

Village woodlots have often been advocated as an efficient means to combat desertization. Indeed, the planting of fast-growing trees in privileged sites may be a very successful way to overcome the firewood crisis. But this is often wishful thinking, because most of the areas where trees could grow are already used for growing food plants. Understandably, crops have higher priority than firewood for the villager. Woodlots need total protection from man and beast for at least five years, but protection and management are often difficult if not impossible to enforce. Nevertheless, a small number of rural civilizations in arid-zone Africa have shown that agroforestry in cropland or pastures is a useful way of conserving soil productivity while producing higher overall return than conventional open land cropping (Le Houérou, 1978).

More recently, experiments with several shrubs have been successful. Such species as the apple ring acacia *(Faidherbia albida)* in the tropics, and spineless cacti *(Opuntia ficus-indica)*, and saltbushes *(Atriplex* spp.) in the Mediterranean, subtropics and eastern African highlands have proved that aesthetic, conservative and productive artificial or seminatural agro-sylvo-pastoral production systems can be established as a viable alternative to technocratic rural "development" (i.e., conventional agronomic blueprints that are capital- and energy-intensive, often environmentally destructive, and not adapted to the social, cultural, and economic situation of the would-be users) (De Kock, 1980; Monjauze and Le Houérou, 1965; Le Houérou and Froment, 1966; Franclet and Le Houérou, 1971).

There are other constraints, conditions, and limits to "biological recovery." It would seem that on shallow soils where average annual precipitation is less than 100–150 mm, recovery is virtually impossible, even under conditions of total protection (Le Houérou, 1959, l974). On sealed soils, recovery is extremely difficult unless the seals are broken and the soil surfaces roughened artificially in order to increase water intake, storage, seed trapping, and seedling emergence and establishment. On deep, sandy soils, on the contrary, biological recovery has been naturally achieved with rainfalls as low as 60–80 mm, under conditions of total protection and over periods of 5–10 years or more (Le Houérou, 1976a). Generally speaking, biological recovery is more difficult as the climate is more arid, the soil more shallow and desertization more advanced; it obviously is a matter of water availablity in the soil and of the "entropy" of the system—how far it has already deteriorated. For example, the rate of recovery would depend on the availability of spores and seeds from pioneering, perennial species.

The return of desertized areas to nearly pristine conditions is no problem where the annual rainfall is more than 150–200 mm, assuming that there is protection from depredation over time spans of 5–10 years or more. Many exclosures of various size $(0.01–1,000 \text{ km}^2)$ throughout the whole African arid zone, from the Mediterranean to the Cape of Good Hope, have shown that flora and vegetation can, in many cases, recover from desertization or, still better, be protected from it (Le Houérou, 1959, 1962, 1968, 1969, 1975a,b, 1977b, 1979; Boudet 1977; Adam, 1967; Lamprey, 1983).

Regeneration of tree and shrub cover is of paramount importance in the recovery process. It has been shown that photosynthetic efficiency and productivity of the combined tree–shrub ("trub") and grass layers is 2–3 times higher than that of a grass layer alone (Bille, 1977; Le Houérou, 1980a; Pratchett, personal communication). Therefore, the practice of managing for a more or less homogeneous grass layer is probably short-sighted. In addition, this practice prevents efficient nutrient cycling and does not provide a balanced diet for ruminants throughout the year. Where dry seasons are long, as in the Sahel or the Horn, woody vegetation is a "must" in the diets of grazers and browsers because it is the only source of protein, carotene, and often of phosphorus and other minerals (Le Houérou, 1973, 1978, 1980a; Granier, 1977; Gillet, 1984; Boudet, 1975).

In theory, protein, carotene, and minerals could be added to the grazing diet, but this is not possible under current socio-economic conditions. Our conclusion is simple: when the "trub" layer is destroyed, livestock cannot be maintained in the Sahel outside of the rainy season (Granier, 1977; Le Houérou, 1980a). Desertization thus seriously threatens the survival of the main regional industry.

Though recovery is possible in principle, demographic, political, cultural, and economic forces have militated against its widespread success. For the most part, the small-scale development schemes carried out so far, though often technically successful, have been financial and economic failures. The record is worse for medium- and large-scale projects; it would be a simple matter to compile a long list of disastrous agricultural development projects in arid and semi-arid Africa. In the past 25 years alone, at least a billion dollars (much of it from international agencies) has been spent on such projects, but not one, to our knowledge, has been an economic success or has become self-sustaining. Moreover, few if any of these projects have included a distinct conservation element. In many cases, the only visible result has been a heavy burden of debt for the countries receiving the aid.

The reasons for these failures are various and complex. One is that these projects have been imposed upon rural populations without their voluntary participation. Others are poor management, corruption, nepotism, and excessive and apathetic bureaucracies. In addition, these projects often use technologies that are inappropriate to the prevailing socio-economic conditions.

CONSERVATION PROBLEMS AND CONSTRAINTS

Impacts on wildlife

The need for conservation in North Africa to salvage rare and endangered species was recently reviewed (Le Houérou, 1984b). A number of plant species and ecotypes with high ecological and economic value have already become rare or endangered in the African arid zone, including *Calligonum arich, Cupressus dupreziana, Hedysarum argentatum,* and *Digitaria nodosa.*

Before the Second World War, wildlife could roam freely over immense territories in pursuit of unpredictable and far-flung feeding opportunities. Until about 1960 the scimitar-horned oryx, *Oryx dammah,* a shy and gregarious desert antelope weighing up to 200 kg, was found in herds of several hundreds in Chad (Gillet, 1965). In 1963 the estimated number for that country was 4000 to 5000. The oryx was still common in Western Sahara and Mauritania in the 1950s but became extinct in the early 1960s (Valverde, 1969).

The addax, *Addax nasomaculatus,* the most drought-tolerant ungulate on earth, was not uncommon in the driest parts of the Sahara until World War II; some individuals are still occasionally reported. The travel writer Ph. Diole saw three of them in February 1956 in the dunes of the Edeyen of Murzuk, and four or five were reported from the southern fringe of the Sahara between 1979 and 1985 in Mauritania, Niger and Chad. Monod (1958) saw some 12,500 fresh tracks in the empty quarter of southeast Mauritania in the winter of 1952, but those have been destroyed since that time.

The dorcas and dama gazelles *(Gazella dorcas* and *G. dama),* once very common (particularly the former) in large herds on both sides of the Sahara and in the Sahel until the 1960s, are now very rare— especially the latter which is believed to be extinct in the northern and western Sahara (Valverde, 1969; Panouse, 1968; Dekeyser, 1955; Dekeyser and Derivot, 1969; Hufnagl, 1972; Grenot, 1969).

Today the Sahelian large mammals are on the verge of extinction. The main areas where the oryx and addax used to be found in sizeable number (Chad, Western Sahara and Mauritania) have been the theater of warfare for the past 10 years, and the antelopes have been chased by all warring parties with vehicle-mounted machine guns. European dwellers of the uranium mining town or Arlit are still chasing down the last gazelles with four-wheel drive vehicles. Only a few individuals may have survived the massacre by retreating to inaccessible rocky and hilly terrain or to the main dune systems (although hunting by helicopter has been reported as well). Some salvaging attempts have been made in Chad; the game reserve of Wadi Achem/Wadi Rime was set up in 1970, but survived only five years because of the political situation. The day is rapidly coming when the only surviving addax and scimitar-horned oryx will live in zoological gardens (as happened to the Arabian oryx, *O. leucoryx).*

The impact of desertization on wildlife is amplified as a result of a drying environment and intensive poaching by a hungry human population. Large mammals, still common in the Sahel until the Second

World War, have become very rare, even in the most remote places, or are now extinct in the Sahel and the Sahara where there are no national parks or game reserves. The anubis baboon *(Papio anubis)* and the papas monkey *(Erythrocebus papas)*, formerly common in the Air and Tibesti mountains, have probably disappeared. Among the extremely rare or recently extinct species are the lion *(Panthera leo)*, leopard *(Panthera pardus)*, cheetah *(Acynonix jubatus)* and smaller cats *(Felis serval, F. caracal, F. margarita)*, hunting dog *(Lycaon pictus)*, aardvark *(Orycteropus afer)*, elephant *(Loxodonta africana)*, hippo *(Hippopotamus amphibius)*, giraffe *(Giraffa camelopardalis)*, addax *(Addax nasomaculatus)*, scimitar-horned oryx *(Oryx dammah)*, the defassa waterbuck *(Kobus defassa)* and kob *(K. kob)*, topi *(Damaliscus korrigum)*, bushbuck *(Tragelaphus scriptus)*, greater kudu *(T. strepticeros)*, bubal hartebeest *(Alcelaphus buselaphus)*, the gazelles *Gazella dorcas* and *G. dama*, rhim *(G. leptoceros)*, red-fronted gazelle *(G. rufifrons)*, eland *(Taurotragus derbyanus)*, buffalo *(Syncerus caffer)*, Barbary sheep *(Ammotragus lervia)*, and Nubian ibex *(Capra ibex)*. Only a few small mammals are still observed.

Crocodiles *(Crocodylus niloticus)*, pythons *(Python saeba)*, and ostriches *(Struthio camelus)* have become very rare, but other reptiles and birds are still fairly common, including some game species such as Guinea fowl *(Numida meleagris)*, quail *(Coturnix coturnix)*, bustard *(Choriotis arabes, C. kori)*, doves *(Streptopelia turtur, Columba livia)*, and francolins *(Francolinus spp., Pernistis spp.)*.

Some economically valuable fish are becoming rarer by the day, notably the Nile perch *(Lates niloticus)*. The traditional fishing industry of the inner delta of the Niger River, which used to catch and export 30,000 tons of dried fish annually to many countries of West Africa, is now on the decline because of overfishing. On the other hand, a small bird called the "millet eater" *(Quelea quelea)* is still considered a serious pest in spite of long-standing international efforts to control it.

Constraints on conservation

Most of the national parks and other conservation areas were established long before African countries became independent and at a time when pressure on the land was far less. An exception, however, is the case of Botswana where the Chobe Game Reserve (1.1 million ha) was set up in 1967 and the Gemsbock National Park (2.5 million ha) was established in 1971. Wildlife are efficiently protected in both.

There never were many national parks in the arid zone north of the equator, however. There are none in the arid zones of northern and western Africa (unless on paper). A relatively small number of game reserves, national parks and game ranches in the arid zones of eastern and southern Africa (Kenya, Tanzania, Ethiopia, Botswana, and South Africa) have shown that wildlife protection and management can be a most important asset in economic development, both as sources of meat and of foreign exchange from tourism.

Range management techniques, such as deferred grazing and matching stocking rates to carrying capacities, are applicable to African arid rangelands, including the ranching industry in southern Africa (Le Houérou and Froment, 1966; Boudet, 1975). Another technique—mixed herding of wild or semi-domesticated ungulates with cattle—also has potential, at least in principle. Ancient Egyptian bas-reliefs show mixed herds of cattle with oryx or addax. Similar mixed herding is still observable in Masailand and is reported to have been practiced by the Moors until recently. Mixed herds of cattle and giraffes were also recently reported from the region of Ayorou in Niger. Because cattle and wild ungulates have different nutritional needs, there is often little competition between them.

Why, then, are so few conservation projects implemented? The answer is a very simple one: it is socially and politically very difficult to withdraw large tracts of land from the present pastoral or cropping land use pattern. In other words, the African arid zone harbors too many people.

In 1980 the carrying capacity for humans in the Sahel countries was only 50 percent of its population, given the levels of natural input and artificial subsidies existing then (Kassam and Higgins, 1980; Le Houérou and Popov, 1981). This means that the Sahel now has 2.5 times more people than it could possibly feed on a long-term, sustained basis. This also applies to most other African arid zones, especially those north of the Sahara.

The Sahel used to be a pastoral region, but as a result of the present desertization disaster, more and more destitute pastoralists and impoverished farmers are moving to towns and cities in search of some means of survival. Cities are burgeoning and growth rates are 10–20 percent per year. They are increasingly surrounded by ever-expanding slums where a miserable population depends to a large extent on foreign food aid for survival.

Some successful conservation measures have been taken and enforced in Kenya, such as the ban on the trade of all wildlife products, but this is an isolated case. Good conservation laws that protect trees and wildlife have been issued in most countries. Enforcement is another matter. Not only are the laws not enforced, but too often they are first infringed by those who, provided with state-owned four-wheel drive vehicles, are supposed to enforce them.

In sum, the prognosis for conservation in general, and wildlife in particular, in the African arid zone is very gloomy. Per capita food production has been declining for the past 25 years. Desert is expanding at an average rate of about 0.5 percent per year as a result of faulty land use and overpopulation. More than one million people starved to death in Ethiopia and the Sahel from 1971–1973 and 1984–1985. Warfare has become endemic in Ethiopia, Somalia, Chad, Mauritania, Western Africa, Angola, and Namibia, not to speak of South Africa. Competition for the land between farmers, pastoralists, and wildlife is acute. The first loser is the wildlife, followed by the pastoralist. Finally, the farmer is under permanent threat of starvation.

This assessment may seem excessively pessimistic, but it reflects our attempt to be as realistic as possible, after having spent a combined total of 60 years doing research in African arid zones.

SECTION VI

INTERACTING WITH THE REAL

WORLD

FLEXIBILITY

The three chapters in this section may appear at first glance to be unrelated, but there is a common denominator: enlightened pragmatism. Each of the authors, in his own way, is saying that the successful application of conservation biology to protecting biological diversity may depend on appreciating the diversity of human motivation and cultures.

Naess (Chapter 25) reminds us that individual humans have a capacity for identification with all forms of life, and that a one-dimensional premise (self-interest) of human motivation demeans our species and artificially constrains the arena of expression and action. Compassion for landscapes and for other forms of life can be a powerful motivator.

Cairns (Chapter 23) helps extend this line of holistic thinking to the level of different subgroups within society. His pluralistic approach to restoration and rehabilitation of habitats is based on the recognition that different parts of the scientific–technocratic community use different rules of evidence and employ different criteria for success. In other words, the rules of academia are different that the rules of bureaucracy; the general public operates with still a different set of rules. Hence, the game of science is different from the game of public policy and administration, which, in turn, is different from the game of public relations. He argues that if one wishes to be effective, one must have the flexibility to mentally take off and put on the appropriate rules, and the skill to alter one's form of communication.

Diamond (Chapter 24) ushers us to the level of culture and politics. He argues that different customs and political systems require vastly different approaches to the design of a national conservation strategy. While granting that the same *biological* rules and criteria apply to both Papua New Guinea and Irian Jaya, Diamond implies that any attempt to impose the same conservation system on the two nations (Irian Jaya is part of Indonesia) would be foolish.

HUMAN NATURE AND THE REAL WORLD

In seeking simplicity, we sometimes dismiss the complexity of human beings as a bothersome irrelevancy. The temptation is to appeal to what people want most (sex, babies, health, security, recognition, control, entertainment) and to disregard the balancing needs for relationships, community, and a transcending connectedness to the universe (which may or may not be "religious" in its expression).

Such a one-sided approach to human nature has led some conservationists to a of tactic of purely materialistic promotion, namely, the substitution of "genetic resources" for biological diversity. There is nothing bad or illogical about promoting nature reserves because the organisms in them may have genes that might someday cure cancer, enhance the productivity of crops, and reduce our dependency on fossil fuels. But the "real world" is not simply an economic world. Humans respond more completely when the appeal is expressed in terms of deeply felt values of life.

Naess (Chapter 25) lists some reasons why many experts are reluctant to adopt these intuitive or experiential lines of persuasion. While acknowledging the great difficulty in overcoming cultural conditioning and disciplinary conventions, he suggests that a lopsided, utilitarian platform is hurting our cause. It also truncates our development as individuals.

I think that the reader will enjoy the many levels of common sense in these chapters.

M.S.

CHAPTER 23

RESTORATION, RECLAMATION, AND REGENERATION OF DEGRADED OR DESTROYED ECOSYSTEMS

John Cairns, Jr.

Ecosystems are exposed to a wide range of disturbances that range from the fall of a single large tree in a forest that opens to sunlight a patch that had previously been shaded, to disturbances covering a substantial portion or possibly all of the globe as described in the "nuclear winter" article by Ehrlich et al. (1983). Small disturbances caused by natural phenomena are often followed by a very rapid recovery process. The disturbances covered in this chapter, however, are entirely the result of human activity and are not so quickly healed. The nature of the disturbance, its duration, scale, and, frequently, selectivity may ensure that recovery to original condition is highly improbable. Since the nature of the disturbance plays a pivotal role in ecosystem recovery, a different management strategy usually must be developed for each type of disturbance. An illustrative list follows that the reader will almost certainly be able to expand.

1. *Sudden and unexpected disturbances.* Examples are derailments of railway cars carrying hazardous chemicals, oil drilling blowouts, ship accidents, or manufacturing facility failures such as Three

465

Mile Island in the United States or the Seveso accident in Italy. As a result of the unexpectedness of the accident, the preaccident condition of the ecosystem will most likely be unknown. In addition, confusion following the accident results in data not gathered, data gathering handled improperly, or data gathering of the wrong kind or in the wrong place. In many instances, spill control teams may contain the deleterious material, immobilize it, or destroy it in situ. In other cases, such as the tetraalkyl lead spill described by Tiravanti and Passino (in press), the material may spread over a substantial area and be difficult or impossible to control.

2. *Disturbances that have been occurring for a substantial period of time but were only recently detected.* Examples of this are industrial discharges that were thought to be harmless to the indigenous biota but were actually more harmful than anticipated. In the absence of careful environmental monitoring, including chemical, physical, and biological information, such predictive errors may go undetected until the disturbance reaches gross levels identifiable by laymen. Such evidence may include fish kills, noxious growths, or direct harmful effects on humans. In these cases it is also quite likely that no substantive ecological background information has been gathered, so the predisturbance condition of the ecosystem is known in only the most general and cursory fashion. If the material being discharged or leaching from a burial site is a persistent chemical at toxic concentrations, reduction to tolerable concentrations in the ecosystem may be difficult. Thus, even if the discharge ceases or is reduced to appropriate levels, recovery of the ecosystem to any significant degree is not likely until the hazardous material is either removed or immobilized. Allowing the waste to move and become less harmful through dilution is a common but unsatisfactory solution to the problem. Decontamination in situ may be both expensive and difficult with present technology as attempts to cope with improperly designed hazardous waste storage sites have shown.

3. *Situations where the disturbance is planned.* If the disturbance is anticipated, a thorough ecological evaluation of the predisturbance characteristics can be undertaken. Examples of this situation are surface mining, discharge of sewage or chemical wastes, construction of a manufacturing plant, or construction of a dam or a highway.

Three major determinations must be made following ecological disturbance before appropriate corrective action can be taken: the degree of change; the area in which change has occurred; and the ecological significance of the change, including the probability that it will adversely affect adjacent ecosystems. Some questions will assist in these determinations; for example: What is (or was) the ecosystem like (including variability)? At what rate does normal change occur? How does one determine a deviation from the nominative (as defined in Odum et al., 1979) state? What parameters provide an early warning of recovery malfunction?

DESIGNATED USE CATEGORIES

Regulatory criteria for judging recovery of damaged ecosystems may be dramatically different from criteria based on ecological principles. Illustrations of this point are based entirely on water, since I know this area best. The Clean Water Act defines acceptable uses for water and mandates continuation of specified uses. These include propagation of fish and wildlife, public water supply, recreation, agricultural, industrial, and navigation (Sec. 303, C, 2 Clean Water Act). The U.S. Environmental Protection Agency (1980), in reviewing the regulations enacted by the 50 states, indicates that the states have at least 15 general use categories for streams. However, only one of these categories—propagation of fish and wildlife—was common to all. This is most fortunate, because it is a category entirely compatible with sound ecological principles, whereas navigation is not. However, the regulatory requirements for a significant number of use categories may be satisfied by water quality most ecologists would not consider ideal for aquatic life. It is important to recognize that both the types and number of specific use categories developed by each state are often driven by the needs of its commerce.

DECIDING ON RECOVERY GOALS

Magnuson et al. (1980) have noted that one might *restore* a displaced ecosystem to its original condition, *rehabilitate* it by restoring some of the most desirable original features, or choose some alternative ecosystem that has been designated *enhanced* (Figure 1). Other alternatives are to let the degradation proceed if the stress has not been removed, or to let it remain in its present condition if the stress has been removed but there are no signs of recovery. There are some instances, such as the use of dispersants in oil spill cleanup, where the management practices proved more harmful to the ecosystem than doing nothing. As a consequence, situations exist where knowledge of the system itself and the recovery process, or of the effects of the

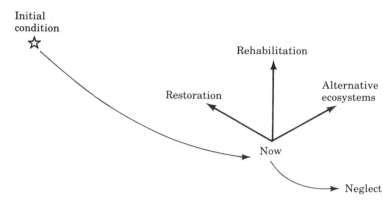

FIGURE 1. Management options for surface-mined lands. (Modified from Magnuson et al., 1980.)

cleanup processes, are so poorly understood that doing something might prove more harmful to the ecosystem than doing nothing. This is repugnant, particularly to Americans who tend to feel that immediate positive action should be taken in almost every situation. (Readers wishing to study some case histories for Europe and North America can find them in some of the books listed under suggested reading.)

When there is insufficient data to decide among alternative recovery goals or actions, one positive course of action is research. If properly designed, such research into alternative recovery projects can proceed in such a way that mistakes can be quickly rectified, and the information generated can be more generally applied to a number of other sites. This will require enormous flexibility and mutual trust on the part of the regulatory agencies, the industry causing the damage, and the academic community or consulting firms carrying out the study. Flexibility of the regulatory agency will be essential because whatever course of action is taken will probably not comply with existing laws and regulations. Flexibility is essential on the part of the industry (or other group causing the damage) because it will be faced with what it will probably regard as unjustifiable research expenses. This group may prefer a court fight to a study that will further understanding of the recovery process if it thinks the fight will be less expensive. The academic community will have to be flexible because, although it is naturally interested in the research aspects of the study, it must also realize that the information generated must be useful in the decision making process and must be, in so far as possible, cost effective. All of this must be communicated effectively to the general public before the study begins.

RECOVERY OF ECOSYSTEMS FOLLOWING DISTURBANCE

Criteria for determining degree of recovery

Criteria selected to characterize ecosystem recovery must not only be scientifically sound but must, of necessity, encompass a broad range of attributes related to the expectations of and intended uses by the general public. Ecologists must recognize that industry is unlikely to take any action not required by law unless there are compelling economic reasons to do so. In the United States, the Environmental Protection Agency and the individual states, in continuing to meet their responsibilities, must promulgate regulations using the best information available to establish degrees of recovery. They are not likely to embrace joyfully the latest research finding, even if published in the most prestigious scholarly journal, until it is widely accepted and used. A number of other factors also influence acceptance.

1. *Reliability of extrapolations from one system to another.* Since ecologists regularly emphasize the uniqueness of each ecosystem, regulators and industry may be wary of claims of general applicability of results and/or methods. As a consequence, the reliability of extrapolations must be adequately documented.
2. *Interpretation.* What does the parameter or characteristic mean in terms of environmental recovery? Do ecologists agree on this interpretation? Will laymen, including courts of law, accept this interpretation and the level of significance assigned to it? In short, the interpretation must have the endorsement of a substantial majority of professionals and, if possible, its significance should be understood and accepted by the public.
3. *Sensitivity.* From a regulatory standpoint, each response or end point should be sufficiently sensitive to avoid excessive false positives (that is, no one should think that the ecosystem has recovered when it has not).[1] At the same time, industry will want an end point that does not produce excessive false negatives (that the system has not recovered when it has). The precise level of sensitivity finally accepted will be a function of an array of factors, including the number and kinds of alternative methods that provide confirming or alternative evidence, the objectives of the study, the correspondence of the end point to ecosystem stability, and the consequences of error in estimating the degree of recovery.

---

[1] For those unaccustomed to these word usages, Cairns (1985) might be helpful.

4. *Variability.* If the response being measured is a discrete variable (such as diversity), the precision of the measurements being made can be more easily documented than is possible for nondiscrete variables (for example, nutrient spiraling). Nevertheless, a rough positive correlation appears to exist between the relative sensitivity of the tests and the degree of variability encountered. It is possible that variability of functional parameters is itself an indicator of ecosystem condition. Both J. Harte (personal communication) and I have noted increased variability of functional attributes in microcosms under stress. However, even if this hypothesis is correct, simultaneous attempts to optimize sensitivity and reduce variability will almost certainly not succeed. Since both are desirable, a compromise between the acceptable degree of sensitivity and reduced variability is the probable outcome. Considerable professional judgment will be required for these decisions because a fixed formula for reaching a decision seems unlikely.

5. *Replicability.* From a regulatory standpoint, methods used for determining degrees of recovery should be sufficiently simple and standardized so that they can be routinely used by a variety of laboratories (consulting firms, state and federal agencies, industry, and academia) that have widely varying capabilities. Since our knowledge of the recovery process in damaged ecosystems is not extensive, it is unlikely that methodology meeting these criteria will be available soon. An interim alternative would be a group of "ombudsmen" or a "science court," composed of respected professionals with experience in these matters, to determine the adequacy of the proposed methods (and staff competency to use them and interpret the results) on a site-specific basis. This approach presents a number of disadvantages (such as finding qualified people willing to serve), but, given the complexity of the problem and our present state of knowledge, there are no other good solutions or alternatives.

Selection of end points to demonstrate a desired state has been achieved

At the population level, end points are relatively well known and fairly generally accepted. Examples include reproductive success, recruitment rate, and age and sex structure. End points for higher levels of biological organization include productivity, diversity, trophic balance, and nutrient spiraling, among others. The probability of achieving anything approaching a professional consensus on the relative importance, reliability, replicability, measurement, interpretation, and a variety of other issues regarding end points at the community and ecosystem level is small. During this period, ecological end points might easily be discarded and replaced with horticultural ones (for example, easily established vegetation) if the discussion is so contentious that laymen lose confidence in ecological end points.

The first step in establishing criteria for ecosystem recovery or achievement of alternative goals (such as rehabilitation) is an investigation into the organization, nature, and function of a number of types of systems needed for the decision making just described. There must also be recognition that changes in ecosystem characteristics are more often step functions than trends, and that ecosystems are highly variable. Recent evidence suggests that species aggregates may be one of the criteria for the recovery of ecosystems. In this case, "modules of species" means interacting subsets of the larger natural community that can be maintained in a natural state in the laboratory. Where various interactions occur, the use of modules of species to predict ecosystem condition has promise. The use of "guilds" (species that exploit the same class of environmental resources in a similar way) for estimating ecosystem condition is another alternative worth investigating.

The "performance" of a system is anything the system does as a whole, including its state and structure. End points should not be based on transitory system performance because these are likely to generate false positives or negatives with regard to condition. However, if one can determine the capacity or potential of the system to perform,[2] this would allow the selection of end points that are most invariant and presumably most fundamental.

Whatever end points are chosen to document ecosystem condition, their relevance must be communicated to the general public and to decision makers in regulatory agencies, industry, and various levels of government. An illustrative checklist follows.

1. *Technical relevance.* Does the end point represent a realistic measure of population, community, or ecosystem condition? Using this information, can one determine the degree of ecosystem change following disturbances based on objective criteria?
2. *Social relevance.* Is the end point meaningful to the public? If not, an effort must be made to communicate the information and reasoning. Our present economic situation requires that any activity requiring substantial investment be understood and supported.
3. *Legal relevance.* One must establish that each end point is useful for establishing that ecosystem condition following disturbance is

---

[2] "Capacity to perform" means processing nutrients, cycling energy, and other functions characteristic of ecosystems (Cairns, 1985).

satisfactory. The information has no legal relevance unless it influences the decision being made.

4. *Cost and timing*. Well-publicized increases in energy and labor costs coupled with tough foreign competition have made industries extremely cost conscious in recent years. Federal deficits and pressures on state tax funds have produced comparable pressures in regulatory agencies. In industry, these pressures are manifested by a reluctance to do anything not required by law. In regulatory agencies, one important manifestation is an attempt to standardize the actions necessary to demonstrate compliance with legislation. In addition, persuasive information must be provided that the cost is reasonable in terms of the objectives. Cost is, as usual, largely a function of the time necessary to carry out the tests, space and facilities required, and the level of professional competence necessary to generate data and interpret them. The decisions regarding acceptable costs are driven primarily by the degree of certainty required about ecosystem condition.

REGULATORY CONDITIONS FOR TERMINATING MANAGEMENT RESPONSIBILITIES

Phase I

The organization responsible for damaging an ecosystem can be identified in a majority of cases. If damage was caused by normal operations, as in strip mining, all necessary information should be part of the permitting process. Where ecosystem damage is the result of an episodic event, such as a catastrophic spill of hazardous chemicals, defining the damaged area and the extent of the damage precisely may be extraordinarily difficult and is often impossible. This is particularly true of spills into the atmosphere. The highest priority should then be given to defining the fate of the material spilled, including touchdown points for atmospheric movement and transformations in water and soil. Simultaneously, biological evidence should be gathered, including estimates of damage through acute toxicity, estimates of bioconcentration and accumulation, and estimates of long-term toxicity, using laboratory and microcosm tests for materials that are persistent or bioconcentrating. In cases where delayed effects are anticipated for any reason, biological monitoring programs should be designed and implemented to document these and predictive models should be developed to aid in this process. In addition, the following institutional and administrative steps should be taken:

1. The organizational structure and responsibilities of various components should be immediately and crisply defined.
2. A means of pooling and sharing information should be immediately developed, preferably utilizing a system already prepared for this purpose.
3. The credentials and qualifications of the organizations and personnel employed in the tasks just described should be carefully examined to eliminate at least the grossly incompetent and to reduce conflict over the quality and interpretation of the data when they are finally analyzed.
4. A quality assurance program, preferably along the lines of a standard practice recommended by the American Society for Testing and Materials or some other standard-setting agency, should be followed.

Phase II

Once the extent of the damage has been defined as precisely as possible, a management decision should be made on which of four options should be used (Cairns, 1983). These are:

1. Restoration to original condition.
2. Rehabilitation of some of the original conditions and possibly some that were not there originally.
3. Development of an alternative ecosystem (e.g., a pond or lake where there was a terrestrial ecosystem previously).
4. Neglect or natural reclamation (doing nothing either because of inadequate information, or because the courses of action available might incur additional damage).

Although ecologists instinctively want to choose restoration to original condition, this is not always possible, either for physical reasons or because the knowledge or funds are not adequate to ensure success. For example, strip mining in the western part of Virginia and most of the state of West Virginia, as well as parts of Kentucky, is carried out in areas where the slope may well be in excess of 20 degrees (Figure 2). Under these conditions, restoring to original contours is often impossible for two reasons. First, even after coal or minerals have been removed, the broken up rock and other material will often not fit in the space from which it was removed. Replacement on the original site most commonly results in a bulge of material that does not closely resemble the original contour. Second, even if the original contours are restored in areas where the slope is steep, it is unlikely

FIGURE 2. Devastated landscape generated by unregulated coal surface mining in the southern Appalachians of Virginia. Ecosystem recovery on such sites is hampered by acid spoil materials, excessive spoil compaction, low water holding capacity, and low levels of nitrogen and phosphorus available for plants. (Courtesy of W.L. Daniels, Powell River Project.)

that they will be maintained long enough for the indigenous vegetation to become fully established.

In the phosphate mining areas of Florida, where large quantities of material are removed from the surface of a relatively flat landscape, a majority of the citizens may prefer lakes to the original condition of the land. If any option other than restoration to original condition is selected, a determination must be made as to whether the level of importance of the ecosystem is national, state, or local. It is important that the citizens (or their representatives) at the level selected be fully informed in lay terms of the reasoning behind the selection of the alternative options, the course of action proposed, the time frame within which it will occur, and that they be given a fairly explicit description of the final product. One or two open public meetings would be the best course of action at the local level. For larger systems, a consensus reached by representatives of various organizations, environmental and commercial as well as public, will probably furnish the most detailed and important commentaries. At any rate, some considerable attention to these matters will be essential.

Phase III

The proposed course of action decided upon in Phase II should be immediately undertaken because there is good evidence that postponement often increases the cost of whatever option is finally chosen. Additionally, the ecosystem will presumably serve a wider variety of public needs after whatever course of action is undertaken than it will in its damaged condition. If the damage was caused by a persistent hazardous chemical, there are public health reasons for expeditious action. The work of the organization(s) chosen to implement the course of action selected should be monitored by a scientific advisory group representing the various interests involved. The advisory board should make periodic reports in lay terms to the representatives of interested organizations and the general public. Detailed evidence should be made available to them on the progress being made as well as any reasons for deciding the performance has been satisfactory or unsatisfactory.

Phase IV

Determining when management responsibilities for implementing the option chosen have been fully met is particularly difficult because of lack of evidence on the reasonableness of the various courses of action. For example, for Virginia strip mine reclamation, all of the practices are implemented immediately (grading, seeding, planting, fertilizing, etc; Figure 3). Then at the end of five years, the results are examined and determined satisfactory or unsatisfactory by a state agency. Since the five-year period has not yet elapsed for any of the sites known to me, it is difficult to determine how successful this particular policy will be. I have chosen it because it illustrates some of the problems involved. The intent is clearly sound—namely, to have vegetation that will survive the climatic conditions of that region without management intervention for a moderate number of years. Since the organization doing the strip mining is bonded and a significant amount of money is tied up while the bond is in effect, the five years represent

(A)

(B)

FIGURE 3. Thick, uncompacted, productive minesoils produced by controlled placement, grading, fertilization, and seeding of nontoxic overburden in southwest Virginia. (A) Approximately two months after seeding; the vigorous growth in the foreground plots is due to sewage sludge applications. (B) Photograph taken after the end of the second growing season, usually the critical period for the establishment of permanent vegetation. In order for this plant–soil system to become fully self-sustaining, however, considerable amounts of atmospheric nitrogen and mineral phosphorus must be fixed into organic form by the vegetative and microbial communities, then decomposers must become established to release the nutrients back to the plant community. To achieve this, a diversity of nitrogen-fixing legumes, grasses, and woody species must be established on the site with appropriate microbial populations. (Courtesy of W.L. Daniels, Powell River Project.)

a compromise between ecologists, who might feel that it is not enough time, and those paying for the bond, who almost certainly feel it is too long. During the five years, no additional seeding, fertilizing, watering, or other management practices are permitted. One might make a good case on ecological grounds for permitting two years of management of any kind and requiring that the system be acceptable after three years of no-management intervention. It seems at least possible that the quality of the ecosystems might be better and that more of them might meet the specifications required if the management period were longer. It might also reduce the cost even if only by spreading it over a greater number of years. More important, the number of total failures should be reduced as should the number of fresh starts required.

To my knowledge, an explicit statement of ecosystem qualities that will be required to determine when management is no longer needed in the rehabilitation process (whether the initial disturbance was strip mining, hazardous chemicals, or anything else) is not available. Furthermore, it seems improbable even if a list were available that qualities such as stability or diversity would be uniformly interpreted by a majority of ecologists (Pimm, Chapter 14). Ecologists should give considerable attention to these matters, not only because they are of considerable national and international importance but because they are an ideal means of testing our understanding of ecosystems and a number of theoretical models. We should also keep in mind that perfection in these activities will probably not be achieved in our lifetimes, but improvement over the present situation is almost guaranteed. Therefore, flexibility in our attitudes, as well as the regulatory and industrial attitudes, should be fostered.

ECOSYSTEMS RISK MATRIX

Every ecosystem in the world is theoretically at risk from an accidental spill of hazardous material or from such widespread problems as acid rain. However, certain types of ecosystems undoubtedly are disturbed more frequently than others, and vulnerability to disturbance is clearly not uniform. A simple association matrix (Table 1) can be used to identify the categories where information should be obtained. Ecologists will undoubtedly have a difficult time determining degrees of vulnerability, but even the crudest estimate will enhance efforts to develop better precision.

Another correlation that will help with decision making relates the severity of disturbance to the time of recovery or the degree of ecosystem displacement. Possibly the latter should be used initially

TABLE 1. Simple association matrix to determine categories of information needed on recovery of ecosystems with different frequencies of disturbance and varying vulnerability.

	Vulnerability to disturbance			
Frequency of disturbance	Extremely vulnerable [1]	Highly vulnerable [2]	Moderately vulnerable [3]	Minor vulnerability [4]
[1] Once a year or more	1	2	3	4
[2] 1–3 years	2	4	6	8
[3] 3–10 years	3	6	9	12
[4] 10–100 years	4	8	12	16
[5] >100 years	5	10	15	20

since there is a toxicological data base for making the displacement estimate but little information for estimating recovery time.

HAZARDOUS WASTE SITE CLOSURE

Although hazardous waste sites are small in total area, their closure will be one of the most difficult problems we face in rehabilitating disturbed ecosystems. These sites were developed for chemicals so hazardous to human health and the environment that immobilization and storage were considered preferable to release into the environment. Although there are some exceptions, most of the material is difficult to transform into less harmful material. Unfortunately, many of these sites were poorly designed—as most of the public now knows. Monitoring of their performance has ranged from nonexistent to exemplary. Because containment of the hazardous wastes has not been as effective as originally predicted, or because new methods for improved containment or transformation of the hazardous materials are now available, closure of many of these sites is now being considered. In many cases, hazardous materials will have seeped into the surrounding area and may be difficult to isolate and immobilize. In such cases, rehabilitation of the disturbed ecosystem will have to be carried out in such a way as to minimize risk to human health and the environment in the surrounding area rather than selecting characteristics matching those of the original ecosystem.

The most important criterion for selecting colonizing species will be reducing transport of hazardous chemicals into the surrounding area. Transport might occur because species from the surrounding area feed in the disturbed site or when the hazardous material is transported as detritus. In addition, the probable effect of any introduced organisms on the geohydrology must be given serious attention. If they alter soil characteristics (such as density, porosity, or permeability), the hydrostatic conductivity or gradient, the infiltrating rainwater rate, or the groundwater or surface water flow rates, the transport of the residual hazardous waste will almost certainly be altered, possibly in an undesirable way. Fortunately, if the preceding information is known, then equations using contamination as a function of depth, soil adhesion properties, distribution coefficients, solubility, and the like will permit relatively accurate predictions of transport potential. These provide estimates of contamination as a function of distance and time, unsaturated and saturated zone transport rates, hydrological flux, surface and subsurface water contamination potential, and material balance. For example, a pathway analysis for metals transport would show the transport rate to be affected markedly by soil adhesion properties. These would in turn be affected by such things as pH and alkali and organic content of the soil. One could then use regulatory standards for the hazardous materials to determine whether or not the measures taken to implement ecosystem recovery from disturbance would keep the concentration of the hazardous substances within acceptable limits.

ARE ECOSYSTEM SERVICES MORE EASILY RESTORED THAN ECOSYSTEM FORM?

An important but essentially unaddressed question is whether ecosystem "services" (functions beneficial to society) are more easily restored than ecosystem form (species diversity, trophic structure, and the like). There are three possible scenarios:

1. Ecosystem form and services are so intimately related that one cannot be restored without inadvertently restoring the other.
2. There is so much functional redundancy in natural systems that services can be restored much more easily than form.
3. Ecosystem form must be restored completely in order that the services be delivered under all seasonal and other changing conditions, as well as to reduce the management costs that are incurred when functional redundancy is minimal.

Magnuson et al. (l980) differentiate between ecosystem *recovery* (a return to original condition following displacement) and *rehabilitation* (restoration of certain desirable attributes that have been altered, but not all of the original attributes). They believe that under certain circumstances certain portions of the original form as well as certain of the original services can be restored independently of some of the other attributes. There seems to be some empirical evidence that this can indeed be accomplished (examples include the restoration of the Thames River in the United Kingdom and Lake Washington in the United States). Additionally, they make a case for the creation of alternative ecosystems in which both form and services (these words were not used but implied) could be totally different from the original condition but designed to meet societal needs using ecological management principles.

I have discussed elsewhere (Cairns, 1983) the difficulties of having an adequate information base, as well as the scientific capability for restoring disturbed ecosystems to their original condition. Environmental impact statements often have exhaustive inventories of indigenous species, but I think even the most charitable ecologist would not accept an inventory alone as an adequate description of ecosystem form. Spatial relationships, recruitment rates, population and community dynamics, predator/prey relationships, trophic interrelationships, and a variety of other attributes would be necessary to describe ecosystem form adequately. As a consequence, even when a complete species inventory is available, restoring to original form will not be easily accomplished.

Most environmental impact statements pay little or no attention to ecosystem services, at least in so far as they are related to ecosystem function. The functional attributes of ecosystems upon which services presumably would depend heavily either are not described at all or are given only cursory attention. In cases of accidental spills where no predevelopment information was gathered (as opposed to surface mining and other types of planned disturbance), one would be forced to use presumably comparable ecosystems as models for restoring either form or services.

Ironically, both form and services are more easily described for alternative ecosystems than for most original ecosystems. This is because the alternative ecosystems in relatively common use are based on well-tried aquacultural or horticultural practices to achieve comparatively limited objectives. For example, a lake or pond that replaces a terrestrial ecosystem consisting of pine, shrubs, etc. would have an ecosystem form designed to optimize recreational fishing services. If an array of restoration options, all attractive to the public and achievable, are to be available, the scientific information base on restoration of ecosystem form and services must be markedly improved.

USING DAMAGED ECOSYSTEMS TO PRESERVE
RARE, ENDANGERED, AND THREATENED SPECIES

The number of rare, endangered, and threatened species is probably increasing exponentially (Myers, Chapter 19). A large number of these will probably be lost forever if protective measures are not implemented soon. Another major problem is the increasing number of damaged ecosystems, many of which cannot be restored to their original condition either because the original condition is unknown or because the science and technology to do so are not presently available (Pimm, Chapter 14).

There is a marvelous opportunity to address both these problems simultaneously when the ecosystem cannot be restored to original condition. In some of these cases, the ecosystem could be restored in such a way as to enhance the chances of survival for one or more rare, endangered, or threatened species. Since a large number of species are in difficulty because of habitat loss, possibly some favorable habitat might be restored on the sites of surface-mined lands or where other types of damage have occurred. Surface-mined lands can be rejuvenated in the ecological sense to either aquatic or terrestrial ecosystems. Since restoration is now mandated by law, the alternative form of rehabilitation just espoused can be implemented through increasing the flexibility of both state and federal regulatory authorities (Cairns, in press).

In this age of computer data sorting and information display, a list could easily be compiled of rare, endangered, and threatened species by region, together with a list of habitat requirements for each one. A printout would be available of the damaged ecosystem sites where rehabilitation is contemplated along with a list of their habitat conditions at present and how well they match those of the rare, endangered, and threatened species. Considerable scientific judgment would be necessary to determine how many of the missing habitat characteristics could be supplemented in an ecologically justifiable way to meet the requirements of one or more species. This idea should be tested in three stages.

1. Stage 1 involves making a damaged ecosystem acceptable to a relatively common species indigenous to the region. The advantage of doing this is that it enables one to work out the ecological prerequisites and to test predictive models without risking a species

already in danger. This design offers a higher probability of success because the habitat requirements are being checked against a large number of existing habitats within the region. In the case of an endangered species, the current habitat is probably very restricted. Such a species, therefore, is not a good candidate for a research program of this magnitude. It is worth noting that Stage 1 should pose no serious regulatory problems because a damaged ecosystem is being restored to provide habitat for a component of the indigenous biota. However, some regulatory flexibility will be required because that species may not have lived on this particular site before it was destroyed.

2. When a sufficient number of successful demonstration projects have materialized and the habitat modeling has become more precise and scientifically sound, Stage 2 can begin. In this stage, a threatened species from the same region will be used, but the species should not be threatened everywhere. This will enable a further check on the rehabilitation model's capabilities and species habitat requirement estimates and, at the same time, introduce another important component: moving a species that is not doing well elsewhere in the same region to a site where ecological conditions are thought to be better as a consequence of the rehabilitation effort. The importance of this exercise is to determine whether a strain or race can be established on a site other than the one it inhabits. No doubt many variations will occur among plants and animals that are on one of the three lists; part of this exercise is to accumulate information on the types of variablity and adaptability that exist. Presumably, most threatened species are in danger because of their limited adaptability, but this may not necessarily be true.

3. For the final and crucial Stage 3, an endangered species likely to become extinct soon will be considered, and regenerated habitat on a damaged ecosystem site will be used for continuing its existence. In a sense, the site would be similar to an outdoor "zoo" or a botanical garden, but with a habitat that would permit natural recruitment of replacements for the species.

Organizational requirements

Perhaps an organization such as the Nature Conservancy could accept the organizational requirements of this new mission. If not, perhaps a new organization, the "Species Conservancy," could be formed. Alternatively, various wildlife organizations already devoted to the preservation of species might form a consortium for this purpose. Among the responsibilities of the organization would be:

1. Identifying species that need help and matching their needs with potential sites.
2. Gathering experts with the appropriate experience to carry out each specific project.
3. Accepting the management responsibilities and stewardship of those sites where the effort is successful. Presumably, a few people in the vicinity of each site might form an organization for this purpose.

Financial support

Since industries and organizations that have damaged ecosystems are now legally responsible for their rehabilitation, in some instances this proposed process could be paid for entirely within these requirements. In others, more than the normal cost of rehabilitation may be necessary; this might be provided by special tax incentives to the industry or other organization responsible for the rehabilitation. In other cases, additional funding will undoubtedly be necessary. A particularly troubling problem will be those instances where the attempt to turn a damaged ecosystem into an acceptable habitat for a particular species has failed and the site must be rehabilitated for other purposes. Presumably, since an ecologically viable system was produced, it could be left alone without any further effort. In some instances, the continuing management requirements for ecological stability might be greater than they would have been with "normal rehabilitation." In such instances, the industry or other organization willing to experiment should not be penalized for doing so, and additional funds should be available for such experimentation. Since the National Science Foundation and other government funded agencies have been instructed to consider some of the practical benefits of research, perhaps some of the funding could be through these traditional sources.

Derelict or abandoned ecosystems, where the people and institutions who should bear the burden of the rehabilitation costs are no longer alive or accessible, pose a particular problem. The cost of such rehabilitation will clearly have to be borne by either state or federal governments. These sites are the ones in which particularly risky experimentation should be carried out, leaving the more straightforward rehabilitation for sites where funding sources other than tax dollars are more clearly available. The opportunities for ecological research on such sites are enormous (Cairns, in press), and perhaps directing some of the purely theoretical projects to sites of this sort might furnish additional evidence that will help in habitat rehabilitation and, at the same time, furnish data that will be useful to theoretical ecologists.

SUGGESTED READINGS

Bradshaw, A.D. and M.J. Chadwick. 1980. *The Restoration of Land*. University of California Press, Berkeley.

Cairns, J. Jr. (ed.). 1980. *The Recovery Process in Damaged Ecosystems*. Ann Arbor Science Publishers, Ann Arbor, MI.

Cairns, J. Jr., K.L. Dickson and E.E. Herricks (eds.). 1977. *Recovery and Restoration of Damaged Ecosystems*. University Press of Virginia, Charlottesville.

Dubos, R. 1980. *The Wooing of Earth*. Charles Scribner's Sons, New York.

Hardin, G. and J. Baden (eds.). 1977. *Managing the Commons*. Freeman, San Francisco.

Holdgate, M.W. and M.J. Woodman (eds.). 1978. *The Breakdown and Restoration of Ecosystems*. Plenum, New York.

Janick, J. 1963. *Horticultural Science*. Freeman, San Francisco.

Lewis, R.R. III (ed.). 1981. *Creation and Restoration of Coastal Plant Communities*. CRC Press, Boca Raton, FL.

Loehr, R.C. (ed.). 1977. *Land as a Waste Management Alternative*. Ann Arbor Science Publishers, Ann Arbor, MI.

Swanson, G.A. (coordinator). 1979. *The Mitigation Symposium*. General Technical Report RM-65, U.S. Government Printing Office, Washington, DC.

CHAPTER 24

THE DESIGN OF A NATURE RESERVE SYSTEM FOR INDONESIAN NEW GUINEA

Jared Diamond

In this chapter I shall illustrate how biological principles have been used in the comprehensive design of a terrestrial nature reserve system for Indonesian New Guinea. I shall draw further examples from Papua New Guinea and other Pacific countries where I have also worked as consultant in the design and establishment of reserve systems.

My presentation will focus on terrestrial vertebrates, the species with which I am most familiar. In practice, considerations of public relations require conservation biology to emphasize vertebrates, because the public cares far more about vertebrate species than about other taxa. For example, the public appeal of tigers, big mammals, and the California condor induced the governments of India, Kenya, and the United States, respectively, to set aside as reserves large areas of habitat that would otherwise surely not have been spared.

Fortunately, there are reasons why reserves designed to protect vertebrates are likely to be valuable for protecting other species. First, distributions of most terrestrial vertebrate species depend on habitats defined by plants, just as do the distributions of invertebrate species. Hence there is often much overlap between distributions of vertebrates, invertebrates, and plants. Second, large-bodied vertebrates tend to be the species with the lowest population densities. Thus, any reserve large enough to contain self-sustaining populations of large vertebrates is likely to contain self-sustaining populations of other species whose distributions are centered on that reserve. Finally, the top predators in an ecosystem are generally vertebrates. For that reason, significant losses of vertebrates may cause losses of many other species (Willis, 1974; Karr, 1982a; cf. Paine, 1966).

Nevertheless, in discussions of reserve design that focus on vertebrates, one must beware of overlooking considerations important for other species. For instance, many species of invertebrates and plants occurring at high population densities may be endemic to a very small area that in itself would not warrant a reserve for vertebrates. Thus, plants and invertebrates may require small "vest-pocket reserves" in addition to the larger reserves needed for vertebrates. To name one example, the world's largest butterfly, the Queen Alexandra Birdwing, is confined to a small area in the lowlands of the northern watershed of New Guinea's southeastern peninsula. The Papua New Guinea government wisely decided to set aside a protected area, the Hurupa-Jagiko Wildlife Management Area, centered on the range of this splendid butterfly. However, the converse is not true: while the large reserves essential for vertebrates are likely to be valuable for other species, the vest-pocket reserves required for local species of plants and invertebrates will play little role in vertebrate conservation and will require more intensive management and habitat manipulation (Wilcove et al., Chapter 11; Lovejoy et al., Chapter 12).

OPPORTUNITIES AND PROBLEMS FOR CONSERVATION IN NEW GUINEA

My work in conservation biology takes place on the islands of the southwest Pacific, primarily for the governments of Papua New Guinea and Indonesian New Guinea, the government of the Solomon Islands, and the South Pacific Commission (the regional consortium of Pacific countries). Among Pacific nations the threats to the biota vary, and so do the conservation programs. At one extreme, the New Zealand Wildlife Service operates one of the most advanced and boldest conservation programs in the world. At the other extreme, some Pacific countries have no effective conservation program. This discussion will emphasize the island of New Guinea.

The New Guinea biota

New Guinea is one of the world's biological treasures (Gressitt, 1982). It is the world's largest island after Greenland, and the largest tropical island. Along with the Andes and the mountains of East Africa, New Guinea's Central Cordillera, rising to 16,500 feet, is one of three places in the world where glaciers occur on the equator. New Guinea actually functions as a miniature continent; it is large enough to have supported numerous cases of speciation in both birds and mammals by the mechanism otherwise virtually confined to the major continents— namely, formation of geographic isolates within a single land mass (Diamond, 1977).

The mammals of New Guinea comprise about 200 species, most of which are endemic to the New Guinea region. These mammals are approximately equally divided between marsupials, rodents, and bats. The best known and largest of the marsupials are tree kangaroos, wallabies, and phalangers. There have been several radiations of rodents, including giant rats. The radiation of bats includes some of the world's largest bats, *Pteropus* fruit bats with a five-foot wing spread.

The breeding avifauna consists of over 500 species. About half of the species and many of the genera are endemic, so that New Guinea has the most distinctive avifauna of any comparable-sized region in the world. Best known of the birds are the birds of paradise, bowerbirds, mound-builders, and cassowaries. Many of the other birds appear superficially similar to Eurasian warblers, flycatchers, nuthatches, and creepers and were long classified as such. However, recent biochemical studies have shown that these New Guinea and related Australian taxa constitute an independent local radiation, an example of convergent evolution similar to that of the marsupials (Sibley and Ahlquist, 1985).

The reptiles on New Guinea include an endemic species of freshwater crocodile, and the world's longest lizard (*Varanus salvadorii*, which can be up to 15 feet long). New Guinea's insects include the bird-wing butterflies *(Ornithoptera)*, among which is the world's largest butterfly; and a group of large flightless beetles, whose backs bear specialized pits in which grow miniature forests of moss, lichens, and liverworts, in which in turn lives a miniature fauna comprising an endemic family of mites plus rotifers and nematodes (Gressitt et al., 1965). New Guinea's plant communities include the world's largest expanses of mangroves, a large variety of orchids and rhododendrons, and large areas of sago swamp *(Metroxylon sagu)*.

At present, the New Guinea biota is still intact, with large expanses of natural habitats and no documented extinctions. However, problems are accumulating.

Socio-political considerations

The island of New Guinea is politically divided into two halves whose differences in government require very different approaches to conservation (Figure 1). The eastern half, the independent country of Papua New Guinea, was formed from the former British colony of

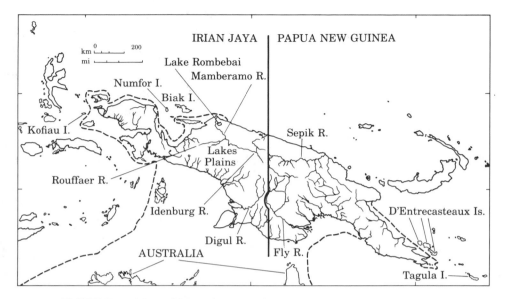

FIGURE 1. Map of New Guinea, showing the international border (heavy, nearly vertical line) between Irian Jaya in the west and Papua New Guinea in the east; the edge of the New Guinea–Australian continental shelf as a dashed line (the area within this dashed line was dry land at low sea level times of the Pleistocene); oceanic islands lying off the continental shelf; some major rivers and lakes; and the inland basin of the Lakes Plains.

Papua and the former German colony of Kaiser-Wilhelms-Land after a long period of Australian colonial mandate government. The population consists of several million New Guineans and only very small European and Chinese minorities. The people are divided into over 400 mutually unintelligible languages, each language being spoken on the average by a few thousand people. In the highlands the languages replace each other approximately every 10 miles. Every bit of land in the country is owned by some clan, and people feel close ties to the land.

Papua New Guinea is a successful democracy. Since independence there have been three elected changes of government, all of them smooth. In addition to the national government, there are strong provincial governments with considerable responsibility for economic development and other local affairs. Related to Papua New Guinea's democratic government and New Guineans' long tradition of freedom from higher authority, the government does not forbid private ownership of firearms or free movement of people throughout the country. Because all land is privately owned and viewed as nearly sacred, the government finds it difficult to acquire land for any purpose, including national parks.

These socio-political considerations determine approaches to conservation biology in Papua New Guinea. Of the two main threats to wildlife, one is the usual threat of habitat destruction from an expanding population and the timber industry. The other threat, and the more acute one, is the shooting of large birds and animals by men carrying shotguns and traveling freely around the country. New Guineans traditionally went to great effort and expense to acquire plumes of birds of paradise and other birds, but former conditions of intertribal warfare restricted each people to their own territory and created extensive no-man's lands. Hunters now charter airplanes to remote areas.

Because government acquisition of land in Papua New Guinea is difficult, conservation will largely depend on an indigenous concept called wildlife management areas. Under this concept, local people manage their wildlife for their own benefit: they are permitted to exclude other people as hunters, but in return must accept restrictions on their harvest designed to ensure self-sustaining populations. Papua New Guinea already has numerous functioning wildlife management areas, plus several small national parks.

The western half of the island of New Guinea is a former Dutch colony that is now a province (Irian Jaya) of the much larger country of Indonesia. The native population of Irian Jaya is as fragmented linguistically as that of Papua New Guinea but is sparser. Irian Jaya's population of about 1 million is dwarfed by Indonesia's total population of over 120 million. Indonesia has a strong centralized government based on Java, several thousand miles west of Irian Jaya, with military participation. The top figures in the provincial government of Irian Jaya are not native New Guineans but instead are mainly people from Java, Celebes, and Sumatra. As one approach to Java's overpopulation problem, Javans are being resettled in Irian Jaya, and eventually native New Guineans will probably constitute a minority. Private ownership of firearms in Irian Jaya is absolutely forbidden and nonexistent; only the army and police have firearms. Visitors from other countries, and Indonesians (including New Guineans) travelling away from home, must register with local police and obtain permission for a stay. Irian Jaya therefore does not have Papua New Guinea's problem of local people travelling around their country with firearms to hunt wildlife for their private purposes. Instead, there is an illegal but extensive trade in fauna for the international market, specializing in parrots, crocodiles, turtles, and reef animals. The other main threat is habitat destruction from logging.

Because Indonesia has a strong central government, and because the ultimate authority over land rests with the government rather than with local tribesmen, the government can plan large national parks whose establishment in Irian Jaya has recently begun. The

Indonesian government plans to devote about 20 percent of the area of Irian Jaya to reserves. Ten individual reserves exceed 1000 square miles in area, and two reserves exceed 5000 square miles. In addition, an approximately equal total area will be set aside as protected forests.

BIOLOGICAL CONSIDERATIONS RELEVANT TO THE DESIGN OF IRIAN JAYA'S RESERVE SYSTEM

The remainder of this chapter emphasizes the reserve system of Irian Jaya, whose design involved compromises between biological and socio-political considerations. The biological input into the selection of protected areas requires considering four questions:

1. What are the likely potential causes of extinctions?
2. What are the major types of habitats that support distinctive biological communities?
3. What are the major biogeographic districts that constitute separate centers of endemism?
4. How much area is required for effective conservation?

Let us now consider these questions in turn.

Modern causes of extinctions

The purpose of a reserve system is to protect species against the risk of extinction. Hence the first biological considerations that require discussion are the potential causes of vertebrate extinctions, which I shall summarize from a worldwide perspective to place New Guinea's problems in context. A conservation biologist, asked to give a brief summary of causes of extinctions in the modern world, finds himself in a situation like that of a physician who has authored one of the usual 2000-page textbooks of medicine and is asked to summarize briefly the causes of disease. Here it goes.

Even under natural conditions, local populations of vertebrates risk extinction. The two main predictors of natural extinction rates are small population size, and large temporal coefficient of variation of population size (MacArthur and Wilson, 1967; Leigh, 1981; Karr, 1982b; Diamond, 1984a). While natural processes are responsible for many witnessed extinctions of local subpopulations on islands and in habitat patches, the modern background rate of natural extinctions for whole species is very low. Instead, most or nearly all recorded modern extinctions of vertebrate species appear to have been caused by man. For example, man is in one way or another responsible for the extermination of about 63 of the 4200 modern species of mammals and about 88 of the approximately 8500 modern species of birds since A.D. 1600. The diverse mechanisms by which man has caused these extinctions (reviewed in Diamond, 1984b) fall into five categories:

1. *Overkill.* This is the sole mechanism by which man has exterminated or decimated marine species, such as Steller's sea cow plus various species of whales, seals, turtles, crocodiles, and fish. It has also been the or a main cause for virtually all exterminations of large terrestrial mammals and giant tortoises, plus about 15 percent of bird extinctions and some extinctions of trees logged for wood or sap (Gentry, Chapter 8). In New Guinea the terrestrial species most under hunting pressure are Goura pigeons, large cockatoos and Pesquet's Parrot, birds of paradise, imperial pigeons *(Ducula)*, hornbills, tree kangaroos, and wallabies.
2. *Habitat destruction.* This has now become the leading cause of extinction worldwide, especially as the destruction of tropical rainforest—the world's most species-rich habitat—has accelerated (Myers, Chapter 19). Two related but distinct effects are involved: reduction of total habitat area, and fragmentation of large chunks of habitat into small pieces. The most frequent form of habitat destruction has been deforestation for timber, agriculture, or stock grazing. Other forms include impacts by introduced grazing and browsing animals, especially goats and rabbits, and wetland drainage and fire. Commercial logging, subsistence agriculture, and commercial oil-palm plantations are currently the main motives for habitat destruction in New Guinea.
3. *Impact of introduced species.* The most spectacular effects of introduced species on native faunas have been the effects of introduced predators on faunas lacking experience of functionally equivalent predators (Pimm, Chapter 14). About half of all extinctions of island birds have been due to introduced mammals (especially rats and cats) on islands lacking native mammals and also lacking native land crabs (the invertebrate equivalent of rats). Native birds of islands with either native rats or native land crabs have been virtually immune to introduced mammals (Atkinson, 1985).

    Predators are not the only introduced species that have contributed to extinctions. As indicated in item two above, introduced mammalian herbivores have eliminated native plants on islands whose plants evolved in the absence of mammalian browsers. Victims of introduced pathogens have included Hawaiian native birds, the American chestnut, and Swayne's hartebeest and other African ungulates, decimated by avian malaria, chestnut blight, and rinderpest, respectively (see also Dobson and May, Chapter 16). Notable impacts of introduced competitors have included effects of introduced fish on native fish throughout the world (often involving mixed predator–competitor effects; see Werner, 1985), effects of introduced placental herbivores on native Australian marsupial herbivores, and the impact of introduced deer on New Zealand's flightless grazing rail, the takahe.

   While introduced predators and herbivores are the main cause of extinctions on Pacific islands from the Solomons eastwards, they are not currently a major threat in New Guinea, whose biota was "immunized" by long exposure to native predators and herbivores. The introduced mammals with feral populations established to date consist of several rat species, pigs, dogs, cats, and rusa deer. Potentially far more dangerous would be the transport of monkeys to New Guinea from other Indonesian islands.

4. *Pollutants*. Chemical toxicants are rising in importance as agents of extinction. Victims to date include barn owl populations decimated by the new generation of superpoisons against rodents, and Guam birds to whose elimination pesticides may have contributed (Jenkins, 1983; but see Grue, 1985; Savidge, 1985).

5. *Secondary effects*. One extinction may lead to other extinctions as a secondary consequence (Pimm, Chapter 14). The best documented examples are the chains of extinctions in rivers or lakes following introduction of piscivorous fish (Zaret and Paine, 1973; Diamond, 1985a; Werner, 1985). Other examples involve plant–pollinator and fruit–frugivore systems.

The relative importance of these modes of extinction varies from case to case. Hence one of the practical tasks of the conservation biologist concerned with preventing extinctions is to determine which modes have been or are likely to be operative in a given situation. For example, on Rennell Island, an island of the Solomon Archipelago with no native mammals and numerous endemic bird taxa, I was especially concerned with the risk from introduced rats and forest destruction (Diamond, 1984e). On Guam, where practically all native forest-dwelling land birds have recently become either extinct or endangered, habitat destruction has been a negligible factor, while introduced diseases, chemical pollutants, and especially an introduced snake appear to be the main factors (Jenkins, 1983; Savidge, 1985). The efforts of the New Zealand Wildlife Service to protect two endangered native bird species, the takahe and saddleback, are focusing on an introduced competitor in the former case, and introduced predators in the latter. For most bird species of the continental tropics today, the main risks come from habitat fragmentation and habitat destruction. For New Guinea today, I consider overkill and habitat destruction to be the main risks.

Habitat considerations

New Guinea's mountains rise to 16,500 feet. From lowland rainforest one ascends through hill forest, oak forest *(Castanopsis)*, southern beech forest *(Nothofagus)*, subalpine forest, and alpine grassland to glaciers on the highest peaks. An essential consideration in reserve design is that almost all species occupy only a fraction of this altitudinal gradient. For instance, 90 percent of New Guinea bird species have altitudinal ranges spanning less than 6000 feet, and many span less than 1000 feet. A further consideration is that even species occupying broad spans, such as certain birds of paradise and bowerbirds, often require the whole span for their life cycle and are unable to complete their life cycle in a fraction of it: they breed at high altitudes but live as immatures at low altitudes. Species diversity decreases with altitude, but the proportion of species endemic to New Guinea increases with altitude. Hence the foremost habitat considerations in reserve design are that reserves must be selected to represent all altitudinal bands, and that the protected altitudinal bands in a given district must be joined as a continuous transect in order to permit seasonal and age-related altitudinal migration of individual animals.

In addition to this vertical sequence of communities, there is a horizontal sequence at the same altitude. While most of New Guinea is covered by forest of various types, extensive dry areas of southern New Guinea are covered by savanna woodlands of *Eucalyptus* and *Melaleuca*, with a biota more similar to that of Australia than to the rest of New Guinea. Other nonforest communities (Figure 1) include several lakes of modest size (Rombebai and Bian in the lowlands, Wissel and Anggi in the mountains), several major rivers [Fly, Digul, Mamberamo and its tributaries the Rouffaer (Tariku) and Idenburg (Taritatu), plus the Sepik in Papua New Guinea], extensive marshes (especially in the basin termed the Meervlakte or Lakes Plains that includes the Rouffaer and Idenburg Rivers, plus the marshes of the Fly River and other areas of southern New Guinea), strand vegetation, and grassland (largely anthropogenic in the mountains). Among specialized forest types New Guinea's mangrove forests are the most extensive in the world, and its swamp forest (especially those dominated by sago, *Metroxylon sagu)* are also extensive. Substrates associated with distinctive forest types include limestone, ultrabasic rocks, and alluvium. On a given substrate, closely related taxa can be found replacing each other between high- and low-rainfall areas (for example, the fruit dove *Ptilinopus pulchellus* versus *P. coronulatus).*

The Irian Jaya reserve system has been designed to incorporate these specialized vertical and horizontal sequences of communities (Figure 2). The system encompasses a complete altitudinal cross sec-

FIGURE 2. Terrestrial reserves established or approved in Irian Jaya as of 1983. Reserves 1 and 2 exceed 5000 square miles in area, while reserves 3, 4, 5, 6, 9, 10, 14, and 15 exceed 1000 square miles. The northern, southern, and Vogelkop lowland centers of endemism are protected by reserves 2 and 3, 1 and 6, and 10 and 14–17, respectively. The main west–east blocs of the Central Cordillera are in reserves 1, 4, and 5. The outlying mountain ranges of the New Guinea mainland are in reserves 2 and 11–17. Reserves 18–22 are on land bridge islands, while reserves 23 and 24 are on oceanic islands. There are large protected areas of savanna in reserves 6–9; mangrove in reserve 10; and marshes, lakes, and major rivers in reserves 2–4 and 7. Eleven smaller reserves are not depicted on this map.

tion of New Guinea, including its highest peaks and equatorial glaciers. There are large protected areas of savanna in the reserves at Kumbe/Merauke and Pulau Dolok, and there is a large mangrove reserve at Teluk Bintuni. The major lakes, most of the Lakes Plains, the whole length of the Mamberamo River, and extensive portions of other marshes and rivers are also included in reserves.

Biogeographic considerations

Let us consider first the differentiation between New Guinea itself and its offshore islands, then the differentiation of biogeographic districts within New Guinea.

New Guinea's offshore islands fall into three categories (Figures 1 and 3). First come those islands of the north and southeast coasts that lie off the shallow shelf surrounding New Guinea (the Sahul Shelf). These lacked Pleistocene land connections to New Guinea and were colonized overwater from New Guinea. Of these, Biak is the largest and has the most endemic species, followed by Kofiau and Numfor of Irian Jaya plus Tagula and the D'Entrecasteaux group of Papua New Guinea. The islands of the Sahul Shelf (Yapen, Salawati, Batanta, Waigeu, and Misol) had intermittent land connections to New Guinea during the Pleistocene and received overland some colonists that subsequently differentiated, especially on Waigeu and Batanta. All these major islands of Irian Jaya have planned reserves except for Kofiau, on which logging had unfortunately proceeded too far before the reserve system was set up (Figure 2). Finally, New Guinea's numerous scattered small islands share some species ("supertramps") absent from the New Guinea mainland, and the small islands serve as important breeding sites for marine turtles and certain pigeons. Many of the small islands, including Mapia, Sayang, Auri, Asia, Aju, and

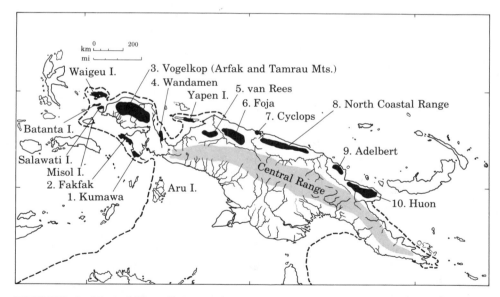

FIGURE 3. Map of New Guinea, showing the Central Cordillera, shaded; the outlying mountain ranges of the New Guinea mainland, numbered and in solid black; the edge of the New Guinea–Australian continental shelf as a dashed line; the five large islands of the continental shelf, which were connected to the New Guinea mainland during the Pleistocene by land bridges; and the mountain ranges on these land bridge islands in solid black.

Sabuda, are included in reserves whose importance for the terrestrial biota is overshadowed by their importance for the reef biota and as turtle nesting beaches.

New Guinea itself contains many centers of endemism, each with its own set of endemic species. For instance, among the ribbon-tailed birds of paradise (genus *Astrapia)*, the species *A. nigra* is confined to the Arfak Mountains, *A. rothschildi* to the mountains of the Huon Peninsula, and *A. splendissima*, *A. mayeri*, and *A. stephaniae* to the western, central, and eastern portions respectively of the Central Cordillera. Thus, it is not enough to have one reserve for each habitat type; there must be multiple reserves in different parts of the New Guinea "continent".

Clearly, information about species distributions and centers of endemism is critical in deciding where reserves are needed to protect local species. Let us briefly summarize New Guinea biogeographic distributions for the Central Cordillera, outlying mountains, and lowlands.

The Central Cordillera runs east–west for 2500 miles, with no passes under 5000 feet (Figure 3). Hence, for those hillforest species confined to elevations under 5000 feet, the populations of the northern and southern watersheds are isolated, and each watershed has endemic taxa (such as the honeyeater *Meliphaga montana* of the northern watershed and *M. mimikae* of the southern watershed). For species living over 5000 feet endemic taxa replace each other east-to-west, as in the example of *Astrapia* species given earlier in this section. The four main divisions of the Central Cordillera are: (in Irian Jaya) the Snow Mountains and Weyland Mountains in the west; the Star Mountains and Jayawijaya Mountains in the east; (in Papua New Guinea) the Central Highlands just east of Strickland Gorge in the west, and the southeast peninsula east of Menyamya in the east. Each of the large blocks in Irian Jaya is the site of a large reserve.

Isolated from the Central Cordillera by a "sea" of lowlands inhospitable to montane species are nine mountain ranges (Figure 3) along the north coast (Huon, Adelbert, and North Coastal Range in Papua New Guinea; Cyclops, Foja, van Rees, Wandamen, Arfak, and Tamrau of Irian Jaya) and two on the south coast (Fakfak and Kumawa in Irian Jaya). All except the van Rees, the lowest of these mountains, have endemic montane species, such as the previously mentioned *Astrapia* species of the Arfak and Huon Mountains (Diamond, 1985b). Of these isolated ranges, the Arfak, Tamrau, and Huon have the highest mountains and are richest in endemics, followed by the Foja Mountains. All these isolated mountain ranges of Irian Jaya have planned reserves except for the van Rees Mountains, which are low, have few inhabitants, and are remote from risk of exploitation in the foreseeable future (Figure 2).

Just as New Guinea's mountains are carved into island-like centers of endemism for montane species by the lowlands, so too New Guinea's lowlands are carved into districts with endemic lowland species by its mountains. The three main lowland biotas are that south of the Central Cordillera in both Irian Jaya and Papua New Guinea, that north of the Central Cordillera in both Irian Jaya and Papua New Guinea, and that of the Vogelkop Peninsula (Kepala Burung) at the west end of Irian Jaya. For instance, each of these three districts has a different species of crowned pigeon (genus *Goura*), large mound-builder *(Talegalla)*, and streaked lory *(Chalcopsitta)*. Irian Jaya has very large lowland blocks reserved in the northern and southern lowlands, and smaller blocks on the Vogelkop Peninsula.

Area considerations

Besides specifying in what parts of New Guinea and in what habitats reserves should be located, biologists can also offer guidelines about how large a reserve needs to be. Reserve area determines whether particular species can maintain self-sustaining populations within the reserve, and how many species the reserve can maintain.

The minimum area requirement varies greatly with species. Common forest flycatchers and rats may be able to persist for decades in a woodlot of a fraction of a square mile in area, while eagles and tigers require hundreds and probably thousands of square miles. The reasons for these widely differing minimum area requirements are discussed in Chapter 2.

The larger a reserve, the more species it will contain. As a rough rule of thumb, a tenfold increase in area doubles the number of species. Part of the reason is that as a reserve gets larger it can sustain populations of species with larger area requirements. Another reason is that larger reserves contain a greater variety of habitats, hence more species limited to habitats absent from smaller reserves.

The New Guinea biota includes many species requiring very large areas for self-sustaining populations. Among these species are ones with huge territories, such as the New Guinea harpy eagle *(Harpyopsis novaeguineae)*; species confined to specialized habitats, such as the bird of paradise *Seleucides melanoleuca*, a denizen of sago swamps; flocking species that wander widely in search of food, such as the parrot *Pseudeos fuscata*; and species whose different age classes live at different altitudes (including many of New Guinea's montane bowerbirds and birds of paradise). If such species, which include many of

New Guinea's most distinctive endemics, are to survive in the wild, it is essential to have reserve areas concentrated in a moderate number of large blocks rather than in many small fragments. Indeed, the extinctions resulting from habitat fragmentation (Lovejoy et al., Chapter 12) are one of the things that a reserve system must prevent.

For this reason, although the Irian Jaya reserve system does contain vest-pocket reserves set up for special purposes, most of the area of the system is concentrated in large reserves. Large reserves have the added advantage of minimizing degradation of reserve usefulness through edge effects. In New Guinea as elsewhere, a forest at its edge becomes transformed in structure and in plant and bird species composition. The edge is of limited value in conservation of forest interior species, which are generally the species most at risk in the New Guinea region, while forest-edge species tend to be ones that also live in second-growth habitats and require no special conservation measures. The edge of a reserve is also the portion most subject to damage by human encroachment. (Chapters 11, 12, and 13 give details of edge effects in temperate and tropical forests.)

In areas of the world where one is designing reserve systems aimed especially at conservation of rainforest, and possibly of climax forest generally, I feel that the Irian Jaya policy of concentrating one's reserve area in large blocs is a wise one (for example, New Zealand forest; see Diamond, 1984c). There may, however, be situations where this is not the best policy—for example, if an entire reserve is at risk of degradation by some factor that can spread readily within a reserve but cannot spread readily between reserves. Countries where the prime risk to the biota comes from fire, disease outbreaks, and locally established introduced species are prime examples of places where one may prefer multiple small reserves over fewer large reserves. However, these latter risks are not the main ones in Irian Jaya.

EVOLUTION OF THE IRIAN JAYA RESERVE SYSTEM

The reserve system began in 1919, when the Dutch colonial administration gazetted a reserve that was later abolished. Efforts resumed under Indonesian sovereignty in 1975. At that time the Forestry Department made an initial proposal consisting of two reserves. More detailed proposals were made by van der Zon and Mulyani (1978) on an FAO-sponsored survey and by Schultze-Westrum (1978) on a World Wildlife Fund-sponsored survey. In 1979 I made my first visit to Irian Jaya and made further suggestions for reserves. FAO consultant John MacKinnon (1980) played an important part in detailed negotiations that began to establish the outlines of the reserve system.

A major practical step came in 1980, when Dr. Ronald Petocz took up residence in Irian Jaya as a consultant for World Wildlife Fund, with the goal of producing a detailed reserve plan for Irian Jaya province. The present status is summarized in reports by Petocz (1983, 1984). Some reserves have already been gazetted, and their boundaries have been delineated. Proposals for the remaining anticipated reserves have been made and are awaiting gazettement. A detailed management plan has been drawn up for one reserve, and others will soon be underway.

My involvement in the planning of the Irian Jaya reserve system has consisted of providing a proposal for the overall design, based on the biological principles discussed earlier in this chapter; making extended visits to Irian Jaya in 1979, 1981, 1983, and 1986 to survey the proposed Mamberamo/Foja, Rouffaer, Fakfak, Kumawa, Wandamen, Yapen, Salawati, Waigeu, Misol, and Batanta reserves; and providing information on bird distributions for other reserves (Diamond, 1980, 1981, 1984d).

At this point, it is still too early to predict what degree of success the Irian Jaya reserve system will meet. The conservation movement does not yet enjoy deep roots or widespread popular understanding in Indonesia. There is not yet a functioning reserve staffed with ground personnel. The threats from settlers from other provinces, the international faunal trade, and economic development (especially logging) remain considerable. At the same time, the Indonesian government's commitment to the reserve system has evidenced itself in the cancellation of timber leases and transmigration projects that would conflict with reserves. If implemented as planned, the Irian Jaya reserve system will consist of one of the largest in the tropical world, occupying a large fraction of the province, with numerous large reserves, and providing coverage of almost all significant habitats and centers of endemism. I am optimistic that the reserve system will become a reality and will be a source of pride to Indonesia and an example to other countries.

LESSONS FROM THE PROCESS

In Indonesia as elsewhere in the Third World, most visiting western biologists whose primary interests lie in research have made disappointingly little contribution to conservation programs, in comparison with the contribution that they could make. The stated and unstated reasons that hold biologists back from involvement tend to fall into four groups of beliefs: the effort seems doomed to futility; far more basic biological information must be gathered first; one person, especially a westerner, can make little impact; and time and effort devoted to conservation in the Third World is not productive in the western academic world. I would like to conclude by examining each of these four points.

"It's hopeless"

Western biologists often assume that conservation programs in the Third World are hopeless efforts doomed to failure. It seems that there is little local appreciation of conservation, that reserves can only be the scraps of land unwanted for development, and that any reserve program will soon be crushed under the grim realities of financial impoverishment, an explosively growing population, and desperate need for economic growth. Why waste time at Third World conservation if the prospects of success are so dim?

My experience in Indonesia and elsewhere in the Pacific fails to support this pessimism. At first, Indonesia does not look like a promising candidate for a successful conservation program. There is not a popular base of environmental awareness and appreciation for conservation on which to draw. The country suffers from the classic problems of expanding population and budgetary limitations. The government has features that seemingly are more closely associated with exploitation than with protection of the biota.

Nevertheless, a small core of Indonesians devoted to conservation, aided by strong biological arguments, has been able to accomplish a good deal. An important way in which we biologists can contribute is by offering *specific* biological arguments why reserves are required in particular areas, why they should be of at least a certain size, and what species are most likely to go extinct if the reserves are of lesser size. Increasingly over the last decade, biogeography has offered such specific arguments. Until recently, biologists intuitively felt strongly that it was important for reserves to be of adequate size, but they could not justify this feeling by specific arguments. When an unsympathetic government official wanted to trim reserve areas by 50 percent, it was hard to offer an effective reply. Increasingly nowadays, we can offer a reply: examination of species distributions and population viability analysis (Gilpin and Soulé, Chapter 2) indicate that such a smaller reserve would almost surely lose species X, Y, and Z and would have only a 50 percent chance of retaining species A, B, and C. When I made my first visit to Papua New Guinea after independence and sat down for my first discussion with the leaders of the country's new Wildlife Division, I was pleasantly surprised to find that they had read and could quote from papers on application of biogeographic principles to reserve design. Thus, we biologists can offer ammunition in the form of specific arguments that will help Third World conservationists to convince reluctant governments.

"We shouldn't act until we have detailed biological information"

Scholars working in academic disciplines are driven by the quest for knowledge, independent of external deadlines. Hence one of the cardinal sins of pure science is to leap to a conclusion prematurely, before adequate supporting evidence has been assembled. Thus when pure scientists without practical experience of conservation programs write about conservation, one often hears the caveat that decisions must be based on detailed autecological studies of individual species.

However, conservation biology is an applied science and is subject to different rules from those governing pure science (Soulé, 1985 and Chapter 1). The goal of conservation biology is to solve problems, constrained by externally imposed deadlines and budgets. One has to make one's recommendations on the basis of whatever knowledge is available on the deadline date. Occasionally, in well-funded programs focussing on single species not yet on the verge of extinction, the conservation biologist has the time and money to gather knowledge that would satisfy a pure scientist. More often, the available knowledge is ludicrously inadequate. One is dealing with communities of hundreds or thousands of species, most or all of whose distributions, life histories, and requirements are unknown. One is operating on a shoestring budget, under a tight deadline imposed by development projects that are about to start and to produce irreversible consequences.

The development of the Irian Jaya reserve system illustrates this point. Irreversible decisions about the locations of reserves in unknown or poorly known terrain had to be made in the early 1980s so that land not set aside for reserves could be made available for timber leases and other development programs. When the Kumawa Mountains were recommended as a reserve, no biologist had ever entered those mountains; we could not name a single species of animal that was definitely known to occur in them; and the locations and elevations of the highest peaks were unknown. Instead, the recommendation was based on the isolated location of the Kumawa Mountains, which leads one to expect a moderate degree of endemism. This expectation was confirmed by my survey of the mountains in 1983, and the results of that survey will be used in eventual decisions about a management plan for the reserve; but the reserve could not have been set aside at all unless a decision had been taken in the total absence of specific knowledge.

No one denies that conservation biology would be more successful if it could operate with generous budgets and extensive lead-times to gather ample biological information. Under prevailing Third World conditions, however, this expectation reminds me of Andrew Marvell's poem "Had we but world enough, and time." Starting with this title, the poet tells his coy mistress that he would woo her by spending 100 years praising each of her eyes, 200 years adoring each breast. . . but he and she would both be dead by then, so she should decide quickly. Like Marvell's mistress, conservation biologists will usually have to decide quickly, based on whatever they know as of the deadline date.

"What can one person do?"

The magnitude of the conservation problems facing Third World countries often seems such that one scientist could not begin to make a useful contribution. In fact, the reverse is true: precisely because in many cases so little biological information and scientific expertise are available to Third World countries, one person can readily make a big contribution. The following example is an illustration.

In the 1970s the government of the Solomon Islands was negotiating with a Japanese mining company over a plan to mine bauxite on Rennell Island. The proposed project would have constituted the largest development project in the Solomon Islands. The plan called for construction of a road network on Rennell, preliminary processing of ore on the island, logging of forest as a side-benefit made possible by the road system, and maintenance of a large work force on Rennell to carry out these operations. However, the Solomon Islands government was in the position of having to assess this project in the virtual absence of relevant biological and environmental information.

In 1976 the South Pacific Commission and the Planning Office of the Solomon Islands asked me to carry out a week-long environmental impact study on Rennell (Diamond, 1984e). It quickly became clear to me that the project would be disastrous for both biological and economic reasons. Biological considerations included that the island is still almost entirely forested and lacks introduced European rats; it has a highly endemic avifauna that is adapted to forest and that evolved in the absence of rats; fruits of most of the major timber tree species are bird-dispersed; Rennell is one of the world's largest raised atolls, with the former lagoon converted into the largest freshwater lake in the Pacific, harboring a distinctive ecosystem at risk from effects of pollutants; and only slight regeneration of vegetation had taken place on sites subjected to trial mining six years previously. From an economic point of view, the proposal would have meant attempting to process ore and build a city on an outlying coral island with little surface water except at the lake, low agricultural potential to feed a large work force, and negligible economic prospects for the city after exhaustion of the bauxite. A short discussion with the Planning Office sufficed to make most of these problems obvious, and the negotiations between the government and the mining company were not consummated.

The point of this episode is that budget limitations and lack of trained scientists had prevented the Solomon Islands government from obtaining even the minimum biological input into their largest proposed development project.

"It's not academically profitable"

It is difficult to persuade western scientists to make a major investment of their time in practical conservation studies in the Third World. There is no problem understanding why. Such an investment of time yields no academic credit; it may even yield academic scorn, because of the need to make decisions on the basis of limited scientific knowledge. Practical conservation work competes for time with field work that could yield research publications. The people with whom one is having discussions are generally not fellow scientists but nonscientists who don't share one's interests, and may even oppose one's interests. Finally, it is almost unbearably painful to have to watch the destruction of biological communities that one values and loves.

All these perceived drawbacks of doing conservation work in Third World countries are real, even though the objections described in the preceding sections are not. Nevertheless, doing practical conservation work is an obligation that pure scientists owe to Third World countries in return for the opportunity to do research there. Scientists often complain about the bureaucratic hostility that they encounter in doing field research abroad. In the Pacific countries where I work, this hostility is well deserved; many western scientists make no contribution to the pressing needs of their host countries, and don't even bother to send reprints or reports describing their findings. On a less selfish level, a person who is willing to do practical conservation work can contribute a great deal, in ways that will have a long-lasting beneficial influence on the earth's future.

SUGGESTED READINGS

Gressitt, J. L. 1982. *Ecology and Biogeography of New Guinea.* Junk, The Hague.

CHAPTER 25

INTRINSIC VALUE: WILL THE DEFENDERS OF NATURE PLEASE RISE?

Arne Naess

The venerable German philosopher Immanuel Kant insisted that we never use a human being *merely* as a means to an end. But why should this philosophy apply only to human beings? Are there not other beings with intrinsic value? What about animals, plants, landscapes, and our very special old planet as a whole?

I hope you all answer "yes." Is it my privilege as a philosopher to announce what *is* of intrinsic value, whereas scientists, as such, must stick to theories and observations? No, it is not—because you are not scientists as such; you are autonomous, unique persons, with obligations to *announce* what has intrinsic value without any cowardly subclass saying that it is just your subjective opinion or feeling. On the other hand, it does not follow that you are entitled to "beat up on" those who disagree with you. The rational solution of value conflicts is not something that is impossible to achieve.

But what is intrinsic value?

Expressions such as "this should be protected *for its own sake*" are very common; but pseudoscientific philosophers and scientists find them objectionable when they are applied to natural phenomena. They insist that there must be a being valuing things—that is, there must be humans in the picture. In a sense this is true. Theories of value, like theories of gravity and rules of logical or methodological infer-

Keynote Address, Second International Conference on Conservation Biology, University of Michigan, May 1985.

ences, are human products. But this does not leave out the possibility of truth or correctness. The positions in philosophy often referred to as "value nihilism" and "subjectivity of value" reject the concept of valid norms. Other positions accept the concept. I accept it.

The world of experience is the only world with which we are firmly acquainted. The world as spontaneously experienced, including intrinsic values, cannot be denounced as less real than that of scientific theory, because we always ultimately refer back to the immediate reality.

Recent developments in physics substantiate the primacy of immediate experience. As long as atoms were conceived of as small, hard things, physical reality could be conceived of as the real world. But quantum theory now offers us a picture so abstract, so mathematical, that it is reasonable to see it as furnishing only the abstract structure and outline of the real world, not its content. Color hues are real in their own way, just as electromagnetic "waves" are real in their function as abstract entities. We experience good old friends as values in themselves on a par with ourselves, and we do things for their sakes as naturally as for our own. Our friends may be useful to us, but that is not all. Why shouldn't this also apply to other living beings than humans? We are forced by modern science back to nature, basically as the naturalists conceived it. And it is, in its essential features, worth protecting for its own sake.

The position that nothing in the natural world has intrinsic value, that the whole conservation movement is motivated only by narrow utilitarian aims centered on human health and prosperity, corrodes in the long run the public image of the movement. Highly dedicated persons who cannot help but work for conservation and for whom it is a vital need to live with nature are confused by what they take to be the utter cynicism of scientists and experts who use purely utilitarian, flat language in their assessment of environmental risks,"genetic resources," and extinction. These experts are often seen as traitors.

There is an important philosophical argument against talking about protecting natural entities for their own sake. Is there not always, in any sort of valuation, a human subject that projects value into an object? Therefore, is not everything we do basically something we do for our own sake? I may answer "yes" insofar as we may use the expression "for our own sake" in a very abstract way. But everyday use of the expression is also legitimate. We undertake a hike for the sake of ourselves and our dog, but sometimes, in bad weather and having pressing things to do, we take the dog for a walk for its own sake. There are cases of doubt, but to announce that we do everything for our own sake, that is, that *we*, each of us, are the sole intrinsic value, is plain rubbish from a semantic point of view. In short, the argument against the *possibility* of doing things for the sake of others is untenable.

Spontaneous value experience is something to be conveyed to others even in our capacity as scientists. What we feel spontaneously has weight when we decide how to act, for instance, in regard to conservation policies. And the public and politicians should know what has weight for biologists.

Let me mention a rather touching sentence I found in a standard handbook of how to treat our domestic animals. It was extensively used in the 1920s and 1930s. The author talks about caressing pigs. The sentence reads approximately like this: "Those who have experienced the satisfaction of pigs stroked this way, cannot but do it." How can *the author* experience the satisfaction of a *pig?* The question is badly posed. It assumes a cleavage between the human subject and animal object. Actually such a cleavage does not belong to spontaneous experience, and should not be introduced in order to make the sentence more scientific. Much that passes for objectivity in scientific talk is really pseudoscientific and renders the language of scientists gray and flat!

The quoted sentence is instructive in another way. The last part, about compulsion, is marvelous: "Those who have experienced the satisfaction of pigs stroked in this way, *cannot but do it."* The farmer may say to himself, "Dear pig, I don't have time to stroke you today," but in vain. He just goes on stroking the pig. Here also a so-called scientific textbook writer would object; of course the farmer *can* refrain from stroking the pig.

Of less importance but perhaps worth mentioning is the way the sentence reminds us that when we talk about technical progress in the agricultural sector, we do not include techniques for caressing. Why? Is it better to be incomplete than to be accused of being sentimental or unscientific?

Back to Immanuel Kant and the use of a human being merely as a means to an end. What makes possible a vivid experience of intrinsic value corresponding to a vastly generalized Kantian maxim? In short, what makes intense personal appreciation of diversity of life forms and the whole ecosphere possible?

There is one process which perhaps is more important in this respect than any other: the process of so-called *identification* (Naess, 1985). We tend to see ourselves in everything alive. As scientists we observe the death struggle of an insect, but as mature human beings we spontaneously also experience our own death in a way, and feel sentiments that relate to struggle, pain, and death. Spontaneous identification is of course most obvious when we react to the pain of persons we love. We do not observe that pain and by reflecting on it decide that it is bad. What goes on is difficult to describe; it is a task of philosophical phenomenology to try to do the job. Here it may be sufficient to give some examples of the process of identification, or "seeing oneself in others." A complete report on the death struggle of an insect as some of us experience such an event must include the positive and negative values which are attached to the event as firmly as the duration, the movements, and the colors involved.

There is nothing unduly romantic or poetic here. Given our biological endowment each of us has the capacity to identify with all living beings. In addition, given the physiological, psychological, and social basis of gestalt perception and apperception, humans have the capacity of experiencing the intimate relations between organisms and the nonorganic world—that is, between the biosphere and the ecosphere in general. So we have natural expressions such as "living landscapes" and "the living planet." There is nothing here that goes against the scientific attitude.

I take it therefore to be an empirically testable hypothesis that the attainment of well-rounded human maturity leads to *identification with all life forms* in a wide sense of "life" and including the acknowledgment of the intrinsic value of these forms. The process of maturation is here conceived as something different from the mere learning of new skills. It encompasses the realization of different kinds of capabilities inherent in human nature. These capabilities are not necessarily related to increasing one's biological fitness. Through this conception of identification and maturity, ecologically sound policies gain a basis of justification that is not entirely homocentric, but is biocentric in the wide sense of *bios*.

Let me eliminate a couple of misunderstandings of this hypothesis about maturity. Even if 90 percent of humanity developed a high degree of identification with other life forms and openly acknowledged their intrinsic value, this might not stop governments from implementing policies resulting in large-scale extinctions and further destruction of wilderness and habitats. Social conditions such as hunger, war, and power conflicts on individual or group levels, along with mismanagement, may override considerations that spring from the genuine feelings and value priorities of the majority. A tragic situation may arise: attitudes towards nature show a high degree of maturity, but social and political chaos make the forceful expression of these attitudes impossible. Or, biologists may conclude that the biological programming of *Homo sapiens* simply is such that human overpopulation and cruel dominance of other life forms are inevitable. In other words, mature realization of the social and political potential may not make a "live and let live" attitude realizable.

Let us not take such a pessimistic view too seriously; yet it points up the fact that an increase in the breadth and intensity of the identification process in large numbers of individuals does not automatically increase the political strength of the conservation movement. The fight for basic policy changes is a necessary corollary of the effort to assist the development of identification through education and otherwise.

Let us look more closely at the complex relationship between basic value positions and concrete environmental policies. It has been encouraging for me to lead a project of systematic interviews of so-called "ordinary people" on the rights of animals, plants, and landscapes, and on their intrinsic value. In spite of what one would guess from the way they vote (and I am now speaking as a Scandinavian) there is a substantial majority with quite far-reaching ideas about the rights and value of life forms, and a conviction that *every life form has its place in nature* which we must respect.

On the other hand, there is one widespread opinion among ordinary people which should be disturbing to biologists. This is the opinion that scientists and so-called environmental experts have largely deserted them. It is not difficult to see some of the causes of this feeling. For example, people read about an expert who favors Plan A over Plan B as to where to place a dam, or how to increase the production of energy. The public has no available information about what the expert thinks in his heart. He may be of the opinion that both plans are irresponsible, that to increase energy production is sheer nonsense, and that the implied interference in natural processes is a calamity. But he or she (rarely a she, I am sorry to say) has been asked only to compare plans. If the expert publishes his real opinions, he will not be asked by authorities to function as an expert in the future.

In any case, whatever the causes of widespread silence, many people would give up their own passivity if offered more support or leadership by those whom they consider experts, or at least more knowledgable and articulate than themselves.

Let me mention the questions raised in the interviews. The first part asks whether the interviewee thinks we have duties or obligations toward animals, plants, rivers, and landscapes. The great majority of persons interviewed maintain that we do have duties and obligations towards the nonhuman world, organic and nonorganic. The second part asks analogous questions about rights. Answers have the same positive character. The third part concerns intrinsic value, or value in itself. Most people think they understand these expressions. Things may have value, people say, without having value for humans.

The last questions concern population, extinction, and territorial conflicts between humans and nonhumans. "Animals *have* equal rights but humans take away the right" is a common answer. When asked what they think about the prediction that a million species may be wiped out if policies are not changed, it is pathetic to see how this idea elicits horror, indignation, and despair. In short, the answers of so-called ordinary people were such that one might, perhaps naively, expect them to press for substantially different conservation policies.

Are the experts really narrowly utilitarian in their views, and are they really in favor of present environmental policies? In an attempt to find out I recently sent a long personal letter to 110 people who influence national environmental policy in Norway. About one out of three has responded, some with long, interesting essays. The respondents include high-level personnel in the departments of finance, justice, and energy—persons with comprehensive educations in various branches of natural science and technology.

The experts were asked to react to the following eight points, which, incidentally, I call "the platform of the deep ecology movement," or rather, one formulation of such a platform.[1]

1. "The flourishing of human and nonhuman life on Earth has inherent value. The value of nonhuman life forms is independent of the usefulness of the nonhuman world for human purposes." The great majority indicated their agreement.
2. "Abundance and diversity of life forms are values in themselves and contribute to the flourishing of human and nonhuman life on Earth." The great majority agree.
3. "Humans have no right to reduce this abundance and diversity except to satisfy vital needs." The great majority *tend* to agree. Many comment on the term "vital."
4. "The flourishing of human life and cultures is compatible with a substantial decrease in the human population, and the flourishing of nonhuman life requires such a decrease." The great majority agree.
5. "Present human interference with the nonhuman world is excessive, and the situation is rapidly worsening." The great majority agree.
6. "Policies must therefore be changed. The changes in policies affect basic economic, technological, and ideological structures. The resulting state of affairs would be deeply different from the present and would make possible a more joyful experience of the connectedness of all things." The great majority tend to agree. Some find the last sentence rhetorical and doubtful.

[1] A more thorough treatment of these points is presented in Chapter 5 of the volume *Deep Ecology: Living As If Nature Mattered*, edited by B. Devall and G. Sessions (1985).

7. "The ideological change is mainly that of appreciating *life quality* rather than adhering to an increasingly higher standard of living. There will be a profound awareness of the difference between big and great." The great majority tend to agree.

8. "Those who subscribe to the foregoing points have an obligation, directly or indirectly, to participate in the attempt to implement the necessary changes." The great majority agree.

The eight points are of course in need of clarification, elaboration, and comment. In the letter to the experts more than two pages of comments were added. One may suspect that some of those who did not answer my letter largely disagreed or at least did not find the formulations acceptable as expressions of their personal views. But the main point is clear: a tendency among respondents to agree.

These results do not confirm the belief of ordinary people that "the experts" have deserted them in their basic views on the man/nature relationship and in their views as to the necessity of fundamental changes in policy. There is no pronounced technocratic philosophy of value. The concrete governmental decisions which ordinary people find lamentable are rather based on priorities of a nonfundamental character plus the conviction that "people" *must* be promised *both* a steadily higher (material) standard of living *and* unspoiled nature. At least this is a general impression, though there is no scientific evidence that it is the case.

It is easy to accept lofty principles verbally. Hundreds of wars which we now consider more or less crazy were glorified by reference to God, patriotism, love of mankind, and supreme justice. Likewise it is easy to agree upon the intrinsic value of the richness and diversity of life on the planet Earth. What is needed is a methodology of *persistently connecting basic value judgments and imperative premises with decisions in concrete situations* of interference or noninterference in nature. What I therefore suggest is that those who are thought to be experts and scientists repeatedly and persistently deepen their arguments with reference to basic value judgments and imperative premises. That is, they should announce their normative philosophy of life and discuss environmental problems in their most comprehensive time and space frame of reference—ultimately in terms of millions of years of evolution and intimate global interactions.

Neither the so-called deep ecology movement nor any other movement which attributes intrinsic value to the richness and diversity of life forms today meets antagonists with an opposite *articulated* philosophy.[2] The movement to institute responsible, respectful treatment

---

[2] Various forms of Hegelianism conceive of mankind as "spiritualizing" nature, and

of nature is up against much more formidable forces than well articulated antagonistic philosophies. Strong economic, social, and political forces are operating against it, as well as lifestyles, habitual attitudes, and the preferences of individuals who are encouraged to adopt a consumerist style. Old habitual attitudes find expression in such phrases as "fight nature;" "improved land;" "push back the jungle;" or "conquer Mt. Everest."

To some it may appear that there is a conflict between what I have said about identification and what might be called fundamental human nature as it is understood by evolutionary biologists. Let us consider the hypothesis that all-around maturity among human beings inevitably fosters a high level of identification with all life forms. It is justifiable to doubt the compatibility of such a view with the contemporary theories of evolution and ethology. One may ask whether a rather impartial "live and let live" attitude implies a kind of altruism incompatible with the egoism or very narrow altruism of the human genetic heritage. Before answering we should agree that the concepts of egoism and altruism as understood by evolutionary and genetic theorists are not those of the concepts shared by most people in everyday life. No criticism of either concept is implied in saying this; there is room for many concepts, each useful within its limits.

Without going into detail I will venture to say that deep changes of environmental policies in favor of nonhumans do not imply anything definite about any particular view of biological fitness. What is good for nonhumans obviously may also be good for humans; furthermore, there is nothing in the current theories about the evolution of behavior that contradicts the possibility of human management of the human population size.

In his concluding article in the anthology *Natural Selection and Social Behavior,* Richard D. Alexander says:

We are, then, hedonistic or selfish individualists to the extent that such behavior maximizes the survival by reproduction of those copies of our genes residing in our own bodies; and we are group altruists to the extent that this behavior maximizes the survival by reproduction of the copies of our genes residing in the bodies of others. At least this is what we have evolved to be— and to all accounts it is all that we have evolved to be.

---

allow that only through the agency of mankind can nature partake in the realm of intrinsic values. Such views are today still a step removed from representing an articulated, antagonistic philosophy; for instance, they don't directly negate any of the eight points formulated on pages 509–510. Another potentially antagonistic philosophy is the Marxist "labor theory of value," which has been interpreted to negate the intrinsic value of life forms. Only things transformed through labor acquire value is one interpretation. But today Marxists who work for the protection of wilderness propose a much wider concept of "labor" than that of gross material transformation.

It is paramount to realize, however, that—as opposed to what we have evolved to be—what we actually are or become is whatever we can make of ourselves, given our history, and our propensities and talents, which are great, for creating novelty in our environments, at rates and of kinds that the process of genetic evolution has no possibility of controlling or keeping up with. Nowadays, we are closer than ever before to being able to become what we wish to be, if for no other reason than because we know about ourselves the things that I have just mentioned.

Personally I would perhaps be a little more cautious about whether we are today more able than before to become what we wish. But at least we have more knowledge about the practical obstacles we face, and one obstacle is lack of articulate leadership.

How do environmental experts express themselves when they are hired to take part in vital environmental decisions, or when they are asked their views about what is going on on this planet? Not like the poets, and I think that is good. But the predominant way the experts express themselves in public needs some comment.

The so-called language of metaphors, used to a certain extent within the old naturalist tradition, is not competing, or should not be competing, with the language of tough modern biology. On the other hand, there is not a single theorem in natural science which can undermine the reality of a person's wealth of spontaneous experience. But biologists have the precious privilege to be acquainted with worlds, and in a vital sense to feel at home in worlds, largely outside the experience of others—worlds of microscopic living beings, and of life processes which amaze us all. These are sources of joy and wonder that the biologist should be able to expose and communicate, not only in the form of textbooks, but also through the direct language of spontaneous experience.

Spontaneous experience is not limited to so-called pure sense experience. It has cognitive elements, but elements of acquaintance and insight rather than of abstract knowledge. These must not be left unexpressed in the name of science. There may of course be biologists who suspect that an expert who uses spontaneous language is incapable of scientific rigor. But we should not neglect our linguistic abilities from fear of such misunderstandings. When biologists refrain from using the rich and flavorful language of their own spontaneous experience of all life forms—not only of the spectacularly beautiful but of the mundane and bizarre as well—they support the value nihilism which is implicit in outrageous environmental policies.

The high-level experts I asked to comment on the eight points of a so-called deep ecology platform answered in a way favorable to a remarkably strong conservation policy. But they answered a personal letter from me, and I guaranteed not to reveal their names. These experts and most others do not propagate their strong views on conservation in public. Why? Here are some possible reasons:

1. Time taken away from professional work.
2. Consequent adverse effects of this on promotion and status.
3. Feeling of insufficient competence outside their "expertise."
4. Lack of training in the use of mass media and in facing nonacademic audiences.
5. Negative attitude towards expressing "subjective" opinions and valuations, or violating norms of "objectivity;" reluctance to enter controversial issues.
6. Fear that colleagues or bosses think that they dabble in irrelevant, controversial fields, and that their going public is due to vainglory and publicity seeking.
7. Fear of fellow researchers, institution personnel or administrations; fear of the stigma "unscientific."

For example, when Barry Commoner says "Nature knows best!" and explains what he means thereby, some philosophers who like to be scientific tend, nevertheless, to class this as rubbish because nature, not being an organism, cannot "know" anything. For them, to speak of "destruction" is taken to be unscientific because ecosystems only *change,* man's interaction with the system being a factor of change on a par with all others. And one should not in serious discussion use a slogan such as "Let the river live!" because it affirms old superstitions such as those about river elves. "Mysticism" is bad: when somebody exclaims "this place is part of me," and points to a place along a river, it is assumed to be nonsense from the strict point of view of logic and physics. But why choose this way of interpreting these exclamations?

Let us recognize that whereas some terms and phrases are scientific and others unscientific, most are neither. That is, the context is such that the distinction is irrelevant. The expression "the life of a river" *may* be introduced in a scientific text by using the terminology of ecosystems. The slogan "Let the river live!" has had an important function in a situation of social conflict. When biologists participate in such conflicts they generally use the language of social conflict as others do, and if challenged they are willing and able to clarify their use of this language. They may point out that exclaiming "Let the river live!" does not imply that the river is a biological organism, and that "Nature knows best!" does not imply that nature "knows" in the same sense that a person "knows." It is up to the challenger to justify the claim that there is a relation of implication.

What can be done to counteract the tendency to public silence among the experts who are in sympathy with strong conservation measures? Here are some suggestions:

1. Find them.
2. Listen to their explanations about why they are silent.
3. Find out whether they are willing in principle to expose their views publicly. If they are,
4. Help find suitable occasions at which they can enter public discussion; or,
5. Suggest themes of articles they can write or how they might add certain paragraphs to what they are writing. (We presuppose that they, as experts, touch upon problems of relevance to the intrinsic value of natural richness and diversity.)
6. Propose that they talk to their colleagues about the social and political issues discussed at their professional meetings.
7. Propose special sessions on these issues at their professional meetings.

In environmental conflicts today, deep motivation is necessary for dedication and persistence. Philosophical views encompassing ethical views are therefore relevant. The way we experience reality largely determines our ethical norms, including our environmental ethics.

What I have said so far is not meant to invalidate or make unnecessary narrow, utilitarian, short-range arguments. We need them in order to get things done. Let me end by mentioning an example of the use of narrowly utilitarian proposals.

There are wolves in many European countries. Controversies abound. Protection of all habitats is extremely costly and is often energetically fought against by the local communities affected. Recently wolf specialists have started to work out a plan for the conservation of wolves within the framework of market economy. One simply tries to *sell* the idea that protection of wolves in large, thinly populated areas in many countries has local commercial advantages. Hunting, photography, and viewing safaris are recommended. The chorus of howling wolves should be "sold" as nature's most intense, most wild wilderness experience. (Actually, musical composers are already studying and using the howling structure.) An essential point is that the local population, not urban capitalists or the state, should earn the money. At least in my century, local people will shoot the wolves if they are not part of a positive economy. In a market economy there must be a product to sell. The wolf specialist Ivar Mysterud with whom I cooperate uses the term "experience product," and the source of the product is termed "experience resource."

In addition, however, the people's fear of wolves must be addressed. In Norway small buses are available for children going to school in wolf country, even if the school is very near their homes. It costs a great deal of money, but unless their children are 100 percent protected, local people will shoot the wolves.

It is the idea of Mysterud and others that we must adapt to the prevalent market economy, and not rely indefinitely on public funds. Wolves require extensive territories and in Europe there are no large uninhabited areas. The fight against shrinking habitats must use commercialism as a regrettable but necessary means.

I have mentioned this in order to emphasize that "romanticism" and "cynicism" must be combined. By this I mean that philosophical or religious views, labeled romantic and sentimental by some, must be combined with what others will label cynicism and opportunism. But even the use of all these means may prove insufficient, and we may face the decision to give up protecting some wolf territories. There is much work to be done.

In short, biologists endowed with the necessary energy and enthusiasm will talk and act on the basis of a normative total view—what I call an ecosophy, or *wisdom* of household, not mere ecology, or knowledge of household. The biologist will combine philosophical, including ethical, fundamental positions with practical arguments. If they do this full of trust in doing the right thing, and without too many negative utterances against the opponents of strong conservation, they are unlikely to be hurt in their capacities as experts and scientists.

Socrates was not popular among the people in power. He pestered them, but in a way that made them respect him. That is all we can hope for as devoted conservationists: to be pests, but *respected* pests.

Acknowledgments

Michael E. Soulé: Chapter 1

I thank Michael Gilpin, Stuart Pimm, Hal Salwasser, and Daniel Simberloff for their words of caution and their many suggestions. I want also to acknowledge my teachers, Paul Ehrlich and Arne Naess, who have been supportive and inspirational to me and to many others.

Michael E. Gilpin and Michael E. Soulé: Chapter 2

The authors thank Mark Shaffer for his critique of an early draft.

Katherine Ralls, Paul H. Harvey, and Anna Marie Lyles: Chapter 3

We thank those people who provided us with unpublished data and manuscripts: K. Arnold, A Bittles, J. Berger, A. Cockburn, S. Emlen, J. Hoogland, W. Koenig, J. Malcolm, G. McCracken, A. van Noordijk, C. Packer, A. Pusey, J. Reese, J. Rood, I. Rowley, P. Sherman, P. Sikes, R. Schreiber, G. Svendsen, and A. Zahavi. We are also grateful to those people who discussed ideas or commented on parts or all of our draft manuscript: A. Bittles, W. Koenig, D. Mech, R. May, G. McCracken, E. Morton, M. Nelson, L. Partridge, J. Seger, and W. Shields.

Fred W. Allendorf and Robb F. Leary: Chapter 4

F.W.A. was supported by National Science Foundation Grant BSR-8300039 while this manuscript was written. R.F.L. was supported by a student stipend from the Montana Department of Fish, Wildlife, and Parks. We thank the following, who provided us with reprints and preprints of their work: Y. P. Altukhov, R. Angus, D. W. Foltz, R. K. Koehn, J. B. Mitton, J. L. Patton, R. J. Shultz, M. H. Smith, M. Turelli, W. B. Watt, and E. Zouros.

Alan R. Templeton: Chapter 6

This work was supported by N.I.H. Grant R01 GM31571 and the Missouri Conservation Commission Contract for Reintroduction of Eastern Collared Lizards. My special thanks go to Scott Davis, who performed the genetic survey of the collared lizards as well as much field work, and to Scott McWilliams, the director of the Peck Ranch, who provided critical help and support in turning the reintroduction program into a reality. Finally, I wish to thank Bruce Read for the many exciting discussions we have had on the problems and solutions of managing endangered species.

Alwyn H. Gentry: Chapter 8

I thank P. Ashton and W. D'Arcy for reviewing this manuscript and D. Boufford, T. Croat, W. D'Arcy, S. McDaniel, L. Morse, W. Wagner, and G. Wilhelm for providing data. Field work in Ecuador was done in collaboration with C.

Dodson and funded by National Science Foundation grants INT-790-6850 and BSR-834-2764; field work in western Colombia was done in collaboration with E. Forero and funded by Colciencias and by N.S.F.R. grant INT-792-0783; field work in Amazonian Peru was done in collaboration with F. Ayala, D. Smith, R. Vasquez, and others, and was funded by the National Geographic Society, N.S.F. and U.S.A.I.D. (DAN-5542-G-SS-1086-00, Proj. No. 936-5542). Cerro Tacarcuna was explored with a grant from the National Geographic Society.

Deborah Rabinowitz, Sara Cairns, and Theresa Dillon: Chapter 9

We thank our colleagues for giving us their time to fill out the survey: Joy Belsky, Philip Dixon, David Ellsworth, Sana Gardescu, David Gill, Laura Jackson, Rhonda Janke, Sandra Knapp, Martin Lechowicz, James Mallet, Peter Marks, Charles Mohler, Nathan Stephenson, Marcia Waterway, and Peter Wimberger. Suzanne Jones assisted with preparation and data analyses, and Peter Marks, Peter Ewell, Chris Thomas, Jonathan Silvertown, and Heather Robertson made helpful comments on the manuscript.

Stephen P. Hubbell and Robin B. Foster: Chapter 10

We thank the National Science Foundation, the Smithsonian Tropical Research Institute, the World Wildlife Fund, the Center for Field Research, the Guggenheim Foundation, the Geraldine R. Dodge Foundation, the Walcott Fund, and numerous other private donors for supporting this research. We also thank the more than 100 field asistants from more than 10 countries who have generously contributed their labor over the past 5 years for inadequate pay to make this project a reality. Finally, we thank Joe Wright for reading the manuscript and making helpful suggestions for its improvement.

Thomas E. Lovejoy et al.: Chapter 12

This study was supported by the World Wildlife Fund—U.S., the Instituto Nacional de Tesquisas da Amazônia (INPA), the Instituto Brasileiro de Desenvolvimento Florestal (IBDF), the Andrew W. Mellon Foundation, the Tinker Foundation, the Sequoia Foundation, and a grant from the National Park Service, Cooperative Agreement CX-0001-9-0041. It represents publication number 23 in the Minimum Critical Size of Ecosystems Project (Dinâmica Biológica de Fragmentos Florestais) Technical Series.

Work of this scope inevitably involves the help of many individuals, to all of whom we are most grateful. We are also grateful for the counsel and encouragement of Brazil's Special Secretary for the Environment, Paulo Nogueira-Neto.

Andrew P. Dobson and Robert M. May: Chapter 16

This work was supported in part by National Science Foundation grant DEB83-03772 (R.M.M.) and by a NATO Postdoctoral Fellowship (A.P.D.). We thank R. M. Anderson, C. R. Kennedy, D. H. Molyneux, and A. M. Lyles for many helpful discussions.

David C. Culver: Chapter 21

A number of colleagues generously supplied unpublished manuscripts, reprints, and answers to a large number of questions. These include W. M.

Elliott, F. G. Howarth, T. M. Iliffe, J. F. Jacobs, V. M. Tipton, M. D. Tuttle, and J. Yager. Dr. Virginia Tipton of Radford University was indispensable in providing me with references to and reprints of the bat conservation literature.

John Cairns Jr.: Chapter 23

This paper was written at the Rocky Mountain Biological Laboratory, Crested Butte, Colorado. I am indebted to colleagues at R.M.B.L. for useful comments and discussion about the recovery process. I am also indebted to participants in the Water Quality Criteria Workshop organized by Abel Wolman and M. Gordan Wolman and held at the Belmont Conference Center, January 15–18, 1984, for views on criteria and end points. This manuscript was typed by Betty Higginbotham and prepared for publication by Darla Donald, Editorial Assistant, University Center for Environmental Studies, Virginia Polytechnic Institute and State University, Blacksburg, Virginia.

Jared Diamond: Chapter 24

It is a pleasure to record my debt to Ardy Irwanto, John MacKinnon, Hans Makabory, Jeffrey McNeely, Ronald Petocz, John Ratcliffe, Franz Sadsuitubun, Achmad Sounuddin Solihin, and many other colleagues during my work in Irian Jaya; and to the National Geographic Society, World Wildlife Fund, and Lievre Fund, for support.

Literature Cited

Numbers in parentheses at the end of each reference indicate the chapter or chapters in which the work is cited.

Abele, L.G. and E.F. Connor. 1979. Application of island biogeography theory to refuge design: Making the right decision for the wrong reasons. In Proc. 1st Conf. on Scientific Research in National Parks, R.M. Linn (ed.). U.S.D.I. National Park Service, Washington, D.C. pp. 89–94. (11)

Abrokwa-Ampadu, R. 1984. The Volta River hydroelectric project in Ghana. In Hydroenvironmental Indices: A Review and Evaluation of their Use in the Assessment of the Environmental Impacts of Water Projects. Technical Documentation in Hydrology, J.R. Card (ed.). UNESCO, Paris. pp. 57–77. (18)

Adam, J.G. 1967. Evolution de la végétation dans les sous-parcelles protégées de l'UNESCO-IFAN à Atar. Bull. Inst. Fon. Afr. Noire, Sér A 29 (1), 92–106. (22)

Adamkewicz, L., S.R. Raub, and J.R. Wall. 1984. Genetics of the clam *Mercenaria mercenaria*. II. Size and genotype. Malacologia 25, 525–533. (4)

Adis, J. 1984. Seasonal igapó forests of Central Amazonian blackwater rivers and their terrestrial arthropod fauna. In *The Amazon: Limnology and Landscape Ecology of a Mighty Tropical River and its Basin*, H. Sioli (ed.). Junk, Dordrecht. pp. 245–268. (18)

Ågren, G. 1981. Two laboratory experiments on inbreeding avoidance between siblings in gerbils. Behav. Process. 6, 292–297. (3)

Ågren, G. 1984. Incest avoidance and bonding between siblings in gerbils. Behav. Ecol. Sociobiol. 14, 161–169. (3)

Ahearn, G.A. and F.G. Howarth. 1982. Physiology of cave arthropods in Hawaii. J. Exper. Zool. 222, 227–238. (21)

Allard, R.W. 1960. *Principles of Plant Breeding*. Wiley, New York. (5)

Allen, P.H. 1961. Natural selection in loblolly pine stands. J. For. 59, 598–599. (5)

Allendorf, F.W., K.L. Knudsen and G.M. Blake. 1982. Frequencies of null alleles at enzyme loci in natural populations of ponderosa and red pine. Genetics 100, 497–504. (5)

Allison, A.C. 1955. Aspects of polymorphism in man. Cold Spring Harbor Symp. Quant. Biol. 20, 239–255. (4)

Altukhov, Y.P. and N.V. Varnavskaya. 1983. Adaptive genetic structure and its relationship to intrapopulation sex, age, and growth rate differentiation in sockeye, *Oncorhynchus nerka* (Walb.). Genetika 19, 796–806 (Transl. 614–624). (4)

Amerson, A.B., Jr., R.C. Clapp and W.O. Wirtz II. 1974. The natural history of Pearl and Hermes Reef, northwestern Hawaiian Islands. Atoll Research Bulletin 174. Smithsonian Institution, Washington, D.C. (3)

Anderson, R.M. and R.M. May. 1978. Regulation and stability of host-parasite population interactions. I. Regulatory processes. J. Anim. Ecol. 47, 219–247. (16)

Anderson, R.M. and R.M. May. 1979. Population biology of infectious diseases: Part I. Nature 280, 361–367. (16)

Anderson, R.M. and R.M. May. 1981. The population dynamics of microparasites and their invertebrate hosts. Phil. Trans. Roy. Soc. Lond. B. 291, 451–524. (16)

Anderson, R.M., H.C. Jackson, R.M. May and A.D.M. Smith. 1981. Population dynamics of fox rabies in Europe. Nature 289, 765–771. (16)

Andersson, M. 1984. Brood parasitism within species. In *Producers and Scroungers: Strategies of Exploitation and Parasitism*, C.J. Barnard (ed.). Croom Helm, London. pp. 195–227. (3)

Andjelkovic, M., D. Marinkovic, N. Tucic and M. Tosic. 1979. Age-affected changes in viability and longevity loads of *Drosophila melanogaster*. Amer. Natur. 114, 915–920. (5)

521

Annest, J.L. and A.R. Templeton. 1978. Genetic recombination and clonal selection in *Drosophila mercatorum*. Genetics 89, 193–210. (6)

Armitage, K.B. 1984. Recruitment in yellow-bellied marmot populations: kinship, philopatry, and individual recognition. In *Biology of Ground- Dwelling Squirrels*, J.O. Murie and G.R. Michener (eds.). University of Nebraska Press, Lincoln. pp. 377–403. (3)

Ashton, P.C. 1984. Long-term changes in dense and open inland forests following herbicidal attack. In *Herbicides in War: The Long-term Ecological and Human Consequences*, A.H. Westing (ed.). Taylor and Francis, London. pp. 33–37. (20)

Aspelin, P.S. and S.C. dos Santos. 1981. Indian areas threatened by hydroelectric projects in Brazil. IWGIA Document, Copenhagen. 201 pp. (18)

Assman, E. 1970. *The Principles of Forest Yield Study*. Pergamon, Oxford. (5)

Atkinson, I.A.E. 1985. Effects of rodents on islands. In *Conservation of Island Birds*, P.J. Moors (ed.). International Council for Bird Preservation, Cambridge, England. pp. 35–81. (24)

Atlas of Hawaii, 2nd Ed. 1983. University of Hawaii Press, Honolulu. (14)

Aubert, G. and Maignien, R. 1949. L'érosion éolienne dans le Nord du Sénégal. Bull. Agr. Congo Belge 40, 1309–1316. (22)

Aubréville, A. 1949. *Climats, Forêts et Désertification de l'Afrique Tropicale*. Soc. Edit. Géogr., Mar. & Col., Paris. 351 pp. (22)

Ayala, F.J. 1965. Evolution of fitness in experimental populations of *Drosophila serrata*. Science 150, 903–905. (4)

Ayala, F.J. 1969. Evolution of fitness. V. Rate of evolution in irradiated populations of *Drosophila*. Proc. Nat. Acad. Sci. USA 63, 790–793. (4)

Ayala, F.J. 1984. Molecular polymorphism: How much is there and why is there so much? Develop. Genet. 4, 379–391. (4)

Baccus, R., H.O. Hillestad, P.E. Johns, M.N. Manlove, R.L. Marchinton and M.H. Smith. 1977. Prenatal selection in white-tailed deer. Proc. Annual Conf. S.E. Assoc. Fish and Wildlife Agencies 31, 173–179. (4)

Bagley, F.M. 1984. A Recovery Plan for the Ozark Big-eared Bat and the Virginia Big-eared Bat. U.S. Fish and Wildlife Service, Rockville, Md. (21)

Baker, A.E.M. 1981a. Gene flow in house mice: behavior in a population cage. Behav. Ecol. Sociobiol. 8, 83–90. (3)

Baker, A.E.M. 1981b. Gene flow in house mice: introduction of a new allele into free-living populations. Evolution 35, 243–258. (3)

Baker, C.B. and S.F. Fox. 1978. Dominance, survival, and enzyme polymorphism in dark-eyed juncos, *Junco hyemalis*. Evolution 32, 697–711. (4)

Bakus, G.J. 1982/83. The selection and management of coral reef preserves. Ocean Management 8, 305–316. (17)

Balon, E.K. and A.G. Coche (eds.). 1974. *Lake Kariba: A Man-made Tropical Ecosystem in Central Africa*. Junk, The Hague. 767 pp. (18)

Bannister, M.G. 1965. Variation in the breeding system of *Pinus radiata*. In *The Genetics of Colonizing Species*, H.G. Baker and G.L. Stebbins (eds.). Academic Press, New York. pp. 353–372. (5)

Barber, H.N. 1965. Selection in natural populations. Heredity 20, 551–572. (5)

Barbira-Scazzocchio, F. (ed.). 1980. *Land, People and Planning in Contemporary Amazonia*. Centre of Latin American Studies, Cambridge University, Cambridge, U.K. (19)

Barr, T.C. and J.R. Holsinger. 1985. Speciation in cave faunas. Ann. Rev. Ecol. Syst. 16, 313–337. (21)

Barral, H. et al. 1983. Systèmes de Production d'Elevage au Sénégal, dans la Région du Ferlo. GERDAT/ORSTOM, Paris. 172 pp. (22)

Barry, J.P., G. Boudet, A. Bourgeot, J.C. Celles, A.M. Coulibally, J.C. Leprun, and R. Manière. 1983. Études des Potentialités Pastorales et du leur Évolution en milieu Sahélien au Mali. GERDAT/ORSTOM, Paris. 116 pp. (22)

Baskin, J.M. and C.C. Baskin. 1978. The seed bank in a population of an endemic plant species and its ecological significance. Biol. Conserv. 14, 125–130. (13)

Bateson, P.P.G. 1978. Early experience and sexual preferences. In *Biological Determinants of Sexual Behavior*, J.B. Hutchinson (ed.). Wiley, London. pp. 29–53. (3)

Bateson, P. 1982. Preferences for cousins in Japanese quail. Nature 295, 236–237. (3)

Bateson, P. 1983. *Mate Choice*. Cambridge University Press, Cambridge. (3)

Baur, G.N. 1964. The ecological basis of rainforest management. Forestry Commission, New South Wales. (20)

Bawa, K.S. 1974. Breeding systems of tree species of a lowland tropical community. Evolution 28, 85–92. (5)

Beardmore, J.A. 1960. Developmental stability in constant and fluctuating temperatures. Heredity 14, 411–422. (4)

Beardmore, J.A. and S.A. Shami. 1976. Parental age, genetic variation and selection. In *Population Genetics and Ecology*, S. Karlin and E. Nevo (eds.). Academic Press, New York. pp. 3–22. (4)

Beardmore, J.A. and S.A. Shami. 1979. Heterozygosity and the optimum phenotype under stabilizing selection, Aquilo Ser. Zool. 20, 100–110. (4)

Beardmore, J.A. and R.D. Ward. 1977. Polymorphism, selection, and multi-locus heterozygosity in the plaice, *Pleuronectes platessa* L. In *Measuring Selection in Natural Populations*, S. Levin (ed.). Lecture Notes in Biomathematics 19. Springer Verlag, New York. pp. 207–221. (4)

Beaumont, A.R., C.M. Beveridge and M.D. Budd. 1983. Selection and heterozygosity within single families of the mussel *Mytilus edulis* L. Mar. Biol. Lect. 4, 151–161. (4)

Beddington, J.R. 1979. Harvesting and population dynamics. In *Population Dynamics*, R.M. Anderson, B.R.Turner and L.R. Taylor (eds.). Blackwell, Oxford. (14)

Beddington, J.R. and R.M. May. 1980. Maximum sustainable yield in systems subject to harvesting at more than one trophic level. Math. Biosc. 51, 261–281. (14)

Beecher, I.M. and M.D. Beecher. 1983. Sibling recognition in bank swallows (*Riparia riparia*). Z. Tierpsychol. 62, 145–150. (3)

Beecher, M.D. 1982. Signature system and kin recognition. Amer. Zool. 22, 477–490. (3)

Beecher, M.D., I.M. Beecher and S. Hahn. 1981a. Parent–offspring recognition in bank swallows (*Riparia riparia*): II. Developmental and acoustic basis. Anim. Behav. 29, 95–101. (3)

Beecher, M.D., I.M. Beecher and S. Lumpkin. 1981b. Parent–offspring recognition in bank swallows (*Riparia riparia*): I. Natural History. Anim. Behav. 29, 86–94. (3)

Bekoff, M. 1977. Mammaliam dispersal and the ontogeny of individual behavioral phenotypes. Amer. Natur. 111, 715–732. (3)

Bekoff, M. 1981. Mammalian sibling interactions: genes, facilitative environments and the coefficient of familiarity. In *Parental Care in Mammals*, D.J. Gubernick and P.H. Klopfer (eds.). Plenum, New York. pp. 307–346. (3)

Benthem, R.J., S. Crowe, D.W. Goode, K.L. Lahiri and H. Sioli. 1971. The northern Salt Lake Zone of Calcutta, West Bengal, with special reference to the proposed establishment of a bird sanctuary. Paper and Proceedings, IUCN Eleventh Technical Meeting, New Delhi, 25–28 November 1969. pp. 125–128. (18)

Berger, A.J. 1981. *Hawaiian Birdlife*. University Press of Hawaii, Honolulu. (14)

Berger, J. 1986. *Wild Horses of the Great Basin: Social Competition and Population Size*. University of Chicago Press, Chicago. (3)

Bergmann, F. 1973. Genetische Untersuchungen bei *Picea abies* mit Hilfe der Isoenzym-Identifizierung. II. Genetisch Kontrolle von Esterase- und Leucinaminopeptidase-Isoenzymen in haploiden Endosperm ruhender Samen. Theor. Appl. Genet. 43, 222–225. (5)

Bergmann, F. and H.R. Gregorius. 1979. Comparison of the genetic diversities of various poulations of Norway spruce (*Picea abies*). In Proc. Conf. Biochem. Genet. For. Trees, D. Rudin (ed.). Dep. For. Genet. Plant Physiol., Swed. Univ. Agr. Sci., Umea, Sweden. pp. 99–107. (5)

Bernstein, I.S., P. Balcaen, L. Dresdale, H. Gouzoules, M. Kavanagh, T. Patterson and P. Neyman-Warner. 1976. Differential effects of forest degradation on primate populations. Primates 17, 410–411. (12)

Bernus, E. 1971. Possibilities and limits of pastoral watering plans in the Nigerian Sahel. FAO Seminar on Nomadism, Cairo. (22)

Bernus, E. 1974. Les Illabaken, une tribu touarègue sahélienne et son aire de nomadisation. *Atlas des structures agraires au sud du Sahara.* Mouton & Co., La Haye, and ORSTROM, Paris. 116 pp. (22)

Bernus, E. 1981. Touaregs Nigériens: Unité Culturelle et Diversité Régeionale d'un Peuple Pasteur. Mém. No. 94. ORSTOM, Paris. 508 pp. (22)

Bernus, E. and G. Savonnet. 1973. Les problèmes de la lècheresse dans l'Afrique de l'ouest. Presence Africaine 88, 4, 113–138. (22)

Berry, R.J. 1978. Genetic variation in wild house mice: where natural selection and history meet. Amer. Sci. 66, 52–60. (3)

Bertram, B.C.R. 1973. Sleeping sickness survey in the Serengeti Area (Tanzania). Acta Tropica 30, 36–48. (16)

Beven, S.F., E.F. Connor and K. Beven. 1984. Avian biogeography in the Amazon Basin and the biological model of diversification. J. Biogeog. 11, 383–399. (8)

Bider, J.R. 1968. Animal activity in uncontrolled terrestrial communities as determined by a sand transect technique. Ecol. Monogr. 38, 269–308. (11)

Biémont, C. 1983. Homeostasis, enzymatic heterozygosity, and inbreeding depression in natural populations of *Drosophila melanogaster.* Genetica 61, 179–189. (4)

Bierregaard, R.O., Jr. in preparation. A mist-netting study of community structure in understory birds of the central Amazonian terra firme forests. (12)

Bierregaard, R.O., Jr. and T.E. Lovejoy. in preparation. Preliminary effects of isolation on understory bird communities in Amazonian forest fragments. (12)

Bille, J.C. 1977. Étude de la production primaire nette d'un écosystème sahélien. Trav. et Doc. No. 65, ORSTOM, Paris. (22)

Bille, J.C. 1978. Woody forage species in the Sahel: their biology and use. Proceed. 1st Internat. Range Congr., Denver, Colorado, pp. 392–395. (22)

Bingham, R.T. and A.E. Squillace. 1955. Self-compatibility and effects of self-fertility in western white pines. For. Sci. 1, 121–129. (5)

Bingham, R.T. and G.E. Rehfeldt. 1970. Cone and seed yields in young western white pines. USDA For. Serv. Res. Paper INT-79. 12 pp. (5)

Birdsall, D.A. and D. Nash. 1973. Occurrence of successful multiple insemination in natural populations of deer mice (*Peromyscus maniculatus*). Evolution 27, 106–110. (3)

Bishop, J.A. and L.M. Cook. 1980. Industrial melanism and the urban environment. Adv. Ecol. Res. 11, 373–404. (6)

Bishop, J.P. 1979. Family agricultural–swine–forestry production in the Spanish American humid tropics. In Agro-Forestry Systems in Latin America, Workshop Proceedings. G. de las Salas (ed.). Centro Agronómico Tropical de Investigación y Enseñanza, Turrialba, Costa Rica. pp. 140–144. (20)

Bittles, A.H. 1979. Incest re-assessed. Nature 280, 107. (3)

Black, F.L. and F.M. Salzano. 1981. Evidence for heterosis in the HLA system, Am. J. Hum. Genet. 33, 894–899. (4)

Blackwood, J.W. and C.R. Tubbs. 1970. A quantitative survey of chalk grassland in England. Biol. Conserv. 3, 1–5. (11)

Blouin, M.S. and E.F. Connor. 1985. Is there a best shape for nature reserves? Biol. Conserv. 32, 277–288. (11)

Boag, P.T. and P.R. Grant. 1981. Intense natural selection on a population of Darwin's finches (Geospizinae) in the Galapagos. Science 214, 82–85. (4)

Boag, P.T. 1983. The heritability of external morphology in Darwin's ground finches (*Geospiza*) on Isla Daphne Major, Galapagos. Evolution 37, 877–894. (4)

Boecklen, W.J. and N.J. Gotelli. 1984. Island biogeography theory and conservation practice: species–area or specious area relationships? Biol. Conserv. 29, 63–80. (11)

Bögli, A. 1980. *Karst Hydrology and Physical Speleology.* Springer Verlag, Berlin. (21)

Bond, P. and P. Goldblatt. 1984. Plants of the Cape flora: a descriptive catalogue. J. South African Bot. Soc., Suppl. Vol. 13, 1–455. (8)

Bond, R.R. 1957. Ecological distribution of breeding birds in the upland forests of southern Wisconsin. Ecol. Monogr. 27, 351–384. (11)

Bond, W. 1983. On alpha diversity and the richness of the Cape flora: a study in southern Cape fynbos. In *Mediterranean Type Ecosystems: The Role of Nutrients*, F.J. Kruger, D.T. Mitchell and J.U.M. Jarvis (eds.). Springer Verlag, New York. pp. 337–356. (7)

Bond, W. and P. Slingsby. 1984. Collapse of an ant–plant mutualism: the Argentine ant (*Iridomyrmex humilis*) and myrmecochorous Proteaceae. Ecology 65, 1031–1037. (7)

Bonnell, M.L. and R.K. Selander. 1974. Elephant seals: genetic variation and near extinction. Science 184, 908–909. (5)

Boorman, L.A. 1967. *Limonium vulgare* Mill. and *L. humile* Mill. J. Ecol. 55, 221–232. (9)

Bottini, E., F. Gloria-Bottini, P. Lucarelli, A. Polzonetti, F. Santoro and A. Varveri. 1979. Genetic polymorphisms and intrauterine development: evidence of decreased heterozygosity in light-for-dates human newborn babies. Experientia 35, 1565–1567. (4)

Bou, C. and R. Rouch. 1967. Un nouveau champ de recherches sur la faune aquatique souterraine. Compt. Rend. Acad. Sci. Paris 265, 369–370. (21)

Boudet, G. 1972. Desertification de l'Afrique tropicale sèche. Andansonia, Sér. 2, 12 (4), 505–524. (22)

Boudet, G. 1975. The inventory and mapping of rangelands in West Africa. Proceed. Sympos. on Evaluation and Mapping of Tropical African Rangelands, ILCA, Addis-Ababa. pp. 57–77. (22)

Boudet, G. 1977. Désertification ou remontée biologique au Sahel? Cah. ORSTOM, Sér. Biol., XII, 4. 293–300. (22)

Boughton, R.V. 1932. The influence of helminth parasitism on the abundance of the snowshoe rabbit in western Canada. Canad. J. Research 7, 524–547. (16)

Boulière, F. and M. Hadley. 1970. The ecology of tropical savannahs. Ann. Rev. Ecol. Syst. 1, 125–152. (15)

Boyd, S.K. and A.R. Blaustein. 1985. Familiarity and inbreeding avoidance in the gray-tailed vole (*Microtus canicaudus*). J. Mammal. 66, 348–352. (3)

Braakhekke, W.G. and C.A. Drijver. 1984. Wetlands. WWF-India 5, 4. Newsletter 51, 4–5. (18)

Bradbury, R.H. and R. Reichelt. 1981. The reef and man: rationalizing management through ecological theory. In Proc. IV Int'l. Coral Reef Symp. Univ. Philippines, Manila, 1, 219–224. (17)

Bradley, B.P. 1980. Developmental stability of *Drosophila melanogaster* under artificial and natural selection in constant and fluctuating environments. Genetics 95, 1033–1042. (4)

Bradshaw, A.D. 1971. Plants and industrial waste. Trans. Bot. Soc. Edinburgh 41, 71–84. (5)

Brady, J., T. Kunz, M.D. Tuttle and D. Wilson. 1982. Gray Bat Recovery Plan. U.S. Fish and Wildlife Service, Rockville, Md. (21)

Bramlett, D.L. and T.W. Popham. 1971. Model relating unsound seed and embryonic lethal alleles in self-pollinated pines. Silvae Genet. 20, 192–193. (5)

Bramwell, D. and Z.I. Bramwell. 1974. *Wildflowers of the Canary Islands*. Thornes, London. (7)

Branch, L.C. 1983. Seasonal and habitat differences in the abundance of primates in the Amazon (Tapajos) National Park, Brazil. Primates 24, 424–431. (12)

Brewbaker, J.L. 1964. *Agricultural Genetics*. Prentice-Hall, Englewood Cliffs, N.J. 156 pp. (5)

British Ecological Society. 1975. The biological flora of the British Isles. J. Ecol. 63, 335–344. (9)

Brittingham, M.C. and S.A. Temple. 1983. Have cowbirds caused forest songbirds to decline? BioScience 33, 31–35. (11)

Brody, A.K. and K.B. Armitage. in press. The effects of adult removal on dispersal of yearling yellow-bellied marmots. Canad. J. Zool. (3)

Brokaw, N.V.L. 1982. Treefalls: frequency, timing, and consequences. In *The Ecology of a Tropical Forest: Seasonal Rhythms and Long-Term Changes*, E.G. Leigh, Jr, A.S. Rand and D.M. Windsor (eds). Smithsonian Institution Press, Washington D.C. pp. 101–108 (10)

Brokaw, N.V.L. 1985. Gap-phase regeneration in a tropical forest. Ecology 66, 682–687. (10)

Brooke, R.K. 1984. South African red data book—birds. So. Afr. Sci. Prog. Rep. 97, 1–213. (7)

Brower, L.P., B.E. Horner, M.A. Marty, C.M. Moffitt and B. Villa R. 1985. Mice (*Peromyscus maniculatus, P. spicilegus*, and *Microtus mexicanus*) as predators of overwintering monarch butterflies (*Danaus plexippus*) in Mexico. Biotropica 17, 89–99. (13)

Brown, A.G. 1972. The role of the hybrid in forest tree breeding. In Proc. Joint Symp. For. Tree Breeding, Genet. Subject Group, IUFRO and Sect. 5, For. Trees, SABRAO. Govt. For. Exp. Sta. Jap., Tokyo. pp. C-1 (I). (5)

Brown, J.H. 1984. On the relationship between abundance and distribution of species, Amer. Natur. 124, 255–279. (II,9)

Brown, J.H. 1985. Experiments on desert rodents. In *Community Ecology*, T.J. Case and J.M. Diamond (eds.). Harper & Row, New York. (14)

Brown, J.H. and A. Kodric-Brown. 1977. Turnover rates in insular biogeography: effect of immigration on extinction. Ecology 58, 445–449. (11)

Brown, J.R. and E.R. Brown. 1981. Extended family system in a communal bird. Science 211, 959–960. (3)

Brown, K.S., Jr. in preparation. Comparisons between Lepidoptera faunas in reserves of different size and degree of isolation. (12)

Brussard, P.F. 1985. The current status of conservation biology. Bull. Ecol. Soc. Amer. 66, 9–11. (1)

Buchanan, D.B., R.A. Mittermeier and M.G.M. van Roosmalen. 1981. The saki monkeys genus *Pithecia*. In *Ecology and Behavior of Neotropical Primates,I*, A.F. Coimbra-Filho and R.A. Mittermeier (eds.). Academia Brasileira de Ciências, Rio de Janeiro. pp. 391–417. (12)

Burger, G.V. 1973. *Practical Wildlife Management*. Winchester Press, New York. (11)

Burgess, R.C. and D.M. Sharp (eds.). 1981. *Forest Island Dynamics in Man-dominated Landscapes*. Springer Verlag, New York. (11)

Buroker, N. E. 1979. Overdominance of a muscle protein (Mp-1) locus in the Japanese oyster, *Crassostrea gigas* (Ostreidae). J. Fish. Res. Board Canad. 36, 1313–1318. (4)

Burton, R.S. 1983. Protein polymorphisms and genetic differentiation of marine invertebrate populations. Mar. Biol. Lett. 4, 193–206. (4)

Buschbacher, R.J., C. Uhl and E.A.S. Serrao. 1984. Forest development following pasture use in the north of Para, Brazil. Proc. First Symp. on the Humid Tropics. Empresa Brasileira de Pesquisa Agropecuaria, Belem, Brazil. (20)

Bush, R., F.T. Ledig and P.E. Smouse. in preparation. The fitness consequences of multiple-locus heterozygosity: the relationship between heterozygosity and growth rate in pitch pine (*Pinus rigida* Mill.). (5)

Cain, S.A. 1944. *Foundations of Plant Geography*. Harper & Brothers, New York. (9)

Cairns, J., Jr. (ed.). 1985. *Multispecies Toxicity Testing*. Pergamon, New York. (23)

Cairns, J., Jr. in press. Disturbed ecosystems as opportunities for research in restoration ecology. In *Restoration Ecology: Progress Toward a Science and Art of Ecological Healing*, W.R. Jordan, J.D. Abert and M.E. Gilpin (eds.). (23)

Cairns, J., Jr. 1983. Management options for the rehabilitation and enhancement of surface-mined ecosystems. Miner. Environ. 5, 32–38. (23)

California Department of Parks and Recreation. 1975. Torrey Pine State Reserve and State Beach. Sacramento, Calif. (5)

California Gene Resources Program. 1982. Douglas-fir Genetic Resources: An Assessment and Plan for California. National Council of Gene Resources, Berkeley, Calif. 274 pp. (5)

Camara, I. de G. 1983. Tropical moist forest conservation in Brazil. In *Tropical Rain Forest: Ecology and Management*, S.L. Sutton, T.C. Whitmore and A.C. Chadwick (eds.). Special Publication No. 2, British Ecological Society, pp. 413–421. (13)

Campbell, R.K. 1979. Genecology of Douglas-fir in a watershed in the Oregon Cascades. Ecology 60, 1036–1050. (5)

Carlquist, S. 1974. *Island Biology*. Columbia University Press, New York. (8)

Carpenter, I.W. and A.T. Guard. 1950. Some effects of cross-pollination on seed production and hybrid vigor of tuliptree. J. For. 48, 852–855. (5)

Carter, G.S. undated (ca. 1954). *The Papyrus Swamps of Uganda*. Heffer & Sons, Cambridge, 25 pp. (18)

Case, T.J. and J.M. Diamond (eds). 1985. *Community Ecology*. Harper & Row, New York. (14)

Cates, R.G. and G.H. Orians. 1975. Successional status and the palatability of plants to generalized herbivores. Ecology 56, 410–418. (12)

CESP (Centrais Elétricas de São Paulo). Annual Report 1979. São Paulo, Brazil, CESP. (18)

Chaisson, R.E., L.A. Serunian and T.J.M. Schopf. 1976. Allozyme variation between two marshes and possible heterozygote superiority within a marsh in the bivalve *Modiolus demissus*. Biol. Bull. 151, 404. (4)

Chakraborty, R. and N. Ryman. 1983. Relationship of mean and variance of genotypic values with heterozygosity per individual in a natural population. Genetics 103, 149–152. (4)

Chalmers, D.J. and B. van den Ende. 1975. Productivity of peach trees: factors affecting dry weight distribution during tree growth. Ann. Bot. 39, 423–432. (5)

Chapin, F.S. 1980. The mineral nutrition of wild plants. Ann. Rev. Ecol. Syst. 11, 233–260. (20)

Chapin, G. and R. Wasserstrom. 1981. Agricultural production and malaria resurgence in Central America and India. Nature 293, 181–185. (16)

Chapman, F.M. 1907. *The Warblers of North America*. D. Appleton and Company, New York. (11)

Chasko, G.G. and J.E. Gates. 1982. Avian habitat suitability along a transmission-line corridor in an oak-hickory forest region. Wildl. Monogr. 82. (11)

Chepko-Sade, B.D. and D.S. Sade. 1979. Patterns of group splitting within matrilineal kinship groups. Behav. Ecol. Sociobiol. 5, 56–86. (3)

Chesser, R.K. 1983. Genetic variability within and among populations of the black-tailed prairie dog. Evolution 37, 320–331. (3)

Chesser, R.K., M.H. Smith, P.E. Johns, M.N. Manlove, D.O. Straney and R. Baccus. 1982. Spatial, temporal, and age-dependent heterozygosity of beta-hemoglobin in white-tailed deer. J. Wildl. Manage. 46, 983–990. (4)

Christiansen, F.B., O. Frydenberg and V. Simonsen. 1977. Genetics of *Zoarces* populations. X. Selection component analysis of the *Est*III polymorphism using samples of successive cohorts. Hereditas 87, 129–150. (4)

Christiansen, K.A. and D.C. Culver. 1968. Geographical variation and evolution in *Pseudosinella hirsuta*. Evolution 22, 237–255. (21)

Cicmanec, J.C. and A.K. Campbell. 1977. Breeding the owl monkey (*Aotus trivirgatus*) in a laboratory environment. Lab. Anim. Sci. 27, 517. (6)

Cicmanec, J.L. 1978. Medical problems encountered in a Callitrichid colony. In *The Biology and Conservation of the Callitrichidae*, D.G. Kleiman (ed.). Smithsonian Institution Press, Washington, D.C. (16)

Clarke, B. 1979. The evolution of genetic diversity. Proc. Roy Soc. Lond. B. Biol. Sci. 205, 453–474. (4)

Clutton-Brock, T.H. and P.H. Harvey. 1977. Species differences in feeding and ranging behavior in primates. In *Primate Ecology: Studies of Feeding and Ranging Behaviour in Lemurs, Monkeys and Apes*, T.H. Clutton-Brock (ed.). Academic Press, London. pp. 557–584. (5)

Cockburn, A., M.P. Scott and D. Scotts. 1985. Inbreeding avoidance and male-biased natal dispersal in *Antechinus* spp. Anim. Behav. 33, 908–915. (3)

Cody, M.L. 1974. *Competition and the Structure of Bird Communities*. Princeton University Press, Princeton, N.J. (7)

Cody, M.L. 1975. Towards a theory of continental species diversities: bird distributions over Mediterranean habitat gradients. In *Ecology and Evolution of Communities*, M.L. Cody and J.M. Diamond (eds.). Belknap Press of Harvard University Press, Cambridge, Mass. pp. 214–257. (7)

Cody, M.L. 1983a. Continental diversity patterns and convergent evolution in bird communities. In *Mediterranean-Type Ecosystems*, F.J. Kruger, D.T. Mitchell and J.U.M. Jarvis (eds.). Ecological Studies V. 43, Springer Verlag, Berlin. pp. 357–402. (7)

Cody, M.L. 1983b. Bird diversity and density in South African forests. Oecologia 59, 201–215. (7)

Cody, M.L. 1985a. Habitat selection in the sylviine warblers of western Europe and North Africa. In *Habitat Selection in Birds*, M.L. Cody (ed.). Academic Press, New York. (7)

Cody, M.L. 1985b. Structural niches in plant communities. In *Community Ecology*, T.J. Case and J.M. Diamond (eds.). Harper & Row, New York. (7)

Cody, M.L. 1985c. Convergent evolution, habitat area, species diversity and distribution. Res. Rep. 18, 233–241. Natl. Geogr. Soc., Washington, D.C. (7)

Cody, M.L. and J.M. Diamond. 1975. *Ecology and Evolution of Communities*. Harvard University Press, Cambridge, Mass. (14)

Cody, M.L. and H.A. Mooney. 1978. Convergence versus nonconvergence in Mediterranean-climate ecosystems. Ann. Rev. Ecol. Syst. 9, 265–321. (7)

Coles, J.F. and D.F. Fowler. 1976. Inbreeding in neighboring trees in two white spruce populations. Silvae Genet. 25, 29–34. (5)

Colgan, P. 1983. *Comparative Social Recognition*. Wiley, New York. (3)

Colwell, R.N. 1951. The use of radioactive isotopes in determining spore distribution patterns. Am. J. Bot. 38, 511–523. (5)

Conkle, M.T. 1974. Enzyme polymorphism in forest trees. In Proc. Third North Am. For. Biol. Workshop, C.P.P. Reid and G.H. Fechner (eds.). Colorado State University, Fort Collins. pp. 95–105. (5)

Conkle, M.T. 1981. Isozyme variation and linkage in six conifer species. In Proc. Symp. Isozymes of North Am. For. Trees and For. Insects, M.T. Conkle (tech. coord.). USDA For. Serv. Gen. Tech. Rep. PSW-48. pp. 11–17. (5)

Connor, E.F. and E.D. McCoy. 1979. The statistics and biology of the species–area relationship. Amer. Natur. 113, 791–833. (11)

Cooke, A.S. 1979. Changes in egg shell characteristics of the Sparrowhawk (*Accipiter nisus*) and Peregrine (*Falco peregrinus*) associated with exposure to environmental pollutants during recent decades. J. Zool. 187, 245–263. (16)

Correll, D.S. and M.C. Johnston. 1970. *Manual of the Vascular Plants of Texas*. Texas Research Foundation, Renner. (8)

Cothran, E.G., R.K. Chesser, M.H. Smith and P.E. Johns. 1983. Influences of genetic variability and maternal factors on fetal growth in white-tailed deer. Evolution 37, 282–291. (4)

Cram, W.H. 1984. Some effects of self-, cross-, and open-pollinations in *Picea pungens*. Canad. J. Bot. 62, 392–395. (5)

Cramp, S. and K.E.L. Simmons (eds.). 1979. *The Birds of the Western Palearctic*, Vol. II. Oxford University Press, Oxford. (11)

Critchfield, W.B. 1984. Impact of the Pleistocene on the genetic structure of North American conifers. In Proc. Eighth North Amer. For. Biol. Workshop, R.M. Lanner (ed.). Logan, Utah. pp. 70–118. (5)

Croat, T.B. 1975. Phenological behavior of habit and habitat classes on Barro Colorado Island (Panama Canal Zone). Biotropica 7, 270–277. (15)

Croat, T.B. 1978. *Flora of Barro Colorado Island*. Stanford University Press, Stanford, Calif. (8, 10, 15)

Croat, T.B. 1985. A revision of *Anthurium* for Mexico and Central America. Part 2: Panama. Monogr. Missouri Bot. Gard. (8)

Cronquist, A., A.H. Holmgren, N.H. Holmgren and J.L. Reveal. 1972. Intermountain Flora 1, 1–270. (8)

Crow, J.F. 1948. Alternative hypotheses of hybrid vigor. Genetics, 33, 477–487. (4)

Crow, T.R. 1980. A rainforest chronicle: a 30-year record of changes in structure and composition at El Verde, Puerto Rico. Biotropica 12, 42–55. (20)

Culver, D.C. 1982. *Cave Life: Evolution and Ecology*. Harvard University Press, Cambridge, Mass. (21)

Curl, R. 1966. Caves as a measure of karst. J. Geol. 74, 798–830. (21)

Daly, J.C. 1981. Effects of social organization and environmental diversity on determining the genetic structure of a population of the wild rabbit, *Oryctolagus cuniculus*. Evolution 35, 689–706. (3)

Dancik, B.P. and F.C. Yeh. 1983. Allozyme variability and evolution of lodgepole pine

(*Pinus contorta* var. *latifolia*) and jack pine (*P. banksiana*) in Alberta. Canad. J. Genet. Cytol. 25, 57–64. (5)

Danielopol, D. 1981. Distribution of ostracods in the groundwater of the northwestern coast of Euboea (Greece). Int. J. Speleology 11, 91–103. (21)

Danzmann, R.G. and G.P. Buchert. 1983. Isozyme variability in Central Ontario jack pine. In Proc. Twenty-eighth Northeast. For. Tree Improv. Conf. Inst. Natur. Environ. Resources, University of New Hampshire, Durham. pp. 232–248. (5)

Danzmann, R.G., M.M. Ferguson, F.W. Allendorf and K.L. Knudsen. 1986. Heterozygosity and developmental rate in a strain of rainbow trout (*Salmo gairdneri*). Evolution 40, 86–93. (4)

Dasmann, R.F. 1964. *Wildlife Biology*. Wiley, New York. (11)

Dasmann, W. 1971. *If Deer Are To Survive*. Stackpole Books, Harrisburg, Pa. (11)

Daveau, S. 1969. La découverte du climat d'Afrique tropicale au cours des navigations portugaises (XVème siècle et début du XVIème siècle). Bull. IFAN B 31, 953–988. (22)

Davis, J.W., L.H. Karstad and D.O. Trainer (eds.). 1970. *Infectious Diseases of Wild Mammals*. Iowa State University Press, Ames. (16)

Davis, L.S. 1982. Sibling recognition in Richardson's ground squirrels (*Spermophilus richardsonii*). Behav. Ecol. Sociobiol. 11, 65–70. (3)

Davis, L.S. 1984. Behavioral interactions of Richardson's ground squirrels: asymmetries based on kinship. In *The Biology of Ground-Dwelling Squirrels*, J.O. Murie and G.R. Michener (eds.). University of Nebraska Press, Lincoln. pp. 424–444. (3)

Dawson, G. 1979. The use of time and space by the Panamanian tamarin, *Saguinus oedipus*. Folia Primatol. 31, 253–284. (12)

De Boer, L.E.M. 1982. Karyological problems in breeding owl monkeys, *Aotus trivirgatus*. Internatl. Zoo Yearbook 22, 119–124. (6)

De Kock, G.C. 1965. The management and utilization of spineless cactus. Proceed. 9th Internat. Grassland Congr., São Paulo, Brazil. (22)

De Wispelaere, G. 1980. Les photographies aeriennes témoins de la dégradation du couvert ligneux dans un écosystème sahélien sénégalais. Influence de la proximité d'un forage. Cah. ORSTOM Sér. Sc. Human., XVII, 3–4. (22)

De Wispelaere, G. and B. Toutain. 1976. Estimation de l'évolution du couvert végétal en 20 ans, consécutivement à la sécheresse dans le Sahél Voltaique. Rev. Photinterpr. 76, 3/2. (22)

De Wispelaere, G. and B. Toutain. 1981. Étude diachronique de quelques géosystèmes sahéliens en Haute Volta septentrionale. Rev. Photointerpr. 81, 1/1–1/5. (22)

Degener, O., I. Degener, K. Sunada and M.K. Sunada. 1976. *Argyroxiphium kauense*, the key silversword. Phytologia 33, 173–177. (9)

Degos, L., J. Colomboni, A. Chaventre, B. Bengston and A. Jacguard. 1974. Selective pressure on HLA polymorphism. Nature 249, 62–63. (4)

Dekeyser, P.L. 1955. *Les Mammifères de l'Afrique Noire Française*, 2nd Ed. Inst. Fond. d'Afrique Noire, Dakar. 426 pp. (22)

Dekeyser, P.L. and J. Derivot. 1969. *La vie animale au Sahara*. Colin, Paris. 220 pp. (22)

Denevan, W.M. 1981. Swiddens and cattle versus forest: the imminent demise of the Amazon rain forest reexamined. In *Where Have All the Flowers Gone? Deforestation in the Third World*, V.H. Sutlive, N. Altshuler, and M.D. Zamora (eds.). Pub. 13, Studies in Third World Societies, Dept. of Anthropology, College of William and Mary, Williamsburg, Va. pp. 25–44. (20)

Devall, B. and G. Sessions. 1985. *Deep Ecology: Living As If Nature Mattered*. Peregrine Smith Books, Layton, Ut. (25)

Devi, A.R.R. and N.A. Rao. 1981. Consanguinity, fecundity and post-natal mortality in Karnataka, South India. Ann. Hum. Biol. 8, 469–472. (3)

Devi, A.R.R., N.A. Rao and A.H. Bittles. 1982. Inbreeding in the State of Karnataka, South India. Hum. Heredity 32, 9–10. (3)

Dewsbury, D.A. 1982. Avoidance of incestuous breeding between siblings in two species of *Peromyscus* mice. Biol. Behav. 7, 157–169. (3)

Diamond, J.M. 1975a. Assembly of species communities. In *Ecology and Evolution of Communities*, M.L. Cody and J.M. Diamond (eds.). Harvard University Press, Cambridge, Mass. pp. 342–444. (11)

Diamond, J.M. 1975b. The island dilemma: lessons of modern biogeographic studies for the design of natural reserves. Biol. Conserv. 7, 129–146. (11)

Diamond, J.M. 1977. Continental and insular speciation in Pacific land birds. Syst. Zool. 26, 263–268. (24)

Diamond, J.M. 1980. Proposal for a reserve in the Mamberamo region, Irian Jaya. World Wildlife Fund Indonesia Program Special Report No. 3. (24)

Diamond, J.M. 1981. Surveys of proposed reserves in the Foja and Fakfak Mountains, Irian Jaya, Indonesia. World Wildlife Fund Indonesia Program Special Report. (24)

Diamond, J.M. 1984a. "Normal" extinctions of isolated populations. In *Extinctions*, M.H. Nitecki (ed.). University of Chicago Press, Chicago. pp. 191–246. (11, 14, 24)

Diamond, J.M. 1984b. Historic extinctions: a Rosetta Stone for understanding prehistoric extinctions. In *Quaternary Extinctions, A Prehistoric Revolution*, P.S. Martin and R.G. Klein (eds.). University of Arizona Press, Tucson. pp. 824–862. (14, 24)

Diamond, J.M. 1984c. Distributions of New Zealand birds on real and virtual islands. New Zealand J. Ecol. 7, 37–55. (24)

Diamond, J.M. 1984d. Surveys of five proposed reserves in Irian Jaya, Indonesia: Kumawa Mts., Wandamen Mts., Yapen Island, Salawati Island, and Batanta Island. World Wildlife Fund Indonesia Program Special Report. (24)

Diamond, J.M. 1984e. The avifaunas of Rennell and Bellona Islands. Nat. Hist. Rennell Island, Br. Solomon Isls. 8, 127–168. (24)

Diamond, J.M. 1985a. Introductions, extinctions, exterminations, and invasions. In *Community Ecology*, T.J. Case and J.M. Diamond (eds.). Harper & Row, New York. pp. 65–79. (14, 24)

Diamond, J.M. 1985b. New distributional records and taxa from the outlying mountain ranges of Irian Jaya. Emu 85, 65–91. (24)

Dice, L.R. and W.E. Howard. 1951. Distances of dispersal by prairie deermice from birthplace to breeding sites. Contr. Lab. Vert. Biol. Univ. Mich. 50, 1–15. (3)

Diehl, W.J., P.M. Gaffney, J.H. McDonald and R.K. Koehn. 1985. Relationship between weight-standardized oxygen consumption and multiple-locus heterozygosity in the marine mussel *Mytilus edulis* L. (Mollusca), Proc. XIX European Mar. Biol. Symp. (4)

Dietz, K. 1975. Transmission and control of arbovirus diseases. In *Epidemiology*, D. Ludwig and K.L. Cooke (eds.). Proceedings of a SIMS Conference on Epidemiology, Alta, Utah. (16)

DiMichele, L. and D.A. Powers. 1982a. LDH-B genotype-specific hatching times of *Fundulus heteroclitus* embryos. Nature 296, 563–565. (4)

DiMichele, L. and D.A. Powers. 1982b. Physiological basis for swimming endurance differences between LDH-B genotypes of *Fundulus heteroclitus*. Science 216, 1014–1016. (4)

DiMichele, L. and M.H. Taylor. 1980. The environmental control of hatching in *Fundulus heteroclitus*. J. Exper. Zool. 214, 181–187. (4)

DiMichele, L. and M.H. Taylor. 1981. The mechanism of hatching in *Fundulus heteroclitus*: development and physiology. J. Exper. Zool. 217, 73–79. (4)

Dobson, A.P. and A.E. Keymer. 1985. Life history models. In *Biology of the Acanthocephala*, D.W.T. Crompton and B. Nickol (eds.). Cambridge University Press, Cambridge. (16)

Dobson, A.P. and R.M. May. 1985. Patterns of invasions by pathogens and parasites. In *Ecology of Biological Invasions of North America and Hawaii*, H.A.Mooney (ed.). Springer Verlag, New York. (16)

Dobson, F.S. 1979. An experimental study of dispersal in the California ground squirrel. Ecology 60, 1103–1109. (3)

Dobson, F.S. 1982. Competition for mates and predominant juvenile male dispersal in mammals. Anim. Behav. 30, 1183–1192. (3)

Dobson, F.S. and W.T. Jones. 1985. Multiple causes of dispersal. Amer. Natur. 126, 855–858. (3)

Dobzhansky, Th. 1948. Genetics of natural populations. XVIII. Experiments on chromosomes of *Drosophila pseudoobscura* from different geographical regions. Genetics 33, 588–602. (6)

Dodd, J., E.M. Heddle, J.S. Pate and K.W. Dixon. 1984. Rooting patterns of sandplain plants and their functional significance. In *Kwongan: Plant Life in the Sandplain*,

J.S. Pate and J.S. Beard (eds.). University of Western Australia Press, Nedlands. pp. 146–177. (7)

Dodson, C. and A.H. Gentry. 1978. Flora of the Rio Palenque Science Center. Selbyana 4, 1–628. (8, 19)

Dodson, C., A.H. Gentry and F.M. Valverde. 1985. Flora of Jauneche. Los Rios, Ecuador. *Selbyana.* (8)

Dogiel, V.A. 1966. *General Parasitology.* Academic Press, New York. (16)

Dollar, S.J. 1982. Wave stress and coral community structure in Hawaii. Coral Reefs 1, 71–81. (17)

Dorn, R.D. 1977. *Manual of the Vascular Plants of Wyoming,* Vol. 2. Garland, New York. (8)

Doubleday, W.C. 1976. Environmental fluctuations and fisheries management. Int. Comm. Northw. Atl. Fish. Sel. Pap. 1, 141–150. (14)

Dover, M. and B. Croft. 1984. Getting Tough. Public Policy and the Management of Pesticide Resistance. World Resources Institute, Study 1. WRI Pubs. (16)

Drake, J. A. 1985. Some theoretical and experimental explorations of the structure of food webs. Ph.D. dissertation, Purdue University, West Lafayette, Ind. (14)

Dregne, H.E. 1983. *Desertification of Arid Lands.* Academic Press, New York. 242 pp. (22)

Dressler, R.L. 1982. Biology of orchid bees *(Euglossini).* Ann. Rev. Ecol. Syst. 13, 373–394. (12)

Drummond, D.C. 1970. Variation in rodent populations in response to control measures. Symp. Zool. Soc. Lond. 26, 351–367. (4)

Duba, S.E. 1985. Polymorphic isozymes from megagametophytes and pollen of longleaf pine: characterization, inheritance, and use in analyses of genetic variation and genotype verification. In Proc. Eighteenth Southern For. Tree Improv. Conf. Long Beach, Miss. pp. 88–98. (5)

Duncan, P., C. Feh, J.C. Gleize, P. Malkas and A.M. Scott. 1984. Reduction of inbreeding in a natural herd of horses. Anim. Behav. 32, 520–527. (3)

Eanes, W.F. 1978. Morphological variance and enzyme heterozygosity in the monarch butterfly. Nature 276, 263–264. (4,5)

Eckert, R.T., R.J. Joly and D.B. Neale. 1981. Genetics of isozyme variants and linkage relationships among allozyme loci in 35 eastern white pine clones. Canad. J. For. Res. 11, 573–579. (5)

Edmisten, J. 1970. Soil studies in the El Verde rain forest. In *A Tropical Rain Forest,* H.T. Odum (ed.). Division of Technical Information, U.S. Atomic Energy Commission, Washington, D.C. pp. H79–H87. (20)

Ehler, L.E. and R.W. Hall. 1982. Evidence for competitive exclusion of introduced natural enemies in biological control. Environ. Entomol. 11, 1–4. (14)

Ehrlich, P.R. and A.H. Ehrlich. 1981. *Extinction: the Causes and Consequences of the Disappearance of Species.* Random House, New York. (14)

Ehrlich, P.R. 1983. Genetics and extinction of butterfly populations. In *Genetics and Conservation: A Reference for Managing Wild Animal and Plant Populations,* C.M. Schonewald-Cox, S.M. Chambers, F. MacBryde and L. Thomas (eds.) Benjamin-Cummings, Menlo Park, Calif. pp. 164–184. (2)

Ehrlich, P.R., J. Harte, M.A. Harwell, R.H. Raven, C. Sagan, G.M. Goodwell, J. Berry, E.S. Ayensu, A.H. Ehrlich, T. Eisner, S.J. Gould, H.D. Grover, R. Herrera, R.M. May, E. Mayr, C.P. McKay, H.A. Mooney, N. Myers, D. Pimental and J.M. Teal. 1983. Long-term biological consequences of nuclear war. Science 222, 1293–1300. (23)

Eisenberg, J.F. and R.W. Thorington. 1973. A preliminary analysis of a Neotropical mammal fauna. Biotropica 5, 150–161. (15)

Eisner, T. 1983. Chemicals, genes, and the loss of species. Nature Conservancy News 33, 23–24. (19)

El-Kassaby, Y.A. 1983. Repeated relation between allozyme variation and a quantitative trait in Douglas-fir. Egypt. J. Genet. Cytol. 12, 329–344. (5)

Eldridge, K.G. 1970. Breeding system of *Eucalyptus regnans.* In IUFRO Sect. 22 Working Group on Sexual Reproduction of Forest Trees, Vol. 1. Finn. For. Res. Inst., Helsinki. 13 pp. (5)

Eldridge, K.G. and A.R. Griffin. 1983. Selfing effects in *Eucalyptus regnans*. Silvae Genet. 32, 216–221. (5)

Elton, C.S. 1958. *The Ecology of Invasions by Animals and Plants*. Chapman and Hall, London. (14)

Emlen, S.T. 1984. Cooperative breeding in birds and mammals. In *Behavioural Ecology: An Evolutionary Approach*, 2nd Ed., J.R. Krebs and N.B. Davies (eds.). Sinauer Associates, Sunderland, Mass. pp. 305–309. (3)

Emmerson, D.K. 1980. Rethinking artisanal fisheries development: Western concepts, Asian experiences. World Bank Staff Work. Pap. 324, 1–97. (17)

Emmons, L.H. 1984. Geographic variation in densities and diversities of non-flying mammals in Amazonia. Biotropica 16: 210–222. (12)

Emmons, L.H., A. Gautier-Hion and G. Dubost. 1983. Community structure of the furgivorous-folivorous forest mammals of Gabon. J. Zool. Lond. 199, 209–222. (15)

Endler, J.A. 1982. Pleistocene forest refuge: fact or fancy? In *Biological Diversification in the Tropics*, G. Prance (ed.). Columbia University Press, New York. pp. 179–200. (8)

Ennos, R.A. 1983. Maintenance of genetic variation in plant populations. Evol. Biol. 16, 129–155. (5)

Erickson, A.B. 1944. Helminth infections in relation to population fluctuations in snowshoe hares. J. Wildl. Manage. 8, 134–153. (16)

Erwin, T.L. 1983. Tropical forest canopies: the last biotic frontier. Bull. of the Entomol. Soc. Amer. (Spring), 14–19. (19)

Esman, M.J. 1979. *Landlessness and Near Landlessness in Developing Countries*. Cornell University Press, Ithaca, N.Y. (19)

Estrada, A., R. Coates-Estrada and C. Vazquez-Yanes. 1984. Observations on fruiting and dispersers of *Cecropia obtusifolia* at Los Tuxtlas, Mexico. Biotropica 16, 315–318. (13)

Evans, R.M. 1970. Imprinting and mobility in young ring-billed gulls, *Larus delawarensis*. Anim. Behav. Monogr. 3, 193–248. (3)

Falconer, D.S. 1981. *Introduction to Quantitative Genetics*. Longman, New York. (4)

FAO. 1980. Proceedings of World Conference in Agrarian Reform and Rural Development. Food and Agriculture Organization, Rome. (19)

FAO/UNEP. 1981. Conservation of the genetic resources of fish: problems and recommendation. Report of the Expert Consultation on the genetic resources of fish. Rome, 9–13 June, 1980. FAO Fish. Tech. Pap. 217. 43 pp. (18)

FAO/UNEP. 1982. Tropical Forest Resources. Food and Agriculture Organization, Rome, and United Nations Environment Programme, Nairobi. (19)

Farris, M.A. and J.B. Mitton. 1984. Population density, outcrossing rate, and heterozygote superiority in ponderosa pine. Evolution 38, 1151–1154. (5)

Fearnside, P.M. 1985. Human-use systems and the causes of deforestation in the Brazilian Amazon. Paper presented at UNU International Conference on Climatic, Biotic and Human Interactions in the Humid Tropics, São Jose, dos Campos, São Paolo, Brazil, 25th February–1st March, 1985. (19)

Fearnside, P.M. and J. Rankin. 1980. Jari and development in the Brazilian Amazon. Interciencia 5, 146–156. (20)

Fearnside, P.M. and J. Rankin. 1982. The new Jari: risks and prospects of a major Amazonian development. Interciencia 7, 329–339. (20)

Feder, J.L., M.H. Smith, R. K. Chesser, M.J.W.Godt and K. Asbury. 1984. Biochemical genetics of mosquitofish. II. Demographic differentiation of populations in a thermally altered reservoir. Copeia 1984, 108–119. (4)

Feinsinger, P. 1980. Asynchronous migration patterns and the coexistence of tropical hummingbirds. In *Migrant Birds in the New World Tropics*, E.S. Morton and A. Keast (eds.). Smithsonian Institution Press, Washington, D.C. pp. 411–419. (15)

Feller, W. 1957. *An Introduction to Probability Theory and its Applications*, Vol. 1. Wiley, New York. (2)

Fienberg, S.E. 1981. *The Analysis of Cross-classified Categorical Data*, 2nd Ed. M.I.T. Press, Cambridge, Massachusetts. (9)

Fins, L. and W.J. Libby. 1982. Population variation in Sequoiadendron: seed and seeding studies, vegetative propagation, and isozyme variation. Silvae Genet. 31, 101–148. (5)

Fins, L. and L.W. Seeb. in preparation. Genetic variation in allozymes of western larch. Canad. J. For. Res. (5)

Fisher, H.I. 1966. Aerial census of Laysan albatrosses breeding on Midway Atoll in December, 1962. Auk 83, 670–673. (3)

Fisher, H.I. 1967. Body weights in Laysan albatross *Diomedea immutabilis*. Ibis 109, 373–382. (3)

Fisher, H.I. and M.L. Fisher. 1969. The visits of Laysan albatross to the breeding colony. Micronesica 5, 173–221. (3)

Fisher, R.A. 1930. *The Genetical Theory of Natural Selection*. Dover Publications, New York. (4)

Fisher, R.A. 1941. Average excess and average effect of a gene substitution. Ann. Eugen. 11, 53–63. (5)

Fleagle, J.G. and R.A. Mittermeier. 1980. Locomotor behaviour, body size and comparative ecology of seven Surinam monkeys. Amer. J. Phys. Anthro. 52, 301–314. (12)

Fleischer, R.C., R.F. Johnston and W.J. Klitz. 1983. Allozymic heterozygosity and morphological variation in house sparrows. Nature 304, 628–630. (4, 5)

Fleming, G.L. 1916. Rediscovering Torrey Pines. Calif. Gard. 8(1), 8–10. (5)

Floret, C. 1981. The effects of protection on steppic vegetation in the Mediterranean arid zone of southern Tunisia. Vegetatio 46, 117–129. (22)

Floret, C. and E. Le Floch. 1973. Production, sensibilité et évolution de la végétation et du milieu en Tunisie présaharienne. Doc. No. 71, CEPE/CNRS, Montpellier. 45 pp. (22)

Floret, C. and R. Pontanier. 1982. L'aridité en Tunisie présaharienne. Trav. & Doc. No. 150, ORSTOM, Paris. 544 pp. (22)

Floret, C., E. Le Floch and R. Pontanier. 1983. Phytomasse et production végétale en Tunisie preésaharienne. Oecol. Plant., Oecol. Applic. 4, 18, 2, 133–152. (22)

Flyger, V. 1970. Urban grey squirrels: problems, management and comparisons with forest populations. Transactions of Northeast Fish and Wildlife Conference 27, 107–113. (11)

Fogden, M.P. 1972. The seasonality and population dynamics of tropical forest birds in Sarawak. Ibis 114, 307–343. (15)

Foltz, D.W. 1981. Genetic measures of inbreeding in *Peromyscus*. J. Mammal. 62, 470–476. (3)

Foltz, D.W. and J.L. Hoogland. 1983. Genetic evidence of outbreeding in the black-tailed prairie dog (*Cynomys ludovicianus*). Evolution 37, 273–281. (3)

Foltz, D.W. and E. Zouros. 1984. Enzyme heterozygosity in the scallop *Placopecten magellanicus* (Gmelin) in relation to age and size. Mar. Biol. Lett. 5, 255–263. (4)

Foltz, D.W., G.F. Newkirk and E. Zouros. 1983. Genetics of growth rate in the American oyster: absence of interactions among enzyme loci. Aquaculture 33, 157–165. (4)

Forbes, S.A. 1887. The lake as a microcosm. Bull. Peoria (Ill.) Sci. Assoc., reprinted in Bull. Ill. Nat. Hist. Surv., 1925, 15, 537–550. (18)

Ford, J. 1971. The role of the trypanosomiases in African ecology. Clarendon Press, Oxford. (16)

Fosbrooke, H. 1963. The stomoxys plague in Ngorongoro, 1962. East African Wildlife Journal 1, 124–126. (16)

Fossey, D. 1983. *Gorillas in the Mist*. Houghton Mifflin, Boston. (3)

Foster, R.B. 1982a. The seasonal rhythm of fruitfall on Barro Colorado Island. In *The Ecology of a Tropical Forest: Seasonal Rhythms and Long-term Changes*, E.G. Leigh, Jr., A.S. Rand and D.M. Windsor (eds.). Smithsonian Institution Press, Washington, D.C. pp. 151–172. (15)

Foster, R.B. 1982b. Famine on Barro Colorado Island. In *The Ecology of a Tropical Forest: Seasonal Rhythms and Long-term Changes*, E.G. Leigh, Jr., A.S. Rand and D. Windsor (eds.). Smithsonian Institution Press, Washington, D.C. pp. 201–212. (15)

Fowler, D.P. 1964. Pre-germination selection against a deleterious mutant in red pine. For. Sci. 10, 335–336. (5)

Fowler, D.P. 1965a. Effects of inbreeding in red pine, *Pinus resinosa* Ait. Silvae Genet. 13, 170–177. (5)

Fowler, D.P. 1965b. Effects of inbreeding in red pine, *Pinus resinosa* Ait. II. Pollination studies. Silvae Genet. 14, 12–23. (5)

Fowler, D.P. 1965c. Effects of inbreeding in red pine, *Pinus resinosa* Ait. III. Factors affecting natural selfing. Silvae Genet. 14, 37–46. (5)

Fowler, D.P. 1965d. Effects of inbreeding in red pine, *Pinus resinosa* Ait. IV. Comparison with other northeastern *Pinus* species. Silvae Genet. 14, 76–81. (5)

Fowler, D.P. and R.W. Morris. 1977. Genetic diversity in red pine: evidence for low genic heterozygosity. Can. J. For. Res. 7, 343–347. (5)

Fowler, D.P. and Y.S. Park. 1983. Population studies of white spruce. I. Effects of self-pollination. Can. J. For. Res. 13, 1133–1138. (5)

Fox, J.E.D. 1976. Constraints on the natural regeneration of tropical moist forest. For. Ecol. and Manage. 1, 37–65. (20)

Franclet, A. and H.N. Le Houérou. 1971. Les atriplex en Tunisie et en Afrique du Nord. FAO, Rome. 244 pp. (English translation on micofiche No. 17909/E1, FAO, Rome). (22)

Frankel, O.H. and M.E. Soulé. 1981. *Conservation and Evolution.* Cambridge University Press, Cambridge. (1, 2, 4, 11, 16)

Frankel, R. (ed.). 1983. *Heterosis: Reappraisal of Theory and Practice.* Springer Verlag, Berlin. (4)

Frankham, R. 1980. The founder effect and response to artificial selection in *Drosophila.* In *Selection Experiments in Laboratory and Domestic Animals,* A. Robertson (ed.). Commonwealth Agricultural Bureaux, Farnham Royal, Slough, U.K. pp. 87–90. (4)

Frankie, G.W., H.G. Baker and P.A. Opler. 1974. Comparative phenological studies of trees in tropical wet and dry forests in the lowlands of Costa Rica. J. Ecol. 62, 881–919. (15)

Franklin, E.C. 1969. Mutant forms found by self-pollination in loblolly pine. J. Hered. 60, 315–320. (5)

Franklin, E.C. 1970. Survey of mutant forms and inbreeding depression in species of the family Pinaceae. USDA For. Serv. Res. Pap. SE-61. 21 pp. (5)

Franklin, E.C. 1972. Genetic load in loblolly pine. Amer. Natur. 106, 262–265. (5)

Freeman, C.C. 1980. Annotated list of the vascular flora of Konza Prairie Research Natural Area, Div. Biol. Kansas State University, Manhattan. pp. 1–27. (8)

Frelinger, J.A. 1971. Maternally derived transferrin in pigeon squabs. Science 171, 1260–1261. (4)

Frelinger, J.A. 1972. The maintenance of transferrin polymorphism in pigeons. Proc. Nat. Acad. Sci. USA 69, 326–329. (4)

Fretwell, S. 1972. *Populations in a Seasonal Environment.* Princeton University Press, Princeton, N.J. (11)

Friedman, S.T. and W.T. Adams. 1985. Estimation of gene flow into two seed orchards of loblolly pine (*Pinus taeda* L.). Theor. Appl. Genet. 69, 609–615. (5)

Fryer, J.H. and F.T. Ledig. 1972. Microevolution of the photosynthetic temperature optimum in relation to the elevational complex gradient. Canad. J. Bot. 50, 1231–1235. (5)

Fujino, K. and T. Kang. 1968. Transferrin groups in tunas. Genetics 59, 79–91. (4)

Furnier, G.R. 1984. Population genetic structure of Jeffrey pine. Ph.D. dissertation, Oregon State University, Corvallis. 59 pp. (5)

Futuyma, D.J. 1979. *Evolutionary Biology.* Sinauer Associates, Sunderland, Mass. (I)

Futuyma, D.J. 1983. Interspecific interactions and the maintenance of genetic diversity. In *Genetics and Conservation: A Reference for Managing Wild Animal and Plant Populations,* C.M. Schonewald-Cox, S.M. Chambers, F. MacBryde, and L. Thomas (eds.). Benjamin/Cummings, Menlo Park, Calif. pp. 364–373. (2)

Gaines, M.S. and L.R. McClenaghan, Jr. 1980. Dispersal in small mammals. Ann. Rev. Ecol. Syst. 11, 163–196. (3)

Game, M. 1980. Best shapes for nature reserves. Nature 287, 630–632. (11)

Gardner, M.R. and W.R. Ashby. 1970. Connectance of large, dynamic, (cybernetic) systems: critical values for stability. Nature 228, 784. (14)

Garton, D.W. 1984. The relationship between multiple locus heterozygosity and the physiological energetics of growth in the estuarine gastropod *Thais haemostoma*. Physio. Zool. 57, 530–543. (4)

Garton, D.W., R.K. Koehn and T.M. Scott. 1984. Multiple locus heterozygosity and the physiological energetics of growth in the coot clam *Mulinia lateralis* from a natural population. Genetics 108, 445–455. (4)

Garwood, N.C., D.P. Janos and N. Brokow. 1979. Earthquake-caused landslides: a major disturbance to tropical forests. Science 205, 997–999. (20)

Gaston, A. 1981. La végétation du Tchad: évolutions récentes sous influences climatiques et humaines. D.Sc. thesis, University of Paris XII. IEMVT, Maisons-Alfort. (22)

Gates, J.E. and L.W. Gysel. 1978. Avian nest dispersion and fledging success in field–forest ecotones. Ecology 59, 871–883. (11)

Gauthreaux, S.A., Jr. 1978. The ecological significance of behavioral dominance. In *Perspectives in Ethology*, Vol. 3, P.P.G. Bateson and P.H. Klopfer (eds.). Plenum, New York. pp. 17–54. (3)

Gautier-Hion, A. 1980. Seasonal variations of diet related to species and sex in a community of *Cercopithecus* monkeys. J. Anim. Ecol. 49, 237–269. (15)

Gautier-Hion, A., J.-M. Duplantier, L. Emmons, F. Feer, P. Hecketsweiler, A. Moungazi, R. Quris and C. Sourd. In press. Coadaptation entre rythmes de fructification et frugivorie en forêt tropicale humide du Gabon: mythe ou realité. Rev. Ecol. (Terre et Vie). (15)

Gavish, L., J.E. Hofmann and L.L. Getz. 1984. Sibling recognition in the prairie vole, *Microtus ochrogaster*. Anim. Behav. 32, 362–366. (3)

Gentille, J. (ed.). 1971. *Climates of Australia and New Zealand*. World Survey of Climatology, Vol. 13. Elsevier, Amsterdam. (7)

Gentry, A.H. 1977. Endangered plant species and habitats of Ecuador and Amazonian Peru. In *Extinction is Forever*, G. Prance and T. Elias (eds.). New York Bot. Gard. pp. 136–149. (8)

Gentry, A.H. 1978. A new *Freziera* (Theaceae) from the Panama/Colombia border. Ann. Missouri Bot. Gard. 63, 773–774. (8)

Gentry, A.H. 1979. Extinction and conservation of plant species in tropical America: a phytogeographical perspective. In *Systematic Botany, Plant Utilization, and Biosphere Conservation*, I. Hedberg (ed.). Almquist & Wiksell, Stockholm. pp. 110–126. (8)

Gentry, A.H. 1980. Sabiaceae. In *Flora of Panama*, Ann. Missouri Bot. Gard. 67, 949–963. (8)

Gentry, A.H. 1981a. Distributional patterns and an additional species of the *Passiflora vitifolia* complex: Amazonian species diversity due to edaphically differentiated communities. Plant Syst. Evol. 137, 95–105. (8)

Gentry, A.H. 1981b. Inventario floristico de Amazonia Peruana: estado y perspectivas de conservación. In *Seminario Sobre Proyectos de Investigación Ecológica para el Manejo de los Recursos Naturales Renovables del Bosque Tropical Húmido*, T. Gutierrez G. (ed.). Dirección General Forestal y de Fauna, Ministerio de Agricultura, Lima, Peru. pp. 36–44. (8)

Gentry, A.H. 1982a. Neotropical floristic diversity: phytogeographical connections between Central and South America, Pleistocene climatic fluctuations, or an accident of the Andean orogeny? Ann. Missouri Bot. Gard. 69, 557–593. (8)

Gentry, A.H. 1982b. Patterns of neotropical plant species diversity. Evol. Biol. 15, 1–84. (8, 19)

Gentry, A.H. 1982c. Phytogeographic patterns in northwest South America and southern Central America as evidence for a Choco refugium. In *Biological Diversification in the Tropics*, G. Prance (ed.). Columbia University Press, New York. pp.112–136. (8)

Gentry, A.H. 1983a. *Alstonia* (Apocynaceae): another paleotropical genus in Central America. Ann. Missouri Bot. Gard. 70, 206. (8)

Gentry, A.H. 1983b. Dispersal and distribution in Bignoniaceae. Sonderbd. Naturwiss. Ver. Hamburg 7, 187–199. (8)

Gentry, A.H. 1984. *An Overview of Neotropical Phytogeographic Patterns with an Emphasis on Amazonia*. Missouri Botanical Garden, St. Louis. (19)

Gentry, A.H. 1985a. Contrasting phytogeographic patterns of upland and lowland Panamanian plants. In *Natural History of Panama*, W. D'Arcy and M. Correa (eds.). Missouri Botanical Garden, St. Louis. pp. 214–237. (8)

Gentry, A.H. 1985b. Exploring the mountain of the mists. World Book Science Year, pp. 124–139. (8)

Gentry, A.H. in press. An overview of neotropical phytogeographic patterns with an emphasis on Amazonia. Proceedings of 1st Simposio do Tropico Humedo, Belem, Brazil. (8, 15)

Gentry, A.H. and C.H. Dodson. 1986. Tropical rain forests without trees: still the world's most species-rich plant communities. Biotropica 18. (8)

Gentry, A.H. and R. Foster. 1981. A new Peruvian *Styloceras* (Buxaceae): discovery of a phytogeographical missing link. Ann. Missouri Bot. Gard. 68, 122–124. (8)

George, A.S. 1984. *The Banksia Book*. Kangaroo Press, Kenhurst, N.S.W. (7)

Gibb, J.A. and J.E.C. Flux. 1973. Mammals. In *The Natural History of New Zealand*, G.R. Williams (ed.). Reed, New Zealand. (14)

Gilbert, L.E. 1980. Food web organization and conservation of neotropical diversity. In *Conservation Biology*, M.E. Soulé and B.A. Wilcox (eds.). Sinauer Associates, Sunderland, Mass., pp. 11–34. (11, 15, 19)

Gillet, H. 1965. L'Oryx la gazell et l'addax au Tchad. La Terre et la Vie 3, 257–272. (22)

Gillet, H. 1984. La chèvre et la gazelle: exploitation comparée de pâturages par la faune et le bétail en Afrique tropicale sèche. Le Courrier de la Nature 90, 17–26. (22)

Gilpin, M.E. 1986a. Patchily distributed populations. In *Viable Populations*, M.E. Soulé (ed.). Cambridge University Press, Cambridge. (2)

Gilpin, M.E. 1986b. Effective population size under range shift: an explanation for the genetic depaupercy in cheetahs. Evolution. (2)

Gilpin, M.E. and R.A. Armstrong. 1981. On the concavity of island biogeographic rate functions. J. Theor. Biol. 20, 209–217. (11)

Gilpin, M.E. and T.J. Case. 1976. Multiple domains of attraction in competition communities. Nature 261, 40–42. (14)

Gilpin, M.E. and J.M. Diamond. 1976. Calculation of immigration and extinction curves from the species-area-distance relation. Proc. Nat. Acad. Sci. USA 73, 4130–4134. (2, 11)

Gilpin, M.E. and J.M. Diamond. 1981. Immigration and extinction probabilities for individual species: relation to incidence functions and species colonization curves. Proc. Nat. Acad. Sci. USA 78, 392–396. (11)

Glanz, W.E. 1982. The terrestrial mammal fauna of Barro Colorado Island: censuses and long-term changes. In *The Ecology of a Tropical Forest: Seasonal Rhythms and Long-term Changes*, E.G. Leigh, Jr., A.S. Rand and D.M. Windsor (eds.). Smithsonian Institution Press, Washington, D.C. pp. 455–568. (15)

Glanz, W.E., R.W. Thorington, Jr., J. Giacalone-Madden and L.R. Heaney. 1982. Seasonal food use and demographic trends in *Sciurus granatensis*. In *The Ecology of a Tropical Forest: Seasonal Rhythms and Long-term Changes*, E.G. Leigh, Jr., A.S. Rand and D.M. Windsor (eds.). Smithsonian Institution Press, Washington, D.C. pp. 239–252. (15)

Goeden, G. B. 1979. Biogeographic theory as a management tool. Environ. Conserv. 6, 27–32. (17)

Goldblatt, P. 1978. An analysis of the flora of Southern Africa: its characteristics, relationships, and origins. Ann. Missouri Bot. Gard. 65, 369–436. (8)

Gomez, E. D. 1982/83. Perspectives on coral reef research and management in the Pacific. Ocean Management 8, 281–295. (17)

Goodland, R.J.A. 1980. Environmental ranking of Amazonian development projects in Brazil. Environ. Conserv. 7, 9–26. (20)

Goodland, R.J.A. and H.S. Irwin. 1975. *Amazon Jungle: Green Hell to Red Desert?* Elsevier, Amsterdam. (20)

Goodman, D. 1975. The theory of diversity-stability relationships in ecology. Quart. Rev. Biol. 50, 237–266. (14)

Goodman, D. 1986. The demography of chance extinction. In *Viable Populations*, M.E. Soulé (ed.). Cambridge University Press, Cambridge. (2)

Gore, A.J.P. 1983. Introduction. In Ecosystems of the World, Vol. 4A, *Mires: Swamp, Bog, Fen and Moor*, A.J.P. Gore (ed.). Elsevier, Amsterdam. pp. 1–34. (18)

Gould, S.J. 1981. *The Mismeasure of Man*, Norton, New York. (9)

Gowaty, P.A. and A.A. Karlin. 1984. Multiple maternity and paternity in single broods of apparently monogamous eastern bluebirds (*Sialia sialis*). Anim. Behav. 30, 497–505. (3)

Graham, J.H. and J.D. Felley. 1985. Genomic coadaptation and developmental stability within introgressed populations of *Enneacanthus gloriosus* and *E. obesus* (Pisces, Centrarchidae). Evolution 39, 104–114. (5)

Granier, P. 1977. Rapport d'activités agropastorales en république du Niger. IEMVT, Maisons-Alfort. 140 pp. (22)

Grant, P.R. 1972. Centripetal selection and the house sparrow. Syst. Zool. 21, 23–30. (4)

Grau, H.J. 1982. Kin recognition in white-footed deermice (*Peromyscus leucopus*). Anim. Behav. 28, 1140–1162. (3)

Greaves, A. 1979. *Gmelina* large scale planting, Jarilandia, Amazon Basin. Commonwealth Forestry Review 58, 267–269. (20)

Greaves, J.H. and P. Ayres. 1967. Heritable resistance to warfarin in the rat. Nature 215, 877–878. (4)

Greaves, J.H. and P. Ayres. 1969. Linkages between genes for coat colour and resistance to warfarin in *Rattus norvegicus*. Nature 224, 284–285. (4)

Greaves, J.H., R. Redfern, P.B. Ayres and J.E. Gill. 1977. Warfarin resistance: a balanced polymorphism in the Norway rat. Genet. Res., Camb. 30, 257–263 (4)

Greenway, J.C., Jr. 1967. *Extinct and Vanishing Birds of the World*. Dover, New York. (14)

Greenwood, P.J. 1980. Mating systems, philopatry and dispersal in birds and mammals. Anim. Behav. 28, 1140–1162. (3)

Greenwood, P.J. and P.H. Harvey. 1982. The natal and breeding dispersal of birds. Ann. Rev. Ecol. Syst. 13, 1–21. (3)

Greenwood, P.J., P.H. Harvey and C.M. Perrins. 1978. Inbreeding and dispersal in the great tit. Nature 271, 52–54. (3)

Greenwood, P.J., P.H. Harvey and C.M. Perrins. 1979. The role of dispersal in the great tit (*Parus major*): the causes, consequences and heritability of natal dispersal. J. Anim. Ecol. 48, 123–142. (3)

Greig, J.C. 1979. Principles of genetic conservation in relation to wildlife management in Southern Africa. S. African J. Wildlife Res. 9, 57–78. (6)

Grenot, Cl. 1979. Les mammifères du désert. Science et Vie 742, 32–38. (22)

Gressitt, J.L. 1982. *Ecology and Biogeography of New Guinea*. Junk, The Hague. (24)

Gressitt, J.L., J. Sedlacek and J.J.H. Szent-Ivany. 1965. Flora and fauna on backs of large Papuan moss-forest weevils. Science 150, 1833–1835. (24)

Grier, J.W. 1982. Ban of DDT and subsequent recovery of reproduction in Bald Eagles. Science, 218, 1232–1235. (16)

Grigg, R. W., 1979. Coral reef ecosystems of the Pacific islands: issues and problems for future management and planning. In *Literature Review and Synthesis of Pacific Island Ecosystems*, J.E. Byrne (ed.). U.S. Fish. Wildlife Serv., Off. Biol. Serv., Washington, D.C. FWS-OBS, 79-35,6-1-6-17 (17)

Grime, J.P. 1979. *Plant Strategies and Vegetation Processes*. Wiley, Chichester. (7)

Grubb, P.J. 1977. The maintenance of species richness in plant communities: the importance of the regeneration niche. Biol. Rev. 52, 107–145. (7, 10)

Grue, C.E. 1985. Pesticides and the decline of Guam's native birds. Nature 316, 301. (24)

Gulland, J.A. 1983. World resources of fisheries. In *Marine Ecology*, Vol. V, part 2, O.Kinne (ed.). Wiley, Chichester. (14)

Gullberg, U., R. Yazdoni, R. Rudin and N. Ryannen. in press. Allozyme variation in Scots pine (*Pinus silvestris* L.) native to Sweden. Silvae Genet. (5)

Guries, R.P. and F.T. Ledig. 1982. Genetic diversity and population structure in pitch pine (*Pinus rigida* Mill.). Evolution 36, 387–402. (5)

Haber, W.A. 1984. Pollination by deceit in a mass-flowering tropical tree *Plumeria rubra* L. (Apocynaceae). Biotropica 16, 269–275. (13)

Haber, W.A. and G.W. Frankie. 1982. Pollination of *Luehea* (Tiliaceae) in the Costa Rican deciduous forest. Ecology 63, 1740–1750. (13)

Habtemariam, T., R. Howitt, R. Ruppanner and H.P. Rieman. 1984. Application of a linear programming model to the control of African trypanosomiasis. Preventive Veterinary Medicine 3, 1–14. (16)

Habtemariam, T., R. Ruppanner, H.P. Rieman and J.H. Theis. 1982. Evaluation of trypanosomiasis control alternatives using an epidemiological simulation model. Preventive Veterinary Medicine 1, 147–156. (16)

Haffer, J. 1969. Speciation in Amazonian forest birds. Science 165, 131–137. (8)

Hagman, M. 1973. The Finnish standard stands for forestry research. In *Proc. Thirteenth Meet. Comm. For. Tree Breeding Can. Part 2. Symp. Conserv. For. Gene Resources*, D.P. Fowler and C.W. Yeatman (eds.). Prince George, British Columbia. pp. 67–76. (5)

Haila, Y. and I. Hanski. 1984. Methodology for studying the effect of fragmentation on land birds. Annales Zoologica Fennici 21, 393–397. (11)

Haldane, J.B.S. 1957. The cost of natural selection. J. Genet. 55, 511–524. (5)

Hall, A.V., B. de Winter, S.P. Fourie and T.H. Arnold. 1984. Threatened plants in southern Africa. Biol. Conserv. 28, 5–20. (7)

Hall, R.W. and L.E. Ehler. 1979. Rate of establishment of natural enemies in classical biological control. Bull. Ent. Soc. Amer. 25, 280–282. (14)

Hammond, N. 1982. The exploration of the Maya world. Amer. Sci. 70, 482–495. (20)

Hamrick, J.L., Y.B. Linhart and J.B. Mitton. 1979. Relationships between life history characteristics and electrophoretically detectable genetic variation in plants. Ann. Rev. Ecol. Syst. 10, 173–200. (5)

Handford, P. 1980. Heterozygosity at enzyme loci and morphological variation Nature 286, 261–262. (4)

Handford, P. 1983. Age-related allozymic variation in the cyprinid fish *Alburnus alburnus*. Canad. J. Zool. 61, 2844–2848. (4)

Hanken, J. and P.W. Sherman. 1981. Multiple paternity and Belding's ground squirrel litters. Science 212, 351–353. (3)

Harberd, D.J. 1963. Observations on natural clones of *Trifolium repens* L. New Phytol. 62, 198–204. (5)

Harcombe, P.A. 1977a. The influence of fertilization on some aspects of succession in a humid tropical forest. Ecology 58, 1375–1383. (20)

Harcombe, P.A. 1977b. Nutrient accumulation by vegetation during the first year of recovery of a tropical forest ecosystem. In *Recovery and Restoration of Damaged Ecosystems*, J. Cairns, K.L. Dickson and E.E. Herricks (eds.). University of Virginia Press, Charlottesville. pp. 347–378. (20)

Hargreaves, G.H. 1977. World water for agriculture. Utah State University Contract 1103, US AID, Washington, D.C. (22)

Harper, J.L. 1977. *Population Biology of Plants*. Academic Press, New York. (7)

Harper, L. in preparation. Birds and army ants (*Eciton burchelli*), observations on their ecology in undisturbed forest and isolated reserves. (12)

Harper, R.M. 1946. Preliminary list of endemic flowering plants of Georgia. J. Tennessee Acad. Sci. 21, 225. (8)

Harper, R.M. 1947. Preliminary list of Southern Appalachian endemics. Castanea 12, 100–112. (8)

Harper, R.M. 1948. More about Southern Appalachian endemics. Castanea 13, 124–127. (8)

Harrington, J.F. 1970. Seed and pollen storage for conservation of plant gene resources. In *Genetic Resources in Plants—Their Exploration and Conservation*. IBP Handbook No. 11, O.H. Frankel and E. Bennett (eds.). Blackwell, Oxford. pp. 501–521. (5)

Harris, L.D. 1984. *The Fragmented Forest: Island Biogeography Theory and the Preservation of Biotic Diversity*. University of Chicago Press, Chicago. 211 pp. (5, 11, 13)

Harry, D.E. 1984. Genetic structure of incense-cedar (*Calocedrus decurrens*) populations. Ph.D. dissertation. University of California, Berkeley. 163 pp. (5)

Hartshorn, G.S. 1978. Tree falls and tropical forest dynamics. In *Tropical Trees as Living Systems*, P.B. Tomilson and M.H. Zimmer (eds.). Cambridge University Press, Cambridge. pp. 617–638. (20)

Hartshorn, G.S. 1981. Report to Institute of Current World Affairs, December, 1981, on activities as a Forest and Man Fellow, sponsored by that Institute. Institute of Current World Affairs, Wheelock House, Hanover, New Hampshire. (20)

Harvey, P.H. 1985. Raising the wrong children. Nature 313, 95–96. (3)

Harvey, P.H., P.J. Greenwood and C.M. Perrins. 1979. Breeding area fidelity of the great tit (*Parus major*). J. Anim. Ecol. 48, 305–13. (3)

Hatcher, B.G. 1984. A maritime accident provides evidence for alternate stable states in benthic communities on coral reefs. Coral Reefs 3, 199–204. (17)

Hawkesworth, D.L. (ed.). 1974. *The Changing Flora and Fauna of Britain*. Academic Press, London. (11)

Haywood, M. 1981. Évolution de l'utilisation des terres et de la végétation dans la zone Soudano-Sahélienne du projet CIPEA au Mali. Doc. Trav. No. 3, CIPEA/ILCA, Addis Ababa. 187 pp. (22)

Heald, E.J. and W.E. Odum. 1970. The contribution of mangrove swamps to Florida fisheries. Proc. Gulf Caribb. Fish. Inst. 23rd Ann. Sess., 130–135. (17)

Hecht, S.B. 1981. Deforestation in the Amazon basin: magnitude, dynamics and soil resource effects. Studies in Third World Societies 13, 61–108. (19)

Hedrick, P.W. 1982. Genetic hitchhiking: a new factor in evolution? BioScience 32, 845–853. (5)

Hedrick, P.W., M.E. Ginevan, and E.P. Ewing. 1976. Genetic polymorphism in heterogeneous environments. Ann. Rev. Ecol. Syst. 7, 1–32. (5)

Hemming, J. (ed.). 1985. *Change in the Amazon Basin*. University of Manchester Press, Manchester. (19)

Henderson-Sellars, A. 1982. The effects of land clearance and agricultural practices on climate. Studies in Third World Societies 14, 443–485. (19)

Henry, J.P. 1978. Observations sur les peuplements de Crustaces Aselloides des milieux souterrains. Bull. Soc. Zool. France 103, 491–497. (21)

Hepper, P.G. 1983. Sibling recognition in the rat. Anim. Behav. 31, 1177–1191. (3)

Hermodson, M.A., J.W. Suttie and K.P. Link. 1969. Warfarin metabolism and vitamin K requirement in the warfarin resistant rat. Amer. J. Physiol. 217, 1316–1319. (4)

Hiebert, R.D. and J.L. Hamrick. 1983. Patterns and levels of genetic variation in Great Basin bristlecone pine, *Pinus longaeva*. Evolution 37, 302–310. (5)

Hierneaux, P. 1980. Inventory of the browse potential of bushes, trees, and shrubs in an area of the Sahel in Mali: Methods and initial results. In *Browse in Africa*, H.N. Le Houérou (ed.). International Livestock Center for Africa, Addis Ababa. pp. 197–205. (22)

Higgs, A.J. and M.B. Usher. 1980. Should nature reserves be large or small? Nature 285, 568–569. (11)

Hill, J.L. 1974. *Peromyscus:* effect of early pairing on reproduction. Science 186, 1042–1044. (3)

Hime, J.M., I.F. Keymer, and C.J. Baxter. 1975. Measles in recently imported colobus monkeys (*Colobus guereza*). Vet. Res. 97, 392. (16)

Hitchcock, A.S. and A. Chase. 1950. *Manual of the Grasses of the United States*, 2nd Ed. U.S. Government Printing Office, Washington, D.C. (9)

Hobdy, R.W. 1984. A re-evaluation of the genus *Hibiscadelphus* (Malvaceae) and the description of a new species. Bishop Mus. Occ. Papers 25(11), 1–7. (8)

Hoeck, H.N. 1982. Population dynamics, dispersal and genetic isolation in two species of Hyrax (*Heterohyrax brucei* and *Procavia johnstoni*) on habitat islands in the Serengeti. Z. Tierpsychol. 59, 177–210. (16)

Hoffman, C.O. and J.L. Gottschang. 1977. Numbers, distribution and movements of a raccoon population in a suburban residential community. J. Mammal. 58, 623–636. (11)

Holdgate, M.W. 1960. The fauna of the mid-Atlantic islands. Proc. Roy Soc.Lond. B 152, 550–567. (14)

Holekamp, K.E. 1984. Dispersal in ground-dwelling sciurids. In *The Biology of Ground-Dwelling Squirrels*, J.O. Murie and G.R. Michener (eds.). University of Nebraska Press, Lincoln. pp. 295–320. (3)

Holmes, W.G. 1984. Sibling recognition in thirteen-lined ground squirrels: effects of genetic relatedness, rearing association, and olfaction. Behav. Ecol. Sociobiol. 14, 225–233. (3)

Holmes, W.G. and P.W. Sherman. 1982. The ontogeny of kin recognition in two species of ground squirrels. Amer. Zool. 22, 491–517. (3)

Holmes, W.G. and P.W. Sherman. 1983. Kin recognition in animals. Amer. Sci. 7, 46–55. (3)

Holsinger, J.R. 1966. A preliminary study of the effects of organic pollution of Banners Corner Cave, Virginia. Int. J. Speleology 2, 75–89. (21)

Holsinger, J.R. and D.C. Culver. 1985. The invertebrate cave fauna of Virginia and a part of eastern Tennessee: zoogeography and ecology. Brimleyana. (21)

Holsinger, J.R., J.S. Mort and A.D. Recklies. 1983. The subterranean crustacean fauna of Castleguard Cave, Columbia Icefields, Alberta, Canada, and its zoogeographic significance. Arctic Alpine Res. 15, 543–549. (21)

Honegger, R.E. 1979. The IUCN Red Data Book, Vol. 3: Amphibia and Reptilia. International Union for the Conservation of Nature and Natural Resources, Morges, Switzerland. (21)

Hoogland, J.L. 1982. Prairie dogs avoid extreme inbreeding. Science 215, 1639–1641. (3)

Hopkins, A.J.M. and E.A. Griffin. 1984. Floristic patterns. In *Kwongan: Plant Life in the Sandplain*, J.S. Pale and J.S. Beard (eds.). University of Western Australia Press, Nedlands. pp. 69 -83. (7)

Hopper, S.D., N.A. Campbell and G.F. Mason. 1982. *Eucalyptus caesia,* a rare mallee of granite rocks from southwestern Australia. In *Species at Risk: Research in Australia,* R.H. Graves and W.D.L. Ride (eds.). Springer Verlag, New York. (7)

Horn, H.S. 1978. Optimal tactics of reproduction and life history. In *Behavioural Ecology: An Evolutionary Approach,* J.R. Krebs and N.B. Davies (eds.). Sinauer Associates, Sunderland, Mass. pp. 411–429. (3)

Horn, H.S. and D.I. Rubenstein. 1984. Behavioural adaptations and life history. In *Behavioural Ecology: An Evolutionary Approach,* 2nd Ed., J.R. Krebs and N.B. Davies (eds.). Sinauer Associates, Sunderland, Mass. pp. 279–298. (3)

Houghton, R.A. et al. 1985. Net flux of carbon dioxide from tropical forests in 1980. Nature 317, 617–620. (19)

Howard, W.E. 1949. Dispersal, amount of inbreeding and longevity in a local population of prairie deermice on the George Reserve, southern Michigan. Contrib. Lab. Vert. Biol. Univ. Mich. 43, 1–50. (3)

Howarth, F.G. 1973. The cavernicolous fauna of Hawaiian lava tubes. I. Introduction. Pacific Ins. 15, 139–151. (21)

Howarth, F.G. 1980. The zoogeography of specialized cave animals: a bioclimatic model. Evolution 34, 394–406. (21)

Howe, H.F. 1982. Fruit production and animal activity in two tropical trees. In *The Ecology of a Tropical Forest: Seasonal Rhythms and Long-term Changes,* E.G. Leigh, Jr., A.S. Rand and D.M. Windsor (eds.). Smithsonian Institution Press, Washington, D.C. pp. 189–199. (15)

Howe, H.F. and G.A. vande Kerckhove. 1979. Fecundity and seed dispersal of a tropical tree. Ecology 60, 180–189. (15)

Howe, H.F. and G.A. vande Kerckhove. 1981. Removal of wild nutmeg *(Virola surinamensis)* crops by birds. Ecology 62, 1093–1106. (15)

Howe, R. W. 1984. Local dynamics of bird assemblages in small forest habitat islands in Australia and North America. Ecology 65, 1585–1601. (13)

Hrdy, S.B. 1977. *The Langurs of Abu.* Harvard University Press, Cambridge. (3)

Hubbell, S.P. 1984. Methodologies for the study of the origin and maintenance of tree diversity in tropical rainforest. In Biology International (IUBS) Special Issue No. 6, G. Maury-Lechon, M. Hadley and T. Younes (eds.). pp. 8–13 (10)

Hubbell, S.P. and R.B. Foster. 1983. Diversity of canopy trees in a neotropical forest and implications for conservation. In *Tropical Rain Forest: Ecology and Management,* S.L. Sutton, T.C. Whitmore and A.C. Chadwick (eds.). Blackwell, Oxford. (10)

Hubbell, S.P. and R.B. Foster. 1985a. Canopy gaps and the dynamics of a neotropical forest. In *Plant Ecology,* M.J. Crawley (ed.). Blackwell, Oxford (10)

Hubbell, S.P. and R.B. Foster.1985b. Biology, chance, and history, and the structure of tropical rain forest tree communities. In *Community Ecology*, T. J. Case and J. Diamond (eds.). Harper & Row, New York. pp. 314-329. (10)

Hubbell, S.P. and R.B. Foster. 1985c. La estructura en gran escala de un bosque neo-tropical. Revista Biologia Tropical. (10)

Hubbell, S.P. and R.B. Foster. 1985d. The spatial context of regeneration in a neotropical forest. In *Colonization, Succession, and Stability*, M. Crawley, P. J. Edwards, and A. Gray (eds.). Blackwell, Oxford. (10)

Huck, U.W. and E.M. Banks. 1979. Behavioral components of individual recognition in the collared lemming (*Didrostonyx groenlandicus*). Behav. Ecol. Sociobiol. 6, 85–90. (3)

Hudson, P. and A. Watson. 1985. The red grouse. Biologist 32, 13–18. (16)

Hudson, P.J., A.P. Dobson and D. Newborn. 1985. Cyclic and noncyclic populations of red grouse: a role for parasitism? In *Host Parasite Populations: Ecology and Genetics*, D. Rollinson and R.M. Anderson (eds.). Symposia of the Linnean Society of London. Wiley, New York. pp. 207–221. (4)

Hufnagl, E. 1972. *Libyan Mammals*. The Oleander Press, Stoughton, Wisc. and Harrow, England. 85 pp. (22)

Hughes, C.E. and B.T. Styles. 1984. Exploration and seed collection of multiple-purpose dry zone trees in Central America. International Tree Crops Journal 3. (13)

Huiskes, A.H.L. 1979. *Ammophila arenaria* L. Link. J. Ecol. 67, 363–382. (9)

Hulten, E. 1968. *Flora of Alaska and Neighboring Territories*. Stanford University Press, Stanford, Calif. (8)

Hutchinson, G.E. 1959. Homage to Santa Rosalia, or why are there so many kinds of animals? Amer. Natur. 93, 145–159. (14)

Iliffe, T.M., T.D. Jickells and M.S. Brewer. 1984. Organic pollution of an inland marine cave from Bermuda. Mar. Envir. Res. 12, 173–189. (21)

International Labour Organization. 1977. Poverty and Landlessness in Rural Asia. International Labour Office, Geneva, Switzerland. (19)

Irion, G. and J. Adis. 1979. Evolucão de florestas amazónicas inundadas, de igapó—um exemplo do rio Tarumã Mirim. Acta Amazonica 9, 299–303. (18)

Itamies, J., E.T. Valtonen and H.P. Fagerholm. 1980. *Polymorphus minutus* infestation in eiders and its role as a possible cause of death. Ann. Zool. Fennici. 17, 285–289. (16)

Izawa, K. 1979. Foods and feeding behavior of wild black-capped capuchin (*Cebus apella*). Primates 20, 57–76. (15)

Jackson, D.D. 1982. *Underground Worlds*. Time-Life Books, Alexandria, Va. (21)

Jacobs, B.F., C.R. Werth and S.I. Gutman. 1984. Genetic relationships in *Abies* (fir) of eastern United States: an electrophoretic study. Canad. J. Bot. 62, 609–616. (5)

Jacquard, A. 1974. *The Genetic Structure of Populations*. Springer Verlag, New York. (3)

James, C.W. 1961. Endemism in Florida. Brittonia 13, 225–244. (8)

James, J.W. 1971. The founder effect and response to artificial selection. Genet. Res., Camb. 12, 249–266. (4)

Janson, C.H., J. Terborgh and L.H. Emmons. 1981. Non-flying mammals as pollinating agents in the Amazonian forest. Biotropica Suppl. 1–6. (15)

Jantz, R.L. and R.S. Webb. 1982. Interpopulation variation in fluctuating asymmetry of the palmar A-B ridge-count. Amer. J. Physiol. Anthro. 57, 253–259. (4)

Janzen, D.H. 1967. Synchronization of sexual reproduction of trees within the dry season in Central America. Evolution 23, 620–637. (15)

Janzen, D.H. 1971. Seed predation by animals. Ann. Rev. Ecol. Syst. 2, 465–492. (13)

Janzen, D.H. 1974. Tropical blackwater rivers, animals, and mast fruiting by the Dip-terocarpaceae. Biotropica 6, 69–103. (13, 15)

Janzen, D.H. 1975. *Ecology of Plants in the Tropics*. Arnold, London. (19)

Janzen, D.H. 1978. Seeding patterns of tropical trees. In *Tropical Trees as Living Systems*, P.B. Tomlinson and M.H. Zimmerman (eds.). Cambridge University Press, New York. pp. 83–128. (13)

Janzen, D.H. 1979. How to be a fig. Ann. Rev. Ecol. Syst. 10, 13–51. (15)

Janzen, D.H. 1983. No park is an island: increase in interference from outside as park size increases. Oikos 41, 402–410. (10, 11, 12, 13)

Janzen, D.H. 1984a. Two ways to be a tropical big moth: Santa Rosa saturniids and sphingids. Oxford Surveys in Evolutionary Biology 1, 85–140. (13)

Janzen, D.H. 1984b. Dispersal of small seeds by big herbivores: foliage is the fruit. Amer. Natur. 123, 338–353. (13)

Janzen, D.H. in press. Biogeography of an unexceptional place: what determines the saturniid and sphingid moth fauna of Santa Rosa National Park, Costa Rica, and what does it mean to conservation biology? Proc. 1984 Merida Symposium in Mesoamerican Biogeography. (13)

Janzen, D.H. and P.S. Martin. 1982. Neotropical anachronisms: the fruits the gomphotheres ate. Science 215, 19–27. (13)

Jenkins, J.M. 1983. The Native Forest Birds of Guam. Amer. Ornithol. Union Ornithol. Monogr. No. 31. (24)

Jennings, A.R., E.J.L. Soulsby and C.B. Wainwright. 1961. An outbreak of disease in Mute Swans at an Essex reservoir. Bird Study 8, 20–25. (16)

Jimenez, J.A. 1984. A hypothesis to explain the reduced distribution of the mangrove Pelliciera rhizophorae Tr. & Pl. Biotropica 16, 304–308. (8)

Jinks, J.L. 1983. Biometrical genetics of heterosis. In Heterosis—Reappraisal of Theory and Practice, Monogr. Theor. Appl. Genet., Vol. 6, R. Frael (ed.). Springer Verlag, Berlin. pp. 1–46. (5)

Johannes, R.E. 1975. Pollution and degradation of coral reef communities. In Tropical Marine Pollution, E.J.F. Wood and R.E. Johannes (eds.). Elsevier, Amsterdam. pp. 13–51. (17)

Johannes, R.E. 1978a. Reproductive strategies of coastal marine fishes in the tropics. Env. Biol. Fish. 3, 65–84. (17)

Johannes, R.E. 1978b. Traditional marine conservation methods in Oceania and their demise. Ann. Rev. Ecol. Syst. 9, 349–364. (17)

Johannes, R.E. and S. Betzer. 1975. Marine communities respond differently to pollution in the tropics than at higher latitudes. In Tropical Marine Pollution, E.J.F. Wood and R.E. Johannes (eds.). Elsevier, Amsterdam. pp. 1–12. (17)

Jokiel, P.L. 1980. Solar ultraviolet radiation and coral reef epifauna. Science 207, 1069–1071. (17)

Jolly, A., P. Oberle and E.R. Albignac (eds.). 1984. Madagascar. Pergamon, Oxford. (19)

Jones, C.A. 1973. The conservation of chalk downland in Dorset. Dorset County Council, Dorchester, England. (11)

Jones, D.A. 1977. Population structure and local differentiation in Liatris cylindraceae: a criticism. Amer. Natur. 111, 608–610. (5)

Jones, D.D. and C.J. Walters. 1976. Catastrophe theory and fisheries regulation. J. Fish. Res. Board Canad. 33, 2829–2833. (14)

Jones, D.M. 1982. Conservation in relation to animal disease in Africa and Asia. In Animal Disease in Relation to Animal Conservation, M.A. Edwards and U. McDonnell (eds.). Symposia of the Zoological Society of London 50, 271–285. (16)

Jones, G.W. and H.V. Richter. 1981. Population mobility and development: Southeast Asia and the Pacific. Development Studies Centre, Canberra, Australia. (19)

Jones, W.K. 1973. Hydrology of limestone karst in Greenbrier County, West Virginia. W. Va. Geol. Econ. Surv. Bull. 36, 1–49. (21)

Jones, W.T. 1984. Natal philopatry in bannertailed kangaroo rats. Behav. Ecol. Sociobiol. 15, 151–155. (3)

Jones, W.T. in preparation. Dispersal patterns in kangaroo rats (Dipodomys spectabilis). (3)

Jordan, C.F. 1984. Nutrient regime in the wet tropics: physical factors. In Physiological Ecology of Plants in the Wet Tropics, E. Medina, H.A. Mooney and C. Vazquez-Yanes (eds.). Junk, The Hague. pp. 3–12. (20)

Jordan, C.F. 1985. Nutrient Cycling in Tropical Forest Ecosystems. Wiley, Chichester. 190 pp. (20)

Jordan, W.R. and M.E. Gilpin (eds.). in press. Restoration Ecology. Cambridge University Press, Cambridge. (1)

Juberthie, C. 1983. Le milieu souterrain: etendue et composition. Mem. Biospeologie 10, 17–65. (21)

Junk, W.J. 1980. Die Bedeutung der Wasserstandsschwankungen fuer die Oekologie von Ueberschwemmungsgebieten, dargestellt an der Várzea des mittleren Amazonas. Amazoniana 7, 19–29. (18)

Junk, W.J. 1983. Aquatic habitats in Amazonia. In *Ecological Structures and Problems of Amazonia*. IUCN Commisson on Ecology, No. 5, 24–34. (18)

Junk, W.J. 1984. Ecology of the várzea, floodplain of Amazonian whitewater rivers. In *The Amazon: Limnology and Landscape Ecology of a Mighty Tropical River and Its Basin*, H. Sioli (ed.). Junk, The Hague. pp. 215–243. (18)

Junk, W.J. and C. Howard-Williams. 1984. Ecology of aquatic macrophytes in Amazonia. In *The Amazon: Limnology and Landscape Ecology of a Mighty Tropical River and Its Basin*, H. Sioli (ed.). Junk, The Hague. pp. 269–293. (18)

Junk, W.J., B.A. Robertson, A.J. Darwich and I. Vieira. 1981. Investigaçóes limnologicas e ictiológicas em Curúa-Una, a primeira represa hidrelétrica na Amazônia Central. Acta Amazonica 11, 689–716. (18)

Kallen, A., P. Arcuri and J.D. Murray. 1985. A simple model for the spatial spread of rabies. J. Theor. Biol. 116, 700–721. (16)

Kareem, A.M. 1983. Effect of increasing periods of familiarity on social interactions between male sibling mice. Anim. Behav. 31, 919–926. (3)

Kareem, A.M. and C.J. Barnard. 1982. The importance of kinship and familiarity in social interactions between mice. Anim. Behav. 30, 594–601. (3)

Karn, M.N. and L.S. Penrose. 1951. Birth weight and gestation time in relation to maternal age, parity and infant survival. Ann. Eugen. Lond. 16, 147–164. (4)

Karr, J.R. 1982a. Avian extinction on Barro Colorado Island, Panama: a reassessment. Amer. Natur. 119, 220–239. (11, 12, 24)

Karr, J.R. 1982b. Population variability and extinction in the avifauna of a tropical land bridge island. Ecology 63, 1975–1978. (24)

Kassam, A.H. and G.M. Higgins. 1980. Land Resources for Populations of the Future. FAO/UNFA, Rome. 369 pp. (22)

Kat, P.W. 1982. The relationship between heterozygosity for enzyme loci and developmental homeostasis in peripheral populations of aquatic bivalves (Unionidae). Amer. Natur. 119, 824–832. (4, 5)

Keiding, H. 1968. Preliminary investigations of inbreeding and outcrossing in larch. Silvae Genet. 17, 159–165. (5)

Kenchington, R.A. and B.E.T. Hudson. 1984. *Coral Reef Management Handbook*. UNESCO, Jakarta. (17)

Kermack, W.O. and A.G. McKendrick. 1927. A contribution to the mathematical theory of epidemics. Proc. Roy. Soc. Lond. A 115, 700–721. (16)

Khattali, H. 1983. Contribution à l'étude de l'érosion éolienne dans la Jeffara tunisienne. Bull. Techn. No. 3, Instit. Rég. Arides, Médenine, Tunisia. 48 pp. multigraph. (22)

Kiltie, R.A. 1981. Distribution of palm fruits on a rain forest floor: why white-lipped peccaries forage near objects. Biotropica 13, 141–145. (15)

Kiltie, R.A. 1982. Bite force as a basis for niche differentiation between rain forest peccaries *(Tayassu tajacu* and *T. pecari)*. Biotropica 14, 188–195. (15)

Kiltie, R.A. and J. Terborgh. 1983. Observations on the behavior of rain forest peccaries in Peru: why do white-lipped peccaries form herds? Z. Tierpsych. 62, 241–255. (15)

Kimura, M. and T. Ohta. 1971. *Theoretical Aspects of Population Genetics*. Princeton University Press, Princeton, N.J. 219 pp. (5)

King, A.W. and S.W. Pimm. 1983. Complexity, diversity and stability; a reconciliation of theoretical and empirical results. Amer. Natur. 122, 229–239. (17)

King, J.E. 1981. Late Quaternary vegetational history of Illinois. Ecol. Monogr. 51, 43–62. (6)

King, J.N. and B.P. Dancik. 1983. Inheritance and linkage of isozymes in white spruce *(Picea glauca)*. Canad. J. Genet. Cytol. 25, 430–436. (5)

King, J.N., B.P. Dancik and N.K. Dhir. 1984. Genetic structure and mating system of white spruce (*Picea glauca*) in a seed production area. Canad. J. For. Res. 14, 639–643. (5)

Kinkead, G. 1981. Trouble in D.K. Ludwig's jungle. Fortune, April 20, 1981, 102–117. (20)

Kinloch, B.B., R.D. Westfall and G.I. Forrest. in preparation. Caledonian Scots pine: origins and genetic structure. (5)

Kinzey, W. and A.H. Gentry. 1979. Habitat utilization in two species of *Callicebus*. In *Primate Ecology: Problem-oriented Field Studies*, R. Sussman (ed.). Wiley, New York. pp. 89–100. (8)

Kinzey, W.G. 1974. Ceboid models for the evolution of hominoid dentition. J. Hum. Evol. 3, 193–203. (15)

Kleiman, D.G. 1977. Monogamy in mammals. Quart. Rev. Biol. 52, 39–69. (3)

Klekowski, E.J., Jr. 1976. Mutational load in a fern population growing in a polluted environment. Am. J. Bot. 63, 1024–1030. (5)

Knowles, P. 1984. Genetic variability among and within closely spaced populations of lodgepole pine. Canad. J. Genet. Cytol. 26, 177–184. (5)

Knowles, P. and M.C. Grant. 1981. Genetic patterns associated with growth variability in ponderosa pine. Am. J. Bot. 68, 942–946. (5)

Kochan, I., C.A. Golden and J.A. Bukovic. 1969. Mechanism of tuberculostasis in mammalian serum. II. Induction of serum tuberculostasis in guinea pigs. J. Bacteriol. 100, 64–70. (4)

Koehn, R.K. and P.M. Gaffney. 1984. Genetic heterozygosity and growth rate in *Mytilus edulis*. Mar. Biol. 82, 1–7. (4)

Koehn, R.K. and S.R. Shumway. 1982. A genetic/physiological explanation for differential growth rate among individuals of the American oyster *Crassostrea virginica* (Gmelin). Mar. Biol. Lett. 3, 35–42. (4)

Koehn, R.K., R. Milkman and J.B. Mitton. 1976. Population genetics of marine pelecypods. IV. Selection, migration and genetic differentiation in the blue mussel *Mytilus edulis*. Evolution 30, 2–32. (4)

Koehn, R.K., J.E. Perez and R.B. Merritt. 1971. Esterase enzyme function and genetical structure of populations of the freshwater fish *Notropis stramineus*. Amer. Natur. 105, 51–69. (4)

Koehn, R.K., F.J. Turano and J.B. Mitton. 1973. Population genetics of marine pelecypods. II. Genetic differences in microhabitats of *Modiolus demissus*. Evolution 27, 100–105. (4)

Koeman, J.H., M. Rijksen, B. Smies, B. Na'Isa and K.J.R. MacLennan. 1971. Faunal changes in a swamp habitat in Nigeria sprayed with insecticide to exterminate *Glossina*. Neth. J. Zool. 21, 443–463. (16)

Koenig, W.D., R.L. Mumme and F.A. Pitelka. 1984. The breeding system of the acorn woodpecker in central coastal California. Z. Tierpsychol. 65, 289–308. (3)

Kojima, K. 1971. Is there a constant fitness value for a given genotype? No! Evolution 25, 281–285. (4)

Kormutak, A., F. Bencat, D. Rudin and R. Seyedyazdani. 1982. Isoenzyme variation in the four Slovakian populations of *Abies alba* Mill. Biologia 37, 433–440. (5)

Koski, W. 1970. A study of pollen dispersal as a mechanism of gene flow in conifers. Commun. Inst. For. Fenniae 70.4. Helsinki. 78 pp. (5)

Koski, V. 1971. Embryonic lethals of *Picea abies* and *Pinus sylvestris*. Commun. Inst. For. Fenniae 75.3. Helsinki. 30 pp. (5)

Koski, V. 1973. On self-pollination, genetic load, and subsequent inbreeding in some conifers. Commun. Inst. For. Fenniae 78.10. Helsinki. 43 pp. (5)

Krebs, C.J. 1971. Territory and breeding density in the great tit, *Parus major* L. Ecology 52, 2–22. (3)

Krebs, C.J. 1984. *Ecology: The Experimental Analysis of Distribution and Abundance*, 3rd Ed. Harper & Row, New York. (14)

Kreitman, M. 1983. Nucleotide polymorphism at the alcohol dehydrogenase locus of *Drosophila melanogaster*. Nature 304, 412–417. (4)

Kruckeberg, A.R. 1951. Intraspecific variability in the response of certain native plant species to serpentine soil. Am. J. Bot. 38, 408–419. (5)

Kruckeberg, A.R. 1954. Plant species in relation to serpentine soils. Ecology 35, 267–274. (9)

Kruckeberg, A.R. and D. Rabinowitz. 1985. Biological aspects of endemism in higher plants. Ann. Rev. Ecol. Syst. 16, 447–479. (9)

Kruger, F.J. and H.C. Taylor. 1979. Plant species diversity in Cape fynbos: gamma and delta diversity. Vegetation 41, 85–93. (7)

Lack, D. 1968. *Ecological Adaptations for Breeding in Birds.* Methuen, London. (3)

Ladd, D. and P. Nelson. 1982. Ecological synopsis of Missouri glades. Mo. Acad. Sci. Occasional Paper 7, 1–20. (6)

Laing, C.D., G.R. Carmody and S.B. Peck. 1976. How common are sibling species in cave-inhabiting invertebrates? Amer. Natur. 110, 184–189. (21)

Lamprey, H.F. 1975. Report on the desert encroachment reconnaissance in northern Sudan. UNEP, Nairobi. 14 pp. mimeo. (22)

Lamprey, H.F. 1983. Pastoralism yesterday and today: the overgrazing problem. In *Tropical Savannas*, F. Bourliére (ed.). Vol. 13 of *Ecosystems of the World.* Elsevier, Amsterdam. pp. 643–666. (22)

Lande, R. and D.W. Schemske. 1985. The evolution of self-fertilization and inbreeding depression in plants. I. Genetic models. Evolution 39, 24–40. (5)

Langner, W. 1970. Some cases of inbreeding and hybrid weakness. In Second World Consult. For. Tree Breeding, Vol. 2. FAO, U.N., Rome. pp. 903–911. (5)

Lanner, R.M. 1966. Needed: a new approach to the study of pollen dispersion. Silvae Genet. 15, 50–52. (5)

Larkum, A.W.D. and R. West. 1983. Stability, depletion and restoration of seagrass beds. Proc. Linn. Soc. N.S.W. 106, 201–212. (17)

Le Houérou, H.N. 1959. Recherches écologiques et floristiques sur la végétation de la Tunisie méridionale. Inst. Rech. Sahar., University d'Alger. 510 pp. (22)

Le Houérou, H.N. 1962. Les pâturages naturels de la Tunisie aride et désertique. Instit. Sciences Econ. Appl., Paris. 106 pp. (22)

Le Houérou, H.N. 1968. La désertisation du Sahara septentrional et des steppes limitrophes. Ann. Algér de Géogr. 6, 2–27. (22)

Le Houérou, H.N. 1969. La végétation de la Tunisie steppique, avec référance aux végétations analogues de l'Algérie de la Libye et du Maroc. Ann. Inst. Nat. Rech. Agron. de Tunisie 42, 5, 1–624. (22)

Le Houérou, H.N. 1972. An assessment of the primary and secondary production of the arid grazing lands ecosystems of North Africa. AGPC, Misc. 5, FAO, Rome. 25 pp. [Abbreviated version in Ecophysiological Foundations of Arid Lands, L.E. Rodin (ed.). Nauka Publ., Leningrad. pp. 168–172.] (22)

Le Houérou, H.N. 1973. Écologie, démographie et production agricole dans les pays méditerranéens du tiers monde. Options Méditerranéenes 17, 53–61. (22)

Le Houérou, H.N. 1974. Detrioration of the ecological equilibrium in the arid zones of North Africa. Proceed. France–Israel Sympos. on Ecological Research on Development of Arid Zones with Winter Precipitation., Spec. Publ. No. 39, Organiz. of Agric. Research, Volcani Center, Bet-Dagan, Israel. pp. 45–56. (22)

Le Houérou, H.N. 1975a. The rangelands of North Africa: typology, yield, productivity, and development. In Proceed. Sympos. on Evaluation and Mapping of African Tropical Rangelands, ILCA, Addis Ababa. pp. 41–56. (22)

Le Houérou, H.N. 1975b. Problémes et potentialités des zones arides Nord-Africaines. Options Méditerranéenes 26, 17–35. (22)

Le Houérou, H.N. 1976a. Peut-on lutter contre la désertisation? In Cpte Rend. Coll. Nouakchott sur la désertification aud sud du Sahara, Th. Monod (ed.). Niles Edit. Afric., Dakar/Abidjan, pp. 158–163. (22)

Le Houérou, H.N. 1976b. Nature et désertisation. Cpte Rend. Consultation sur la Foresterie au Sahel. CILSS/UNSO/FAO, FO:RAF/305/3, FAO, Rome. 21 pp. (22)

Le Houérou, H.N. 1977a. The nature and causes of desertization. In *Desertification: Environmental Degradation in and around Arid Lands*, M.H. Glantz (ed.). Westview Press, Boulder, Colo. pp. 17–38. (22)

Le Houérou, H.N. 1977b. Biological recovery versus desertization. Econ. Geogr. 53, 4, 413–420. (22)

546 LITERATURE CITED

Le Houérou, H.N. 1977c. The grasslands of Africa: classification, production, international evolution and development outlook. Proc. XIII Grassland Congress, Vol. 1. Akademic Verlag, Berlin, pp. 99–116. (22)

Le Houérou, H.N. 1978. The role of shrubs and trees in the management of natural grazing lands. Position paper No. 10, Eighth World Forestry Congress, Jakarta, Indonesia. 24 pp. (22)

Le Houérou, H.N. 1979. La désertisation des régions arides. La Recherche 99, 336–334. (22)

Le Houérou, H.N. 1980a. Browse in Africa (Les fourrages ligneux en Afrique). International Livestock Center for Africa, Addis Ababa. 491 pp. (22)

Le Houérou, H.N. 1980b. The rangelands of the Sahel. J. Range Mgt. 33, 41–46. (22)

Le Houérou, H.N. 1984a. Rain use efficiency: a unifying concept in arid land ecology. J. of Arid Envir. 7, 213–247. (22)

Le Houérou, H.N. 1984b. Fuel and forage species in the Mediterranean arid and semi-arid zone. In Proceed. Kew Internat. Conf. on Economic Plants for Zones, G.E. Wickens (ed.). Kew Royal Botanic Gardens, Richmond, England. (22)

Le Houérou, H.N. 1985a. The desert and arid zones of North Africa. In *Hot Deserts and Arid Shrublands*, M. Evenari and D.W. Goodall (eds.). Vol. 12 of Ecosystems of the World. Elsevier, Amsterdam. pp. 66–97. (22)

Le Houérou, H.N. 1985b. The impact of climate on pastoralism. In *Climate and Societies*, W. Kates (ed.). Wiley, New York. (22)

Le Houérou, H.N. and D. Froment. 1966. Définition d'une doctrine pastorale pour la Tunisie steppique. Bull. Ec. Nat. Sup. Agron. de Tunis 10–12, 72–152. (22)

Le Houérou, H.N. and C.H. Hoste. 1977. Rangeland production and annual rainfall relations in the Mediterranean Basin and in the African Sahelian and Sudanian zones. J. Range. Mgt. 30, 181–189. (22)

Le Houérou, H.N. and G.F. Popov. 1981. An eco-climatic classification of Intertropical Africa. Plant Prod. and Protect. Paper No. 31, FAO, Rome. 40 pp. (22)

Le Houérou, H.N., R.L. Bingham and W. Sherbek. 1984. Towards a probabilistic approach to rangeland development planning. Plenary Lecture, Second International Rangeland Congress, Australian Range Management Society, Adelaide. (22)

Leary, R.F., F.W. Allendorf and K.L. Knudsen. 1983. Developmental stability and enzyme heterozygosity in rainbow trout. Nature 301, 71–72. (4)

Leary, R.F., F.W. Allendorf and K.L. Knudsen. 1984. Superior developmental stability of heterozygotes at enzyme loci in salmonid fishes. Amer. Natur. 124, 540–551. (4, 5)

Leary, R.F., F.W. Allendorf and K.L. Knudsen. 1985. Inheritance of meristic variation and the evolution of developmental stability in rainbow trout. Evolution 39, 308–314. (4)

Leck, C.F. 1972. Seasonal changes in feeding pressures of fruit- and nectar-eating birds in Panama. Condor 74, 54–60. (15)

Ledig, F.T. 1974. An analysis of methods for the selection of trees from wild stands. For. Sci. 20, 2–16. (5)

Ledig, F.T. and M.T. Conkle. 1983. Gene diversity and genetic structure in a narrow endemic, Torrey pine (*Pinus torreyana* Parry ex Carr.). Evolution 37, 79–85. (5)

Ledig, F.T., R.P. Guries and B.A. Bonefeld. 1983. The relation of growth to heterozygosity in pitch pine. Evolution 37, 1227, 1238. (5)

Lee, A.K. and A. Cockburn. 1985. *Evolutionary Ecology of Marsupials*. Cambridge University Press, Cambridge. (3)

Leentvaar, P. 1967. The artificial Brokopondo Lake of the Surinam River, its biological implications. Atas do Simpósio sobre a Biota. Amazônica 3 (Limnologia), 127–140. (18)

Leentvaar, P. 1973. Lake Brokopondo. In *Man-made Lakes: The Problems and Environmental Effects*, W.C. Ackermann, G.F. White and E.B. Worthington (eds.). Geophysical Union, Washington, D.C. pp. 186–196. (18)

Leentvaar, P. 1984. The Brokopondo Barrage Lake in Suriname, South America, and the planned Kabalebo Project in West Suriname. In *Hydro-Environmental Indices: A Review and Evaluation of their Use in Assessment of the International Conclusions*, J.R. Card (ed.). Rombach, Freiburg. pp. 9–33. (18)

Leigh Brown, A.J. 1977. Physiological correlates of an enzyme polymorphism. Nature 269, 803–804. (4)

Leigh, E.G. 1975. Population fluctuations, community stability and environmental variability. In *Ecology and Evolution of Communities*, M.L. Cody and J.M. Diamond (eds.). Harvard University Press, Cambridge. pp. 51–73. (2, 11)

Leigh, E.G., Jr. 1981. The average lifetime of a population in a varying environment. J. Theor. Biol. 90, 213–239. (2, 11, 24)

Leigh, E.G., Jr. and N. Smythe. 1978. Leaf production, leaf consumption, and the regulation of frugivory on Barro Colorado Island. In *Ecology of Arboreal Folivores*, G.G. Montgomery (ed.). Smithsonian Institution Press, Washington, D.C. pp. 51–73. (15)

Leigh, J.H., J.D. Briggs and W. Hartley. 1982. The conservation status of Australian plants. In *Species at Risk: Research in Australia*, R.H. Groves and W.D. L. Ride (eds.). Springer Verlag, New York. pp. 13–25. (7)

Leigh, E.G., Jr., A.S. Rand and D.M. Windsor (eds.). 1982. *The Ecology of a Tropical Forest: Seasonal Rhythms and Long-Term Changes*. Smithsonian Institution Press, Washington, D.C. (10)

Leighton, M. and D.R. Leighton. 1983. Vertebrate responses to fruiting seasonality within a Bornean rain forest. In *Tropical Rain Forest: Ecology and Management*, S.L. Sutton, T.C. Whitmore and A.C. Chadwick (eds.). Blackwell, Oxford. pp. 181–196. (15)

Leopold, A. 1933. *Game Management*. Scribners, New York. (11, 12)

Lerner, I.M. 1954. *Genetic Homeostasis*. Oliver & Boyd, London. (4, 5)

Leroy, J.F. 1978. Composition, origin and affinities of the Madagascar vascular flora. Ann. Missouri Bot. Gard. 65, 535–589. (19)

Lessios, H.A., D.R. Robertson, and J.D. Cubit. 1984a. Spread of *Diadema* mass mortality through the Caribbean. Science 226, 335–337. (16)

Lessios, H.A., J.D. Cubit, D.R. Robertson, M.J. Shulman, M.R. Parker, S.D. Garrity and S.C. Levings. 1984b. Mass mortality of *Diadema antillarum* on the Caribbean coast of Panama. Coral Reefs 3, 173–182. (16)

Levin, D.A. 1981. Dispersal versus gene flow in plants. Ann. Missouri Bot. Gard. 68, 233–253. (5)

Levin, D.A. 1984. Inbreeding depression and proximity-dependent crossing success in *Phlox drummondii*. Evolution 38, 116–127. (5)

Levin, D.A. and H.W. Kerster. 1974. Gene flow in seed plants. Evol. Biol. 7, 139–220. (5)

Levins, R. and D. Culver. 1971. Regional coexistence of species and competition between rare species. Proc. Nat. Acad. Sci. USA 68, 1246–1248. (11)

Lewin, R. 1983a. Santa Rosalia was a goat. Science 221, 636–639. (14)

Lewin, R. 1983b. Predators and hurricanes change ecology. Science 221, 737–740. (14)

Lewin, R. 1984. Parks: How big is big enough? Science 225: 611–612. (12)

Lewontin, R.C. 1974. *The Genetic Basis of Evolutionary Change*. Columbia University Press, New York. 346 pp. (4, 5)

Lewontin, R.C. and J.L. Hubby. 1966. A molecular approach to the study of genic heterozygosity in natural populations. II. Amount of variation and degree of heterozygosity in natural populations of *Drosophila pseudoobscura*. Genetics 54, 595–609. (5)

Libby, W.J., B.G. McCutchan and C.I. Millar. 1981. Inbreeding depression in selfs of redwood. Silvae Genet. 30, 15–25. (5)

Lidicker, W.Z. 1962. Emigration as a possible mechanism permitting the regulation of population density below carrying capacity. Amer. Natur. 96, 29–33. (3)

Lidicker, W.Z., Jr. 1985. Dispersal. In Biology of New World *Microtus*, R.H. Tamarin (ed.). Spec. Publ. Amer. Soc. Mamm. No. 8. (3)

Lidicker, W.Z., Jr. and J.L. Patton. in preparation. Pattern of dispersal and genetic structure in populations of small rodents. (3)

Lill, B.S. 1976. Ovule and seed development in *Pinus radiata*: postmeiotic development, fertilization, and embryogeny. Canad. J. Bot. 54, 2141–2154. (5)

Linhart, Y.B., J.B. Mitton, K.B. Sturgeon and M.L. Davis. 1981. Genetic variation in space and time in a population of ponderosa pine. Heredity 46, 407–426. (5)

Liogier, A.H. 1981. Flora of Hispaniola: Part 1, Phytologia Mem. 3, 1–218. (8)

Livshits, G. and E. Kobyliansky. 1984. Biochemical heterozygosity as a predictor of developmental homeostasis in man. Ann. Hum. Gen. 48, 173–184. (4)

Lloyd, R.M. 1974. Mating systems and genetic load in pioneer and non-pioneer Hawaiian Pteridophyta. Bot. J. Linnean Soc. 69, 23–35. (5)

Long, G.A., B. Lacaze, G. Debussche, E. Le Floch and R. Pontanier. 1978. Contribution à l'analyse écologique des zones arides de Tunisie avec l'aide de la télédetection spatiale. CEPE/CNRS, Montpellier, France. 222 pp. (22)

Long, J. 1981. *Introduced Birds of the World*. David and Charles, London. (14)

Long, R.W. and O. Lakela. 1971. *A Flora of Tropical Florida*. University of Miami Press, Coral Gables, Fla. pp. 1–962. (8)

Lovejoy, T.E. 1980. Discontinuous wilderness: minimum areas for conservation. Parks 5(2), 13–15. (12)

Lovejoy, T.E., R.O. Bierregaard, J.M. Rankin and H.O.R. Schubart. 1983. Ecological dynamics of tropical forest fragments. In *Tropical Rain Forest: Ecology and Management*, S.L. Sutton, T.C. Whitmore and A.C. Chadwick (eds.). Blackwell, Oxford. pp. 377–384. (12, 13)

Lovejoy, T.E., J.M. Rankin, R.O. Bierregaard, K.S. Brown, L.H. Emmons and M.E. Van der Voort. 1984. Ecosystem decay of Amazon forest fragments. In *Extinctions*, M.H. Nitecki (ed.). University of Chicago Press, Chicago. pp. 295–325. (11, 12, 13)

Loveless, M.D. and J.L. Hamrick. in press. Distribution of genetic variation in tropical tree species. Rev. Biol. Trop. (5)

Lowe, R.G. 1977. Experience with the tropical shelterwood system of regeneration in natural forest in Nigeria. Forest Ecology and Management 1, 193–212. (20)

Loya, Y. and B. Rinkevich. 1980. Effects of oil pollution on coral reef communities. Mar. Ecol. Prog. Ser. 3, 167–180. (17)

Lucas, G. and H. Synge. 1978. Red Data Book. I.U.C.N., Morges, Switzerland. (7)

Luckinbill, L.S. and M. Fenton. 1978. Regulation and environmental variability in experimental populations of protozoa. Ecology 59, 1271–1276. (14)

Lugo, A.E. 1980. Mangrove ecosystems: successional or steady state? Biotropica 12, 65–72. (17)

Lugo, A.E., M. Applefield, D.J. Pool and R.B. McDonald. 1983. The impact of hurricane David on the forests of Dominica. Canad. J. For. Res. 13, 201–211. (20)

Lundholm, B. 1977. Ecology and dust transport. In *Saharan Dust: Mobilization, Transport and Deposition*, C. Morales (ed.). Wiley, New York. pp. 61–68. (22)

Lynch, J.F. and D.F. Whigham. 1984. Effects of forest fragmentation on breeding bird communities in Maryland, USA. Biol. Conserv. 28, 287–324. (11)

MacArthur, R.H. 1957. Fluctuations of animal populations and a measure of community stability. Ecology 36, 533–536. (14)

MacArthur, R.H. 1965. Patterns in species diversity. Biol. Rev. 40, 510–533. (7)

MacArthur, R.H. and R. Levins. 1967. The limiting similarity, convergence and divergence of coexisting species. Amer. Natur. 101, 377–385. (14)

MacArthur, R.H. and J. MacArthur. 1961. On bird species diversity. Ecology 42, 594–598. (7)

MacArthur, R.H. and E.O. Wilson. 1967. *The Theory of Island Biogeography*, Princeton University Press, Princeton, N.J. (2, 11, 12, 24)

MacClintock, L., R.F. Whitcomb and B.L. Whitcomb. 1977. Island biogeography and "habitat islands" of eastern forest. II. Evidence for the value of corridors and minimization of isolation in preservation of biotic diversity. Amer. Birds 31, 6–16. (11)

MacDonald, D.W. 1980. *Rabies and Wildlife: A Biologist's Perspective*. Oxford University Press, Oxford. (16)

MacKinnon, J. 1980. The province of Irian Jaya. FAO Field Report, Bogor. (24)

Mack, R. 1970. The great African cattle plague epidemic of the 1890s. Trop. Anim. Hlth. Prod. 2, 210–219. (16)

Magnuson, J.J., H.A. Regier, W.J. Cristie and W.C. Sonzongi. 1980. To rehabilitate and restore Great Lakes ecosystems. In *The Recovery Process in Damaged Ecosystems*, J. Cairns, Jr. (ed.). Ann Arbor Science Publishers, Ann Arbor, Mich. pp. 95–112. (23)

Maguire, B. 1970. On the flora of the Guayana Highland. Biotropica 2, 85–100. (8)

Main, A. 1982. Rare species: precious or dross? In *Species at Risk: Research in Australia*, R.H. Graves and W.D.L. Ride (eds.). Springer Verlag, New York. pp. 163–174. (7)

Malcolm, J. in preparation. Small mammal abundances in isolated and nonisolated primary forest reserves near Manaus, Brazil. (12)

Malcolm, J.R. and K. Marten. 1982. Natural selection and the communal rearing of pups in African wild dogs (*Lycaon pictus*). Behav. Ecol. Sociobiol. 10, 1–13. (3)

Maley, J. 1981. Etudes palynologiques dans le bassin du Tchad et paleo-climatologie de l'Afrique Nord-Tropicale de 30,000 ans à l'époque actuelle. Trav. & Doc. No. 129, ORSTOM, Paris. 586 pp. (22)

Manley, S.A.M. 1975. Genecology of Hybridization in Red Spruce (*Picea rubens* Sarg.) and Black Spruce (*Picea mariana* Mill., B.S.P.). Ph.D. dissertation, Yale University, New Haven, Conn. 154 pp. (5)

Manley, S.A.M. and F.T. Ledig. 1979. Photosynthesis in black and red spruce and their hybrid derivatives: ecological isolation and hybrid adaptive inferiority. Canad. J. Bot. 57, 305–314. (5)

Manwell, C. and C.M.A. Baker. 1982. Alcohol dehydrogenase in the marine polychaete *Hyalinoecia tubicola*: heterozygote deficit and growth. Comp. Biochem. Physio. 733, 411–416. (4)

Margalef, R. 1968. *Perspectives in Ecological Theory*. University of Chicago Press, Chicago. (14)

Margules, C., A.J. Higgs and R.W. Rafe. 1982. Modern biogeographic theory: Are there any lessons for nature reserve design? Biol. Conserv. 24, 115–128. (11)

Marks, P.L. 1983. On the origin of the field plants of the northeastern United States. Amer. Natur. 122, 210–228. (13)

Martin, T.E. 1981. Species area slopes and coefficients: a caution on their interpretation. Amer. Natur. 118, 823–837. (11)

Maruyama, M. and K. Kimura. 1980. Genetic variability and effective population size when local extinction and recolonization of subpopulations are frequent. Proc. Nat. Acad. Sci. USA 77, 6710–6714. (2)

Maslin, B.R. and S.D. Hopper. 1982. Phytogeography of *Acacia* (Leguminoseae: Minosoideae) in Central Australia. In *Evolution of the Flora and Fauna of Arid Australia*, W.R. Barker and P.J.M. Greenslade (eds.). Peacock, Frewville, So. Aust. pp. 301–315. (7)

Maslin, B.R. and L. Pedley. 1982. The distribution of *Acacia* (Leguminoseae: Minosoideae) in Australia. Pt. 1. Species distribution maps. W.A. Herbar. Res. Notes No. 6, 1–128. (7)

Matheny, R.T. and D.L. Gurr. 1979. Ancient hydraulic techniques in the Chiapas highlands. Amer. Sci. 67, 441–449. (20)

Mather, K. 1953. The genetical structure of populations. Symp. Soc. Exper. Biol. 7, 66–95. (5)

Mather, K. 1953. Genetical control of stability in development. Heredity 7, 297–336. (4)

Mattheissen, P. 1981. Haematological changes in fish following aerial spraying with endosulfan insecticide for tsetse fly control in Botswana. J. Fish. Biol. 18, 461–469. (16)

Matthiae, P.E. and F. Stearns. 1981. Mammals in forest islands in southeastern Wisconsin. In *Forest Island Dynamics in Man-Dominated Landscapes*, R.L. Burgess and D.M. Sharpe (eds.). Springer Verlag, New York. pp. 55–66. (11)

May, R.M. 1973. *Complexity and Stability in Model Ecosystems*. Princeton University Press, Princeton, N.J. (14)

May, R.M. 1975. Patterns of species abundance and diversity. In *Ecology and Evolution of Communities*, M.L. Cody and J.M. Diamond (eds.). Harvard University Press, Cambridge, Mass. (9)

May, R.M. 1975. The tropical rain forest. Nature 257, 737–738. (19)

May, R.M. 1976. Models for two interacting populations. In *Theoretical Ecology: Principles and Applications*, R.M. May (ed.). Saunders, Philadelphia. pp. 49–70. (21)

May, R.M. 1977. Thresholds and breakpoints in ecosystems with a multiplicity of stable states. Nature 269, 471–477. (14)

May, R.M. 1979. The structure and dynamics of ecological communities. In *Population Dynamics*, R.M. Anderson, B.R. Turner and L.R. Taylor (eds.). Blackwell, Oxford. (14)

May, R.M. 1981. *Theoretical Ecology: Principles and Applications*, 2nd Ed. Blackwell, Oxford. (16)

May, R.M. 1984. Ecology and population biology of parasites. In *Tropical and Geographical Medicine*, K.S. Warren and A.F. Mahmoud (eds.). McGraw-Hill, New York. (16)

May, R.M. 1985. Evolution of pesticide resistance. Nature 315, 12–13. (16)

May, R.M. and R.M. Anderson. 1978. Regulation and stability of host-parasite population interactions. II. Destabilizing processes. J. Anim. Ecol. 47, 249–267. (16)

May, R.M. and R.M. Anderson. 1979. Population biology of infectious diseases: Part II. Nature 280, 455–461. (16)

May, R.M. and Dobson, A.P. 1985. Population dynamics and the rate of evolution of pesticide resistance. *Pesticide Resistance Management*, NAS-NRC Publications. in press. (16)

May, R.M. and R.H. MacArthur. 1972. Niche overlap as a function of environmental variability. Proc. Nat. Acad. Sci. USA 69, 1109–1112. (14)

May, R.M. and S.K. Robinson. 1985. Population dynamics of avian brood parasitism. Amer. Natur. 126, 475–494. (11)

May, R.M., J.R. Beddington, J.W. Horwood and J.G. Shepherd. 1978. Exploiting natural populations in an uncertain world. Math. Biosc. 42, 219–252. (14)

Mayfield, H. 1977. Brown-headed cowbird: agent of extermination? Amer. Birds 31, 107–113. (11)

Maynard Smith, J. 1982. *Evolution and the Theory of Games*. Cambridge University Press, New York. (3)

Mayr, E. 1963. *Animal Species and Evolution*. Belknap Press of Harvard University Press, Cambridge, Mass. (6)

McAndrew, B.J., R.D. Ward and J.A. Beardmore. 1982. Lack of relationship between morphological variance and enzyme heterozygosity in the plaice, *Pleuronectes platessa*. Heredity 48, 117–125. (4)

McArthur, P.D. 1982. Mechanisms and development of parent–young vocal recognition in the pinon jay (*Gymnorhinus cyanocephalus*). Anim. Behav. 30, 62–74. (3)

McClure, H.E. 1966. Flowering, fruiting and animals in the canopy of a tropical rain forest. Malay Forester 29, 182–203. (15)

McCracken, G.F. 1984. Social dispersion and genetic variation in two species of emballonurid bats. Z. Tierpsychol. 66, 55–69. (3)

McCracken, G.F. in press. Genetic structure of bat social groups. In *Recent Advances in the Study of Bats*, M.B. Fenton, P.A. Racey and S.M.V. Rayner (eds.). Cambridge University Press, Cambridge. (3)

McCracken, G.F. and J.W. Bradbury. 1977. Paternity and genetic heterogeneity in the polygynous bat *Phylostomus hastatus*. Science 198, 320–321. (3)

McFarquhar, A.M. and F.W. Robertson. 1963. The lack of evidence for co-adaptation in crosses between races of *Drosophila subobscura* Coll. Genet. Res. 4, 104–131. (6)

McGuire, M.R. and L.L. Getz. 1981. Incest taboo between sibling *Microtus ochrogaster*. J. Mammal. 62, 213–215. (3)

McLellan, C.H., A.P. Dobson, D.S. Wilcove and J.M. Lynch. 1986. Effects of forest fragmentation on New and Old World bird communities: empirical observations and theoretical implications. In *Modeling Habitat Relationships of Terrestrial Vertebrates*, J. Verner, M. Morrison and C.J. Ralph (eds.). University of Wisconsin Press, Madison. (11)

McNaughton, S.J. 1977. Diversity and stability of ecological communities: a comment on the role of empiricism in ecology. Amer. Natur. 111, 515–525. (14)

McNaughton, S.J. 1985. Ecology of a grazing ecosystem: the Serengeti. Ecol. Monogr. 55, 259–294. (14)

McWilliam, J.R. and B. Griffing. 1965. Temperature-dependent heterosis in maize. Aust. J. Biol. Sci. 18, 569–583. (5)

Mech, L.D. 1977. Wolf-pack buffer zones as prey reservoirs. Science 198, 320–321. (3)

Mech, L.D. in preparation. Age, season, distance, direction, and social aspects of wolf dispersal from a Minnesota pack. (3)

Medway, L. 1972. Phenology of a tropical rainforest in Malaya. Biol. J. Linn. Soc. 4, 117–146. (15)

Medway, L. and D.R. Wells. 1976. *The Birds of the Malay Peninsula*, Vol. 5. Broadwater Press, Welwyn Garden City, Herts. (15)

Meigs, P. 1953. World distribution of arid and semi-arid homoclimate. In Review of Research 33 on Arid Zone Hydrology, Arid Zone Programme 1: 203–209, UNESCO, Paris. (22)

Melillo, J.M. 1985. A comparison of recent estimates of disturbance in tropical forests. Environ. Conserv. 12, 37–40. (19)

Melnick, D.J. 1982. Are social mammals really inbred? Genetics 100, s46. (3)

Melnick, D.J., C.J. Jolly and K.K. Kidd. 1984. The genetics of wild populations of rhesus monkeys (*Macaca mulatta*). I. Genetic variability within and between social groups. Am. J. Phys. Anthropol. 63, 341–360. (3)

Michel, J.F. 1969. The epidemiology and control of some nematode infections of grazing animals. Advances in Parasitology 7, 211–282. (16)

Michel, J.F. 1976. The epidemiology and control of some nematode infections in grazing animals. Advances in Parasitology 14, 355–397. (16)

Millar, C.I. 1985. Genetic Studies of Dissimilar Parapatric Populations in Northern Bishop Pine (*Pinus muricata*). Ph.D. dissertation, University of California, Berkeley. 188 pp. (5)

Miller, R.R. 1969. The IUCN Red Data Book. Vol. 4: Pisces. Int. Union. Cons. Nat. Nat. Res., Morges, Switzerland. (21)

Milton, K. 1980. *The Foraging Strategy of Howler Monkeys: A Study in Primate Economics*. Columbia University Press, New York. (15)

Milton, K. 1982. Dietary quality and demographic regulation in a howler monkey population. In *The Ecology of a Tropical Forest: Seasonal Rhythms and Long-term Changes*. E.G. Leigh, Jr., A.S. Rand and D.M. Windsor (eds.). Smithsonian Institute Press, Washington D.C. (16)

Missakian, E.A. 1973. Genealogical mating activity in free-ranging groups of rhesus monkeys (*Macaca mulatta*) on Cayo Santiago. Behaviour 45, 225–241. (3)

Mitchell, R.S. 1963. Phytogeography and floristic survey of a relic area in the Marianna Lowlands, Florida. Amer. Midl. Natural. 69, 328–366. (8)

Mittermeier, R.A. and M.G.M. van Roosmalen. 1981. Preliminary observations on habitat utilization and diet in eight Surinam monkeys. Folia Primatol. 36, 1–39. (12)

Mitton, J.B. 1978. Relationship between heterzygosity for enzyme loci and variation of morphological characters in natural populations. Nature 273, 661–662. (4, 5)

Mitton, J.B. and M.C. Grant. 1980. Observations on the ecology and evolution of quaking aspen, *Populus tremuloides*, in the Colorado Front Range. Am. J. Bot. 67, 202–209. (5)

Mitton, J.B. and M.C. Grant. 1984. Associations among protein heterozygosity, growth rate, and developmental homeostasis. Ann. Rev. Ecol. Syst. 15, 479–499. (4)

Mitton, J.B. and R.K. Koehn. 1975. Genetic organization and adaptive response of allozymes to ecological variables in *Fundulus heteroclitus*. Genetics 79, 97–111. (4)

Mitton, J.B., P. Knowles, K.B. Sturgeon, Y.B. Linhart and M. Davis. 1981. Associations between heteroyzgosity and growth rate variables in three western forest trees. In Proc. Symp. Isozymes of North Am. For. Trees and For. Insects, M.T. Conkle (tech. coord.). USDA Forest Serv. Gen. Tech. Rep. PSW-48. pp. 27–34. (5)

Mogensen, H.L. 1975. Ovule abortion in *Quercus* (Fagacaaeae). Am. J. Bot. 62, 160–165. (5)

Molofsky, J., C.A.S. Hall and N. Myers. in press. *A Comparison of Tropical Forests Surveys.* (19)

Molyneux, D.H. 1982. Trypanosomes, trypanosomiasis and tsetse control: impact on wildlife and its conservation. In *Animal Disease in Relation to Animal Conservation*, M.A. Edwards and U. McDonnell (eds.). Symp. Zool. Soc. Lond. 50, 29–56. (16)

Molyneux, D.H. and R.W. Ashford. 1983. *The Biology of Trypanosoma and Leishmania, Parasites of Man and Domestic Animals.* Taylor and Francis, London. (16)

Monjuaze, A. and H.N. Le Houérou. 1965. Le rôle des Opuntia dans l'économie agricole nord-africaine. Bull. Ec. Nat. Sup. Agron. de Tunis 8–9, 85–164. (22)

552 LITERATURE CITED

Monod, Th. 1957. Les grandes divisions chorologiques de l'Afrique. Scientific Council for Africa, Publication No. 24, Brussels. 147 pp. (22)

Monod, Th. 1958. Majabat Al-Koubra: Contribution à l'étude de l'"empty quarter" ouest-saharien. Instit. François d'Afrique Noire, Dakar. 406 pp. (22)

Mooney, H. (ed.). 1985. *Ecological Consequences of Biological Invasions*. Springer Verlag, New York. (1)

Moore, D. 1983. *Flora of Tierra del Fuego*. Nelson, Shropshire, and Missouri Botanical Garden, St. Louis. (8)

Moore, J. and R. Ali. 1984. Are dispersal and inbreeding avoidance related? Anim. Behav. 32, 94–112. (3)

Moore, N.W. 1962. The heaths of Dorset and their conservation. J. Ecol. 50, 369–391. (11)

Moore, N.W. and M.D. Hooper. 1975. On the number of bird species in British woods. Biol. Conserv. 8, 239–391. (11)

Morales, C. 1979. *Saharan Dust: Mobilization, Transport and Deposition*. Wiley, New York. (22)

Moran, G.F. and J.C. Bell. 1983. Eucalyptus. In *Isozymes in Plant Genetics and Breeding*, Part B, D.S. Tanksley and T.J. Orton (eds.). Elsevier, Amsterdam. pp. 423–441. (5)

Moran, G.F. and A.H.D. Brown. 1980. Temporal heterogeneity of outcrossing rates in alpine ash (*Eucalyptus delegatensis* R.T. Bak.). Theor. Appl. Genet. 57, 101–105. (5)

Moran, G.F. and S.D. Hopper. 1983. Genetic diversity and the insular population structure of the rare granite rock species, *Eucalyptus caesia* Benth. Aust. J. Bot. 31, 161–172. (5)

Morat, P., J.-M. Veillon and H.S. MacKee. 1984. Floristic relationships of New Caledonian rain forest phanerogams. In *Biogeography of the Tropical Pacific*, F. Radovsky, P. Raven and S. Sohmer (eds.). Bishop Museum Spec. Publ. 72, pp. 70–128. (8)

Mori, S.A., B.M. Boom, A.M. de Carvalho and T.S. dos Santos. 1983. Southern Bahian moist forests. Bot. Rev. 49, 155–232. (19)

Morris, R.F. 1951. The effects of flowering on the foliage production and growth of balsam fir. For. Chron. 27, 40–57. (5)

Morrison, D.W. 1978. Foraging ecology and energetics of the frugivorous bat *Artibeus iamaicansis*. Ecology 5, 716–723. (15)

Morton, N.E., J.F. Crow and H.J. Muller. 1956. An estimate of the mutational damage in man from data on consanguineous marriages. Proc. Nat. Acad. Sci. USA 42, 855–863. (5)

Moulton, M.P. and S.L. Pimm. 1983. The introduced Hawaiian avifauna: biogeographic evidence for competition. Amer. Natur. 121, 669–690. (14)

Moulton, M.P. and S.L.Pimm. 1985. The extent of competition in shaping an experimental fauna. In *Community Ecology*, T.J. Case and J.M. Diamond (eds.). Harper & Row, New York. (14)

Moulton, M.P. and S.L. Pimm. 1986. Species introductions to Hawaii. In *Ecology of Biological Invasions to North America and Hawaii*, H. Mooney, (ed.). Springer Verlag, Berlin. (14)

Mourant, A.E., A.C. Kopec and K. Domaniewska-Sobczak. 1978. *Blood Groups and Diseases*. Oxford University Press, London. (4)

Moynihan, M. 1970. Some behaviour patterns of Platyrrhine monkeys. II, *Saguinus geoffroyi* and some other tamarins. Smiths. Contr. Zool. 28, iv + 77. (12)

Mulcahy, D.L. 1975. Differential mortality among cohorts in a population of *Acer saccharum* (Aceraceae) seedlings. Am. J. Bot. 62, 422–426. (5)

Munz, P.A. 1968. *Flora of California*. Stanford University Press, Stanford, Calif. (7)

Murray, M. 1967. The pathology of some diseases found in wild animals in East Africa. East Afr. Wildl. J. 5, 37–45. (16)

Murray, R.D. and E.O. Smith. 1982. The role of dominance and intrafamilial bonding in the avoidance of close inbreeding. J. Hum. Evol. 12, 481–486. (3)

Myers, N. 1980. *Conversion of Tropical Moist Forests*. National Research Council, Washington, D.C. (19)

Myers, N. 1984. *The Primary Source: Tropical Forests and Our Future*. Norton, New York. (9)

Myers, N. 1985. The end of the lines. Natural History 94 (February), 2–12. (13)
Myers, N. 1985. Tropical Moist Forests: Over-Exploited and Under-Utilized? Report to IUCN/World Wildlife Fund, Gland, Switzerland. (19)

Naess, A. 1985. Identification as a source of deep ecological attitudes. In *Deep Ecology*, M. Tobias (ed.). Avant Books, San Diego. (25)
Nagy, K.A. and K. Milton. 1979. Energy metabolism and food consumption by wild howler monkeys (*Alouatta palliata*). Ecology 60, 475–480. (15)
Nair, P.K.R. 1979. Agroforestry research: a retrospective and prospective appraisal. International Cooperation in Agroforestry. Proceedings of an International Conference, T. Chandler and D. Spurgeon (eds.). International Council for Research in Agroforestry. Nairobi, Kenya. pp. 275–296. (20)
National Academy of Sciences. 1972. *Genetic Vulnerability of Major Crops*. Nat. Acad. Sci., Washington, D.C. 307 pp. (5)
National Research Council. 1981. *The Winged Bean: A High Protein Crop for the Tropics*. National Academy Press, Washington, D.C. (19)
Naveh, Z. and R.H. Whittaker. 1980. Structural and floristic diversity of shrublands and woodlands in northern Israel and other Mediterranean areas. Vegetatio 41, 171–190. (7)
Neale, D.B. and W.T. Adams. 1981. Inheritance of isozyme variants in seed tissues of balsam fir (*Abies balsamea*). Canad. J. Bot. 59, 1285–1291. (5)
Nelson, M.E. and L.D. Mech. 1984. Home-range formation and dispersal of deer in northeastern Minnesota. J. Mammal. 65, 567–575. (3)
Nelson, M.E. and L.D. Mech. in preparation. Demes within a northeastern Minnesota deer population. (3)
Nevo, E., A. Beiles and R. Ben-Shlomo. 1984. The evolutionary significance of genetic diversity: ecological, demographic and life history correlates. Lect. Notes Biomath. 53, 13–213. (4)
New Mexico Native Plant Protection Advisory Committee. 1984. *Handbook of Rare and Endemic Plants of New Mexico*. University of New Mexico Press, Albuquerque. (8)
Ng, F.S.P. 1981. Vegetative and reproductive phenology of dipterocarps. Malay. Forester 44, 197–221. (15)
Nicholson, S.E. 1978. Climatic variation in the Sahel and other African regions during the past five centuries. J. Arid Envir. 1, 3–34. (22)
Nicholson, S.E. 1981. Climate and man in the Sahel during the historical period. In *Environmental Change in the Sahel*. National Academy of Sciences, Washington, D.C. (22)
Nicolic, D. and N. Tucic. 1983. Isoenzyme variation within and among populations of European black pine (*Pinus nigra* Arnold). Silvae Genet. 32, 80–88. (5)
Nienstaedt, H. (Chm.) 1980. Issue Group 1: categories of resource management. In Proc. Servicewide Workshop on Gene Resource Management. Timber Manage. Staff, USDA For. Serv. Sacramento, Calif. pp. 315–322. (5)
Nilsson, B. 1973. Recent Results of Inter-provenance Crosses in Sweden and its Implications for Breeding. Dep. For. Genet., Roy. Coll. For., Res. Note No. 12, Stockholm. 14 pp. (5)
Norman, M.J.T. 1979. *Annual Cropping Systems in the Tropics*. University of Florida Press, Gainesville. (20)
Norton-Griffiths, M. 1979. The influence of grazing, browsing, and fire on the vegetation dynamics of the Serengeti. In *Serengeti, Dynamics of an Ecosystem*, A.R.E. Sinclair and M. Norton-Griffiths (eds.). University. of Chicago Press, Chicago. pp. 310–352. (20)
Novikoff, G. 1983. Desertification by overgrazing. Ambio 12, 102–105. (20)
Noy-Meir, I. 1975. Stability of grazing systems: an application of predator-prey graphs. J. Ecol. 63, 459–481. (14)
Nye, P.H., and D.J. Greenland. 1960. The Soil Under Shifting Cultivation. Technical Communication 51. Commonwealth Bureau of Soils. Commonwealth Agricultural Bureaux, Farnham Royal, England. (20)

O'Brien, S.J., M.E. Roelke, L. Marker, A. Newman, C. A. Winkler, D. Meltzer, L. Colly, J.F. Evermann, M. Bush and D.E. Wildt. 1985. Genetic basis for species vulnerability in the cheetah. Science 227, 1428–1434. (4, 5, 16)

O'Brien, S.J., D.E. Wildt, D. Goldman, D.R. Merril and M. Bush. 1983. The cheetah is depauperate in genetic variation. Science 221, 459–462. (5, 16)

O'Malley, D.M., F.W. Allendorf and G.M. Blake. 1979. Inheritance of isozyme variation and heterozygosity in *Pinus ponderosa*. Biochem. Genet. 17, 233–250. (5)

Odum, E.P., J.T. Finn and E.H. Franz. 1979. Perturbation theory and the subsidy–stress gradient. BioScience 29, 349–352. (23)

Odum, H.T. 1970. Summary: an emerging view of the ecological system at El Verde. In *A Tropical Rain Forest*, H.T. Odum (ed.). Division of Technical Information, U.S. Atomic Energy Commission, Washington, D.C. pp. I-191–I-289. (20)

Odum, H.T. and G. Drewry. 1970. The cesium source at El Verde. In *A Tropical Rain Forest*, H.T. Odum (ed.). Division of Technical Information, U.S. Atomic Energy Commission, Washington, D.C. pp. C-23–C-36. (20)

Odum, W.E. and E.J. Heald. 1975. The detritus-based food web of an estuarine mangrove community. In *Estuarine Research*, L.E. Cronin (ed.). Academic Press, New York. (17)

Odum, W.E. and E.J. Heald. 1972. Trophic analyses of an estuarine mangrove community. Bull. Mar. Sci. 22, 671–738. (17)

Ogden, J.C. and E.H. Gladfelter. 1983. Coral reefs seagrass beds and mangroves: their interaction in the coastal zone of the Caribbean. UNESCO Repts. Mar. Sci. No. 23. 123 pp. (17)

Oliveira, J.M. and M.G. Lima. 1980. Observações preliminares sobre o parauacú: Cebidae, Primata (*Pithecia pithecia* Desmarest, 1804). Ciência e Cultura, Suppl. 33, 499–500. (12)

Oliver, E.G.H., H.P. Linder and J.P. Rourke. 1983. Geographical distribution of present-day Cape taxa and their phytogeographical significance. Bothalia 14, 427–440. (8)

Olson, S.L. and H.F. James. 1982. Podromus of the fossil avifauna of the Hawaiian Islands. Smithsonian Contrib. Zool. 365. Smithsonian Institution Press, Washington, D.C. (14)

Ong, J.E. 1982. Mangroves and aquaculture in Malaysia. Ambio 11, 252–257. (17)

Opler, P.A. 1978. Interaction of plant life history components as related to arboreal folivory. In *The Ecology of Arboreal Folivores*, G.G. Montgomery (ed.). Smithsonian Institution Press, Washington, D.C. pp. 23–31. (12)

Opler, P.A., H.G. Baker and G.W. Frankie. 1980. Plant reproductive characteristics during secondary succession in neotropical lowland forest ecosystems. Biotropica 12, 40–46, Special Suppl. Tropical Succession, J. Ewel (ed.). (12)

Ormerod, W.E. 1976. Ecological effect of control of African trypanosomiasis. Science 191, 815–821. (16)

Orr-Ewing, A.L. 1976. Inbreeding Douglas fir to the S3 generation. Silvae Genet. 25, 179–183. (5)

Osbourne, P. 1982. Some effects of Dutch elm disease on nesting farmland birds. Bird Study 29, 2–16. (16)

Ovington, D. 19787. *Australian Endangered Species*. Cassell, Stanmore, N.S.W. (7)

Packer, C. 1977. Inter-troop Transfer and Inbreeding Avoidance in *Papio anubis*. Ph.D. thesis, University of Sussex. (3)

Packer, C. 1979. Inter-troop transfer and inbreeding avoidance in *Papio anubis*. Anim. Behav. 27, 1–36. (3)

Packer, C. 1985. Dispersal and inbreeding avoidance. Anim. Behav. 33, 666–668. (3)

Paine, R.T. 1966. Food web complexity and species diversity. Amer. Natur. 100, 65–75. (24)

Panero, R.B. 1969. A dam across the Amazon. Science Journal, Sept. 1969, 56–60. (18)

Panouse, J.B. 1968. La faune menacée du Maroc. Cpte. Rend. Coll. P.B.I., Sect. C.T., Hammamet, Tunisie. 8 pp. (22)

Park, Y.S. and D.P. Fowler. 1982. Effects of inbreeding and genetic variances in a natural population of tamarack, *Larix laricina* (Du Roi), in Eastern Canada. Silvae Genet. 31, 21–26. (5)

Park, Y.S. and D.P. Fowler. 1984. Inbreeding in black spruce (*Picea mariana* Mill. B.S.P.): self-fertility, genetic load, and performance. Canad. J. For. Res. 14, 17–21. (5)

Park, Y.S., D.P. Fowler and J.F. Coles. 1984. Population studies of white spruce. II. Natural inbreeding and relatedness among neghboring trees. Canad. J. For. Res. 14, 909–913. (5)

Parker, G.A. 1979. Sexual selection and sexual conflicts. In *Sexual Selection and Reproductive Competition in Insects*, M.S. Blum and N.A. Blum (eds.). Academic Press, New York. pp. 123–166. (3)

Parker, G.G. 1985. The Effects of Size of Disturbance on Water and Solute Budgets of Hillslope Tropical Rainforest in Northeastern Costa Rica. Ph.D. dissertation. University of Georgia, Athens. (20)

Parry, M., A. Bruce and C. Harkness. 1981. The plight of the British Moorland. New Scientist 90, 550–551. (11)

Parsons, J.J. 1966. Los Campos de Cultivos Prehispanicos del Bajo San Jorge. Rev. Acad. Colombiana de Ciencias Exactas, Físicas y Naturales (9) 48, 449–460. (18)

Parsons, P.A. 1971. Extreme-environment heterosis and genetic loads. Heredity 26, 479–483. (5)

Pastorok, R.A. and G.R. Bilyard. 1985. Effects of sewerage pollution on coral-reef communities. Mar. Ecol. Prog. Ser. 21, 175–189. (17)

Pate, J.S. and J.S. Beard. 1984. *Kwongan: Plant Life of the Sandplain*. University W. A. Press, Nedlands. (7)

Patton, J.L. and J.H. Feder. 1981. Microspatial genetic heterogeneity in pocket gophers: nonrandom breeding drift. Evolution 35, 912–920. (3)

Paul, A. and J. Kuester. 1985. Intergroup transfer and incest avoidance in semifree-ranging Barbary macaques (*Macaca sylvanus*) at Salem (FRG). Amer. J. Primat. 8, 317–322. (3)

Pauly, D. and G.I. Murphy. 1982. Theory and management of tropical fisheries. ICLARM Conference Proceedings. Intern. Center Living Aquat. Res. Management, Manila, Philippines and Division of Fisheries Research, CSIRO, Cronulla, Australia. (17)

Peck, S.B. 1980. Climatic change and the evolution of cave invertebrates in the Grand Canyon, Arizona. Bull. Nat. Speleological Soc. 42, 53–60. (21)

Pence, D.B., L.A. Windberg, B.C. Pence and R. Sprowls. 1983. The epizootiology and pathology of sarcoptic mange in coyotes, *Canis latrans*, from south Texas. J. Parasitol. 69, 1100–1115. (16)

Pennington, W. 1974. *The History of British Vegetation*, 2nd Ed. English Universities Press, London. (9)

Perring, F.H. 1974. Changes in our native vascular plant flora. In *The Changing Flora and Fauna of Britain*, D.L. Hawksworth (ed.). Academic Press, New York. (9)

Perring, F.H. and P.D. Sell (eds.). 1978. *Critical Supplement to the Atlas of the British Flora*, Botanical Society of the British Isles, EP Publishing, East Ardsley. (9)

Perring, F.H. and S.M. Walters (eds.). 1976. *Atlas of the British Flora*, 2nd Ed. Botanical Society of the British Isles, EP Publishing, East Ardsley. (9)

Persson, R. 1977. *Forest Resources of Africa*. Royal College of Forestry, Stockholm. (19)

Peterken, G.F. and M. Game. 1981. Historical factors affecting the distribution of *Mercurialis perennis* in central Lincolnshire. J. Ecol. 69, 781–796. (11)

Peterman, R.M., W.C. Clark and C.S. Holling. 1979. The dynamics of resilience: shifting domains in fish and insect systems. In *Population Dynamics*, R.M. Anderson, B.R. Turner and L.R. Taylor (eds.). Blackwell, Oxford. (14)

Peters, R.H. 1983. *The Ecological Implications of Body Size*. Cambridge University Press, Cambridge. (15)

Peters, R.L. and J.D.S. Darling, 1985, The greenhouse effect and nature reserves. BioScience 35, 707–717. (2)

Peterson, R.O., J.D. Woolington and T.N. Bailey. 1984. Wolves of the Kenai Peninsula, Alaska. Wildl. Monog. 88. (3)

Petocz, R.G. 1983. Recommended reserves for Irian Jaya Province: statements prepared for the formal gazettement of 31 conservation areas. World Wildlife Fund Special Report. (24)

Petocz, R.G. 1984. Conservation and development in Irian Jaya. World Wildlife Fund Special Report. (24)

Peyre de Fabrègues, B. 1985. Quel avenir pour l'élevage au Sahel. Rev. Elev. et Medec. Vét. Pays Tropic 3. (22)

Peyre de Fabrègues, B. and G. De Wispelaere. 1984. Sahel: La fin d'un mode pastoral? Marchés Tropicaux 12 (October 1984). pp. 2488–2491. (22)

Pierce, B.A. and J.B. Mitton. 1982. Allozyme heterozygosity and growth in the tiger salamander, *Ambystoma tigrinum.* J. Hered. 73, 250–253. (4)

Pigott, C.D. 1955. *Thymus* L. J. Ecol. 43, 365–387. (9)

Pimm, S.L. 1979. Complexity and stability: another look at MacArthur's original hypothesis. Oikos 33, 351–357. (14)

Pimm, S.L. 1980. Food web design and the effects of species deletion. Oikos 35, 139–149. (14)

Pimm, S.L. 1982. *Food Webs.* Chapman and Hall, London. (14)

Pimm, S.L. 1984a. The complexity and stability of ecosystems. Nature 307, 321–326. (14)

Pimm, S.L. 1984b. Food chains and return times. In *Ecological Communities: Conceptual Issues and the Evidence,* D.R. Strong, D. Simberloff, L.G. Abele and A.B. Thistle (eds.). Princeton University Press, Princeton, N.J. (14)

Pimm, S.L. and J. B. Hyman. in press. Stability concepts in multispecies fisheries. Canad. J. Fish. Aquatic Sci. (14)

Pimm, S.L. and J.W. Pimm. 1982. Resource use, competition and resource availability in Hawaiian honeycreepers. Ecology 63, 1468–1480. (14)

Pinker, R.T., O.E. Thompson and R.F. Eck. 1980. The albedo of a tropical evergreen forest. Quart. J. Roy. Meteorol. Soc. 106, 551–558. (19)

Piot, J.J.P. Nebout, R. Nanot and B. Toutain. 1980. Utilisation des ligneux sahéliens par les herbivores domestiques. Centre Technique Forestier Tropical, Nogent/Marne. 213 pp. (22)

Plowright, W. 1982. The effects of rinderpest and rinderpest control on wildlife in Africa. In Animal Disease in Relation to Animal Conservation, M.A. Edwards and U. McDonnell (eds.). Symposia of the Zoological Society of London, 50, 1–28. (16)

Plowright, W. 1985. La peste bovine aujourd'hui dans le monde. Controle et posssibilite d'eradication par la vaccination. Ann. Med. Vet. 129, 9–32. (16)

Pohl, R.W. l983. *Hyparrhenia rufa (jaragua).* In *Costa Rican Natural History,* D.H. Janzen (ed.). University of Chicago Press, Chicago. p. 256. (13)

Porchester, Lord. 1977. *A Study of Exmoor.* HMSO, London. (11)

Porter, R.H. and M. Wyrick. 1979. Sibling recognition in spiny mice (*Acomys cahirinus*): influence of age and isolation. Anim. Behav. 27, 761–766. (3)

Porter, R.H., M. Wyrick and J. Pankey. 1978. Sibling recognition in spiny mice (*Acomys cahirinus*). Behav. Ecol. Sociobiol. 3, 61–68. (3)

Posey, C.E. 1980. Statement of Dr. Clayton E. Posey on Jari Forestal. In Hearings Before the Subcommittee on Foreign Affairs. House of Representatives, 96th Congress, Second Session, May 7, June 19, and Sept. 18, 1980. Tropical Deforestation. U.S. Government Printing Office, Washington, D.C. pp. 428–433. (20)

Potts, G.R., S.C. Tapper and P.J. Hudson. 1984. Population fluctuations in red grouse: analysis of bag records and a simulation model. J. Anim. Ecol. 53, 21–36. (16)

Poulson, T.L. 1964. Animals in aquatic environments: animals in caves. In *Handbook of Physiology,* D.B. Dill (ed.). American Physiological Society, Washington, D.C. pp. 749–771. (21)

Poulson, T.L. 1972. Bat guano ecosystems. Bull. Nat. Speleological Soc. 34, 55–59. (21)

Poulson, T.L. and W.B. White. 1969. The cave environment. Science 165, 971–981. (21)

Poupon, H. 1980. Structure et dynamique de la strate ligneuse d'une steppe sahélienne au Nord Sénégal. Trav. and Doc. No. 115, ORSTOM, Paris. 351 pp. (22)

Powell, A.H. and G.V.N. Powell. in press. Population dynamics of male euglossine bees in Amazonian forest fragments. Biotropica. (12)

Powell, W.R. (ed.). 1974. *Inventory of Rare and Endangered Vascular Plants of California,* 2nd Ed. 1980, revised by J.P. Smith Jr., R.J. Cole and J.O. Sawyer Jr. Spec. Publ. No. 1 (2nd Ed.), Calif. Native Plant Soc., Berkeley, Calif. (7)

Prance, G. 1973. Phytogeographic support for the theory of Pleistocene forest refuges in the Amazon Basin based on evidence from distribution patterns in Caryocaraceae, Chrysobalanaceae, Dichapetalaceae, and Lecythidaceae. Acta Amaz. 3, 5–28. (8)

Prance, G. 1977. The phytogeographic subdivisions of Amazonia and their influence on the selection of biological reserves. In *Extinction is Forever*, G. Prance and T. Elias (eds.). New York Botanical Garden, Bronx, New York. pp. 195–213. (8)

Prance, G.T. 1978. The origin and evolution of the Amazon flora. Interciencia 3, 207–222. (5)

Prance, G.T. 1980. Probable pollination of *Combretum* by monkeys. Biotropica 12, 239. (15)

Prance, G. 1982a. *Biological Diversification in the Tropics*. Columbia University Press, New York. 714 pp. (8, 19)

Prance, G. 1982b. Forest refuges: evidence from woody angiosperms. In *Biological Diversification in the Tropics*, G. Prance (ed.). Columbia University. Press, New York. pp. 137–157. (8)

Prance, G.T. and T.S. Elias. 1977. *Extinction is Forever*. New York Botanical Garden, Bronx, New York. 437 pp. (18)

Preston, F.W. 1962. The canonical distribution of commonness and rarity (2 parts). Ecology 43, 185–215; 410–432. (11)

Price, M.V. and N.M. Waser. 1979. Pollen dispersal and optimal outcrossing in *Delphinium nelsoni*. Nature 277, 294–297. (5)

Pusey, A.E. 1980. Inbreeding avoidance in chimpanzees. Anim. Behav. 28, 543–552. (3)

Pusey, A.E. and C. Packer. 1986a. Dispersal and philopatry. In *Primate Societies*, B.B. Smuts, D.L. Cheney, R.M. Seyfarth, R.W. Wrangham and T.T. Struhsaker (eds.). University of Chicago Press, Chicago. (3)

Pusey, A.E. and C. Packer. 1986b. Philopatry and dispersal in lions. Behaviour. (3)

Pusey, A.E. and C. Packer. in preparation. Dispersal patterns in lions. (16)

Rabinowitz, D. 1981. Seven forms of rarity. In *The Biological Aspects of Rare Plant Conservation*, H. Synge (ed.). Wiley, Chichester. (9)

Radesater, T. 1976. Individual sibling recognition in juvenile Canada geese (*Branta canadensis*). Canad. J. Zool. 54, 1069–1072. (3)

Raemaekers, J.J., F.P.G. Aldrich-Blake and J.B. Payne. 1980. The forest. In *Malayan Forest Primates: Ten Years' Study in Tropical Rain Forest*, D.J. Chivers (ed.). Plenum, New York. pp. 29–61. (15)

Ralls, K. and J. Ballou. 1983. Extinction: lessons from zoos. In *Genetics and Conservation: A Reference for Managing Wild Animal and Plant Populations*, C.M. Schonewald-Cox, S.M. Chambers, B. MacBryde and L. Thomas (eds.). Benjamin-Cummings, Menlo Park, Calif. pp. 164–184. (2, 3, 4)

Ralls, K. and J. Ballou (eds.). in press. Genetic management of captive populations. Zoo Biol. (1)

Ranney, J.W. 1977. Forest island edges—their structure, development, and implication to regional forest ecosystem dynamics. EDFB/IBP-77-1. Oak Ridge National Laboratory, Oak Ridge, Tenn. (11)

Ranney, J.W., M.C. Bruner and J.B. Levenson. 1981. The importance of edge in the structure and dynamics of forest islands. In *Forest Island Dynamics in Man-Dominated Landscapes*, R.L. Burgess and D.M. Sharpe (eds.). Springer Verlag, New York. pp. 67–95. (11, 13)

Rapport, D.J., H.A. Regier and T.C. Hutchinson. 1985. Ecosystem behavior under stress. Amer. Natur. 125, 617–640. (14)

Ratcliffe, D. 1960. *Draba muralis* L. J. Ecol. 48, 737–744. (9)

Ratcliffe, D. 1979. The end of the large blue butterfly. New Scientist 8, 457–458. (11, 16)

Rauh, W. 1979. Problems of biological conservation in Madagascar. In *Plants and Islands*, D. Bramwell (ed.). Academic Press, New York. pp. 405–421. (19)

Raven, P. and D. Axelrod. 1974. Angiosperm biogeography and past continental movements. Ann. Missouri Bot. Gard. 61, 539–673. (8)

Raven, P. and D. Axelrod. 1978. Origin and relationships of the California flora. Univ. Calif. Publ. Bot. 72. 1–134. (8)

Raven, P.H. 1980. Research Priorities in Tropical Biology. National Research Council, Washington, D.C. (19)

Redfield, J. A. 1974. Genetics and selection at the *Ng* locus in the blue grouse (*Dendragapus obscurus*). Heredity 33, 69–78. (4)

Reese, J.G. 1980. Demography of European mute swans in Chesapeake Bay. Auk 97, 449–464. (3)

Regal, P. 1985. Models of genetically-engineered organisms and their ecological impact. In *Ecological Consequences of Biological Invasions*, H. Mooney (ed.). Springer Verlag, New York. (1)

Regier, H.A. and W.L. Hartman. 1973. Lake Erie's fish community: 150 years of cultural stress. Science 180, 1248–1255. (14)

Reichelt, R.E., D.G. Green and R.H. Bradbury. 1983. Explanation, prediction and control in coral reef ecosystems: I. Models for explanation. In Proc. Great Barrier Reef Conf., J.T. Baker, R.M. Carter, P.W. Sammarco and K.P. Stark (eds.). James Cook University, Townsville, Australia, pp. 231–236. (17)

Remson, J.V., Jr. and T.A. Parker III. 1983. Contribution of river-created habitats to bird species richness in Amazonia. Biotropica 15, 223–231. (8)

Rendel, J.M. 1943. Variations in the weights of hatched and unhatched ducks' eggs. Biometrika 33, 48–58. (4)

Richardson, R.H. and K. Kojima. 1965. The kinds of genetic variability in relation to selection responses in *Drosophila fecundity*. Genetics 52, 583–598. (6)

Richdale, L.E. 1957. *A Population Study of Penguins*. Clarendon Press, Oxford. (3)

Richmond, M. and R. Stehn. 1976. Olfaction and reproductive behavior in microtine rodents. In *Mammalian Olfaction, Reproductive Processes, and Behavior*, R.L. Doty (ed.). Academic Press, New York. pp. 197–217. (3)

Richter-Dyn, N. and N.S. Goel. 1972. On the extinction of a colonizing species. Theor. Pop. Biol. 3, 406–433. (2, 11)

Righter, F.I. 1945. *Pinus*: the relationship of seed size and seedling size to inherent vigor. J. For. 43, 131–137. (5)

Ripley, S.D. 1959. Competition between sunbird and honeyeater species in the Moluccan Islands. Amer. Natur. 93, 127–132. (12)

Ripley, S.D. 1961. Aggressive neglect as a factor in interspecific competition in birds. Auk 78, 366–371. (12)

Ritchie, J.C. 1954. *Primula scotica* Hook. J. Ecol. 42, 623–628. (9)

Robbins, C.S. 1979. Effect of forest fragmentation on bird populations. In Workshop Proceedings: Management of North Central and Northeastern Forests for Nongame Birds, R.M. DeGraaf (coord.). USDA Forest Service General Technical Report NC-51. pp. 198–212. (11)

Robbins, C.S. 1980. Effect of forest fragmentation on breeding bird populations in the piedmont of the Mid-Atlantic region. Atlantic Naturalists 33, 31–36. (11)

Robbrecht, E. 1982. The identity of the Panamanian genus *Dressleriopsis* (Rubiaceae). Ann. Missouri Bot. Gard. 69, 427–429. (8)

Rodhouse, P.G. and M. Gaffney. 1984. Effect of heterozygosity on metabolism during starvation in the American oyster *Crassostrea virginica*. Mar. Biol., 80, 179–187. (4)

Rodin, L.E. (ed.). 1970. *Études géobotaniques des pâturages du secteur ouest du département de Medea (Algérie)*. Ed. Nauka, Leningrad. 124 pp. (22)

Rodman, P.S. 1977. Feeding behavior of Orangutans of the Kutai Nature Reserve, East Kalimantan. In *Primate Ecology: Studies of Feeding and Ranging Behaviour in Lemurs, Monkeys and Apes*, T.H. Clutton-Brock (ed.). Academic Press, London. pp. 383–413. (15)

Rogers, C.S., H.C. Fitz, H.C., M. Gilnack, T. Beets and J. Hardin. 1984. Scleractinian coral recruitment patterns at Salt River Canyon, St. Croix, U. S. Virgin Islands. Coral Reefs 3, 69–76. (17)

Rogers, D.J. and Randolph, S.E. 1985. Population ecology of tsetse. Ann. Rev. Entomol. 30, 197–216. (16)

Romashov, V.A. 1977. Helminths of wild ungulates and dynamics of intensity of infestation in relation to the density and number of host species in the Voroneshsky reserve. In XIIIth International Congress of Game Biologists, T.J. Peterle (ed.). Corporate Press, Washington, D.C. (16)

Rood, J.P. in preparation. Dispersal and intergroup transfer in the dwarf mongoose. (3)

Root, R.B. 1967. The niche exploitation pattern of the blue-gray gnatcatcher. Ecol. Monogr. 378, 317-350. (11)

Rose, F. 1948. *Orchis purpurea* Huds. J. Ecol. 36, 366–377. (9)

Rose, M. 1982. Antagonistic pleiotropy, dominance, and genetic variation. Heredity 48, 63–78. (4)

Ross, M.D. 1980. Components of fitness and frequency-dependent overdominance in hermaphrodite plants. Evolution 34, 765–778. (5)

Rossiter, P.B., D.M. Jessett, J.S. Wafula, L. Karstad, S. Chema, W.P. Taylor, L.W. Rowe, J.C. Nyange, M. Otaru, M. Mumbala and G.R. Scott. 1983. The re-emergence of rinderpest as a threat in East Africa. 1979–1982. Veterinary Record 113, 459–461. (16)

Rothstein, S.I. 1975. An experimental and teleonomic investigation of avian brood parasitism. Condor 77, 250–271. (11)

Rouch, R. 1977. Considerations sur l'ecosysteme karstique. Compt. Rend. Acad. Sci. Paris 284, 1101–1103. (21)

Roughgarden, J. 1979. *Theory of Population Genetics and Evolutionary Ecology: An Introduction.* Macmillan, New York. (19)

Rourke, J.P. 1980. *The Proteas of Southern Africa.* Purnell, Cape Town, Johannesburg. (7)

Rummel, J.D. and J. Roughgarden. 1983. Some differences between invasion-structured and coevolution-structured competitive communities: a preliminary theoretical analysis. Oikos 41, 477–486. (14)

Russell, C.E. 1983. Nutrient Cycling and Productivity in Native and Plantation Forests at Jari Florestal, Para, Brazil. Ph.D. dissertation, University of Georgia, Athens. (20)

Rylands, A.B. 1981. Preliminary field observations on the marmoset, *Callithrix humeralifer intermedius* (Hershkovitz, 1977) at Dardanelos, Rio Aripuanã, Mato Grosso. Primates 22, 46–59. (12)

Rylands, A.B. 1982. The behaviour and ecology of three species of marmosets and tamarins (Callitrichidae, Primates) in Brazil. Unpubl. doctoral dissertation, University of Cambridge, Cambridge. (12)

Rylands, A.B. and A. Keuroghlian. 1985. Primate survival in forest fragments in Central Amazonia: preliminary results. Paper presented at the 2nd Brazilian Primatology Congress, State University of Campinas, São Paulo, 29th January to 2nd February, 1985. (12)

Ryu, J.B. and R.T. Eckert. 1983. Foliar isozyme variation in twenty-seven provenances of *Pinus strobus* L.: genetic diversity and population structure. In Proc. Twenty-eighth Northeast. For. Tree Improv. Conf. Inst. Natur. Environ. Resources, Univ. New Hampshire, Durham. pp. 249–261. (5)

Rzóska, J., 1974. The Upper Nile swamps, a tropical wetland study. Freshwat. Biol. 4, 1–30. (18)

Sade, D.S. 1968. Inhibition of son–mother mating among free-ranging rhesus monkeys. Science and Psychoanalysis 12, 18–38. (3)

Sade, D.S., D.L. Rhodes, J. Loy, G. Hausfater, J.A. Breuggeman, J.R. Kaplan, B.D. Chepko-Sade and K. Cushing-Kaplan. 1984. New findings on incest among free-ranging rhesus monkeys. Amer. J. Phys. Anthro. 63, 212–213. (3)

Sagan, C., O.B. Toon and J.B. Pollack. 1979. Anthropogenic albedo changes and the Earth's climate. Science 206, 1363–1368. (19)

Sailer, R.I. 1978. Our immigrant insect fauna. Bull. Ent. Soc. Amer. 23, 3–11. (14)

Sailer, R.I. 1983. History of insect introductions. In *Exotic Plant Pests and North American Agriculture*, C. Graham and C. Wilson (eds.). Academic Press, New York. (14)

Salati, E. and P.B. Vose. 1984. Amazon basin: a system in equilibrium. Science 224, 129–138. (19)

Sale, P.F. 1980. The ecology of fishes on coral reefs. Oceanogr. Mar. Biol. Ann. Rev. 18, 367–421. (17)

Salm, R.V. 1984. Ecological boundaries for coral reef reserves: principles and guidelines. Environ. Conserv. 11, 209–215. (17)

Salwasser, H., S.P. Mealey and K. Johnson. 1984. Wildlife population viability: a question of risk. Trans. N. Am. Wildl. Nat. Resource Conf. 49. (2)

Samollow, P.B. and M.E. Soulé. 1983. A case of stress related heterozygote superiority in nature. Evolution 37, 646–648. (4)

Sanchez, P.A. 1976. *Properties and Management of Soils in the Tropics*. Wiley, New York. (20)

Sanders, R.W. and A.A. Nightingale. 1982. *Juniperius ashei* in the Ozarks: diagnosis and significance. Mo. Acad. Sci. Occasional Paper 7, 71–79. (6)

SAS Institute Inc. 1982. SAS User's Guide: Statistics, 1982 Edition, SAS Institute, Cary, North Carolina. (9)

Sassaman, C. 1978. Dynamics of a lactate dehydrogenase polymorphism in the wood louse *Porcellio scaber* Latr.: evidence for partial assortative mating and heterosis in natural populations. Genetics 88, 591–609. (4)

Saus, G.L. and R.M. Lloyd. 1976. Experimental studies on mating systems and genetic load in *Onoclea sensibilis* L. (Aspleniacaeae: Athyrioideae). Bot. J. Linnean Soc. 72, 101–113. (5)

Savidge, J. 1985. Pesticides and the decline of Guam's native birds. Nature 316, 301. (24)

Scandalios, J.G., E.H. Liu and M.A. Campeau. 1972. The effects of intragenic and intergenic complementation on catalase structure and function in maize: a molecular approach to heterosis. Arch. Biochem. Biophys. 153, 696–705. (5)

Schaal, B.A. 1975. Population structure and local differentiation in *Liatris cyclindracea*. Amer. Natur. 109, 511–528. (5)

Schaal, B.A. and D.A. Levin. 1977. The demographic genetics of *Liatris cylindracea* Michx. (Compositae). Amer. Natur. 110, 191–206. (5)

Schade, A.L. and L. Caroline. 1944. Raw hen egg white and the role of iron in growth inhibition of *Shigella dysenteriae, Staphylococcus aureus, Escherichia coli*, and *Saccharomyces cerevisiae*. Science 100, 14–15. (4)

Schaller, G.B. 1972. *The Serengeti Lion: a Study of Predator–prey Relations*. University of Chicago Press, Chicago. (16)

Schemske, D.W. 1980. Floral ecology and hummingbird pollination of *Combretum farinosum* in Costa Rica. Biotropica 12, 169–181. (15)

Schemske, D.W. and R. Lande. 1985. The evolution of self-fertilization and inbreeding depression in plants. II. Empirical observations. Evolution 39, 41–52. (5)

Schiller, G., M.T. Conkle and C. Grunwald. in press. Isozyme variation among mediterranean populations of Aleppo pine. Silvae Genet. (5)

Schmidt, R. 1981. Summary of Jari Plantation data, supplied by Inventory and Control Manager, Jari Florestal, National Bulk Carriers, New York, N.Y. (20)

Schmitt, D. 1968. Performance of Southern Pine Hybrids in South Mississippi. USDA For. Serv. Res. Pap. SO-36. 15 pp. (5)

Schmitt, D. 1969. Nanism in slash × shortleaf pine hybrids. For. Sci. 15, 174–175. (5)

Schoener, T.W. 1976. The species–area relation within archipelagoes: models and evidence from island land birds. In Proceedings of the 16th International Ornithological Congress, H.J. Frith and J.H. Calaby (eds.). Australian Academy of Science, Canberra. pp. 629–642. (11)

Schoener, T.W. 1983. Field experiments on interspecific competition. Amer. Natur. 122, 240–285. (14)

Schonewald-Cox, C.M., S.M. Chambers, F. MacBryde and L. Thomas (eds.). 1983. *Genetics and Conservation: A Reference for Managing Wild Animal and Plant Populations*. Benjamin/Cummings, Menlo Park, Calif. (2)

Schultze-Westrum, T. 1978. Conservation in Irian Jaya, Indonesia. World Wildlife Fund Special Report. (24)

Schwaar, D. 1965. L'évolution du milieu entre 1949 et 1963 dans une région test de la Tunisie centrale. Etude détaillée réalisée par photointerprétation. UNDP/TUN/I, FAO, Rome. 11 pp. (22)

Schwaegerle, K.E. and B.A. Schaal. 1979. Genetic variability and founder effect in the pitcher plant *Sarracenia purpurea* L. Evolution 33, 1210–1218. (5)

Schwartz, D. and W.J. Laughner. 1969. A molecular basis for heterosis. Science 166, 626–627. (5)

Schwartz, O.A. and K.B. Armitage. 1980. Genetic variation in social mammals: the marmot model. Science 207, 665–667. (3)

Scott, G.R. 1970. Rinderpest. In *Infectious Diseases of Wild Mammals*, J.W. Davis, L.H. Karsted and D.O. Trainer (eds.). Iowa State University Press, Iowa. (16)

Seemanova, E. 1971. A study of children of incestuous matings. Human Heredity 21, 108–128. (3)

Seidensticker, J.C., M.G. Hornocker, W.V. Wiles and J.P. Messick. 1973. Mountain lion social organization in the Idaho primitive area. Wildl. Monogr. 35. (11)

Sekulic, R. 1982. Daily and seasonal patterns of roaring and spacing in four red howler, *Alouatta seniculus*, troops. Folia Primatol. 39, 22–48. (12)

Sexton, O.J. and R.M. Andrews. in preparation. Characteristics of individual growth in a peripheral population of the lizard *Crotaphytus collaris* (Sauria; Iguanidae). (6)

Shaffer, M.L. 1978. Determining minimum viable population sizes: a case study of the grizzly bear. PhD. dissertation, Duke University, Durham, N.C. (2)

Shaffer, M.L. 1981. Minimum population sizes for species conservation. BioScience 31, 131–134. (2)

Shaffer, M.L. 1983. Determining minimum viable population size for the grizzly bear. Int. Conf. Bear Res. Manage. 5, 133–139. (2)

Sharpe, D.M., F.W. Stearns, R.C. Burgess and W.C. Johnston. 1981. Spatio-temporal patterns of forest ecosystems in man-dominated landscapes of the eastern United States. In *Perspectives in Landscape Ecology*, S.P. Tjallingii and A.A. de Veer (eds.). Centre for Agricultural Publication and Documentation, Wageningen, Netherlands. pp. 109–116. (11)

Shick, J.M., R.J. Hoffman and A.N. Lamb. 1979. Asexual reproduction, population structure, and genotype-environment interactions in sea anemones. Amer. Zool. 19, 699–713. (4)

Shields, W.M. 1982. *Philopatry, Inbreeding, and the Evolution of Sex*. State University of New York Press, Albany. (3)

Shields, W.M. in preparation. Dispersal and mating systems: their joint effects on mammalian evolution. (3)

Shmida, A. and S.P. Ellner. 1984. Coexistence of plant species with similar niches. Vegetatio 58, 29–55. (9)

Shull, G.H. 1952. Beginnings of the heterosis concept. In *Heterosis*, J.W. Gowen (ed.). Iowa State College Press, Ames. pp. 14–48. (5)

Sibley, C.G. and J.E. Ahlquist. 1985. The phylogeny and classification of the Australo-Papuan passerine birds. Emu 85, 1–14. (24)

Siegel, M.I. and W.J. Doyle. 1975a. The differential effects of prenatal and postnatal audiogenic stress on fluctuating dental asymmetry. J. Exper. Zool. 191, 211–214. (4)

Siegel, M.I. and W.J. Doyle. 1975b. The effects of cold stress on fluctuating asymmetry in the dentition of the mouse. J. Exper. Zool. 193, 385–389. (4)

Siegel, M.I. and W.J. Doyle. 1975c. Stress and fluctuating limb asymmetry in various species of rodents. Growth 39, 363–369. (4)

Siegel, S. 1956. *Nonparametric Statistics for the Behavioral Sciences*. McGraw-Hill, New York. (9)

Siegfried, W.R., P.G.H. Frost, J. Cooper and A.C. Kemp. 1976. South African Red Data Book: Aves. So. Afr. Sci. Frog. Rep. No. 7, 1–108. (7)

Silander, S. 1984. Succession at the El Verde Radiation Site: A 17-Year Record. Center for Energy and Environmental Research, Rio Piedras, Puerto Rico. (20)

Silen, R.R. 1962. Pollen dispersal considerations for Douglas-fir. J. For. 60, 790–795. (5)

Silvertown, J.W. 1980. The evolutionary ecology of mast seeding in trees. Biol. J. Linn. Soc. 14, 235–250. (13)

Simberloff, D. 1981. Community effects of introduced species. In *Biotic Crises in Ecological and Evolutionary Time*, M.H. Nitecki (ed.). Academic Press, New York. (14)

Simberloff, D. 1986. Introduced insects: a biogeographic and systematic perspective. In *Ecology of Biological Invasions to North America and Hawaii*, H. Mooney (ed.). Springer Verlag, Berlin. (14)

Simberloff, D.S. and L.G. Abele. 1976. Island biogeography theory and conservation practice. Science 191, 285–286. (11)

Simon, N. 1962. *Between the Sunlight and the Thunder. The Wild Life of Kenya.* Collins, London. (16)

Sinclair, A.R.E. 1977. *The African Buffalo.* The University of Chicago Press, Chicago. (16)

Sinclair, A.R.E. and M. Norton-Griffiths (eds.). 1979. *Serengeti. Dynamics of an Ecosystem.* The University of Chicago Press, Chicago. (16)

Singh, B.N. 1972. The lack of evidence for coadaptation in geographic populations of *Drosophila ananassai.* Genetica 43, 582–598. (6)

Singh, S.M. 1982. Enzyme heterozygosity associated with growth at different developmental stages in oysters. Canad. J. Genet. Cytol. 24, 451–458. (4)

Singh, S.M. 1984. Allozyme heterozygosity as a probe for genetic variation associated with fitness characters in marine molluscs. Biol. Morya 1, 27–39. (4)

Singh, S.M. and R.G. Green. 1984. Excess of allozyme heterozygosity in marine molluscs and its possible significance. Malacologia 25, 569–581. (4)

Singh, S.M. and E. Zouros. 1978. Genetic variation associated with growth rate in the American oyster (*Crassostrea virginica*). Evolution 32, 342–353. (4)

Singleton, G.R. 1983. The social and genetic structure of a natural colony of house mice, *Mus musculus*, at Healesville Wildlife Sanctuary. Aust. J. Zool. 31, 155–166. (3)

Sinha, S.K. and R. Khanna. 1975. Physiological, biochemical, and genetic basis of heterosis. Adv. Agron. 27, 123–174. (5)

Sioli, H. 1968. Principal biotopes of primary production in the waters of Amazonia. In Proc. Symp. Recent Adv. Trop. Ecol., R. Misra and B. Gopal (eds.). Dept. Botany, Banaras Hindu University, Varanasi, India. pp. 591–600. (18)

Sioli, H. 1973. Introduction into the problems: The situation of modern civilization in the light of the ecological aspect of life. In *Oekologie und Lebensschutz in internationaler Sicht (Ecology and Bioprotection: International Conclusions)*, H. Sioli (ed.). Rombach, Freiburg. pp. 9–33. (18)

Sioli, H. 1982. Tropische Fluesse in ihren Beziehungen zu terrestrischen Umgebung und im Hinblick auf menschliche Eingriffe. Arch. Hydrobiol. 95, 463–485. (18)

Sioli, H. 1984. Das Leben und die "Entwicklung" der feucht-tropischen Waldgebiete (Beispiel Amazonien). Ber. Sitz. Joachim Jungius Gesellschaft der Wissenschaften, Hamburg, Jg. 2, Heft 2, 17 pp. (18)

Sissenwine, M.P. 1977. The effect of random fluctuations on a hypothetical fishery. Int. Comm. Northw. Atl. Fish. Sel. Pap. 29, 137–144. (14)

Skog, L.E. 1978. Gesneriaceae. In *Flora of Panama*. Ann. Missouri Bot. Gard. 63, 783–996. (8)

Smith, A.T. and B.L. Ivins. 1983. Colonization in a pika population: dispersal vs. philopatry. Behav. Ecol. Sociobiol. 13, 37–47. (3)

Smith, A.T. and B.L. Ivins. 1984. Spatial relationships and social organization in adult pikas: a facultatively monogamous mammal. Z. Tierpsychol. 66, 289–308. (3)

Smith, C.F. 1976. A Flora of the Santa Barbara Region, California. Santa Barbara Mus. Nat. Hist., Santa Barbara, Calif. (8)

Smith, M.H., R.K. Chesser, E.G. Cothran and P.E. Jones. 1982. Genetic variability and antler growth in a natural population of white-tailed deer. In *Antler Development in Cervidae*, R.D. Brown (ed.). Caesar Kleberg Wildlife Research Institute, Kingsville, Tex. pp. 365–387. (4)

Smith, R.H. 1979. On selection for inbreeding in polygynous animals. Heredity 43, 205–211. (3)

Smith, S.V., W.J. Kimmerer, E.A. Laws, R.E. Brock, and T.W. Walsh. 1981. Kaneohe Bay sewage diversion experiment: perspectives on ecosystem responses to nutritional perturbation. Pac. Sci. 35, 279–397. (17)

Smythe, N. 1970. Relationships between fruiting seasons and seed dispersal methods in a Neotropical forest. Amer. Natur. 104, 25–35. (15)

Smythe, N. 1978. The natural history of the Central American agouti (*Dasyprocta punctata*). Smithsonian Contributions to Zoology 257, 1–52. (15)

Smythe, N., W.E. Glanz and E.G. Leigh, Jr. 1982. Population regulation in some terrestrial frugivores. In *The Ecology of a Tropical Forest: Seasonal Rhythms and Long-term Changes*, E.G. Leigh, Jr., A.S. Rand and D.M. Windsor (eds.). Smithsonian Institution Press, Washington. pp. 227–238. (15)

Snyder, E.G., A.E. Squillace and J.M. Hamaker. 1966. Pigment inheritance in slash pine seedlings. In Proc. Eighth Southern Conf. For. Tree Improv. Savannah, Ga. pp. 77–85. (5)

Sokal, R.R. and F.J. Rohlf. 1981. *Biometry: The Principles and Practice of Statistics in Biological Research*, 2nd Ed. W.H. Freeman, San Francisco. (9)

Sokolov, V.E., S.I. Isaev and E.Y. Pavlova. 1981. A study of mechanisms regulating growth and sexual maturation in great gerbils (*Rhombomys opimus*). Zool. Zh. 60, 579–586. (3)

Solbrig, O.T. 1981. Studies on the population biology of the genus *Viola*. II. The effect of a plant size and fitness in *Viola sororia*. Evolution 35, 1080–1093. (5)

Sorensen, F.C. 1969. Embryonic genetic load in coastal Douglas-fir, *Pseudotsuga menziesii var menziesii*. Amer. Natur. 103, 389–398. (5)

Sorenson, F.C. 1970. Self-fertility of a Central Oregon Source of Ponderosa Pine. USDA For. Serv. Res. Pap. PNW-109. 9 pp. (5)

Sorenson, F.C. 1982. The roles of polyembryony and embryo viability in the genetic system of conifers. Evolution 36, 725–733. (5)

Sorenson, F.C., J.F. Franklin and R. Woolard. 1976. Self-pollination effects on seed and seedling traits in noble fir. For. Sci. 22, 155–159. (5)

Soulé, M. 1967. Phenetics of natural populations. II. Asymmetry and evolution in a lizard. Amer. Natur. 101, 141–160. (5)

Soulé, M. 1973. The epistasis cycle: a theory of marginal populations. Ann. Rev. Ecol. Syst. 4, 165–187. (5)

Soulé, M. 1976. Allozyme variation: its determinants in time and space. In *Molecular Evolution*, F.J. Ayala (ed.). Sinauer Associates, Sunderland, Mass. pp. 60–77. (4)

Soulé, M. 1979. Heterozygosity and developmental stability: another look. Evolution 33, 396–401. (4,5)

Soulé, M.E. 1985. What is conservation biology? BioScience 35, 727–734. (I, 1, 24)

Soulé, M.E. 1986. *Viable Populations*, Cambridge University Press, Cambridge. (2)

Soulé, M.E. and J. Cuzin-Roudy. 1982. Allomeric variation. 2. Developmental instability of extreme phenotypes. Amer. Natur. 120, 765–786. (4)

Soulé, M.E. and D. Simberloff. 1986. What do genetics and ecology tell us about the design of nature reserves? Biol. Conserv. 35, 19–40. (2,8)

Soulé, M.E. and B.A. Wilcox. 1980. *Conservation Biology: An Evolutionary-Ecological Approach*. Sinauer Associates, Sunderland, Mass. (11)

Spevak, T.A. in preparation. Breeding structure in semi-natural populations of *Peromyscus maniculatus bardi*. (3)

Spitzer, K., M. Rejmanek and T. Soldan. 1984. The fecundity and long term variability in abundance of noctuid moths (Lepidoptera, Noctuidae). Oecologia 62, 91–93. (14)

SPOF (Strategic Planning for Ontario Fisheries), Working Group 15. 1983. The index of overexploitation. Ontario Ministry of Natural Resources. (14)

Start, A.N. and A.G. Marshall. 1976. Nectarivorous bats as pollinators of trees in West Malaysia. In *Tropical Trees: Variation, Breeding and Conservation*, J. Burley and B.T. Styles (eds.). Academic Press, London. pp. 141–150. (19)

Stebbings, R.E. 1969. Observer influence on bat behaviour. Lynx (Prague) 10, 93–100. (21)

Stebbins, G.L. 1974. Flowering plants. In *Evolution Above the Species Level*. Belknap Press, Harvard University Press, Cambridge, Mass. (8)

Stebbins, G.L. and J. Majors. 1965. Endemism and speciation in the California flora. Ecol. Monogr. 35, 1–35. (8,9)

Steele, J.H. 1985. A comparison of terrestrial and marine ecological systems. Nature 313, 355–358. (14)

Steinhoff, R.J., D.G. Joyce and L. Fins. 1983. Isozyme variation in *Pinus monticola*. Canad. J. For. Res. 13, 1122–1131. (5)

Stevenson, M.F. and A.B. Rylands. in press. The marmoset monkeys, genus *Callithrix*. In *Ecology and Behavior of Neotropical Primates*, II, R.A. Mittermeier and A.F. Coimbra-Filho (eds.). Academia Brasileira de Ciências, Rio de Janeiro. (12)

Steyermark, J.A. 1963. *Flora of Missouri*. Iowa State University Press, Ames. pp. 1–1728. (8)

Steyermark, J.A. 1979. Flora of the Guayana Highland: endemicity of the generic flora of the summits of the Venezuelan tepuis. Taxon 28, 45–54. (8)

Stiles, F.G. 1983. Birds. In *Costa Rican Natural History*, D.H. Janzen (ed.). University of Chicago Press, Chicago. pp. 502–530. (13)

Stott, P. 1981. *Historical Plant Geography: An Introduction*, Allen & Unwin, London. (9)

Strauss, S.H. 1985. The Relation of Heterozygosity to Growth Rate and Stability among Inbred and Crossbred Knobcone Pine (*Pinus attenuata* Lemm.). Ph.D. dissertation, University of California, Berkeley. 148 pp. (5)

Strauss, S.H. and M.T. Conkle. in preparation. Inheritance and linkage of allozymes in knobcone pine. Theor. Appl. Genet. (5)

Strong, D.R., D. Simberloff, L.G. Abele and A.B. Thistle. 1984a. *Ecological Communities: Conceptual Issues and the Evidence*. Princeton University Press, Princeton, New Jersey. (14)

Strong, D.R., J.H. Lawton, and T.R.E. Southwood (eds.). 1984b. *Insects on Plants: Community Mechanisms and Patterns*. Blackwell, Oxford. (14)

Struhsaker, T.T. 1978. Food habits of five monkey species in the Kibale Forest, Uganda. In *Recent Advances in Primatology*, D.J. Chivers and J. Herbert (eds.). Academic Press, New York. (15)

Struhsaker, T.T. and L. Leyland. 1977. Palm-nut smashing by *Cebus a. apella* in Colombia. Biotropica 9, 124–126. (15)

Stuber, C.W. 1970. Theory and use of hybrid population statistics. Pap. Presented at the Second Meet. Working Group on Quant. Genet., Sect. 22 IUFRO. Raleigh, N.C. pp. 100–112. (5)

Sukačev, V.N. 1964. *Fundamentals of Forest Biogeography*. Oliver and Boyd, London. (18)

Sutherland, J.P. 1974. Multiple stable points in natural communities Amer. Natur. 108, 859–873. (14)

Sutton, S.L., T.C. Whitmore and A.C. Chadwick (eds.). 1984. *Tropical Rain Forests: Ecology and Management*. Blackwell, Oxford. (19)

Sweeting, M.M. 1973. *Karst Landforms*. Columbia University Press, New York. (21)

Talbot, L.M. and M.H. Talbot. 1963. The wildebeest in western Masailand, East Africa. Wildl. Monogr. 12, 1–88. (16)

Tanner, J.T. 1942. The ivory-billed woodpecker. National Audubon Society Research Report, Number 1. (11)

Tannock, J., W.W. Howells and R.J. Phelps. 1983. Chlorinated hydrocarbon pesticide residues in eggs of some birds in Zimbabwe. Environmental Pollution (Series B) 5, 147–155. (16)

Temple, S.A. 1977. Plant–animal mutualism: coevolution with dodo leads to near extinction of plant. Science 197, 885–886. (2)

Templeton, A.R. 1979. The unit of selection in *Drosophila mercatorum*. II. Genetic revolutions and the origin of coadapted genomes. Genetics 92, 1265–1282. (6)

Templeton, A.R. 1980. The theory of speciation via the founder principle. Genetics 94, 1011–1038. (8)

Templeton, A.R. 1981. Mechanisms of speciation—a population genetic approach. Ann. Rev. Ecol. Syst. 12, 32–48. (6)

Templeton, A.R. 1983. Natural and experimental parthenogenesis. In *The Genetics and Biology of Drosophila*, Vol. 3C, M. Ashburner, H.L. Carson and J.N. Thompson (eds.). Academic Press, London. pp. 343–398. (6)

Templeton, A.R. and B. Read. 1983. The elimination of inbreeding depression in a captive herd of Speke's gazelle. In *Genetics and Conservation: A Reference for Managing Wild Animal and Plant Populations*, C.M. Schonewald-Cox, S.M. Chambers, F. MacBryde and L. Thomas (eds.). Benjamin/Cummings, Menlo Park, Calif. pp. 241–261. (6)

Templeton, A.R. and B. Read. 1984. Factors eliminating inbreeding depression in a captive herd of Speke's gazelle. Zoo Biol. 3, 177–199. (6)

Templeton, A.R. in preparation. Inferences on natural population structure from genetic studies on captive mammalian populations. (3)

Templeton, A.R., T.J. Crease and F. Shah. 1985a. The molecular through ecological genetics of abnormal abdomen. I. Basic genetics. Genetics, in press. (6)

Templeton, A.R., H. Hemmer, G. Mace, U.S. Seal, W.M. Shields and D.S. Woodruff. 1985b. Coadapted gene complexes and population boundaries. Zoo Biol., in press. (6)

Templeton, A.R., C.F. Sing and B. Brokaw. 1976. The unit of selection in *Drosophila mercatorum*. I. The interaction of selection and meiosis in parthenogenetic strains. Genetics 82, 349–376. (6)

Terborgh, J. 1974. Preservation of natural diversity: the problem of extinction prone species. BioScience 24, 715–722. (8, 11)

Terborgh, J. 1983. *Five New World Primates: A Study in Comparative Ecology*. Princeton University Press, Princeton, N.J. (12, 15)

Terborgh, J. and J.M. Diamond. 1970. Niche overlap in feeding assemblages of New Guinea birds. Wilson Bull. 82, 29–52. (15)

Terborgh, J. and B. Winter. 1980. Some causes of extinction. In *Conservation Biology: An Evolutionary-Ecological Perspective*, M.E. Soulé and B.A. Wilcox (eds.). Sinauer Associates, Sunderland, Mass. pp. 119–133. (9, 11, 15)

Terborgh, J. and B. Winter. 1983. A method for siting parks and reserves with special reference to Colombia and Ecuador. Biol. Conserv. 27, 45–58. (19)

Thienemann, A. (ed.). 1931–1957. Tropische Binnengewaesser. Ergebnisse einer von H.J. Feurborn, F. Ruttner und A. Thienemann im Jahre 1928 und 1929 nach Java, Sumatra, und Bali unternommenen Forschungsreise, Bd. 1–11, Arch. Hydrobiol., Suppl. Bde. Thienemann, A. 1953. Fluss und See. Ein limnologischer Vergleich. Gewaesser und Abwaesser 1, 13–30. (18)

Thoday, J.M. 1955. Balance, heterozygosity and developmental stability. Cold Spring Harbor Symp. Quant. Biol. 20, 318–326. (4)

Thoday, J.M. 1958. Homeostasis in a selection experiment. Heredity 12, 401–415. (4)

Thomas, C.D. and H.C. Mallorie. 1985. Rarity, species richness, and conservation: butterflies of the Atlas Mountains of Morocco. Biol. Conserv. 33, 95–117. (9)

Thomas, J.A. 1976. The ecology of the large blue butterfly. Annual Report of the Institute for Terrestrial Ecology. pp. 25–27. (11)

Thomas, J.H. 1961. 1954. *Flora of the Santa Cruz Mountains of California*. Stanford University Press, Stanford, Calif. (8)

Thomson, G. 1977. The effect of a selected locus on linked neutral loci. Genetics 85, 753–788. (4)

Thorington, R.W., Jr. 1968. Observations of the tamarin, *Saguinus midas*. Folia Primatol. 9, 95–98. (12)

Thornback, J. and M. Jenkins. 1982. The IUCN Mammal Red Data Book. Part 1. Int. Union Cons. Nat. Nat. Res., Gland, Switzerland. (21)

Thorne, R.F. 1954. The vascular plants of Southwestern Georgia. Amer. Midl. Natural. 52, 257–327. (8)

Thorpe, H. 1978. The man–land relationship through time. In *Agriculture and Conservation*, J.G. Hawkes (ed.). Duckworth, London. (11)

Tigerstedt, P.M.A. 1983. Genetic mechanisms for adaptation: the mating system of Scots pine. In Genetics: New Frontiers. Proc. XV Int. Congr. Genet. Oxford and IBH Publ., New Delhi-Bombay-Calcutta. pp. 317–322. (5)

Tilman, D. 1982. *Resource Competition in Plant Communities*. Princeton University Press, Princeton, N.J. (7)

Time. 1976. Ludwig's wild Amazon kingdom. Time Magazine, Nov. 15, 59–59A. (20)

Time. 1979. Billionaire Ludwig's Brazilian Gamble. Time Magazine, Sept. 10, 76–78. (20)

Time. 1982. End of a billion-dollar dream. Time Magazine, Jan. 25, 59. (20)

Tinkle, D.W. and R.K. Selander. 1973. Age-dependent allozymic variation in a natural population of lizards. Biochem. Genet. 8, 231–237. (4)

Tipton, V.M. in press. Cave bats, their ecology, identification, and distribution. In Proceedings of the 1982 National Cave Management and Conservation Symposium. (21)

Tiravanti, G. and R. Passino. in press. Tetraalkyl lead accident in sea water. In *Ecoaccidents*, J. Cairns, Jr. (ed.). NATO ASI Series, Plenum, London. (23)

Toledo, V.M. 1977. Pollination of some rain forest plants by non-hovering birds in Veracruz, Mexico. Biotropica 9, 262–267. (15)

Toupet, Ch. 1976. L'évolution du climat de la Mauritanie du Moyne Age jusqu'à nos jours. In *La désertification au sud du Sahara*, Th. Monod (ed.). Cpte Rend. Colloque de Nouakchott, Nelles Edit. Afric., Dakar/Abidjan. pp. 56–63. (22)

Tracey, M.L., N.F. Bellet and C.D. Gravem. 1975. Excess allozyme homozygosity and breeding population structure in the mussel *Mytilus californianus*. Mar. Biol. 32, 303–311. (4)

Tucker, C.J., B.N. Holben and T.E. Goff. 1984. Intensive forest clearing in Rondonia, Brazil, as detected by satellite remote sensing. Rem. Sens. Envir. 15, 255–261. (19)

Tucker, C.J., C. Vanpraet, M.J. Sharman and G. Van Ittersum. 1985. Satellite remote sensing of total herbaceous biomass production in the Senegalese Sahel, 1980–1983. Rem. Sens. of Envir. 17, 223–243. (22)

Turelli, M. 1978. A reexamination of stability in random versus deterministic environments with comments on the stochastic theory of limiting similiarity. Theor. Pop. Biol. 13, 244–267. (14)

Turner, B.L. and P.D. Harrison. 1981. Prehistoric raised-field agriculture in the Maya lowlands. Science 213, 399–405. (20)

Tuttle, M.D. 1977. Gating as a means of protecting cave-dwelling bats. In *Proceedings 1976 National Cave Management Symposium*, T. Aley and D. Rhodes (eds.). Speleobooks, Albuquerque, N.M. pp. 77–82. (21)

Tuttle, M.D. 1979. Status, causes of decline, and management of endangered gray bats. J. Wildl. Manage. 43, 1–17. (21)

Tuttle, M.D. and S.J. Kern. 1981. Bats and public health. Milwaukee Public Mus. Contrib. Biol. Geol. No. 48. 11 pp. (21)

Tuttle, M.D. and D.E. Stevenson. 1978. Variation in the cave environment and its biological implications. In Proceedings 1977 National Cave Management Symposium, J. Chester, S. Gilbert and D. Rhoads (eds.). Adobe Press, Albuquerque, N.M. pp. 108–121. (21)

Uhl, C. and C.F. Jordan. 1984. Vegetation and nutrient dynamics during the first five years of succession following forest cutting and burning in the Rio Negro region of Amazonia. Ecology 65, 1476–1490. (20)

UNESCO-UNEP. 1984. *Conservation, Science and Society*. Vols. I and II. UNESCO, Paris. (1)

UNESCO. 1978. Tropical forest ecosystems: a state of knowledge report. Natural Resources Research XIV. UNESCO, Paris. (19)

Urquhart, F.A. and N.R. Urquhart 1976. The overwintering site of the eastern population of the monarch butterfly (*Danaus p. plexippus*, Danaidae) in southern Mexico. J. Lepid. Soc. 30, 153–158. (13)

U.S. Environmental Protection Agency. 1980. Water Quality Criteria Summaries: A Compilation of State and Federal Criteria, Vol. I. Office of Water Standards and Regulations, Government Printing Office, Washington, D.C. (23)

Valverde, J.A. 1969. Ecological bases for fauna conservation in the Western Sahara. Proceed. IBP Sympos. C.T. Sect., Hammamet, Tunisia, April 1968. 14 pp. mimeo. (22)

van Noordwijk, A.J. and W. Scharloo. 1981. Inbreeding in an island population of the great tit. Evolution 35, 674–688. (3)

van den Berghe, P.L. 1983. Human inbreeding avoidance: culture in nature. Behav. Brain Sci. 6, 91–123. (3)

van der Zon, A.P.M. and Y. Mulyana. 1978. Nature conservation in Irian Jaya. FAO Field Report 9, Bogor. (24)

Van Valen, L. 1962. A study of fluctuating asymmetry. Evolution 16, 125–142. (4)

Vrijenhoek, R.C. and S. Lerman. 1982. Heterozygdevelopmental stability under sexual and asexual breeding systems. Evolution 36, 768–776. (4)

Vu, M.T. and A. Elwan. 1982. Short-term population projection 1980–2000, and long-term projection 2000 to stationary stage, by age and sex for all countries of the world. The World Bank, Washington, D.C. (19)

Waddington, C.H. 1942. Canalization of development and the inheritance of acquired characters. Nature 150, 563–565. (4)

Waddington, C.H. 1948. Polygenes and oligogenes. Nature 156, 394. (4)

Wadsworth, F.H. 1983. Production of usable wood from tropical forests. In *Tropical Rain Forest Ecosystems, Ecosystems of the World* (14A), F.B. Golley (ed.). Elsevier, Amsterdam. pp. 279–288. (20)

Wagner, W. and W. Gagné. in press. Hawaii's native biota: trouble in "paradise." Bull. Amer. Assoc. Bot. Gard. Arboreta. (8)

Wagner, W., D.R. Herbst and R.S. Yee. in press. Status of the flowering plants of the Hawaiian Islands. In *Protection and Management of Native Hawaiian Terrestrial Ecosystems*, C. Stone and M. Scott (eds.). (8)

Wales, B.A. 1972. Vegetation analysis of north and south edges in a mature oak-hickory forest. Ecol. Monogr. 42, 451–471. (11)

Walker, C.H. 1983. Pesticides and birds—mechanisms of selective toxicity. Agriculture, Ecosystems and Environment 9, 211–226. (16)

Wallace, B. 1968. *Topics in Population Genetics*. W.W. Norton, New York. (6)

Waller, D.M. 1984. Differences in fitness between seedlings derived from cleistogamous and chasmogamous flowers in *Impatiens capensis*. Evolution 38, 427–440. (5)

Walter, H. 1979. *Eleonora's Falcon*. University of Chicago Press, Chicago. (7)

Wang, C.W., T.O. Perry and A.E. Johnson. 1960. Pollen dispersion of slash pine (*Pinus elliottii*) with special reference to seed orchard management. Silvae Genet. 9, 78–86. (5)

Warner, R.E. 1968. The role of introduced diseases in the extinction of the endemic Hawaiian avifauna. Condor 70, 101–120. (16)

Waser, P.M. 1977. Feeding, ranging and group size in the mangabey *Cercocebus albigena*. In *Primate Ecology: Studies of Feeding and Ranging Behavior in Lemurs, Monkeys and Apes*, T.H. Clutton-Brock (ed.). Academic Press, London. (15)

Waser, P.M. and T.J. Case. 1981. Monkeys and matrices: coexistence of "omnivorous" forest primates. Oecologia 49, 102–108. (15)

Waser, P.M. and W.T. Jones. 1983. Natal philopatry among solitary mammals. Quart. Rev. Biol. 58, 355–390. (3)

Watson, R.M. and C. Tippett. 1981. Range and livestock surveys in Northern and Central Somalia. Ministry of Livestock Production, Mogadishu. (22)

Watt, K.E.F. 1973. *Principles of Environmental Science*. McGraw Hill, New York. (14)

Watt, W.B. 1977. Adaptation at specific loci. I. selection on phosphoglucose isomerase of *Colias* butterflies: biochemical and population aspects. Genetics 87, 177–194. (4)

Watt, W.B. 1983. Adaptation at specific loci. II. Demographic and biochemical elements in the maintenance of the *Colias PGI* polymorphism. Genetics 103, 691–724. (4)

Watt, W.B. 1985. Bioenergetics and evolutionary genetics: opportunities for new synthesis. Amer. Natur. 125, 118–143. (4)

Watt, W.B., P.A. Carter and S.M. Blower. 1985. Adaptation at specific loci. IV. Differential mating success among glycolytic allozyme genotypes of *Colias* butterflies. Genetics 109, 157–175. (4)

Watt, W.B., R.C. Cassin and M.S. Swan. 1983. Adaptation at specific loci. III. Field behavior and survivorship differences among *Colias PGI* genotypes are predictable from in vitro biochemistry. Genetics 103, 725–739. (4)

Watters, R.F. 1971. Shifting cultivation in Latin America. FAO Forestry Development Paper No. 17. Food and Agriculture Organization, Rome. (20)

Weaver, P. 1979. Agri-silviculture in tropical America. Unasylva 31(126), 2–12. (20)

Webb, N.R. and L.E. Haskins. 1980. An ecological survey of heathlands in the Poole Basin, Dorset, England in 1978. Biol. Conserv. 17, 281–296. (11)

Wells, S.M., R.M. Pyle and N.M. Collins. 1983. The IUCN Invertebrate Red Data Book. Int. Union Conserv. Nat. Nat. Res., Gland, Switzerland. (21)

Wells, T.C.E. 1976. *Hypochoeris maculata* L. J. Ecol. 64, 757–774. (9)

Werner, E.E. 1985. Species interactions in freshwater fish communities. In *Community Ecology*, J.M. Diamond and T.J. Case (eds.). Harper & Row, New York. pp. 344–357. (24)

Westing, A.H. 1984. Herbicides in war: past and present. In *Herbicides in War: The Long-term Ecological and Human Consequences*, A.H. Westing (ed.). Taylor and Francis, London. pp. 3–24. (20)

Wetterberg, G.B. 1976. An Analysis of Nature Conservation Priorities in the Amazon. Brazilian Institute for Forestry Development, Brasilia, Brazil. (8, 19)

Wheeler, N.C. and R.P. Guries. 1982. Population structure, genic diversity, and morphological variation in *Pinus contorta* Dougl. Canad. J. For. Res. 12, 595–606. (5)

Wheelwright, N.T. 1983. Fruits and the ecology of resplendent quetzals. Auk 100, 286–301. (15)

Wherry, E.T., J.M. Fogg, Jr. and H.A. Wahl. 1979. *Atlas of the Flora of Pennsylvania*, Morris Arboretum, Philadelphia. (9)

Whicker, F.W. and V. Schultz. 1982. *Radioecology: Nuclear Energy and the Environment*. Vol. I. CRC Press, Boca Raton, Florida. (20)

Whitaker, J.O., Jr. 1980. *The Audubon Society Field Guide to North American Mammals*. Chanticleer Press, New York. (11)

Whitcomb, R.F., C.S. Robbins, J.F. Lynch, B.L. Whitcomb, M.K. Klimkiewicz and D. Bystrak. 1981. Effects of forest fragmentation on avifauna of the eastern deciduous forest. In *Forest Island Dynamics in Man-Dominated Landscapes*, R.L. Burgess and D.M. Sharpe (eds.). Springer Verlag, New York. (11)

Whitcomb, R.F., J.F. Lynch, P.A. Opler and C.S. Robbins. 1976. Island biogeography and conservation: strategies and limitations. Science 193, 1030–1032. (11)

White, J. 1980. Demographic factors in populations of plants. In *Demography and Evolution in Plant Populations*, O.T. Solbrig (ed.). University of California Press, Berkeley. pp. 21–48. (5)

White, P.S. 1982. The Flora of Great Smoky Mountains National Park: an Annotated Checklist of the Vascular Plants and a Review of Previous Work. National Park Service, Southeast Region, Atlanta, Ga. (8)

Whitmore, T.C. 1978. Gaps in the forest canopy. In *Tropical Trees as Living Systems*, P.B. Tomlinson and M.H. Zimmermann (eds.). Cambridge University Press, Cambridge. pp. 639–655. (20)

Whittaker, R.H. 1972. Evolution and measurement of species diversity. Taxon 21, 213–251. (7)

Whittaker, R.H. 1975. *Communities and Ecosystems*, 2nd Ed. Macmillan, New York. (10)

Whittaker, R.H. 1977. Evolution of species diversity in land communities. Evol. Bio. 10, 1–67. (7)

Wiens, D. 1984. Ovule survivorship, brood size, life history, breeding systems, and reproductive success in plants. Oecologia 64, 47–53. (5)

Wilcove, D.S. 1985a. Forest Fragmentation and the Decline of Migratory Songbirds. Ph.D. thesis, Princeton University, Princeton, N.J. (11)

Wilcove, D.S. 1985b. Nest predation in forest tracts and the decline of migratory songbirds. Ecology 66, 1211–1214. (11)

Wilcox, B.A. 1980. Insular ecology and conservation. In *Conservation Biology: An Evolutionary-Ecological Perspective*, M.E. Soulé and B.A. Wilcox (eds.). Sinauer Associates, Sunderland, Mass. pp. 95–117. (11)

Wilcox, B.A. and D.D. Murphy, 1985, Conservation strategy: the effects of fragmentation on extinction. Amer. Natur. 125, 879–887. (2)

Wiley, G.R. 1982. Maya archaeology. Science 215, 260–267. (20)

Wilhelm, G.S. 1984. Vascular Flora of the Pensacola Region. Ph.D. thesis, Southern Illinois University, Carbondale. (8)

Wilhelmy, H., 1957. Das Grosse Pantanal in Mato Grosso. Deutscher Geographentag Wuerzburg, Tagungsbericht und Wissenschaftliche Abhandlungen, 45–71. (18)

Wilhelmy, H., 1958. Das Grosse Pantanal. Die Umschau 1958, 555–559. (18)

Wilkins, N.P. 1978. Length correlated changes in heterozygosity at an enzyme locus in the scallop (*Pecten maximus* L.). Anim. Blood Groups. Biochem. Gen. 9, 69–77. (4)

Williams, G.C. 1966. *Adaptation and Natural Selection*. Princeton University Press, Princeton, N.J. (3)

Williams, I.J.M. 1972. A revision of the genus *Leucadendron* (Proteaceae). Contrib. Bolus Herbar. No. 3, 1–145. Rondebosch, C.P., So. Africa. (7)

Williams, N.H. and W.M. Dodson. 1972. Selective attraction of male euglossine bees to orchid floral fragrances and its importance in long distance pollen flow. Evolution 26(1), 84–95. (12)

Williams, W. 1959. Heterosis and the genetics of complex characters. Nature 184, 527–530. (5)

Williamson, M. 1981. *Island Populations.* Oxford University Press, Oxford. (11)

Willis, E.O. 1974. Populations and local extinctions of birds on Barro Colorado Island, Panama. Ecol. Monogr. 44, 153–169. (12)

Willis, E.O. 1978. Birds and army ants. Ann. Rev. Ecol. Syst. 9, 243–263. (12)

Willis. E.O. 1979. The composition of avian communities in remanescent woodlots in southern Brazil. Papeis Avulsos Zool. 33(1), 1–25. (12, 15)

Wills, C. 1981. *Genetic Variability.* Clarendon Press, Oxford. (4)

Willson, M.F. 1974. Avian community organization and habitat structure. Ecology 55, 1017–1029. (7)

Willson, M.F. and N. Burley. 1983. *Mate Choice in Plants: Tactics, Mechanisms, and Consequences.* Princeton University Press, Princeton, N.J. 251 pp. (5)

Wilson, A.C., G.L. Bush, S.M. Case and M.C. King. 1975. Social structure of mammalian populations and rate of chromosomal evolution. Proc. Nat. Acad. Sci. USA 72, 5061–5065. (3)

Wilson, E.O. 1985. The biological diversity crisis. BioScience 35, 700–706. (II)

Wilson, E.O. and E.O. Willis. 1975. Applied biogeography. In *Ecology and Evolution of Communities,* M.L. Cody and J.M. Diamond (eds.). Harvard University Press, Cambridge, Mass. pp. 522–534. (11)

Wilson, J. 1985. Colonization in Rondonia: The Case of Ariquemis. Ph.D. dissertation, University of Florida, Gainesville. (19)

Woessner, R.A. 1972. Crossing among loblolly pines indigenous to different areas as a means of genetic improvement. Silvae Genet. 21, 35–39. (5)

Woessner, R.A. 1975. Interprovenance crosses of loblolly pine. In Proc. Fourteenth Meet. Can. Tree Improv. Ass., Part 2. Symp. Interspecific and Interprovenance Hybridization in Forest Trees, D.P. Fowler and C.W. Yeatman (eds.). Fredericton, New Brunswick. pp. 17–23. (5)

Woessner, R.A. 1982. Plantation forestry and natural forest utilization in the Amazon Basin. Paper presented at the American Society of Foresters' meeting, Sept. 19–22, Cincinnati, Oh. (20)

Wolf, L.L. 1970. The impact of seasonal flowering on the biology of some tropical hummingbirds. Condor 72, 1–14. (15)

Woodbury, R. and J. Figueroa. 1985. *Rare and Endangered Plants of Puerto Rico.* (in press). (8)

Woodhead, N. 1951. *Lloydia serotina* (L.) Rchb. J. Ecol. 39, 198–203. (9)

Woods, J.H., G.M. Blake and F.W. Allendorf. 1983. Amount and distribution of isozyme variation in ponderosa pine from eastern Montana. Silvae Genet. 32, 151–156. (5)

Woodwell, G.M. 1967. Toxic substances and ecological cycles. Sci. Amer. 216 (March), 128–135. (20)

Woodwell, G.M. et al. 1983. Global deforestation: contribution to atmospheric carbon dioxide. Science 222, 1081–1086. (19)

Woolfenden, G.E. and J.W. Fitzpatrick. 1984. *The Florida Scrub Jay: Demography of a Cooperative-breeding Bird.* Princeton University Press, Princeton, N.J. (3)

Woolpy, J.H. and I. Eckstrand. 1979. Wolf pack genetics: a computer simulation with theory. In *The Behavior and Ecology of Wolves,* E. Klinghammer (ed.). Garland Press, New York. pp. 206–224. (3)

World Resources Institute. 1985. Accelerated Action Plan for Tropical Forests. World Bank and World Resources Institute, Washington, D.C. (19)

Wright, J.W. 1952. Pollen Dispersal of Some Forest Trees. USDA For. Serv., Northeast. For. Exp. Sta. Pap. 46. Upper Darby, Penn. 42 pp. (5)

Wright, S.J. and S.P. Hubbell. 1983. Stochastic extinction and reserve size: a focal species approach. Oikos 41, 466–476. (10)

Wu, H.M.H., W.G. Holmes, S.R. Medina and G.P. Sackett. 1980. Kin preference in infant *Macaca nemestrina.* Nature 285, 225–227. (3)

Yager, J. 1981. Remipedia, a new class of Crustacea from a marine cave in the Bahamas. J. Crustacean Biol. 1, 328–333. (21)

Yeatman, C.W. 1973. Gene conservation in relation to forestry practice. In Proc. Thirteenth Meet. Comm. For. Tree Breeding Canad. Part 2. Symp. Conserv. For. Gene Resources, D.P. Fowler and C.W. Yeatman (eds.). Prince George, British Columbia. pp. 19–24. (5)

Yeh, F.C. and Y.A. El-Kassaby. 1980. Enzyme variation in natural populations of Sitka spruce (*Picea sitchensis*). I. Genetic variation patterns among trees from 10 IUFRO provenances. Canad. J. For. Res. 10, 415–422. (5)

Yeh, F.C. and C. Layton. 1979. The organization of genetic variability in central and marginal populations of lodgepole pine *Pinus contorta* spp. *latifolia*. Canad. J. Genet. Cytol. 21, 487–503. (5)

Yeh, F.C. and D. O'Malley. 1980. Enzyme variations in natural populations of Douglasfir, *Pseudotsuga menziesii* (Mirb.) Franco, from British Columbia. 1. Genetic variation patterns in coastal populations. Silvae Genet. 29, 83–92. (5)

Yoakum, J. and W.P. Dasmann. 1969. Habitat manipulation practices. In *Wildlife Management Techniques*, R.H. Giles, Jr. (ed.). The Wildlife Society, Washington, D.C. pp. 173–231. (11)

Yoshiyama, R.M. and C. Sassaman. 1983. Morphological and allozymic variation in the stichaeid fish *Anoplarchus purpurescens*. Syst. Zool. 32, 52–71. (4)

Zakharov, V.M. 1981. Fluctuating asymmetry as an index of developmental homeostasis. Genetika (Belgrade) 13, 241–256 (4)

Zaret, T.M. 1984. Central American limnology and Gatun Lake. In Ecosystems of the World, Vol. 23, *Lakes and Reservoirs*, F.B. Taub (ed.). Elsevier, Amsterdam. pp. 447–465. (18)

Zaret, T.M. and R.T. Paine. 1973. Species introductions in a tropical lake. Science 182, 449–455. (24)

Zhu Zenda and Liu Shu. 1983. Combating desertification in arid and semi-arid zones in China. Inst. of Des. Res., Academia Sinica, Lanzhou, China. (22)

Zobel, B.J. and J.T. Talbert. 1984. *Applied Forest Tree Improvement*. Wiley, New York. 505 pp. (5)

Zouros, E. and D.W. Foltz. 1984. Minimal selection requirements for the correlation between heterozygosity and growth, and for the deficiency of heterozygotes, in oyster populations. Devel. Genet. 4, 393–405. (4)

Zouros, E. and D.W. Foltz. in press. The use of allelic isozyme variation for the study of heterosis. Isozymes: Current Topics in Biological and Medical Research. (4)

Zouros, E., S.M. Singh and H.E. Miles. 1980. Growth rate in oysters: an overdominant phenotype and its possible explanations. Evolution 34, 856–867. (4,5)

Zouros, E., S.M. Singh, D.W. Foltz and A.L. Mallet. 1983. Post-settlement viability in the American oyster (*Crassostrea virginica*): an overdominant phenotype. Genet. Res., Camb. 41, 259–270. (4)

Zucchi, R., S.F. Sakagami and J.M.F. de Camargo. 1969. Biological observations on a neotropical parasocial bee, *Eulaema nigrita*, with a review on the biology of Euglossinae (Hymenoptera, Apidae). A comparative study. J. Fac. Sci. Hokkaido University, Series VI. Zoology 17, 271–382. (12)

INDEX

A vortex, 31–33
Abies balsamea, habitat adaptation in, 96
Abortion, and breeding system, 95, 97, 98
Acacia
and African desertization, 452, 456
alpha-diversity in, 134, 135
as anthropogenic vegetation, 123
as threatened species, 125
Acer saccharum, temporal heterogeneity in, 96
Acinonyx jubatus, low heterozygosity and survival of, 14–15, 74–75, 364
Acipenser, parasite epidemic in, 356
Acorn woodpecker, mating behavior of, 49–50
Adaptation vortex, 31–33
Addax nasomaculatus, 458
Africa, *see also* East Africa
beta-diversity in, 128
desertization in, 444–461
major climatic regions of, 445
protected areas in, 408
reservoirs in, 389–390
savannazation in, 422
African buffalo, rinderpest spread in, 346, 349–351, 353
Africanized honeybee, 293
Age, and heterozygosity, 62
Agriculture, *see* Cultivation
Agrius cingulatus, 291
Agrostis setacea, 195, 196
Agrostis tenuis, habitat adaptation in, 96
Alaska, endemism in, 156–157
Albatross, *see also Diomedea immutabilis*
close inbreeding in, 38
Allele frequency, changes in among cohorts, 60–61
Allele linkage, and fitness, 17
Alleles, *see also* Genes
fixation of, 15, 38, 40
recessive deleterious, 14, 15, 82, 104
Allozymes, 84–89
Alouatta seniculus, area fragmentation effects on, 27, 272, 274, 275–276
Alpha-diversity
determinants of, 129–130
ecological effects of reduced, 142–143
and isolation, 145

in Mediterranean climate areas, 137–141
patterns of, 138, 141
in plant communities, 131
as term, 126
and vegetation structure, 135–140
Amazon Basin, shifting cultivation in, 416
Amazonia
and deforestation, 407–408, 424–425
and forest fragmentation, 257–285
local endemism in, 166
plant diversity in, 157
species-specific interaction in, 257
Amazon river system, damming of, 390–392
Ambystoma tigrinum, heterozygosity in, 66
Ammophila arenaria
geographic distribution of, 187, 191
habitat specificity and, 188, 199
Amphitecna, 173
Amytornis goderi, 128
Anatidae, sex-biased dispersal in, 43
Andes, endemism in, 168–176
Antechinus, inbreeding reduction in, 42, 46
Ant-following birds, 277–280
Anthurium, and local endemism, 173–174
Aotus trivirgatus, and coadaptation, 107
Appalachia, endemism in, 156
Aquatic ecosystems
human interference in, 386–392
material metabolisms of, 384–386
Arctostaphylos
and anthropogenic rarity, 124
and species turnover rate, 134
Ardisia, endemism in, 174
Argyroxiphium macrocephalum, as rare species, 183, 184, 185
Arid lands, defined, 444
Army ants, and ant-following birds, 277–280
Asymmetry, and morphological extremity, 71–72
Ateles paniscus, isolation effects on, 271
Atlas of the Flora of the British Isles, 186–204
Atmospheric impacts, 233

Endangered species, in Mediterranean-climate regions, 125
Endemism
anthropogenic, 162–163
caves, 429, 431
island, 124
Pleistocene refugia model of, 167–168
temperate vs. tropical, 176–181
versus rarity, 154
Endothia parasitica, epidemic infection in, 324–325, 346
England, *see also* British Isles
habitat fragmentation in, 241–242
red grouse in, 356–357
Environment, and species viability, 21, 23–24, 28
Environmental Protection Agency, 467
Environmental stochasticity
and environmental quality, 27
and extinction, 246
and genetic effective population size, 28–30
Eonycteris spelaea, extinction susceptibility of, 403
Epidemics, geographic spread of, 348–351
Epistasis, *see also* Outbreeding depression
and fitness, 15, 16–17
Ethical issues, 9, 361–362
Eucalyptus
and beta-rarity, 128
fitness in, 80
inbreeding depression in, 90–91
in Jari project, 423
threatened species of, 125
as threat in Neotropics, 293
Euglossine bees
area fragmentation effects on, 280–283
isolation effects on, 281
Europe, endemism in, 157
European mute swans, close inbreeding in, 38
Evolutionary potential, and heterozygosity, 75–76
Experimental disturbances, in tropical forests, 414–416
Extinction, *see also* Minimum viable population; Species, preservation of
demographic aspects of, 19
deterministic, 24–25
and environmental perturbations, 28
and forest fragmentation, 238–240
genetic aspects of, 19
and home-range size, 246–247, 284
linked, 403
modern causes of, 490–492
and patchiness, 30

probability of, 28–31
processes of, 19–34, 246–252
rate of, 120, 244
secondary, 251–252
stochastic, 24–25
susceptibility to, 402–406
systems perspective on, 20
vortices of, 25–33
Extinction crisis, 4

F vortex, 31
Fagus
fitness in, 80–81
and one-way pollen dispersal, 288
Falco eleonora, alpha-rarity, 126
Fallow deer, *see Dama dama*
Faramea accidentalis, distribution of, 210, 213
Feral horse, kin recognition in, 48
Ferns, genetic loads of, 90
Figs, as keystone resource, 337–340
Filling, and seagrass habitat, 376
Fire, as preserve threat, 296–298
"Firebreak" width, and elimination of infection, 352
Fish
harvest of, and coral reef communities, 374
heterozygosity and growth in, 67
parasite epidemics in, 356
Fisheries
artisanal, 378–379
collapse of, 313–314
tropical, 378–379
Fitness
and coadaptation, 16
components of, and heterozygosity, 58–72
decline of, as intrinsic coadaptation, 108–110
and epistasis, 16–17
and heterosis, 80–81
heterozygosity versus homozygosity and, 72–73, 103
and hybridity, 109
in outbreeding plants, 77–104
Florida
endemism in, 156
recovery from mining damage in, 474–475
Forest fragmentation
adjustment, phases of, 285
in Amazon, 257–285
edge effects of, 257–285
effects of isolation on, 257–285
effects on primates of, 271–274
incidence functions for two species in, 243

Population size, and rarity, 183–185
Population structure, and extinction, 246
Population structure and fitness and species viability, 21, 24
Population viability, *see* Minimum viable population
Population vulnerability analysis (PVA), 19–34
  fields of, 22, 23
Positive epistasis, 16
Preserves, *see also* Reserves
  external biotic threats to, 287–294
  and external habitats, 291–292
  external physical changes and, 258, 260, 294–302
  threats to surroundings from 294
Primates
  breeding dispersal in, 44
  fruit abundance and diet shifts in, 335–340
  in isolated reserves, 271
Primula
  elatior, 187, 191
  scotica, 186, 200
Procavia johnstoni, sarcoptic mange in, 363
Protea, turnover rates, 135
Protead species, and segregation by leaf morphology, 132–133
Pseudanophthalmusm, speciation of, 429–430
Pseudokarst, 433
Pseudooverdominance, 80
Pseudosinella hirsuta, 431
Pseudotsuga menziesii, mutation in, 90
Psidium anglohondurense, as nondeclining rare species, 223–224
Pterocarpus rohrii, distribution of, 214
Puerto Rico, endemism on, 158–159
PVA, *see* Population vulnerability analysis

Quercus
  demise of, in preserve, 299–302
  selective fertilization in, 94

R vortex, 29–30
Rabies, simulated spread of, 351
Racial purity, *see* Outbreeding depression
Radiation, and tropical ecosystems, 414–415
Rainforest, *see also* Tropical forests
  as migratory habitat, 289–291
Ramphocelus carbo, and edge effects, 266–267
Rarity
  analysis of, and sample problem, 118, 119
  anthropogenic, 124

causes of, 221–223
classification of, 185–186, 190–200, 202–204
at community level, 119
components of, 117–118
defined, 210–211
and distribution range, 183
and environmental scale, 119
in Mediterranean-climate regions, 122–152
natural, and high diversity, 124–125
and spatial scales, 186
species diversity and, 124–130
taxonomic approach to, 118–119, 225–226
Rattus norvegicus, heterozygosity in, 59
Reclamation, 465–484
Recovery
  degree of, 469–470
  goals for, 467–468, 470–472
Red grouse, 356–357
Red howler monkeys, *see Alouatta seniculus*
Red pine, as outcrosser, 101
Red-spotted newt, and fragmentation effects, 247
Regeneration
  of ecosystems, 465–484
  of species, 215–218, 226
Relict endemics, 160–162, *see also* Ozarks, relictual species in
Rennell Island, 492
Renosterveld, core species of, 145–147
Reproduction, tree, and high genic diversity, 91–95
Reproductive competition hypothesis, 50
Reptiles, New Guinea, 487
Rescue effect, 243, 253
Reserves, *see also* Edge effects; Preserves
  atmospheric impacts on, 233, 298–299
  biogeographic considerations, 494–497
  disease buffer zones and, 363–364
  and external human activity, 249
  guidelines in temperate zone, 252–256
  habitat considerations, 493–494
  margins of, 258, 261
  marine, 380–381
  in Minimum Critical Size of Ecosystem study, 258–260
  Minimum viable density and, 120–121
  shape of, 255
  size of, 227–230, 253–255, 497–498
  subdivision of, and disease, 363–364
Reservoirs, *see also* Dams
  and tropical river ecology, 389
Resilience, of populations, 313–317
Resistance, to introductions, 312, 317, 324–327
Resonance, of population densities, 316

This book was set in Linotron 202 Century Schoolbook at DEKR Corporation, and manufactured at the Murray Printing Company. Many of the drawings were rendered by Frederic J. Schoenborn and Joseph Vesely. The copy editor was Carol Hines. Joseph Vesely designed the book and supervised its production.